T0140508

Lecture Notes in Networks and Systems

Volume 67

Series Editor

Janusz Kacprzyk, Systems Research Institute, Polish Academy of Sciences, Warsaw, Poland

Advisory Editors

Fernando Gomide, Department of Computer Engineering and Automation—DCA, School of Electrical and Computer Engineering—FEEC, University of Campinas—UNICAMP, São Paulo, Brazil

Okyay Kaynak, Department of Electrical and Electronic Engineering, Bogazici University, Istanbul, Turkey

Derong Liu, Department of Electrical and Computer Engineering, University of Illinois at Chicago, Chicago, USA, Institute of Automation, Chinese Academy of Sciences, Beijing, China

Witold Pedrycz, Department of Electrical and Computer Engineering, University of Alberta, Alberta, Canada, Systems Research Institute, Polish Academy of Sciences, Warsaw, Poland

Marios M. Polycarpou, KIOS Research Center for Intelligent Systems and Networks, Department of Electrical and Computer Engineering, University of Cyprus, Nicosia, Cyprus

Imre J. Rudas, Óbuda University, Budapest, Hungary

Jun Wang, Department of Computer Science, City University of Hong Kong, Kowloon, Hong Kong

The series "Lecture Notes in Networks and Systems" publishes the latest developments in Networks and Systems—quickly, informally and with high quality. Original research reported in proceedings and post-proceedings represents the core of LNNS.

Volumes published in LNNS embrace all aspects and subfields of, as well as new challenges in, Networks and Systems.

The series contains proceedings and edited volumes in systems and networks, spanning the areas of Cyber-Physical Systems, Autonomous Systems, Sensor Networks, Control Systems, Energy Systems, Automotive Systems, Biological Systems, Vehicular Networking and Connected Vehicles, Aerospace Systems, Automation, Manufacturing, Smart Grids, Nonlinear Systems, Power Systems, Robotics, Social Systems, Economic Systems and other. Of particular value to both the contributors and the readership are the short publication timeframe and the world-wide distribution and exposure which enable both a wide and rapid dissemination of research output.

The series covers the theory, applications, and perspectives on the state of the art and future developments relevant to systems and networks, decision making, control, complex processes and related areas, as embedded in the fields of interdisciplinary and applied sciences, engineering, computer science, physics, economics, social, and life sciences, as well as the paradigms and methodologies behind them.

** **Indexing: The books of this series are submitted to ISI Proceedings, SCOPUS, Google Scholar and Springerlink** **

More information about this series at http://www.springer.com/series/15179

Vincenzo Piuri · Valentina Emilia Balas ·
Samarjeet Borah · Sharifah Sakinah Syed Ahmad
Editors

Intelligent and Interactive Computing

Proceedings of IIC 2018

 Springer

Editors
Vincenzo Piuri
Department of Computer Science
University of Milan
Milan, Italy

Samarjeet Borah
Department of Computer Applications,
Sikkim Manipal Institute of Technology
Sikkim Manipal University
Sikkim, India

Valentina Emilia Balas
Department of Automation
and Applied Informatics
Aurel Vlaicu University of Arad
Arad, Romania

Sharifah Sakinah Syed Ahmad
Department of Intelligent Computing
and Analytics
Universiti Teknikal Malaysia Melaka
Durian Tunggal, Melaka, Malaysia

ISSN 2367-3370   ISSN 2367-3389   (electronic)
Lecture Notes in Networks and Systems
ISBN 978-981-13-6030-5   ISBN 978-981-13-6031-2   (eBook)
https://doi.org/10.1007/978-981-13-6031-2

Library of Congress Control Number: 2018966849

This Springer imprint is published by the registered company Springer Nature Singapore Pte Ltd.
The registered company address is: 152 Beach Road, #21-01/04 Gateway East, Singapore 189721, Singapore

Preface

Intelligent and interactive computing is assorted of research outcomes of the 2nd International Conference on Intelligent and Interactive Computing 2018 (IIC 2018). IIC 2018 is jointly organized by the Department of Intelligent Computing and Analytics (ICA) and the Department of Interactive Media (MI), Faculty of Information and Communication Technology (FTMK), Universiti Teknikal Malaysia Melaka (UTeM). It is an academic conference with the theme "Intelligent and Immersive Computing towards Industrial Revolution 4.0". The aim of this conference was to bring together academicians, researchers, developers and practitioners working in this domain of research from academia and industry to share their experiences and to exchange their ideas. IIC 2018 also serves as a platform for the dissemination of state-of-the-art developments in intelligent and interactive computing technology, robotics, automation and its applications.

This volume of proceedings provides a prospect for readers to encompass with a collection of stated papers that were presented during IIC 2018. Divided into four parts, the 46 papers published here address the theme, "Intelligent and Immersive Computing towards Industrial Revolution 4.0". Part I deals with the theory of and applications for computational intelligence using fuzzy system, rough set, neural networks and deep leaning. Part II continues with applications of various soft-computing techniques, such as Bat algorithm, cuckoo search and artificial bee colony algorithm. Part III deals with system and security theories and applications. Part IV focuses on both multimedia and immersive technologies.

Finally, it is hoped that this book will be able to disseminate required knowledge to the researchers, policymakers, students and academicians in the fields of computational intelligence, machine learning, multimedia technologies and immersive technologies. The book will be a valuable resource to graduate students and anyone interested to conduct research in the said domains.

The volume editors want to convey sincere thanks to all the contributors, reviewers and the editorial board members for making this effort a successful one. The editors also appreciate the help and understanding of the editorial staff at Springer, who supported the publication of this volume in the Lecture Notes in Networks and Systems (LNNS) series.

Milan, Italy	Vincenzo Piuri
Arad, Romania	Valentina Emilia Balas
Sikkim, India	Samarjeet Borah
Durian Tunggal, Malaysia	Sharifah Sakinah Syed Ahmad

Contents

Part II   Applications of the Soft-Computing Techniques

About the Editors

Vincenzo Piuri is a Professor at the University of Milan, Italy (since 2000). He was an Associate Professor at Politecnico di Milano, Italy, a Visiting Professor at the University of Texas at Austin, USA, and a Visiting Researcher at George Mason University, USA. He has founded a start-up company in the area of intelligent systems for industrial applications and is active in industrial research projects. He received his M.S. and Ph.D. in Computer Engineering from Politecnico di Milano, Italy. His main research and industrial application interests are: artificial intelligence, intelligent systems, computational intelligence, pattern analysis and recognition, machine learning, signal and image processing, biometrics, intelligent measurement systems, industrial applications, distributed processing systems, Internet-of-Things, cloud computing, fault tolerance, application specific digital processing architectures, and arithmetic architectures. He published over 400 papers in international journals, international conference proceedings, and books. He is Fellow of the IEEE and Distinguished Scientist of ACM. He has been IEEE Vice President for Technical Activities (2015), member of the IEEE Board of Directors (2010-2012, 2015), and President of the IEEE Computational Intelligence Society (2006-2007). He is Editor-in-Chief of the IEEE Systems Journal (2013-2019).

Valentina Emilia Balas is currently a Full Professor at the Department of Automatics and Applied Software at the Faculty of Engineering, "Aurel Vlaicu" University of Arad, Romania. She holds a Ph.D. in Applied Electronics and Telecommunications from the Polytechnic University of Timisoara. Dr. Balas is the author of more than 250 research papers in refereed journals and international conferences. Her research interests are in intelligent systems, fuzzy control, soft computing, smart sensors, information fusion, modeling and simulation. She is the Editor-in-Chief of the International Journal of Advanced Intelligence Paradigms (IJAIP) and the International Journal of Computational Systems Engineering (IJCSysE), a member of the editorial board of several national and international journals and is the Director of the Intelligent Systems Research Centre at

Aurel Vlaicu University of Arad. She is a member of EUSFLAT, SIAM, a senior member of IEEE, and a member of TC – Fuzzy Systems (IEEE CIS), TC – Emergent Technologies (IEEE CIS), and TC – Soft Computing (IEEE SMCS).

Samarjeet Borah is currently working as Professor in the Department of Computer Applications, Sikkim Manipal University (SMU), Sikkim, India. Dr. Borah handles various academics, research and administrative activities. He is also involved in curriculum development activities, board of studies, doctoral research committee, IT infrastructure management etc. along with various administrative activities under SMU. Dr. Borah is involved with various funded projects in the capacity of Principal Investigator/Co-principal Investigator. The projects are sponsored by agencies like AICTE (Govt. of India), DST-CSRI (Govt. of India), Dr. TMA Pai Endowment Fund etc. He is associated with IEEE, ACM (CSTA), IAENG and IACSIT. Dr. Borah organized various national and international conferences in SMU. Some of these events include ISRO Sponsored Training Programme on Remote Sensing & GIS, NCWBCB 2014, NER-WNLP 2014, IC3-2016, ICACCP 2017, IC3-2018 etc. Dr. Borah is involved with various book volumes and journals of repute in the capacity of Editor/Guest Editor such as IJSE, IJHISI, IJGHPC, IJIM, IJVCSN, JISys, IJIPT, IJDS etc.

Sharifah Sakinah Syed Ahmad is currently an Associate Professor in the Department, Intelligent Computing & Analytics (ICA), Faculty of Information & Communication Technology, Universiti Teknikal Malaysia Melaka (UTeM). She received her Bachelors and Masters degrees of Applied Mathematics in School of Mathematics from University Science Malaysia. Following this, she received her Ph.D. from the University Of Alberta, Canada in 2012 in Intelligent System. Her research in graduate school focused on granular computing and fuzzy modeling. Her current research work is on the granular fuzzy rule-based system, Evolutionary Method, Modeling and Data Science. Sakinah has been a member of numerous program committees of National and International conferences in the area of Intelligent System and Soft Computing. Sharifah Sakinah is intensively involved in organizing and reviewing activities. She is a member of IEEE Systems, Man, and Cybernetics and Association for Computing Machinery (ACM). Sharifah Sakinah has co-authored numerous journal publications, conference articles and book chapters on the topics above. She also has received several grants from various funding agencies including the Ministry Of Education Malaysia, and the Ministry of Science, Technology, and Innovation Malaysia.

Part I
Theory of and Applications
for Computational Intelligence

Dynamic Solution Probability Acceptance Within the Flower Pollination Algorithm for Combinatorial t-Way Test Suite Generation

Abdullah B. Nasser, Kamal Z. Zamli and Bestoun S. Ahmed

Abstract In this paper, the enhanced Flower Pollination Algorithm (FPA) algorithm, called imFPA, has been proposed. Within imFPA, the static selection probability is replaced by the dynamic solution selection probability in order to enhance the intensification and diversification of the overall search process. Experimental adoptions on combinatorial t-way test suite generation problem (where t indicates the interaction strength) show that imFPA produces very competitive results as compared to existing strategies.

Keywords Search-based software engineering · Meta-heuristic · Flower pollination algorithm · t-way testing · Test suite generation

1 Introduction

Meta-heuristic is higher level of stochastic methods that attempt to escape from local optimum, by applying intelligent concepts of exploring and exploiting search space. Over the years, meta-heuristic algorithms have proven successful for solving hard optimization problems in many areas of computer science and software engineering. Many of its applications include software design, project planning and cost estimation, requirement engineering, network packet routing, protein structure prediction, software measurement, and software testing [1–6]. In recent years, active research on meta-heuristic algorithms have resulted in much advancement in the literature. Some

A. B. Nasser · K. Z. Zamli (✉)
Faculty of Computer Systems and Software Engineering, Universiti Malaysia Pahang,
26300 Kuantan, Pahang, Malaysia
e-mail: kamalz@ump.edu.my

A. B. Nasser
e-mail: abdullahnasser83@gmail.com

B. S. Ahmed
Department of Computer Science, Faculty of Electrical Engineering,
Czech Technical University, Prague, Czech Republic
e-mail: bestoon82@yahoo.com

© Springer Nature Singapore Pte Ltd. 2019
V. Piuri et al. (eds.), *Intelligent and Interactive Computing*, Lecture Notes in Networks and Systems 67, https://doi.org/10.1007/978-981-13-6031-2_4

of the developed meta-heuristic algorithms include Gray Wolf algorithm (GW), Bat Algorithm (BA), Fruitfly Algorithm (FA), Whale Algorithm (WA), Jaya Algorithm (JA), Differential Evolution (DE), Great Deluge Algorithm (GDA), Symbiotic Optimization Search (SOS), Particle Swarm Optimization (PSO), Sine Cosine Algorithm (SCA) as well as Firefly Algorithm (FA), to name a few [7].

Most of abovementioned algorithms use parameters to guide and control the local and global parts of search process (via intensification and diversification). Here, intensification explores the promising neighboring regions and diversification to ensure that regions of interests in the search space have been sufficiently explored. For example, GA uses crossover, mutation, and selection operators, and TLBO uses teacher and the learner's operator.

Complementing existing work, our work focuses on enhancing the exploration (i.e., ensuring diversity of solution) and exploitation (i.e., manipulating the current best to get better solution) of the Flower Pollination Algorithm (FPA). We argue that existing work on FPA [8, 9] uses a simple probability p_a to control the exploration (i.e., global pollination) and exploitation (i.e., local pollination). As the search space is dynamic (i.e., based on the given configuration), any fixed and preset p_a can be counter-productive as far as exploration and exploitation are concerned. In fact, there should be more exploration (i.e., via global pollination operator) in the early part of the searching process. Then, towards the end, there should be more exploitation (i.e., via local pollination operator). To do so, FPA needs to do away with its p_a (i.e., probability selection of operators) allowing both global and local pollination to run in sequence. Here, the probability is introduced as a dynamic parameter to select the solution itself (rather than the operators). Initially, the probability is high to accept all solution (for diversification). This probability will decrease with time and only allows good solution to be accepted thereafter.

In this paper, the enhanced Flower Pollination Algorithm (FPA) algorithm, called imFPA, has been proposed. As a case study, we adopt the problem from Search-Based Software Engineering domain involving the t-way combinatorial test suite generation problem [6, 10, 11] (where t represents the interaction strength). Experimental results show that imFPA produces very competitive results as compared to existing strategies.

2   Overview on Combinatorial Testing

Combinatorial testing (or t-way testing) is a sampling technique to minimize the interaction test suite size in a systematic manner. Within t-way testing (where t indicates the interaction strength), all t-combinations of system's components must be covered at least once. As illustration, Fig. 1 shows the window of power and sleep setting in Windows 10. The window consists 4 components (i.e., time to turn off screen on battery power, time to turn off screen when plugged in, time to sleep on battery power and time to sleep when plugged in), each component or input has 16

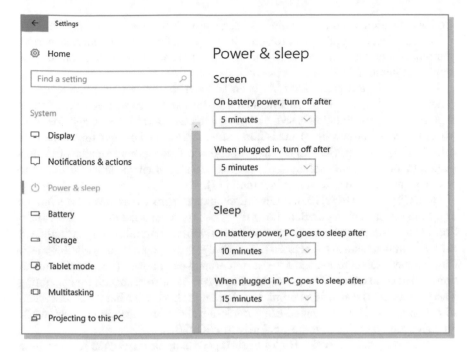

Fig. 1 Customized setting for power and sleep in Windows 10

values (i.e., represented as 1, 2, 3, 5, 10, 15, 20, 25, 30, and 42 min as well as 1, 2, 3, 4, 5 h, and one value termed never).

Ideally, there is a need to test all possible interactions between the inputs' values with 65,536 test cases. However, if only every two pairs (2-way testing) of the components are considered, the final test suite size can be reduced to 310 test cases. Here, the final test suite should include all possibilities of any two inputs at least once, which can detect 76% of software failure [12], while 3-way testing can minimize the test cases to 5300.

3 Related Works

Recently, there are active research on adopting meta-heuristic algorithms for addressing the t-way test suite generation problem. In general, meta-heuristic-based strategies exploit a random set of solutions, called *population*. The population is perturbed using a sequence of mathematical transformation (i.e., population update) in order to find better solution. At each particular iteration, the best candidate is adopted in the final test suite.

GA and ACA [13] are the first population-based algorithm has been used for generating t-way testing. GA adopted three operators, selection, crossover, and mutation, for diversification and intensification. ACA are probability algorithm controlled by 6 parameters (i.e., pheromone control, pheromone persistence, maximum stale period, pheromone amount, elite ants, and heuristic controls) apart from population size and number of iteration [13].

Due to its performance, PSO has been adopted for t-way interaction test suite generation in [14, 15]. The mathematical transformation adopted by PSO controls exploration and exploitation based on three parameters (termed social and cognitive parameter, and inertia weight) and is implicitly implemented by updating its particle velocity [16]. Unlike PSO, HS adopts probabilistic-gradient mathematical transformation to move to better solution. In addition, HS also employs elitism selection to ensure sufficient diversification of solution [17].

Both CS [18] and FPA [19] use Lévy flight in search process as the main technique to update the current population. Lévy flight can be considered as both diversification and intensification depending on generated step size. Addition to Lévy flight, CS uses elitism mechanism to increase the diversity of the population, while FPA uses learning mechanism to intensive the search around each flower (i.e., Local pollination). Apart from CD and FPA, Bat Algorithm (BA) has also been used for generating t-way test suite. The algorithm mimics the hunting behavior of Microbats. The intensification part in is implemented using random walk around the current best, while diversification part shares some similarity to PSO [20].

Unlike many earlier works, High-Level Hyper-Heuristic (HHH) [21] adopts more than one algorithm for generating t-way test suite. In the current implementation, HHH uses five meta-heuristic algorithms; TS as high-level selection algorithm, as well as PSO, CS and GNA as low-level algorithms. Adaptive TLBO (ALTBO) [22] has also been used for generating test suite. ATLBO improves the selection of local and global search operations through the adoption of fuzzy inference rules. Apart from ATLBO, Q-learning with Sine Cosine Algorithm (QSCA) [23] has also recently been adopted to address the t-way generation problem.

4 Proposed Strategy

The standard FPA starts by generating a number of random population, then applies global pollination or local pollination based on the switch probability $p_a \in [0, 1]$, [24]. The complete pseudo code of the standard FPA is presented in Fig. 2.

Unlike FPA, imFPA adopts a new mechanism for selecting local and global pollination. Like FPA, the new generated solution (i.e., either from global pollination or local pollination) will be accepted if it is better than the current solution. If not, the updated solution will only be accepted with probability $P(f(x), f(x_{new}), \text{success\_rate})$ as shown in Fig. 3. In other words, the selection mechanism considers the total success rate of global and local pollinations in order to decide whether to accept or reject the poor current solution. By doing so, imFPA allows a systematic way to get out of local optima.

1. Objective function f(x), x = $(x_1, ..., x_d)$;
2. Initial a population of n flowes x_i (i = 1, 2, ..., n);
3. **while** (t <*MaxGeneration*) or (stop criterion)
4.   **for** i = 1 : n (all n flowers in the population) **Do**
5.   **if** (rand < p)
6.   Draw a (d-dimensional) step vector L which
7.   obeys a L´evy distribution
8.   Global pollination via $x_i^{t+1} = x_i^t + \gamma L\acute{e}vy\,(\lambda)(x_i^t - gbest)$
9.   **else**
10.   Randomly choose j and k among all the solutions
12.   local pollination $x_i^{t+1} = x_i^t + \epsilon(x_j^t - x_k^t)$
13.   **End if**
14.   Evaluate new solutions
15.   If new solutions are better, update them
16.   Find the current best solution *gbest*
17.   **end while**
18.   Postprocess results;
19. **End-Procedure**

Fig. 2 Pseudocode of FPA algorithm

Input: (P, v, t):   Set of parameters P, its parameter-value v and interaction strength
Output: TS final test cases
1. Generate interaction tuples IE list
2. Let *TS* be a set of final test cases
3. Generate random initial population of Pop_{min} pollen
4. **while** the interaction tuples (IE) are not empty **do**
5.   **while** t < *MaxGeneration* or stop criterion is not reached **do**
6.   *for* (each x_t^i in the population)
7.   Perform global pollination, $x_t^{i+1} = x_t^i + L(gbest - x_t^i)$
8.   *if* (x_t^{i+1} better than x_t^i)
9.   Accept x_t^{i+1}
10.   *Global_success_rate*++
11.   **else**
12.   accept x_t^{i+1} with probability P(f(x), f(x_new), *Global_success_rate*)
13.   **end if**
14.   Randomly choose 2 index j and k for all the solutions index where j≠k
15.   Perform local pollination, $x_t^{i+1} = x_t^i + \varrho(x_t^j - x_t^k)$
16.   Perform global pollination, $x_t^{i+1} = x_t^i + L(gbest - x_t^i)$
17.   *if* (x_t^{i+1} better than x_t^i)
18.   Accept x_t^{i+1}
19.   *Local_success_rate*++
20.   **else**
21.   Accept poor solution with probability P(f(x), f(x_new), *Local_success_rate*)
22.   **end if**
23.   **end for**
24.   Find the current best pollen *gbest*
25.   **end while**
26.   Add the best pollen *gbest* into TS.
27.   Remove covered interactions elements from IE list
28. **end while**

Fig. 3 Pseudocode of imFPA for *t*-way test suite generation

5   Results and Evaluation

In this section, imFPA is compared with state-of-the-art t-way strategies including SA, GA, ACA, PSO, HSS, HHH, CS, and FPA. Different system configurations are used as shown in Table 1. The first and second columns, where x^y indicates that the system configuration that has y parameters with x values, and t is the required interaction. Each cell in Table 1 displays the best test case obtained by each strategy, while the darkened cell displays the best minimum size obtained. The NA entries indicate that no published results are available for the case of interest. For experimental setup, the values of population size $= 500$, and iteration $= 500$ have been used. Each configuration is run 30 times, and the best result obtained is recorded.

Table 1 depicts that the results of imFPA are very competitive with the existing strategies. Specifically, imFPA obtains the most minimum results for five cases (S1, S3, S8, S11, and S14), while HHH obtains the best result for six cases. In general, we can observe that meta-heuristic based strategies, such as HS, CS, HHH, FPA, and imFPA, perform better than other strategies.

In order to assess the contribution of imFPA and investigate the behavior of FPA and imFPA, global and local pollination for both FPA and imFPA have been tracked. As Fig. 4 shows, the mean percentage of global and local imFPA dynamically change based on the problem, however, in FPA, it remains the same for all problem instances. Comparatively, the dynamic selection within imFPA can be compared with the selection of candidate solution in SA. In SA, the selection probability (of poor solution) always decreases with time. Unlike SA, the selection probability in imFPA increases

Table 1   Comparison of imFPA with state-of-the-art strategies

Systems			SA	GA	ACA	PSO	HSS	HHH	CS	FPA	imFPA
No.	x^y	t									
S1	3^4	2	9	9	9	9	9	9	9	9	9
S2	3^{13}	2	16	17	17	17	18	NA	NA	18	18
S3	10^{10}	2	NA	157	159	170	155	NA	NA	153	151
S4	5^5	4	NA	NA	NA	779	751	746	776	784	790
S5	5^6	4	NA	NA	NA	1001	990	967	991	988	988
S6	5^7	4	NA	NA	NA	1209	1186	1151	1200	1164	1165
S7	2^{10}	4	NA	NA	NA	34	37	36	28	36	36
S8	3^{10}	4	NA	NA	NA	213	211	207	211	211	205
S9	4^{10}	4	NA	NA	NA	685	691	668	698	661	657
S10	2^{10}	2	NA	NA	NA	8	7	8	8	8	8
S11	2^{10}	3	NA	NA	NA	17	16	16	16	16	16
S12	2^{10}	4	NA	NA	NA	37	37	36	36	35	37
S13	2^{10}	5	NA	NA	NA	82	81	79	79	81	82
S14	2^{10}	6	NA	NA	NA	158	158	153	157	158	153

Fig. 4 Mean of global and local search percentage of FPA and imFPA

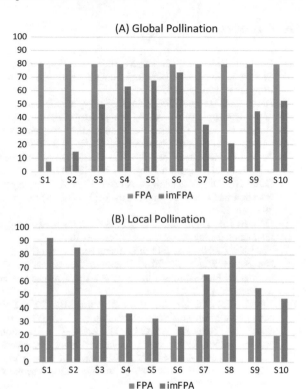

or decreases based on the need of the search process. In this manner, imFPA provides an adaptive way to decide on whether or not to select or discard poor solution.

6 Concluding Remarks

In this paper, an improved algorithm based on Flower Pollination Algorithm (FPA), called imFPA, is proposed. imFPA provides a new mechanism for candidate solution's selection. The experimental results show that imFPA produces competitive results comparing with state-of-the-art strategies. As part of the future work, we are currently investigating imFPA for constraints software product lines test suite generation.

Acknowledgements This work is funded by "FRGS Grant from the Ministry of Higher Education Malaysia titled: A Reinforcement Learning Sine Cosine based Strategy for Combinatorial Test Suite Generation (grant no: RDU170103)".

References

1. Clarke J, Dolado JJ, Harman M, Hierons R, Jones B, Lumkin M, Mitchell B, Mancoridis S, Rees K, Roper M (2003) Reformulating software engineering as a search problem. IEEE Proc-Softw 150:161–175
2. Harman M, Mansouri SA, Zhang Y (2012) Search-based software engineering: trends, techniques and applications. ACM Comput Surv (CSUR) 45:11 (2012)
3. Harman M (2007) The current state and future of search based software engineering. In: Future of software engineering conference (FOSE), pp 342–357
4. Alsewari ARA, Zamli KZ (2012) A Harmony search based pairwise sampling strategy for combinatorial testing. Int J Phys Sci 7:1062–1072
5. Zamli KZ, Din F, Kendall G, Ahmed BS (2017) An experimental study of hyper-heuristic selection and acceptance mechanism for combinatorial t-way test suite generation. Inf Sci 399:121–153
6. Othman RR, Zamli KZ (2011) ITTDG: Integrated t-way test data generation strategy for interaction testing. Sci Res Essays 6:3638–3648
7. Yang X-S (2010) Nature-inspired metaheuristic algorithms. Luniver press
8. Nasser AB, Zamli KZ (2018) Parameter free flower algorithm based strategy for pairwise testing. In: Proceedings of the 2018 7th international conference on software and computer applications, pp 46–50
9. Nasser AB, Zamli KZ, Alsewari AA, Ahmed BS (2018) Hybrid flower pollination algorithm strategies for t-way test suite generation. PLoS ONE 13:e0195187
10. Younis MI, Zamli KZ, Isa NM (2008) MIPOG-modification of the IPOG strategy for t-way software testing. In: Proceeding of the distributed frameworks and applications (DFmA)
11. Younis MI, Zamli KZ, Isa NAM (2008) Algebraic strategy to generate pairwise test set for prime number parameters and variables. In: International symposium on information technology, 2008. ITSim 2008, pp 1–4
12. Kuhn DR, Wallace DR, Gallo AM (2004) Software fault interactions and implications for software testing. IEEE Trans Softw Eng 30:418–421
13. Shiba T, Tsuchiya T, Kikuno T (2004) Using artificial life techniques to generate test cases for combinatorial testing. In: 28th annual international computer software and applications conference, pp 72–77
14. Ahmed BS, Zamli KZ, Lim CP (2012) Constructing a t-way interaction test suite using the particle swarm optimization approach. Int J Innov Comput Inf Control 8:431–452
15. Chen X, Gu Q, Qi J, Chen D (2010) Applying particle swarm optimization to pairwise testing. In: IEEE 34th annual on computer software and applications conference (COMPSAC), pp 107–116
16. Lei L, M Xin, Xiaokui L (2013) Research on hybrid PSO algorithm with appended intensification and diversification. In: International conference on mechatronic sciences, electric engineering and computer (MEC), pp 2359–2363
17. Geem ZW (2006) Optimal cost design of water distribution networks using harmony search. Eng Optim 38:259–277
18. Ahmed BS, Abdulsamad TS, Potrus MY (2015) Achievement of minimized combinatorial test suite for configuration-aware software functional testing using the cuckoo search algorithm. Inf Softw Technol 66:13–29
19. Nasser AB, Alsariera YA, Alsewari ARA, Zamli KZ (2015) Assessing optimization based strategies for t-way test suite generation: the case for flower-based strategy. In: Presented at the 5th IEEE international conference on control systems, computing and engineering, Pinang, Malaysia
20. Alsariera YA, Zamli KZ (2015) A bat-inspired strategy for t-way interaction testing. Adv Sci Lett 21:2281–2284
21. Zamli KZ, Alkazemi BY, Kendall G (2016) A tabu search hyper-heuristic strategy for t-way test suite generation. Appl Soft Comput 44:57–74

22. Zamli KZ, Din F, Baharom S, Ahmed BS (2017) Fuzzy adaptive teaching learning-based optimization strategy for the problem of generating mixed strength t-way test suites. Eng Appl Artif Intell 59:35–50
23. Zamli KZ, Din F, Ahmed BS, Bures M (2018) A hybrid Q-learning sine-cosine-based strategy for addressing the combinatorial test suite minimization problem. PLoS ONE 13:e0195675
24. Yang X-S (2012) Flower pollination algorithm for global optimization. In: International conference on unconventional computing and natural computation, pp 240–249

Aggregating Multiple Decision Makers' Judgement

Jeremy Y. L. Yap, Chiung Ching Ho and Choo-Yee Ting

Abstract Selecting the best location to establish a new business site is very important in order to achieve success. It is therefore one of the most important aspect in any business plan. Multi-criteria decision-making methods such as the Analytic Hierarchy Process (AHP) has been used to elicit information that supports the decision of business site selection. However, AHP often involves multiple decision makers, each with their own opinions and biases. Different decision makers will have different opinions and views on the importance of the criteria and sub-criteria in the AHP model. In this study, three aggregation methods that can be used to carefully aggregate the resultant judgements from the multiple decision makers to form a single group judgement are discussed. The goal of obtaining the single group judgement is to use it as input to the AHP model in order to achieve the goal of selecting the most suitable business location. The study case for this paper is that of the selection of a location for a telecommunication payment point. From this study case, a conclusion can be drawn for the best aggregation method for the selection of the best location to set up a business of the telecommunication nature.

Keywords AHP · Decision support · Location · Site selection · Data analytics · Decision analysis

1 Introduction

Selecting the correct location for a business site is one of the most important strategic decision that decision makers have to face [1]. This decision plays a role in determining the success of the business in monetary terms [2] as well as the image of the

J. Y. L. Yap (✉) · C. C. Ho · C.-Y. Ting
Multimedia University, Cyberjaya, Malaysia
e-mail: jeremyyap.kylmac@gmail.com

C. C. Ho
e-mail: ccho@mmu.edu.my

C.-Y. Ting
e-mail: cyting@mmu.edu.my

© Springer Nature Singapore Pte Ltd. 2019
V. Piuri et al. (eds.), *Intelligent and Interactive Computing*, Lecture Notes in Networks and Systems 67, https://doi.org/10.1007/978-981-13-6031-2_26

13

company [3]. Therefore, the selection needs to undergo thorough judgements that may involve several criteria and a number of decision makers.

When facing complex decision problems, multi-criteria decision methods such as the Analytic Hierarchy Process (AHP) method is often used to aid decision makers in finding a solution to the problem. The resultant ranking from the AHP model is used to aid decision makers in selecting the best site to establish a business. However, AHP often involves multiple decision makers, each with their own opinions and biases. Different decision makers will have different opinions and views on the importance of the criteria and sub-criteria in the AHP model. It is therefore important to carefully aggregate the resultant judgements in order to form a single group judgement that can be used as input to the AHP model in order to achieve the goal of selecting the most suitable business site.

Many studies have been done to propose methods to aggregate the judgements to form a single group judgement. With the availability of these aggregation methods, the selection of the best method to be used in deciding the best site to set up a business will be required. Comparisons of the various aggregation method have been studied in the past [4, 5], however there have not been any studies on the application of these methods on the selection of a telecommunication payment point. A telecommunication payment point in this study is the business site in which customers will go to in order to make payments for their telecommunication services and any other matters that relate to the telecommunication company.

This study looks into the judgements of four different decision makers on which of the four alternative is the best site for the telecommunication payment points, and the judgements are then aggregated to find the ranking of the alternatives. These four alternatives were chosen as they were the top four performing alternative in the scope of Klang Valley. Only four alternatives were chosen to limit the amount of pairwise comparisons to be performed. This ranking is then compared to the real-world rankings of these sites based on their sales performance to validate if the AHP model and the aggregation method.

In Sect. 2, the AHP model that is used to select the best telecommunication payment point is discussed. It is then followed by a discussion on the aggregation methods that were used in this study in Sect. 3. In Sect. 4, the resultant ranking of applying three of the aggregation methods in this study is shown. In Sect. 5, two rank correlation coefficients were used and discussed to determine the degree of similarity of the resultant ranking of the AHP model and the real-world ranking. Finally, in the Sect. 6, a recap of the whole study and discussion on future steps to be undertaken following this study is shown.

2   AHP Model for Telecommunication Payment Point

This study focuses on the selection of the best location for a telecommunication payment point in Selangor, a state in Malaysia. The scope of the case study is in the area of Klang Valley. The case study involved a questionnaire given out to decision

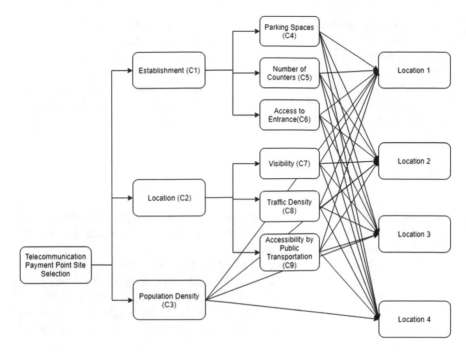

Fig. 1 AHP model for telecommunication payment point

makers of the telecommunication payment point to perform pairwise comparisons between the criteria, sub-criteria, and location. There were four (4) decision makers involved in partaking of filling the questionnaire. The breakdown of the AHP model is: three (3) criteria, six (6) sub-criteria; and four (4) locations. Figure 1 shows the AHP model for telecommunication payment point based on the model proposed by Yap et al. [6].

2.1 Criteria 1: Establishment

To find the most suitable location in order to set up the business site, decision makers have to ensure that the location is appropriate and is able to accommodate and serve its customers. In this case study of selecting the location for a telecommunications payment point, the establishment that is chosen needs to be able to cater for the number of customers that visit the payment point. The sub-criteria that were found to be contributing to the criteria of Establishment are: (i) Parking Spaces, (ii) Number of Counters, and (iii) Access to Entrance.

The number of parking spaces plays a significant role in determining if an establishment is suitable as many customers would go to the site using their own means of transport. The number of parking spaces need to be adequate.

A long waiting time should be avoided in order to satisfy the needs of the customers. Hence, the number of counters that can be set up in the establishment needs to cater to the expected number of customers. More counters would enable a higher number of customers to be serviced hence reducing the waiting time.

Access to entrance refers to how easily a customer would be able to enter or exit the establishment [7]. The access to entrance of the establishment is determined by the presence of automatic sliding doors, ramps, or stairs [6].

2.2 Criteria 2: Location

When determining the location to set up a new business site, there are several aspects of the location that have to be taken into account. The aspects include (i) Visibility, (ii) Traffic Density, and (iii) Accessibility.

Visibility is defined as the ability of customers to view the business site [6]. A site that is easily seen from the walkway or main road will enable the customers to see it. A structure that can be used as a landmark will serve to ease visibility [6].

The traffic density of the location serves to indicate the number of cars passing the business site. When the area is heavily dense, it shows the potential of the site having more customers.

Decision makers must take into account customers that travel using public transportation. Therefore, the selection of site must take into consideration the accessibility of the site by public transportation [6].

2.3 Criteria 3: Population Density

The population density of an area needs to be considered when selecting the location to set up a business site. An area with a higher population density will garner more revenue for the business as the larger number of customers will be in the area.

3 Three Aggregation Methods

Aggregation can be performed at the individual level [8], using techniques such as: (i) aggregation of individual judgement for each pairwise comparison into an aggregate hierarchy; (ii) aggregating each individual's hierarchy and subsequently aggregating the resulting priorities, and (iii) aggregating the individual's derived priorities in every node in the hierarchy. Techniques (i) and (ii) are known as aggregating individual judgements (AIJ) and aggregating individual priorities (AIP) respectively and are techniques which are of interest in this study.

Fig. 2 Flowchart of AIJ
aggregation method

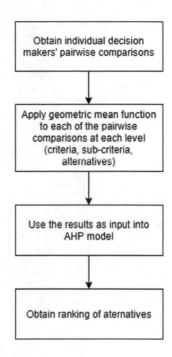

AIJ occurs when individuals in a group agree to forgo their individual preferences and act in tandem—essentially merging their judgements to arrive at a synthetic new "individual". Using AIJ, an individual's judgement may be revised by the group if said judgment results in too high inconsistencies. This technique does not take into account individual priorities, thus ensuring the Pareto principle is not violated. As this new "individual" needs to satisfy the reciprocity requirement for judgments, only geometric mean can be applied when using AIJ. The weighted geometric mean may be applied to AIJ should the individuals in the group be deemed to be not of equal importance. Figure 2 shows the flow of the aggregation method.

AIP is applied when there is interest in each individual's resulting alternative priorities. The aggregation happens at the resulting alternative priorities level, using either geometric or arithmetic mean. Similar to AIJ, AIP does not violate the Pareto principle and individuals involved in AIP can also be weighted to indicate different levels of importance. Figure 3 shows the flow of the AIP aggregation method.

The Borda Count (BC) method is another aggregation method by producing an ordering of N alternative ideas, ranked from best to worst. Once every individual in a group ranks all the options in order of preference, the top-ranked idea is awarded a score of $N - 1$, and the last ranked idea is awarded a score of 0. In summary, the K-th ranked idea is awarded a score of $N - K$. The BC method has been applied in diverse areas ranging from electoral voting, to decision-level aggregation for multi-modal biometric systems.

Fig. 3 Flowchart of AIP
aggregation method

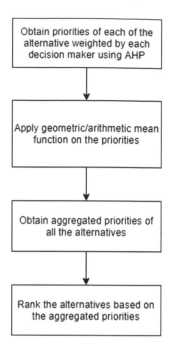

BC has been applied [9, 10], to affect hybrid AHP-Borda approaches resolving preferences during decision-making processes involving multiple people. The hybrid AHP-Borda generic workflow is shown below:

a.   Perform AHP pairwise comparisons at the criteria or sub-criteria level.
b.   Once the AHP weights have been determined, BC is used to rank the alternatives based on the AHP weights given by a person. The alternatives with the highest weight will get $N - 1$ points, while the lowest weight will be awarded zero points. In the event of a tied weight, a tie-breaker of 0.001 is added from one tied weight and 0.001 is subtracted from the other tied weight.
c.   The weights from AHP are multiplied by their corresponding Borda count and are summed up for each alternative. A weight is calculated by dividing the total score for an alternative by the sum of scores.
d.   The scores for each person involved in the decision-making process is added to result in an influence index, the more influence a person has on the group decision-making process. Thus, a person who has a high influence on the wrong set of decisions can be moderated in some ways (by means of weight for example) in order to reduce their influence.

Table 1 Real-world ranking vs AIJ versus AIP versus Borda count

Rank	Real-world ranking	Sales (RM)*	AIJ	AIP	Borda count
1	A1	2,527,619.49	A1	A1	A1
2	A2	2,060,376.51	A2	A2	A2
3	A3	1,942,275.63	A4	A3	A4
4	A4	1,904,286.88	A3	A4	A3

*Sales data are masked

4 Results

Upon receiving the resultant pairwise comparisons from the decision makers, the three aggregation methods discussed above were applied to the judgements. Table 1 presents the ranking of the four alternatives or candidate sites after applying the three different aggregation methods to the decision makers' judgements. The results were also compared to the actual ranking of the alternatives based on their sales performance as a form of validation. The sales performance data was masked for confidentiality purposes but retained their relative ranking.

Based on the comparison of the results of using each aggregation method and real-world sales data of the telecommunication payment points, the aggregating individual priorities (AIP) method is the best aggregation method as the rankings of the telecommunication payment points resulted from this method is the same as the ranking of the payment points based on their sales data.

5 Discussion

In order to determine the rank correlation between the resultant ranking and the real-world ranking of the telecommunication payment points, two rank correlation coefficients were used: (i) Spearman's rho and (iii) Kendall's Tau. These rank correlation coefficients enable us to measure the degree of similarity between the two rankings and therefore determine the significance of the relation between them.

When using the Spearman's Rho coefficient to determine the rank correlation between the resultant ranking using the AIJ and Borda Count methods against the real-world ranking, the degree of similarity is +0.8 which indicates a high degree of similarity with +1 being very similar and −1 being inversely ranked.

When using the Kendall's Tau coefficient on the other hand, the degree of similarity is +0.67 which is still considered a high degree of similarity with the same scale of similarity as that of Spearman's Rho (+1 being very similar and −1 being inversely ranked).

Table 2 Degree of similarity of the result of applying three different aggregation methods

	AIJ	AIP	Borda count
Spearman's rho	+0.8	1	+0.8
Kendall's tau	+0.6666667	1	+0.6666667

Kendall's Tau will often result in a smaller value to that of Spearman's Rho. However, in most situations, the interpretations of the two rank correlation coefficients are very similar and will often lead to the same conclusion.

Results showing the Spearman's rho and Kendall's Tau measures for all three aggregation methods are shown in Table 2.

6   Conclusion

This study discusses the use of three different aggregation methods in order to form a single group judgement from multiple decision makers. The AHP model that was used in this study was proposed in 2017 by Yap et al. [6].

The three aggregation methods that were discussed and applied to the case study of the selection of a site for a telecommunication payment points were (i) Aggregating Individual Judgements (AIJ), (ii) Aggregating Individual Priorities (AIP), and (iii) Borda Count (BC). The results of the ranking of the alternatives from the usage of the three aggregation methods were compared to the real-world ranking of the alternatives based on their sales performance. By using rank correlation coefficients of Spearman's Rho and Kendall's Tau, both rank correlation coefficient resulted in a high degree of similarity between the resultant rankings and real-world rankings.

Based on the comparison and rank correlation coefficient results, the AIP method is considered as the best method of aggregation as it correctly ranked the payment points' performance when compared to real-world ranking of the payment points based on the sales data.

In the future, the case study would be performed again covering a large number of alternatives to test if the result of this study holds true for varying number of alternatives

Acknowledgements This study was supported by Telekom Malaysia.

References

1. Karande K, Lombard JR (2005) Location strategies of broad-line retailers: an empirical investigation. J Bus Res 58:687–695. https://doi.org/10.1016/j.jbusres.2003.09.008
2. Durvasula S, Sharma S, Andrews CJ (1992) STORELOC: a retail store location model based on managerial judgments. J Retail 68.420–444

3. Pope JA, Lane WR, Stein J (2012) A multiple-attribute decision model for retail store location. Southern Bus Rev 37:15
4. Grošelj P, Zadnik Stirn L, Ayrilmis N, Kuzman MK (2015) Comparison of some aggregation techniques using group analytic hierarchy process. Expert Syst Appl 42:2198–2204. https://doi.org/10.1016/j.eswa.2014.09.060
5. Bernasconi M, Choirat C, Seri R (2014) Empirical properties of group preference aggregation methods employed in AHP: theory and evidence. Eur J Oper Res 232:584–592. https://doi.org/10.1016/j.ejor.2013.06.014
6. Yap JYL, Ho CC, Ting C-Y (2017) Analytic Hierarchy process (AHP) for business site selection. In: Proceedings—2017 6th international conference on computer science and computational mathematics
7. Levy M, Weitz B (2001) Retailing management. McGrawHill 688. Retrieved from https://doi.org/10.1057/jors.1992.174
8. Claudio D, Chen J, Okudan GE (2008) AHP based Borda count: a hybrid multi-person decision making method for design concept selection. In: IIE annual conference. Proceedings. Institute of Industrial and Systems Engineers (IISE), p 776
9. Escobar MT, Moreno-Jiménez JM (2007) Aggregation of individual preference structures in AHP-group decision making. Group Decis Negot 16:287–301. https://doi.org/10.1007/s10726-006-9050-x
10. Forman E, Peniwati K (1998) Aggregating individual judgments and priorities with the analytic hierarchy process. Eur J Oper Res 108:165–169. https://doi.org/10.1016/S0377-2217(97)00244-0

A Novel Kohonen Self-organizing Maps Using Exponential Decay Average Rate of Change for Color Clustering

Edwin F. Galutira, Arnel C. Fajardo and Ruji P. Medina

Abstract Kohonen Self-Organizing Maps (KSOM) is the most preferred and commonly used algorithm for clustering high-dimensional or multi-dimensional data. Its success in various domains are undeniable but still, it suffers issues on its convergence which is directly affected by its learning rate decay function. This study adopted Average Rate of Change to modify the learning rate decay function of the algorithm and proposes a new Exponential Decay Average Rate of Change (EDARC) as the modified learning rate decay function of the Enhanced Kohonen Self-Organizing Maps (EKSOM) to address its issue on convergence. The enhanced algorithm and the conventional KSOM was applied to color clustering. The results show that EKSOM outperformed the clustering capability of the conventional KSOM algorithm.

Keywords Learning rate decay · High-dimensional data · Self-organizing maps · Rate of change · Color clustering

1 Introduction

Clustering high-dimensional analyses data from multi-dimensions. It requires a specific result from between clustering algorithms to solve its challenges that could not solve by a conventional algorithm that uses traditional similarity. Clustering has been widely used in different domains like artificial intelligence, bioinformatics, biology, computer vision, data compression, image analysis, information retrieval, machine

E. F. Galutira (✉) · R. P. Medina
Technological Institute of the Philippines, 938 Aurora Blvd., Cubao, Quezon City 1109, Philippines
e-mail: efgalutira@gmail.com

R. P. Medina
e-mail: ruji.medina@tip.edu.ph

A. C. Fajardo
La Consolacion College Manila, Mendiola St., San Miguel Manila, Philippines
e-mail: acfajardo2000@yahoo.com

© Springer Nature Singapore Pte Ltd. 2019
V. Piuri et al. (eds.), *Intelligent and Interactive Computing*, Lecture Notes in Networks and Systems 67, https://doi.org/10.1007/978-981-13-6031-2_28

23

learning, marketing, medicine, pattern recognition, spatial database analysis, statistics, recommendation systems, and web mining [1]. Various clustering algorithms used for high-dimensional or multi-dimensional data have developed such as the K-means, Genetic Algorithm (GA), Self-Organizing Maps (SOM), Multi-Layer Perceptron (MLP), and Particle Swarm Optimization (PSO) [1, 2].

Self-Organizing Maps (SOM) or Kohonen Self-Organizing Feature Maps (KSOM) is the most preferred and commonly used algorithm for clustering high-dimensional or multi-dimensional data. KSOM is an unsupervised neural network which represents high-dimensional or multi-dimensional data into one, two or in particular cases, into three-dimensional space called a map [2, 3]. The elementary of the SOM is the flexible competition of nodes in the output layer; not only one node (the winner) is updated, but also its neighbors are adjusted [4–6]. The Kohonen network can adapt to recognize groups of data and relatively similar classes to the others. The SOM has only two layers: the input layer and the output layer. The input layer is one-dimensional, while the output layer consists of radial units typically organized in two dimensions [4, 5, 7–9]. KSOM is used for a variety of tasks, and it is handy to find regularities in a large volume of data through it [10].

KSOM displayed undeniable success in clustering in the past. However, several papers mentioned some of its drawbacks [6, 11, 12]. These problems point out to the convergence of the network which is directly influenced by the learning rate of the algorithm. The learning rate controls the knowledge of the algorithm to the initial condition of the network and the amount to be learned from the sample data. The lower the learning rate, the slower the convergence, the higher the learning time but will produce the desired result. Whereas too high learning rate the lower the learning time but the data will never converge, and it will result in divergence [13]. Through this paper, it introduces the Exponential Decay Average Rate of Change (EDARC). It enhances the learning rate decay function of the algorithm by using EDARC as its new function. This function facilitates the algorithm in converging to the minimum and produce a more efficient result.

2   Related Literature

2.1   Kohonen Self-organizing Maps

Kohonen Self-Organizing Maps (KSOM) was developed by Professor Teuvo Kohonen of Finland in the early 1980s [12, 14–20]. It adopts unsupervised learning and represents high-dimensional data into low-dimensional data through competitive learning [21]. Figure 1 illustrates a conventional KSOM represented in a 2D lattice with the input vector, winner neuron, and neighboring neurons. The nodes within the network are determined by its specific topological position denoted by (x, y) coordinate sustaining a weights having similar dimensionality to the input data items.

Fig. 1 A SOM with input vectors, a winner cell, and its neighborhood

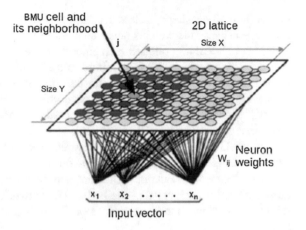

The data in the map or grid is associated with a distinct node within the network denoted as the Best Matching Unit (BMU) and correlated to neighboring nodes. The BMU determines an implicit clustering of data on a graph of similarity. The KSOM model has two interacting subsystems. First, winner-take-all is implemented within the elements of the neural network through competitive learning. Second, elements are regulated by the neural network and correct the local synaptic plasticity of the learning neurons. The learning process is limited to the neighborhood of the most active neurons. The radius of the neighborhood and the learning rate decreases as the algorithm progress as a function of time [11, 21–23]. The Kohonen Self-Organizing Map can be summarized and is broken down into six procedures as follows:

1. Initialize the map's nodes' weight vectors randomly, set the max value for R, set the learning rate
2. Randomly select an input vector and present it to the network
3. Iterate each node in the map and determine BMU based on input vector and all nodes using distance equation (Usually Euclidean Distance)
4. Compute current time learning rate and neighborhood radius
5. Update neighborhood node weights accordingly
6. Increment t and repeat until $t \geq \lambda$ (time constant)

The convergence of a network is greatly influenced by the choice of learning rate. Proper selection of learning rate (step size) schemes results in faster convergence and lower error rates [24, 25]. This will allow the algorithm to converge faster to the desired minimum and is highly influenced by the changes in the learning rate as the training process progresses compared to a constant value for learning rate. At the outset of the process, the algorithm can take larger steps allowing a significant increase in the objective function, but at the later stage of the process, lesser steps are important to settle into refined features of the loss landscape. A needless large learning rate deviates from the right path, moving around a valley of minimum in the error surface affecting the generalization accuracy. In effect, it will take the network to be trained longer because it overshoots the minimum repeatedly or it will not learn at all causing a useless oscillation. This will produce poor generalization accuracy because the weights never converge to the minimum. On the other hand, small enough learning rate allows a smooth descent to the minimum allowing it to converge. After which, it can continue to a smooth path over the error landscape resolving in a minimum. Reducing the learning rate slowly will make the steps finer improving the generalization accuracy positively. Nonetheless, a decrease in learning rate will arrive at a point where the further decay of the learning rate is useless and result in a waste of training time [25, 26].

The BMU in the network has the most significant influence and extending to its adjacent neurons. Upon presentation of the inputs to the nodes it determines the weighted sum using a Euclidian Distance and the nodes having the nearest matching output weight is designated as the BMU. The network will be trained based on the BMU and is extended to its neighborhood. This is accomplished using the following formula:

$$D(j) = \sqrt{\sum_{i=0}^{n}(V_i - W_i)^2}$$ (1)

V = current input vector
W = node's weight vector.

After determining the BMU, the neighborhood radius that shrinks over time must be determined. This is accomplished using an exponential decay formula similar to the function L.

The weights of the vectors inside the neighborhood will be adjusted. This is necessary to mimic the learning process of the algorithm. The process employs an exponential decay to lower down the overall change allowing the network to converge. The weights of the neighborhood are modified using the following formula:

$$W(t + 1) = W(t) + \Theta(t)L(t)(V(t) - W(t))$$ (2)

$W(t)$ = old weight vector
$\Theta(t)$ = neighborhood radius
$L(t)$ = learning rate
$V(t)$ = current input vector.

The current iteration is denoted by t and the learning rate that decreases over time is denoted by the function L. The weights are updated correlative to the distance of the BMU to its neighborhood. The learning rate fades over a range indicated by Θ. This is calculated using the following formula:

$$\Theta(t) = \exp\left(-\frac{\text{dist}^2}{2\sigma^2(t)}\right)$$ (3)

dist^2 = distance from BMU

$\sigma(t) =$ radius of the map.

The learning rate of the conventional KSOM is determined by alpha as a function of t over half. Later, it was enhanced and uses exponential decay learning rate denoted as the function L similar to the neighborhood's radius decrease as follows:

$$L(t) = L_0 \exp\left(-\frac{t}{\lambda}\right) \tag{4}$$

$L_0 =$ initial learning rate
$t =$ current iteration
$\lambda =$ time constant (numIterations/mapRadius).

In the previous formula, L_0 represents the width of the lattice at the time of initialization, λ is a time constant or the maximum iteration and t is the current iteration of the algorithm.

2.2 Average Rate of Change (ARC) Formula

By definition, the average rate of change is the average change in L per one unit of change in t. In determining the amount of change in $L(t)$ between point t_1 and point t_2, divide the change in L over the change in t. Therefore, the relative change of L with respect to t over a change in t is defined as

$$\frac{\Delta L}{\Delta t} = \frac{L(t2) - L(t1)}{t2 - t1} \tag{5}$$

The amount of change in L for a given change in t is computed by dividing ΔL by the Δt [27, 28].

3 Contribution

Kohonen Self-Organizing Maps (KSOM) has been exceptional in contrast to other clustering algorithms. In spite of its success, the performance of KSOM needs to be enhanced by improving its convergence which is greatly affected by the learning rate decay function of the algorithm. Enhancing Kohonen Self-Organizing Maps is conceivable by modifying its learning rate decay function using the new Exponential Decay Average Rate of Change (EDARC) function to be utilized by the Enhanced Kohonen Self-Organizing Maps (EKSOM) in clustering to come up with a more robust and satisfying result.

The Proposed EDARC Function

$$L(t) = L_0 \exp\left(-\frac{\Delta L}{\lambda}\right) \tag{6}$$

$L_0 =$ initial learning rate
$\Delta L =$ change in learning rate with respect to t.

To ensure that the learning rate of the algorithm in a given iteration will not wander within the valley of minimum the ΔL must be divided by the time constant denoted as λ.

To compute for ΔL the following formula is used:

$$\frac{\Delta L}{\Delta t} = \Delta L = \frac{L(t2) - L(t1)}{t2 - t1} \tag{7}$$

$$L(t_n) = t^2 + 1$$

The rate of change ΔL lies between two consecutive two points with a rate of change Δt having a constant value of 1 in any given consecutive two points in the equation. Since the value of Δt is always 1, only ΔL is considered following the rules of division of fractions which produces the same results.

4   Evaluation

The color clustering is one of the most commonly used clustering tests for neural networks. It has been widely used for color feature extraction, and the extracted colors are important in many applications like image analysis and compression [6]. The dataset used for evaluation is a computer-generated color pallets based from the three fundamental color components or the Red, Blue, Green (RGB) components. Following the traditional representation of pixels, data are expressed in vectors with three dimensions $[X_1, X_2, X_3]$ containing three bytes of information. The values of X ranges from 0 to 255 representing no coloration to full coloration of the components. The two networks were run with the same initial parameters of 0.5 initial learning rate and 1000 iterations for a more accurate comparison results. The networks produced a 200×200-pixel image with 50×50 nodes. The individual nodes are scaled within the image size for proper visualization of the result. The KSOM and EKSOM were subjected to two color clustering test. First, it uses predetermined sets of color to facilitate more accurate comparison of test results between the two networks. After which, the two networks were tested using a randomly generated color based on RGB.

As a result, Figs. 2 and 3 illustrates the results when the two networks were subjected for testing using the predetermined colors. Figure 2 shows that there are nodes

(a)   **(b)**

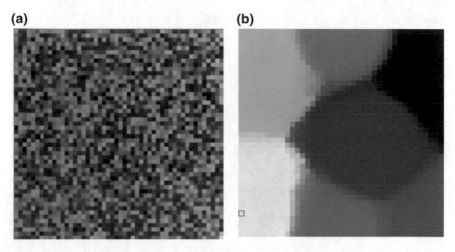

Fig. 2 Color clustering using predetermined color pallets with KSOM; **a** generated base colors; **b** resulting clustered colors

(a)   **(b)**

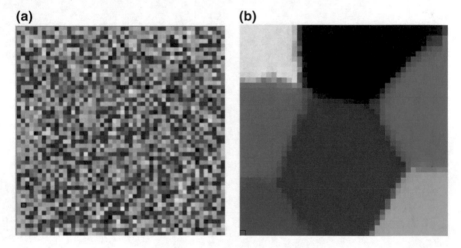

Fig. 3 Color clustering using predetermined color pallets with EKSOM; **a** generated base colors; **b** resulting clustered colors

that do not receive sufficient training which clearly shows by the less prominence of the clusters indicated by a not-clearly defined borders between clusters as visibly seen in (b) after training the generated base colors in (a). Whereas, Fig. 3 evidently indicates a more prominent cluster as shown by a much-defined borders between each cluster in image (b) after 1000 iterations of training the base colors generated in image (a).

(a)   **(b)**

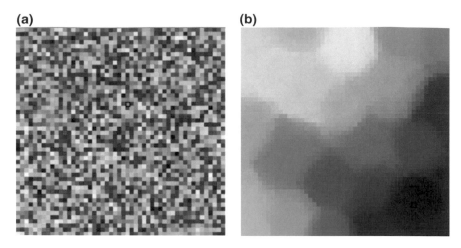

Fig. 4 Color clustering using randomly generated color pallets with KSOM; **a** generated base colors; **b** resulting clustered colors

(a)   **(b)**

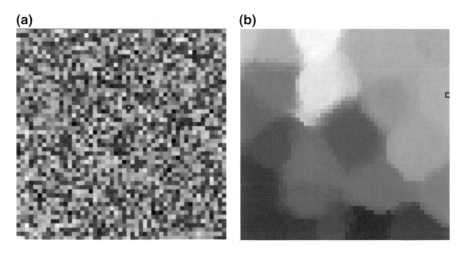

Fig. 5 Color clustering using randomly generated color pallets with EKSOM; **a** generated base colors; **b** resulting clustered colors

The figures apparently show that EKSOM produced a more clustered color compared to the result provided by the conventional KSOM when tested using the predetermined colors.

Further, Figs. 4 and 5 visualizes the test results produced by the two networks when tested using a randomly generated colors. Each test may generate colors randomly as in both image (a) in both figures which produce sets of colors possibly different with the colors made during the other test, but interestingly it provides a remarkable result. Figure 4 depicts the test result after subjecting the KSOM in this test.

Fig. 6 Learning rates of KSOM and EKSOM

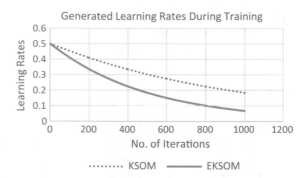

As illustrated in image (b) of Fig. 4, the borders between clusters are less prominent, and the boundaries are vague. This clearly indicates that the nodes did not receive sufficient training.

The image (b) of Fig. 5 illustrates a more trained result and a noticeable border between clusters. This was the result when the EKSOM were subjected to a test using the randomly generated colors.

Visibly, the result produced by EKSOM is more clustered than the result provided by KSOM as indicated by the borders between clusters in the previous results.

These results are strongly backed up by the graph illustrated in Fig. 6 The graph precisely demonstrates the learning rates generated by the two networks during the training process.

Based on the figure, an almost consistent decline of the learning rate was exhibited by KSOM continuously until the end of the iteration as indicated by the dotted line. At an early stage, the EKSOM draws bigger learning rates or steps and smoothly decaying as it reaches the end of the process as marked by the solid line. This evidently implies that the convergence of EKSOM has improved the convergence of KSOM. This further means that the convergence of EKSOM is faster than the convergence of KSOM.

5 Conclusion

This paper has outlined that in spite of the undeniable success of Kohonen Self-Organizing Maps in the field of clustering and its contribution to unsupervised learning its learning rate decay function needs to be enhanced to address its issue on convergence. This paved the way for the introduction of the new Exponential Decay Average Rate of Change (EDARC) for the Enhanced Kohonen Self-Organizing Maps (EKSOM).

The KSOM and EKSOM were subjected to color clustering using a predetermined and randomly generated colors. The results show that KSOM was positively outperformed by EKSOM. The results imply that EKSOM a better and faster in convergence compared to KSOM which makes the enhanced algorithm superior to the conventional algorithm.

6 Future Works

This is an avenue to test further the EKSOM to more sophisticated clustering experiments and applications involving images with different data types, testing alongside or in combination with other clustering algorithms, subjecting to various measurements for clustering tasks results, and exploiting to a real-world application.

References

1. Esmin AAA, Coelho RA, Matwin S (2015) A review on particle swarm optimization algorithm and its variants to clustering high-dimensional data. Artif Intell Rev 44(1):23–45. https://doi.org/10.1007/s10462-013-9400-4
2. Wankhede SB (2014) Analytical study of neural network techniques: SOM, MLP and Classifier—a survey. IOSR J Comput Eng Ver VII 16(3):2278–2661
3. Bourgeois N, Cottrell M, Déruelle B, Lamassé S, Letrémy P (2015) How to improve robustness in Kohonen maps and display additional information in factorial analysis: application to text mining. Neurocomputing 147(1):120–135. https://doi.org/10.1016/j.neucom.2013.12.057
4. Sheela KG (2012) An efficient hybrid neural network model in renewable energy systems. No 978, pp 359–361
5. Gopalakrishnan K, Khaitan S (2008) Enhanced clustering analysis and visualization using Kohone's self-organizing feature map networks. Int J Comput 2(6):64–71
6. Chen LP, Liu YG, Huang ZX, Shi YT (2014) An improved SOM algorithm and its application to color feature extraction. Neural Comput Appl 24(7–8):1759–1770. https://doi.org/10.1007/s00521-013-1416-9
7. Abubaker A, Altayeb A, Kohonen nural network based approach to voltage weak buses/areas identifiers. University of Khartoum
8. Qiang X, Cheng G, Wang Z (2010) An overview of some classical Growing Neural Networks and new developments. In: ICETC 2010—2010 2nd international conference on education technology and computing, vol 3, pp 351–355. https://doi.org/10.1109/icetc.2010.5529527
9. Wehrens R, Buydens LMC (2007) Self- and super-organizing maps in R: the Kohonen package. J Stat Softw 21(5):1–19. https://doi.org/10.18637/jss.v021.i05
10. Khvorostukhina E, L'Vov A, Ivzhenko S (2017) Performance improvements of a Kohonen self-organizing training algorithm. In: Proceedings of the 2017 IEEE Russia section young researches in electrical and electronic engineering conference ElConRus 2017, pp 456–458. https://doi.org/10.1109/eiconrus.2017.7910589
11. Dragomir OE, Dragomir F, Radulescu M (2014) Matlab application of Kohonen Self-organizing map to classify consumers' load profiles. Procedia Comput Sci. 31:474–479. https://doi.org/10.1016/j.procs.2014.05.292
12. Harchli F, Abdelatif ES, Mohamed E (2014) Novel method to optimize the architecture of Kohonen's topological maps and clustering. In: Proceedings of the 2nd international conference on logistics operations management, GOL 2014, pp 8–14. https://doi.org/10.1109/gol.2014.6887439
13. Johnson R, Zhang T (2013) Accelerating stochastic gradient descent using predictive variance reduction. Nips 1(3):315–323

14. Kohonen T (1990) The self-organizing map. Proc IEEE 78(9):1464–1480. https://doi.org/10. 1109/5.58325
15. Ritter H, Kohonen T (1989) Self-organizing semantic maps. Biol Cybern 61(4):241–254. https://doi.org/10.1007/BF00203171
16. Naenna T, Bress RA, Embrechts MJ (2004) A modified Kohonen network for DNA splice junction classification. In: TENCON 2004. 2004 IEEE region 10 conference 2, vol 2729, p 4. https://doi.org/10.1109/tencon.2004.1414570
17. Schlieter T (2016) Application of self-organizing maps in visualization of multi. vol 15, no 1, pp 75–87 (2016)
18. Gorjizadeh S, Pasban S, Alipour S (2015) Noisy image segmentation using a self-organizing map network. Adv Sci Technol Res J 9(26):118–123. https://doi.org/10.12913/22998624/2375
19. Owsley L, Atlas L, Bernard G (1996) Self-organizing feature maps with perfect organization. In: 1996 IEEE international conference on acoustics. speech, signal processing, pp 3557–3560. https://doi.org/10.1109/icassp.1996.550797
20. Flexer A (1997) Limitations of self-organizing maps for vector quantization and multidimensional scaling. In: Advances in neural information processing systems 9, pp 445–451
21. Miljković D (2017) Brief review of self-organizing maps. In: 2017 40th international convention on information and communication technology, electronics microelectronics, pp 1252–1257
22. Guthikonda SM (2005) Kohonen self-organizing maps. Wittenberg University
23. Upadhyay P, Chhabra JK (2015) Modified self organizing feature map neural network (MSOFM NN) based gray image segmentation. Procedia Comput Sci 54:671–675. https://doi.org/10. 1016/j.procs.2015.06.078
24. Senior A, Heigold G, Ranzato M, Yang K (2013) An empirical study of learning rates in deep neural networks for speech recognition. In: 2013 IEEE international conference on acoustics, speech and signal processing, pp 6724–6728. https://doi.org/10.1109/icassp.2013.6638963
25. Wilson D, Martinez T (2001) The need for small learning rates on large problems. In: IJCNN'01. Proceedings of the international joint conference on neural networks (Cat. No. 01CH37222), vol 1, pp 115–119. https://doi.org/10.1109/ijcnn.2001.939002
26. Qiang X, Cheng G, Li Z (2010) A survey of some classic self-organizing maps with incremental learning. In: ICSPS 2010—proceedings of the 2010 2nd international conference on signal processing systems, vol 1, pp 804–809. https://doi.org/10.1109/icsps.2010.5555247
27. Strang G (2001) Calculus. Wellesley-Cambridge Press, Wellesly
28. Calculus online book. Retrieved from http://www.understandingcalculus.com/chapters/04/4-2.php. Accessed on 09 Nov 2017

Affinity Propagation Based on Intuitionistic Fuzzy Similarity Measure

Omar M. Akash, Sharifah Sakinah Syed Ahmad, Mohd Sanusi Azmi and Abd Ulazeez Moh'd Alkouri

Abstract Affinity Propagation (AP) is a recently proposed clustering technique, widely used in literature, which finds the cluster center by exchanging real messages between pairs of data. These messages are calculated on the basis of similarity matrix. Accordingly, the similarity matrix is considered an essential procedure of the AP. Negative Euclidean distance is used as a similarity measure, in order to construct the similarity matrix of the AP. However, most data points lie in the non-Euclidean space which becomes difficult for the Euclidean distance to acquire the real data structure. The performance of AP might be degraded if this drawback occurs. A clustering method is proposed here called Intuitionistic Fuzzy Affinity Propagation (IFAP) that uses an intuitionistic fuzzy similarity measure to construct the similarity matrix among data points. Subsequently, the similarity matrix is fed into the AP procedure, and the cluster center will emerge after a couple of iterations. The numerical experiment is demonstrated. Results show that the IFAP outperforms the other clustering method.

Keywords Affinity propagation · Negative euclidean distance · Intuitionistic fuzzy set · Similarity measures · Hamming distance

1 Introduction

Data clustering is a technique of grouping patterns by separating objects that share the same characteristics. This grouping is based on a mathematical definition of the measure of similarity that calculates the similarity between these data elements, where each group of elements consists of a high degree of correlation [1], where each formed group is called a cluster.

O. M. Akash (✉) · S. S. S. Ahmad · M. S. Azmi
Universiti Teknikal Malaysia Melaka, 76100 Melaka, Malaysia
e-mail: omarakash2010@gmail.com

A. U. M. Alkouri
Ajloun National University, Ajloun, Jordan

© Springer Nature Singapore Pte Ltd. 2019
V. Piuri et al. (eds.), *Intelligent and Interactive Computing*, Lecture Notes in Networks and Systems 67, https://doi.org/10.1007/978-981-13-6031-2_30

Clustering is a widely used technique in data mining, and many techniques of clustering have been proposed lately, such as k-mean clustering [2] and k-medoids method [3]. Among these algorithms, a powerful clustering method is AP by Frey and Dueck [4]. AP is a well-known method that uses the negative Euclidean distance as a similarity measure to find clusters, where each data point is allocated to one cluster only during its iteration. Unlike the K-mean clustering method, setting the number of clusters and defining the initial cluster centers are not needed as it takes into account all data objects as a possible exemplar.

Many machine learning tasks utilize AP, including document summarization and airline routing, clustering face images, and finding genes using microarray data [4], although AP shows its efficiency of finding clusters with less error rate and insensitivity to cluster initialization [5]. However, results may be affected since it primarily relies on the negative Euclidean distance as the similarity measure. This distance is mainly suitable to be used with purely numeric datasets, which can give satisfactory results when all clusters are spheroids with the same size or when all clusters are well-separated [6]. Nevertheless, it is challenging to identify clusters accurately as most of the data points lie in the non-Euclidean space, such that the Euclidean distance is not able to acquire the Euclidean data structure as it is incapable of reflecting the real similarity relations among the data, and thus, the clustering results are consequently affected [7]. Such drawbacks encourage the researcher to utilize different similarity measures with AP.

In some cases, the unsupervised method is affected by uncertainty. Therefore, the introduction of a fuzzy set produced many reliable solutions, and many studies were conducted to combine the clustering method with the theory of fuzzy set. This combination can be specifically advantageous when boundaries between clusters are ambiguous and data points are associated with more than one cluster with a degree of membership [5, 8–10]. In 1965, Zadeh [11] introduced the fuzzy set theory which describes the uncertainty of belongingness by a membership function, where each element in a specific universe is assigned a membership degree, which is in the unit interval [0, 1], a value that indicates the degree of belongness to different clusters [12]. A membership degree can help to discover more sophisticated relationships between a given object and the disclosed clusters [13]. However, due to the lack of knowledge in some datasets, a degree of uncertainty could occur; this uncertainty or hesitation appears during the membership definition. This uncertainty led Atanassov in 1986 [14] to Intuitionistic Fuzzy Set (IFS) concept a generalization of fuzzy sets; the concept shows a great effectiveness in dealing with vagueness, by considering membership and non-membership degrees.

In this paper, an Intuitionistic Fuzzy Affinity Propagation (IFAP) algorithm is proposed which incorporates an intuitionistic fuzzy set to cope with some of the limitations of the AP algorithm. The method integrates elements of IFS to get better clustering results. Section 2 of this paper briefly reviews some necessary background on this research, which proposes a technique to overcome sensitivity of AP. Section 3 overviews the proposed concept and improvements on AP algorithm, while last section describes the conducted experiment and results.

2 Related Work

Affinity propagation method propagates affinity message among objects to detect a good set of clusters. Since AP was published in 2007, it has proven its good performance in image segmentation [15], gene expression [16], and text clustering [17]. Many researchers attempt to improve the AP algorithm by utilizing different similarity measures that can produce better results with datasets applied.

For instance, Geweniger et al. [18] extended the standard AP to a fuzzy variant AP (FAP) and the dissimilarities between the data points were calculated as log-likelihoods of the probability. Li et al. [19] proposed a Dynamic Affinity Propagation (DAP) clustering algorithm for waveforms, based on the Dynamic Time Warping distance (DTW), using the sum of DTW distance instead of similarity as the input data for the AP algorithm. Also, Liu et al. [20] proposed a modified AP using a similarity measure using the negative sine-squared value of the acute angle between discontinuity unit normal vectors.

In text clustering problem, Guan et al. [17] proposed Seeds Affinity Propagation (SAP) for text clustering, utilizing asymmetric similarity metric that takes into account texts structural information called the tri-set method, based on three feature sets. This asymmetric metric is combined with AP. Shrivastava et al. [21] developed a text document AP clustering algorithm based on the vector space model called Phrase Affinity Clustering (PAC). The vector space model calculates the similarity matrix using the cosine similarity measure. Kang and Choi [16] proposed a combination of AP and a new similarity measure called Common Neighborhood Sub-graph Density (CND) based on sub-graph density. This combination shows a higher accuracy finding communities in a different real-world dataset. Jiang et al. [6] introduced an improved AP algorithm relying on path-based similarity (APC_PS). The negative Euclidean distance is utilized to find the exemplars, and then the path-based similarity tries to investigate the basic structure between objects and exemplars, and appoints objects to its related cluster. Hu et al. [17] addressed the problem of Wi-Fi indoor positioning. AP relies on the Jaccard similarity coefficient to build similarity matrix. The fingerprint is measured by merging the similarity measure with radio signal strength.

Many studies are proposed to utilize a different similarity measure with AP. However, throughout conducting literature review, we came across that different similarity measures were used. Several similarity measures were effective and others were modest, while some are unstable in certain cases. This encourages us to study the effectiveness of intuitionistic fuzzy similarity measures.

2.1 Affinity Propagation Clustering

AP is a powerful data clustering technique that finds cluster exemplars with low error rate and fast execution. Briefly, AP takes similarity matrix as an input based on selected similarity measure between data points, to construct a collection of exemplars and identical clusters [4]. In the beginning, data points treated as a possible exemplar. Thereafter, two messages are interchanged, namely, responsibility and availability messages [4].

The following is a brief detail of the AP model. Generally, the used similarity measure is negative squared Euclidean distance $s(i, k)$ between two points. Similarity can be initialized to

$$s(i, k) = -\|x_i - x_k\|^2 \tag{1}$$

Self-similarities are known as preference in [4] and commonly given a single value, which enable point k to become an exemplar. The preference could be set as

$$s(k, k) = \frac{\sum\limits_{i,j=1; i \neq j}^{N} s(i, k)}{N \times (N - 1)} 1 \leq i \leq N \tag{2}$$

Thereafter, the AP approach exchanges two forms of messages between data points, the responsibility $r(i, k)$ and the availability $a(i, k)$. Data point i sends a response message to prospective exemplary point k, displaying how applicable point k is to become an exemplar for point i, taking into account other potential exemplars for point data i. Meanwhile, point k sends availability message to point i, declaring the suitability of point i to choose point k as its exemplar, taking into account the support from other points that point should be an exemplar. The responsibility, $r(i, k)$, and availability, $a(i, k)$, are computed as follows:

Initialization:

$$r(i, k) = s(i, k) - \max\{s(i, k')\}, a(i, k) = 0, \quad \text{Where } k' \neq k \tag{3}$$

Responsibility updates:

$$r(i, k) = s(i, k) - \max\{a(i, k') + s(i, k')\} \tag{4}$$

$$r(i, k) = s(k, k) - \max\{a(k, k') + s(k, k')\} \tag{5}$$

Availability updates:

$$a(i, k) = \min\left(0, r(k, k) + \sum_{i' s.t. i' \notin \{i,k\}} \max\{0, r(i', k)\}\right) \tag{6}$$

Here, $r(k, k)$ is the self-responsibility, which declares that point k at this point suitable to follow another exemplar than becoming an exemplar itself. The "self-availability" $a(k, k)$ presents point k as exemplar, considering the positive responsibilities sent from different points to potential exemplar k. $r(k, k)$ and $a(k, k)$ are updated as follows:

$$a(k, k) = \sum_{i' s.t.i' \notin \{i,k\}} \max\{0, r(i', k)\} \tag{7}$$

$r(k, k)$ and $a(k, k)$ recursively updated, up to specific values or the result remains unchanged for a number of iterations. Finally, $r(k, k)$ and $a(k, k)$ merged to determine the exemplars:

$$c_i^* \leftarrow \arg \max_{1 \leq k \leq n} r(i, k) + a(i, k)$$

3 Proposed Clustering Method

The basic concept of the proposed approach is to extend the AP algorithm to the fuzzy domain. A measure of similarity is required to find the degree of the relationship between objects. The measure of similarity should be determined since it affects the results of the clustering [22]. In this context, introducing IFS to a similarity measure can be used in pattern recognition problems. Generally, a point is assigned to a cluster; however, when calculating the IFS distance measure, a value of a membership and non-membership degree is assigned to a data point that shows how close point x is to an exemplar. Hamming distance [23] is a useful technique to calculate the difference between two parameters such as problems with two elements and sets. The Hamming distance measure is defined in [24] as

$$d(A, B) = \sum_{i=1}^{n} |\mu_A(x_i) - \mu_B(x_i)| \tag{8}$$

In [24], a generalization of distance in Eq. (8) for intuitionistic fuzzy sets is proposed. For two $A, B \in \text{IFS}(X)$, we get

$$d'(A, B) = \frac{1}{2} \sum_{i=1}^{n} |\mu_A(x_i) - \mu_B(x_i)| + |\nu_A(x_i) - \nu_B(x_i)| \tag{9}$$

$d'(A, B)$ denotes the similarity measures from [25] between the membership degree μ_A and μ_B, and the non-membership degree ν_A and ν_B, respectively.

The proposed approach depends on the IFS and AP called Intuitionist Fuzzy Affinity Propagation (IFAP). Unlike the standard AP, it simultaneously utilizes the IFS similarity measure to create the similarity matrix. The technique of the IFAP is as follows:

Step 1: Initially, the crisp dataset is transformed to the fuzzy domain and then into the intuitionistic fuzzy domain. In the beginning, the membership degree $\mu_{ij}(k)$ of IFS is calculated from Eq. (10):

$$\mu_{X'} = \frac{X - X_{\min}}{X_{\max} - X_{\min}} \tag{10}$$

The IFS non-membership function is calculated by Yager [26] from Eq. (11):

$$A = \left\{ x, \mu_A(x), (1 - \mu_A(x)^\alpha)^{1/\alpha} \right\}, 0 < \alpha < 1 \tag{11}$$

Step 2: The intuitionistic fuzzy distance between the data point and the potential clustering exemplars is calculated using Eq. (9).

Step 3: The preference value is updated based on the equation in Geweniger et al. [18]

$$p = \min(\text{sim}) * 2 \tag{12}$$

where sim is the similarity matrix obtained from $d'(A, B)$.

Step 4: Responsibility and availability are initialized according to (3). Then, responsibilities and availability are updated using (4, 5, and 6).

Step 5: Convergence. Repeat steps 4–5 until the results are unchanged for a number of iterations.

4   Results and Discussion

In order to validate the performance of the proposed IFAP technique, IFAP compared with the standard AP, K-AP [27], k-means [2], and k-medoids [28] on several standard datasets. The experiments conducted used three datasets: Iris, Wine, and New-thyroid, obtained from the UCI Machine Learning Repository [29], which are listed in Table 1.

The AP and IFAP are initialized with the damping factor $\lambda = 0.7$. In AP and k-AP, each similarity measure is set to the negative Euclidean distance, and the similarity measure used in IFAP is the intuitionistic fuzzy Hamming distance as in Eq. (9). AP preference values are set to the median of similarities, while in IFAP the preference

Table 1 Details of the datasets applied

Dataset	Cluster	Size	Dimension
Iris	3	150	4
Wine	3	178	13
New-thyroid	3	215	5

Table 2 Clustering results on three UCI datasets

Methods\dataset	NC	Iris	NC	Wine	NC	New-thyroid
AP	12	0.60	12	0.39	28	0.44
K-AP	3	0.76	3	0.66	3	0.85
k-means	3	0.87	3	0.63	3	0.89
k-medoids	3	0.88	3	0.88	3	0.89
IFAP	3	0.94	3	0.91	3	0.96

value is set as in Eq. (12). In IFAP, we investigated the influence of different input values of α in Eq. (11), with the parameter set in the range of $\alpha \in [0.1–20]$ to find the best results. Since k-means and k-medoids are sensitive to initialization, each of them was run for 20 times and the best results were reported. The number of clusters is fixed as $k = 3$, referring to ground truth labels.

We evaluated our experiments with Fowlkes–Mallows index (FM) [30]. The FM index is an evaluation standard that is applied to evaluate the capability of a method to find clusters in a dataset accurately. Whenever FM index value approaches to 1, it shows that the method produced a good set of clusters.

Table 2 shows the clustering results of the methods on different real datasets. It is noticeable that IFAP as well as k-medoids produces a good quality cluster. In addition, IFAP produces the best FM index rates apart from the other four methods in all datasets considering the variation of α; Iris 0.94 with $\alpha = 3$, Wine 0.91 with $\alpha = 4$, and New-Thyroid 0.96 $\alpha = 0.3$. The AP FM index is the lowest in all datasets. NC in Table 2 refers to the resulted number of clusters. It is noticeable that the AP number of clusters for all datasets is different than the real number of clusters, while IFAP could determine the number of clusters accurately.

From these results, we can deduce that the performance of IFAP is the best among the other methods. Intuitionistic fuzzy similarity measure can explore the structures of data, and, consequently, enhances the AP clustering accuracy.

5 Conclusion

In this paper, an improvement of the affinity propagation performance on sample datasets is proposed. The intuitionistic fuzzy hamming distance is proposed to replace the standard negative Euclidean distance. The proposed method converts the data into the fuzzy domain, by calculating the membership and the non-membership degrees. Then, the similarity matrix is constructed using the IFS Hamming distance. The experimental results on the UCI datasets show the effectiveness of the IFAP that outperforms other related algorithms. However, IFAP has the problem of parameter setting, as there are various parameters introduced by user; many experiments needed to be done in order to find a suitable value of a parameter. A further study is going to be focused on the improvement of AP algorithm and applying the method in the diverse field of application.

References

1. Yang J et al (2010) Affinity propagation feature clustering with application to vehicle detection and tracking in road traffic surveillance. In: 2010 Seventh IEEE international conference on advanced video and signal based surveillance (AVSS). IEEE
2. MacQueen J (1967) Some methods for classification and analysis of multivariate observations. In: Proceedings of the fifth Berkeley symposium on mathematical statistics and probability. Oakland, CA, USA
3. Kaufman L, Rousseeuw P (1987) Clustering by means of medoids. North-Holland
4. Frey BJ, Dueck D (2007) Clustering by passing messages between data points. Science 315(5814):972–976
5. Yang C et al (2010) A fuzzy-statistics-based affinity propagation technique for clustering in multispectral images. IEEE Trans Geosci Remote Sens 48(6):2647–2659
6. Jiang Y, Liao Y, Yu G (2016) Affinity propagation clustering using path based similarity. Algorithms 9(3):46
7. Wang X, Wang Y, Wang L (2004) Improving fuzzy c-means clustering based on feature-weight learning. Pattern Recogn Lett 25(10):1123–1132
8. Keller JM, Gray MR, Givens JA (1985) A fuzzy k-nearest neighbor algorithm. IEEE Trans Syst Man Cybern 4:580–585
9. Burrough PA et al (2001) Fuzzy k-means classification of topo-climatic data as an aid to forest mapping in the Greater Yellowstone Area, USA. Landscape Ecol 16(6):523–546
10. Tran D, Wagner H (1999) Fuzzy expectation-maximisation algorithm for speech and speaker recognition. In: Fuzzy information processing society, 1999. NAFIPS. 18th International conference of the North American. IEEE
11. Zadeh LA (1965) Fuzzy sets. Inf Control 8(3):338–353
12. Bezdek JC, Ehrlich R, Full W (1984) FCM: the fuzzy c-means clustering algorithm. Comput Geosci 10(2–3):191–203
13. Tian W et al (2014) Research on clustering based meteorological data mining methods. Adv Sci Technol Lett 79:106–112
14. Atanassov KT (1986) Intuitionistic fuzzy sets. Fuzzy Sets Syst 20(1):87–96
15. Dueck D, Frey BJ (2007) Non-metric affinity propagation for unsupervised image categorization. In: IEEE 11th international conference on computer vision, 2007. ICCV 2007. IEEE
16. Kang Y, Choi S (2009) Common neighborhood sub-graph density as a similarity measure for community detection. In: International conference on neural information processing, Springer
17. Guan R et al (2011) Text clustering with seeds affinity propagation. IEEE Trans Knowl Data Eng 23(4):627–637
18. Geweniger T et al (2009) Fuzzy variant of affinity propagation in comparison to median fuzzy c-means. In: International workshop on self-organizing maps. Springer
19. Li P et al (2018) Dynamic equivalent modeling of two-staged photovoltaic power station clusters based on dynamic affinity propagation clustering algorithm. Int J Electr Power Energy Syst 95:463–475
20. Liu J, Zhao X-D, Xu Z-H (2017) Identification of rock discontinuity sets based on a modified affinity propagation algorithm. Int J Rock Mech Min Sci 94:32–42
21. Shrivastava SK, Rana J, Jain R (2013) Fast affinity propagation clustering based on machine learning
22. Xu Z (2013) Intuitionistic fuzzy aggregation and clustering, vol 279. Springer
23. Kacprzyk J (1997) Multistage fuzzy control: a prescriptive approach. Wiley

24. Szmidt E, Kacprzyk J (2000) Distances between intuitionistic fuzzy sets. Fuzzy Sets Syst 114(3):505–518
25. Grzegorzewski P (2004) Distances between intuitionistic fuzzy sets and/or interval-valued fuzzy sets based on the Hausdorff metric. Fuzzy Sets Syst 148(2):319–328
26. Yager RR (1979) On the measure of fuzziness and negation part I: membership in the unit interval
27. Zhang X et al (2010) K-AP: generating specified K clusters by efficient affinity propagation. In: 2010 IEEE 10th international conference on data mining (ICDM). IEEE
28. Han J, Pei J, Kamber M (2011) Data mining: concepts and techniques. Elsevier
29. Dheeru D, Taniskidou EK (2017) UCI machine learning repository. Retrieved from http://archive.ics.uci.edu/ml
30. Fowlkes EB, Mallows CL (1983) A method for comparing two hierarchical clusterings. J Am Stat Assoc 78(383):553–569

Weight-Adjustable Ranking for Keyword Search in Relational Databases

Chichang Jou and Sian Lun Lau

Abstract Huge volumes of invaluable information are hidden behind web relational databases. They could not be extracted by search engines. The problem is especially severe for long text data, for example, book reviews, company descriptions, and product specifications. Many researches have investigated to integrate information retrieval and database indexing technologies to provide keyword search functionality for these useful contents. Due to diversifying data relationships in application domains and miscellaneous personal preferences, current ranking results of related researches do not satisfy user requirements. We design and implement a Weight-Adjustable Ranking for Keyword Search (WARKS) system to address the issue. Mean average precision (MAP) and mean rank reciprocal difference (MRRD) are proposed as measurements of ranking effectiveness. We use an integrated international trade show database as our experimental domain. User study demonstrates that WARKS performs better than previous practices.

Keywords Search · Information retrieval · Ranking · Mean average precision · Rank reciprocal difference

1 Introduction

Along with the rapid accumulation of Internet contents, search engines have been very important tools in retrieving and collecting information. However, lots of invaluable data are still hidden behind web relational databases. The problem is especially severe for data in long text formats, such as book reviews, company descriptions, and product specifications. They could not be indexed by traditional numeric or short-term indexing of relational databases. Many researchers have tackled the problem by

C. Jou (✉)
Department of Information Management, Tamkang University, New Taipei City, Taiwan
e-mail: cjou@mail.tku.edu.tw

S. L. Lau
Department of Computing and Information Systems, Sunway University, Petaling Jaya, Malaysia
e-mail: sianlunl@sunway.edu.my

© Springer Nature Singapore Pte Ltd. 2019
V. Piuri et al. (eds.), *Intelligent and Interactive Computing*, Lecture Notes in Networks and Systems 67, https://doi.org/10.1007/978-981-13-6031-2_31

integrating information retrieval and database indexing to provide keyword search functionality for these useful contents.

Due to diversifying data relationships in application domains and miscellaneous personal preferences, current ranking results of related researches do not satisfy user requirements. Take the domain of international business trading as an example. All trade show websites store detailed information about participating companies and their products in web relational databases. Their company table includes long text columns for company descriptions, and their product table includes long text columns for product specifications. International traders often need to integrate information about which companies attended which trade shows with which products, especially when they are not certain about the product names. To decide which company to enhance business cooperation, they also need to collect accumulated information across websites about the participating companies, like frequencies, experiences, product categories, exhibition lot sizes, etc. The requirements of personalized keyword search functionality to retrieve company or product information for trade show websites are imminent.

We extend ranking mechanisms of previous researches to design and implement a Weight-Adjustable Ranking for Keyword Search (WARKS) system to meet user requirements. WARKS augments adjustable column weights and tuple weights to rank the retrieved results. Measurements of mean average precision (MAP) and mean rank reciprocal difference (MRRD) are proposed as ranking effectiveness indicators. We use an integrated international trade show database as our experimental domain. User study demonstrates that WARKS performs better than previous practices.

2   Related Work

Bergamaschi et al. [4] surveyed keyword search for relational databases and identified challenging tasks in this research area. Simitsis et al. [14] categorized processing principles for keyword search in relational databases into the following two approaches: based on database schema and based on tuples.

2.1   *Based on Database Schema*

This approach treated database schema as a graph, where each node represents a table, and each edge represents a referential relationship between tables. The following are related researches in this approach:

Discover [8] utilized the join mechanism of relational databases. The system first retrieved those joinable candidate tuple networks (CTN), and then evaluated their execution plans. It used greedy algorithm to restrict number of retrieved CTNs. DiscoverII [9] adopted ranking technology to boost efficiency. It first performed full text retrieving. The obtained tuples were associated with foreign keys to form

CTNs, which were then transformed into SQL syntaxes. The top-ranked results were returned to the user.

DBXplorer [1] performed breadth-first traversal for all tuples in the tuple graph, and built the SQL syntaxes to obtain related tuples. The system considered not only the size of the result sets, but also the relevance of the query results with respect to the keywords.

Keymantic [3] parsed keyword phrases from knowledge bases composed by database schema, data types, data dictionary mapping, and ontology. It listed all possible parsing trees for a query, and let the user select the one that was closest to their intention. The system then transformed the selected parsing into relevant SQL syntaxes.

2.2 Based on Tuples

This approach treated database contents as a directed graph, where each node represents a tuple, and each edge represents a reference between two tuples. Under this principle, these systems directly built tuple trees that satisfy all requirements (AND-semantics). The following are related researches in this approach.

BANKS [6] used backward expanding search for graph traversal to retrieve all keyword-matched tuples in the tuple graph. The search started from each keyword-matched tuple, and then followed the backward direction of an edge. However, for tuples with many keyword matchings, it had a poor search efficiency. BANKSII [11] proposed bidirectional search so that keyword search could start from the root in a tuple graph. This mechanism improved the efficiency shortcoming of BANKS for tuples with many keyword matchings.

Liu et al. [12] proposed a ranking strategy that performed normalizations of tuple tree size, document length, term frequency, and inverse document frequency for candidate tuple trees. Schema terms were given higher scores.

Objectrank [2] applied authority-based ranking. They built a directed graph, where each node was tagged with a node type to represent the tables they were in. Contents of a node are represented by a set of keywords. Weights of all nodes, representing their authorities, were iteratively calculated. The system could support both AND-semantics and OR-semantics for the query keywords.

2.3 Ranking Evaluation and Recent Developments

Coffman and Weaver [7] proposed a framework to quantitatively evaluate the effectiveness of keyword search and their ranking mechanism. Bergamaschi et al. [5] tried to induct the user intentions in choosing keywords. These intentions are then integrated with data representations in the databases to retrieve results satisfying user requirements.

Jabeur et al. [10] proposed a user-centric product search for e-commerce sites by taking user engagement into account. In addition to product features, they included social interactions, namely "like" and "share" tags, in ranking the retrieved results.

Liu et al. [13] extended keyword search to the temporal graphs for users to specify optional predicates and ranking functions related to timestamps. Three types of ranking functions are supported: relevance, time, and duration. They extended Dijkstra's shortest path algorithm to find the best paths between two nodes in each time instant with respect to a ranking function.

Zhu et al. [15] tackled the problem that keyword search results are biased by duplicate tuples that refer to the same real-world entity. They designed a clustering algorithm using the divide-and-conquer mechanism to reduce duplicates in the results.

3 Design and Implementation of WARKS

3.1 Data Model of WARKS

WARKS adopts the data model proposed by Hristidis et al. [9] and Liu et al. [12]:

- Database schema diagram (abbreviated as *DSD*): Suppose there are n tables in the database. A node is designated for each table. If there exists a relationship from table R_i to table R_j, then a directed edge from R_i to R_j is added. Their association degree (1-to-1 or 1-to-many) is augmented to the edge.
- Tuple tree (abbreviated as T): Given a *DSD*, each node in the tree represents a tuple (aka record). A tuple tree is constructed by several related tuples. Suppose $<R_i, R_j>$ is an edge in a *DSD*, if tuple $t_i \in R_i$, $t_j \in R_j$, and $(t_i\ join\ t_j) \in (R_i\ join\ R_j)$, then the directed edge $<t_i, t_j>$ is added.
- Query (abbreviated as Q): WARKS deletes stop words in query terms, and performs stemming on the remaining terms. The remaining terms are treated as a set.

In the integrated trade show database, the following tables are constructed: TradeShow, Company, and Product. Its *DSD* is displayed in Fig. 1.

The bridge tables CompanyProduct, TradeShowCompany, and TradeShowProduct demonstrate foreign key existence in the participating tables. One functionality in WARKS is to use keywords to query the following attributes: the company description of the Company table, the product name of the Product table, and the product specification of the Product table. The product name attribute will be matched using traditional indexing, while the other two attributes will be matched using full-text indexing.

WARKS provides a keyword input text field for searching related companies and their products. We use a company tuple as root of the tuple tree, and build tuple trees as displayed in Fig. 2.

Fig. 1 Example database
schema diagram

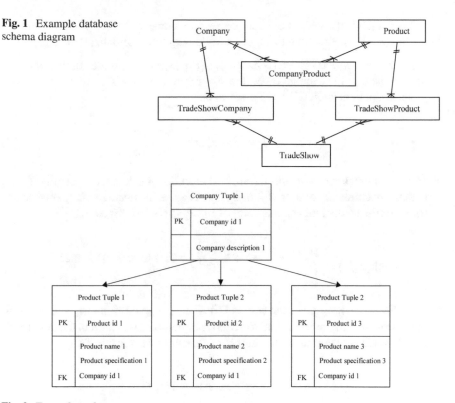

Fig. 2 Example tuple tree

3.2 Ranking Mechanism of Previous Works

This section introduces the ranking mechanism proposed by Hristidis et al. [9] and Liu
et al. [12]. Suppose each tuple is represented as one document D, and the document
collection in a tuple tree T is the set $\{D_1, D_2,..., D_m\}$. For a query phrase Q, the
ranking for a returned T is based on the similarity measure of Q and T, which is
computed by summing the products of the weights of terms in Q and those in T.

$$\text{Sim}(Q, T) = \sum_{k \in Q \cup T} \text{weight}(k, Q) * \text{weight}(k, T) \tag{1}$$

In Eq. (1), weight(k, Q) is the term frequency of k in Q, and weight(k, T) is
computed through the steps expressed in Eqs. (2)–(7):

$$\text{weight}(k, D) = \frac{\text{ntf}(k, D) * \text{idf}^g(k)}{\text{ndl}(k, D) * \text{Nsize}(T)} \tag{2}$$

Equation (2) is the weight of each term k in document D of T. To avoid ranking bias, this weight is computed through the following four normalization steps:

- *Tuple tree size normalization* (Nsize(T)): In Eq. (3), size(T) is the number of tuples in T, avgsize is the average tuple tree size for all tuple trees. Following previous practices, parameter s in Eq. (3) is set as 0.2.

$$\text{Nsize}(T) = (1 - s) * s * \frac{\text{size}(T)}{\text{avgsize}} \tag{3}$$

- *Document length normalization* (ndl(k, D)): In Eq. (4), document length dl(k, D) is to the occurrence number of k in D, and avgdl is the average number of words for all retrieved documents. Parameter s in Eq. (4) is also set as 0.2.

$$\text{ndl}(k, D) = \begin{cases} \left((1 - s) + s * \frac{\text{dl}(k,D)}{\text{avgdl}} \right) * (1 + \ln(\text{avgdl})) & k \in D \\ 0 & k \in D \end{cases} \tag{4}$$

- *Term frequency normalization* (ntf(k, D)): It is against intuition to have the similarity score linearly proportional to occurrence numbers of a term in the document. Equation (5) is to normalize the number of occurrences for k in D:

$$\text{ntf}(k, D) = 1 + \ln(1 + \ln(\text{tf}(k, D))) \tag{5}$$

- *Inverse document frequency normalization* (idf^g (k)): Previously, the normalization computation had been restricted to the local data returned by a search. This was extended to the *global* scale. In Eq. (6), df^g(k) is the number of documents where k appears. N^g is the number of documents in the corpus.

$$\text{idf}^g(k) = \ln \frac{N^g}{\text{df}^g(k) + 1} \tag{6}$$

- *The weight of k in T* (weight(k, T)): In Eq. (7), $\max_{D \in T}$ weight(k, D) is the maximum occurrence number of k for all D in T, and $\sum_{D \in T}$ weight(k, D) is the sum of weights of k for all D in T:

$$\text{weight}(k, T) = \max_{D \in T} \text{weight}(k, D) * \left\{ 1 + \ln \left(1 + \frac{\sum_{DT} \text{weight}(k, D)}{\max_{DT} \text{weight}(k, D)} \right) \right\} \tag{7}$$

The ranking mechanism of previous work has the following shortcomings:

1. Contents of all columns in a tuple were treated equally.
2. All tuples in the database were treated equally.

3.3 Design of WARKS

We propose to increase the weights of keywords in important columns and tuples. Use the trade show database as an example, to increase the weights of keywords in the "product name" column, we could emphasize the tf(k, D) in the term frequency normalization (Eq. 5).

We propose the adjustable ranking mechanism by allowing the user to set up the following two parameters:

1. The weight α for number of occurrences of k in a specific column: To indicate the importance, the value of α must be greater than or equal to 1. WARKS differentiates the importance of individual columns by multiplying occurrence numbers of terms in column i by the weight α_i. Suppose there are l columns in document D. In Eq. (8), n_{ki} denotes k's occurrence numbers in column i.

$$\text{tf}(k, D) = \frac{\sum_{i \in \{1,...,l\}} (n_{k_i} * i)}{\sum_{t \in D} \left(\sum_{i \in \{1,...,l\}} (n_{t_i} * i) \right)} \tag{8}$$

2. The weight β for tuple types: To differentiate the importance of the tuples, WARKS allows the user to adjust the weight of k in a document (Eq. 9). According to the tables that a tuple belongs to, WARKS first categorizes tuples, and then assigns weight β_i to the i-th tuple type. To increase the variation range, the value of β_i will be set as 2^n, where n is between -10 and $+10$.

$$\text{weight}(k, D) = \begin{cases} \dfrac{\text{ntf}(k, D) * \text{idf}^g(k, D)}{\text{ndl}(k, D) * \text{Nsize}(T)} * \beta_i & D \text{ is of the } i\text{-th tuple type} \end{cases} \tag{9}$$

3.4 Implementation of WARKS

WARKS is implemented using PHP 5.3.5 in a Microsoft Server 2012 server using Intel Xeon CPU E5620 2.13 GHz CPU with 12 GB memories. We crawled trade show, company and product data from 10 websites. They are integrated into a MySQL 5.6 database. We utilized full-text indexing of MySQL in creating the related tables to assist answering keyword query. Totally there are 6630 companies and 23,077 products.

WARKS utilized mySQL's "Match Against" function The Match function supports natural language processing. The Against function supports search using the keyword terms. WARKS applies the Match Against syntax to the company description and the product specification columns in the Company and Product tables, respectively.

The term "notebook" is used as a keyword search example to illustrate the adjusted normalization in WARKS. WARKS retrieves 29 company records, and 122 product

Table 1 Statistics of the query term "notebook"

Column	No. of tuples	Total no. of words	Avg tuple size
Company description	29	4011	138.3103
Product specification	133	31,363	235.8158

records for "notebook". Using traditional SQL indexing of the product name column, WARKS retrieves 54 product records. After eliminating duplicated records, 133 product records under 29 companies are returned.

The statistics regarding the query term "notebook" in the Company and Product records are displayed in Table 1. The tuple tree for this company is built by the following foreign keys: C430 → CP1, P1916 → CP1, C430 → CP2, P1917 → CP2, where C430 is the company id, P1916 and P1917 are the product id's, and CP1 and CP2 are record id's in the CompanyProduct bridge table. Table 2 shows the related numbers in the normalization steps for computing the similarity scores for the keyword "notebook" (Q) and the tuple tree (T).

4   Experiments and Discussions

4.1   Environment Setup

From the web directories of industrial products, we collected for the product categories "Consumer Electronics, Electronic Components, and Computer Peripherals" 30 keyword terms, for example, "android", "battery", "camera", "charger", etc.

Using the pooling method, we invited five domain experts to collect query result sets. For each keyword term, most frequently chosen 10 companies and products were selected for the evaluation reference. Then we invited 30 junior and senior university students majored in Information Management as our testers. If a result was chosen by more than half of the testers, then it is chosen as an objective answer.

In our experiments, users could adjust the column weights α only for the product name column, and the tuple weight β only for the product tuples. By fixing α value, users could choose the best β value that fits their preference.

To evaluate the effectiveness of our proposed mechanism, we use the following two indicators: 11-Point mean average precision (MAP) and mean rank reciprocal difference (MRRD). MAP is used to evaluate the precision rate of the retrieved result, and MRDD is used to evaluate the consistency of the retrieved result

1. *MAP*: MAP has been used in many competitions, like TREC, NTCIR, and CLEF. We apply interpolation to obtain the average precision of 11 fixed recalls. In MAP,

Table 2 Computed numbers in the normalization steps

Company	Tuple id	Nsize(T)	ndl(k, D)		ntf(k, D)		idf^g(k)		weight(k, D)	
		Nsize(T)	Company	Product	Company	Product	Company	Product	Company	Product
430	C430	1.406	8.939	0	3.253	0	3.776	0	0.977	0
430	P1916	1.406	0	7.673	0	2.496	0	6.255	0	1.447
430	P1917	1.406	0	8.116	0	2.716	0	6.255	0	1.488

for keyword k, average precision of each result is obtained using Eq. (10), where R_k is the number of query results matching the evaluation reference, P_{kj} is the precision rate for the j-th correct result, and N_Q is number of tested keywords. In other words, MAP is the mean value of the 30 APs.

$$\mathrm{AP}_k(\alpha, \beta) = \frac{\sum_{j=1}^{R_k} P_{k,j}}{R_k} \quad \mathrm{MAP}(\alpha, \beta) = \frac{\sum_{k=1}^{N_Q} \mathrm{AP}_k(\alpha, \beta)}{N_Q} \tag{10}$$

2. *MRRD*: For each query term k, from the top 10 results produced by all combinations of α and β, we identify the company (or product) with the best rank ($F_k(\alpha, \beta)$), and that with the worst rank ($L_k(\alpha, \beta)$). $\mathrm{RRD}_k(\alpha, \beta)$ is defined in Eq. (11). If for some α, β, and k, $\mathrm{RRD}_k(\alpha, \beta)$ is less than 0, it means under that (α, β) setting, the returned query results for k demonstrated a flip-flop phenomenon. For each combination of (α, β), MRRD is the mean value of the 30 RRDs.

$$\mathrm{RRD}_k(\alpha, \beta) = \frac{1}{F_k(\alpha, \beta)} - \frac{1}{L_k(\alpha, \beta)} \quad \mathrm{MRRD}(\alpha, \beta) = \frac{\sum_{k=1}^{N_Q} \mathrm{RRD}_k(\alpha, \beta)}{N_Q} \tag{11}$$

We polled the users with the two questionnaires to proceed with our user study

- For the method proposed by Liu et al. [12]: By setting α as 1, and β as 2^0.
- For a combination setting (α, β): We obtain a best combination setting (α, β) for MAP and MRRD separately.

4.2 Experiment Results and Discussions

For MAP, we fixed the value of β from the set $\{2^{-10}, 2^{-9}, ..., 2^9, 2^{10}\}$ first. Then the value of α was set from 1 to 30. For each pair of (α, β), we calculated their MAP value. The result is shown in Fig. 3. When β is 2^{-7} and α is 7, the MAP measurement achieves the maximum value 0.7244. With fixed α from 1 to 7, the effect of β on the MAP value is displayed in Fig. 4. When β is less than 2^{-7}, the MAP value is almost fixed. The fluctuation periods were when β is from 2^{-2} to 2^4.

For MRRD, we fixed the value of β from the set $\{2^{-10}, 2^{-9}, ..., 2^9, 2^{10}\}$ first. Then the value of α was set from 1 to 30. For each pair of (α, β), we calculated their MRRD value. The result is shown in Fig. 5. When β is 2^1, and α is 2, the MRRD measurement achieves the maximum value 0.7504. There is a trend that when α is greater than 11, the MRRD value becomes very stable for most β values. The line for β equal to 2^2 is a separator. In the above group, the lesser the value of β, the smaller the value of MRRD. In the below group, the more the value of β, the smaller the value of MRRD. With fixed α from 1 to 7, the effect of β on the MRRD value is displayed in Fig. 6. When β is 2^1 and α is 2, the MRRD measurement achieves

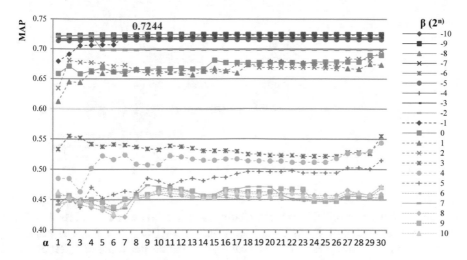

Fig. 3 MAP performance for fixed β

Fig. 4 MAP performance for fixed α

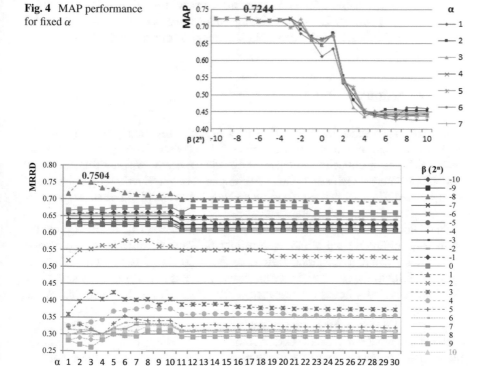

Fig. 5 MRRD performance for fixed β

Fig. 6 MRRD performance
for fixed α

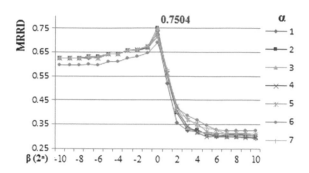

the maximum value 0.7504. In these charts, for a fixed α, when β is less than 2^1, the greater the value of β, the bigger the value of MRRD. For a fixed α, when β is greater than 2^4, the greater the value of β, the smaller the value of MRRD. The fluctuating part is when the value of β is between 2^1 and 2^4.

For user study, we use a fixed pair of α equal to 9 and β equal to 2^{-7}. By checking the average results returned for the 30 keyword terms, we found that using the ranking mechanism proposed by Liu et al. [12], the MAP value is 0.4052, while the MAP value is 0.6471 in WARKS. WARKS has a better performance of 59.70%.

5   Conclusions

There are many valuable data hidden in web databases, and they could not be retrieved through traditional search engines. We crawled and integrated web trade show databases by collecting company and product records. We provided a keyword search user interface with an option of "company" or "product". We designed and implemented WARKS that provided ranked keyword-matching company (or product) records based on user preferences.

In addition to the four normalization steps in Liu et al. [12], WARKS provided adjustable column weight α and tuple weight β. We used mean average precision (MAP) and mean rank reciprocal difference (MRRD) to evaluate ranking effectiveness of WARKS. Our user study for the case when α is 9 and β is 2^{-7} showed that WARKS provided a better performance of 59.70% improvement than that of Liu et al. [12].

We would like to point out the following future work directions: (1) Extend the research into other domains. (2) Consider the relative position of these keywords. (3) Increase the weight for frequent input keywords. (4) Incorporate semantic annotation tool for the user.

References

1. Agrawal S, Chaudhuri S, Das G (2002) DBXplorer: a system for keyword-based search over relational databases. In: Proceedings of the 18th IEEE international conference on data engineering (ICDE 2002), pp 5–16
2. Balmin A, Hristidis V, Papakonstantinou Y (2004) Objectrank: authority-based keyword search in databases. In: Proceedings of the 30th international conference on very large data bases (VLDB'04), pp 564–575
3. Bergamaschi S, Domnori E, Guerra F, Orsini M, Lado RT, Velegrakis Y (2010) Keymantic: semantic keyword-based searching in data integration systems. Proc VLDB Endowment 3(1–2):1637–1640
4. Bergamaschi S, Guerra F, Simonini G (2014) Keyword search over relational databases: issues, approaches and open challenges. In: Lecture notes in computer science, vol 8173, pp 54–73
5. Bergamaschi S, Guerra F, Interlandi M, Lado RT, Velegrakis Y (2016) Combining user and database perspective for solving keyword queries over relational databases. Inf Syst 55:1–19
6. Bhalotia G, Hulgeri A, Nakhe C, Chakrabarti S, Sudarshan S (2002) Keyword searching and browsing in databases using BANKS. In: Proceedings of the 18th IEEE international conference on data engineering (ICDE 2002), pp 431–440
7. Coffman J, Weaver AC (2010) A framework for evaluating database keyword search strategies. In: Proceedings of the 19th ACM international conference on information and knowledge management, pp 729–738
8. Hristidis V, Papakonstantinou Y (2002) DISCOVER: Keyword search in relational databases. In: Proceedings of VLDB'02. VLDB Endowment, Aug 2002, pp 670–681
9. Hristidis V, Gravano L, Papakonstantinou Y (2003) Efficient IR-style keyword search over relational databases. In: Proceedings of VLDB'03, pp 850–861
10. Jabeur LB, Soulier L, Tamine L, Mousset P (2016) A product feature-based user-centric ranking model for e-commerce search. In: Lecture notes in computer science, vol 9822, pp 174–186
11. Kacholia V, Pandit S, Chakrabarti S, Sudarshan S, Desai R, Karambelkar H (2005) Bidirectional expansion for keyword search on graph databases. In: Proceedings of the 31st international conference on very large data bases (VLDB'05), pp 505–516
12. Liu F, Yu C, Meng W, Chowdhury A (2006) Effective keyword search in relational databases. In: Proceedings of ACM SIGMOD'06, pp 563–574
13. Liu Z, Wang C, Chen Y (2017) Keyword search on temporal graphs. IEEE Trans Knowl Data Eng 29(8):1667–1680
14. Simitsis A, Koutrika G, Ioannidis YE (2008) Precis: from unstructured keywords as queries to structured databases as answers. VLDB J 17(1):117–149
15. Zhu L, Du X, Ma Q, Meng W, Liu H (2018) Keyword search with real-time entity resolution in relational databases. In: Proceedings of the 2018 10th international conference on machine learning and computing, pp 134–139

Enhanced Hash Algorithm Using a Two-Dimensional Vector to Improve Data Search Performance

Bobby A. Eclarin, Arnel C. Fajardo and Ruji P. Medina

Abstract Hashing is a procedure that can be used for a program to gain access to data quickly. It is very popular in many IT areas such as route lookup, cryptography development, and string data indexing for fast searching among others. Open Addressing and Separate Chaining are the generally two known collision resolutions if a hash function fails to plot a unique index. Engaging any collision resolution will result in multiple key comparisons and memory accesses causing time to increase linearly thus the lookup latency becomes indeterminate, posing an unpredictable outcome to some time-critical systems. Although the collision problem is inevitable, the proposed hashing algorithm using 2D vector and hash functions on its collision resolution can achieve lookup time at one or two memory accesses while resolving the collision.

Keywords Hashing · Collision resolution · Hash function · Open addressing · Separate chaining · Vector data structure

1 Introduction

Hashing is one of the essential procedures that offer means and ways for a program to gain access to data rapidly [1]. Hashing gains its popularity due to its usefulness in many Information Technology areas [2]. Among which is in the area of networking such as route lookup, packet organization, per-flow state administration, load bal-

B. A. Eclarin (✉) · R. P. Medina
Technological Institute of the Philippines, 938 Aurora Blvd., Cubao,
1109 Quezon City, Philippines
e-mail: bob.eclarin@gmail.com

R. P. Medina
e-mail: ruji.medina@tip.edu.ph

A. C. Fajardo
La Consolacion College Manila, Mendiola St., San Miguel, Manila, Philippines
e-mail: acfajardo2000@yahoo.com

© Springer Nature Singapore Pte Ltd. 2019   59
V. Piuri et al. (eds.), *Intelligent and Interactive Computing*, Lecture Notes in Networks
and Systems 67, https://doi.org/10.1007/978-981-13-6031-2_32

ancing, and network checking for its constant access time [3–5]. It is also being used in cryptography development and string (data) indexing for fast searching [2, 6–11].

Hashing algorithm has two components: the hash function and its collision resolution [12]. A hash function transforms a string of characters into a smaller fixed-length value or key and uses this key to locate and find the original value from a hash table [2, 8, 13]. A hash table is a data structure, also described as an associative array that plots keys to hash values as the index to which a data is stored using a hash function [8, 12, 14]. A good quality hash function provides uniform distribution of elements in the hash table [8], which in turn, by general assumption, delivers O(1) time complexity and a near single-memory access [5].

As data fills up the hash table, a collision may occur [3] even with a good hash function. Collision happens when two or more resulting hashed keys derived by the hash function are equal [9]. Collision resolution techniques are used to resolve such problem and guarantee future key lookup procedures to return the correct data [2]. There are generally two known collision resolutions namely, open addressing and separate chaining. Open addressing technique makes use of probing to find the next empty bucket in the hash table that accommodates the other data [5]. Open addressing employs either or linear probing, quadratic probing, and double hashing methods to determine the number of hops in finding an empty slot to place the data [2]. The same procedure will be used to search for it. On the other hand, separate chaining collision resolution uses linked list data structure to dynamically allocate space for data with the same hashed key [5].

However, engaging any collision resolution can result into multiple key comparisons and memory accesses causing time to increase linearly [5]; thus, the lookup latency becomes indeterminate, posing unpredictable outcome to some time-critical systems [4]. In a linked list chained collision resolution, pointers must be sustained in every bucket to provide associations with other links which make memory overhead in the form of additional space for logging and getting the pointers [13]. The lookup time for the separate chaining collision resolution using linked list depends on the position of the data being searched for and the number of data present in the list. Finding a data involves traversing the list from the beginning, until such item is found or the end of the list is reached [15, 16], resulting in multiple memory access which causes time delays.

The course of direction of this paper is to enhance the hash algorithm by using two hash functions and two-dimensional vector data structure to improve the data search procedure into one or a maximum of two memory accesses. It focuses on the creation of a novel collision resolution of the hash algorithm and may employ existing hash functions.

2 Related Literature

Hashing is the process of locating a record using various calculations to map its key value to a slot in a collection called hash table [15] (Fig. 1).

A hash table is a data structure, also described as an associative array to which a data will be stored using a hash function [8, 12, 14]. As mentioned, the hashing algorithm has two main components: The hash function and its collision resolution.

2.1 Hash Function

A hash function maps each key to an integer in the range $[0, M-1]$, where M is the capacity of the array for the hash table. The main idea is to use the hash value, $h(k)$, as an index into array A, where the data will be stored at $A[h(k)]$ [15]. A good hash function must have the main requirements as follows: must be deterministic, efficient to compute, and must uniformly distribute the keys. According to [14], cryptographic hash functions, like Message Digest Algorithm (MD5) and Secure Hash Algorithm (SHA-1) are thought to offer efficient hash functions for any table sizes. However, [3] argued that such complex hash functions might need more computational time and utilize more CPU resources, especially on frequent execution in a high-speed network. So, choosing a good hash function depends on the application and environment where it is to be implemented (Fig. 2).

In their work [2], listed some essential hashing methodologies to formulate an efficient hash function; this includes the Division Method ($(H(x) = x \bmod m)$), Multiplication Method ($H(x) = (a * b * c * d * ....) \bmod m$), and Random Number Generator. Also, in their work [3], considering the requirement of well load balancing and simplicity, they adopted shift operation and XOR operation in their hash function. According to [15], the last step is to use the modulus operator to the result, using table size M to produce a value within the range. If the sum is not adequately huge, then the modulus operator results in a poor distribution of data in the hash table. Also, [17] claimed that using prime numbers is a good option in coming up with a better hash function since these numbers have some certain dissimilarity among them. This detail can be used to create a uniform distribution of inputs, thus profoundly helps in resolving collisions and clustering problems.

Fig. 1 A hash table with textual data at the different indices

HT index	Data
0	abc
1	cde
-	
-	
20	gth
-	
M-1	lmk

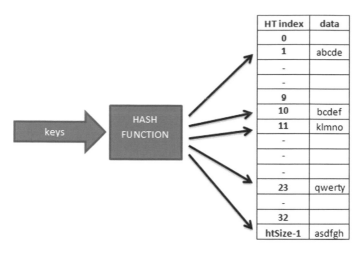

Fig. 2 A hash function that converts the keys and places them in the hash table with their respective indices

2.2 Collision Resolution

Hash algorithm that is well-implemented could reduce the possibility of collision [14]. As the data fills up the hash table, a collision may occur [3] even with a good hash function. Collision happens when two or more resulting hashed keys derived by the hash function are equal [9]. An efficient hash function distributes the data well into the hash table. If that is the case, collision resolution is not needed. However, collision, as mentioned above, is inevitable. There are two options for resolving collision when it occurs. The first one is to find another slot from the table or change the structure of the hash table so that its array indices can hold more than one value [18]. Generally, these are the two known collision resolutions: open addressing and separate chaining.

Open Addressing. Open addressing technique makes use of probing to find the next empty bucket in the hash table to accommodate the other data [5]. Open addressing employs either linear probing, quadratic probing, and double hashing methods to determine the number of hops in finding an empty slot [2]. According to [14], using a prime number in their table size guarantees that any step length will be able to probe all slots in the hash table (Fig. 3).

Reference [13] states that linear probing is the simplest open addressing policy, in which the probing step is usually 1. The key is placed in the first empty table entry from the hashed bucket in a linear order. Thus, given the primary hash function $h': U \to \{0, \ldots, m - 1\}$, linear probing utilizes such hash function to sequentially index the table $h(k, i)$: $(h'(k) + i) \bmod m$; for $i = 0, 1, \ldots, m - 1$. Even though it is easy to implement, such policy suffers from primary clustering; when the number of occupied buckets increases, the average search time also increases [19]. When there are only a few collisions, the probe sequence remains short and data can be

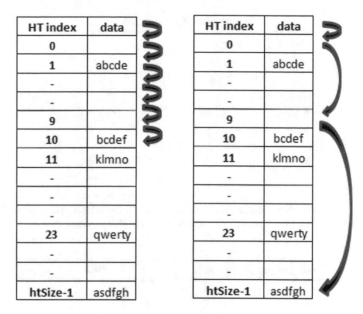

Fig. 3 Linear probing and quadratic probing in open addressing

searched quickly, but when there are lots of collisions, a collision within a cluster increases the size of the cluster [20]. Bigger clusters result in a longer search period, and as the clusters grow in size, they can merge into even larger clusters, making the problem worse [21]. Quadratic probing can solve the problem of primary clustering and iteratively implement $f A[(i + f(j))$ mod $M]$, for $j = 0, 1, 2, \ldots$, where $f(j) = j^2$ until an empty slot is found [15]. However, according to [2] which is also confirmed in [21], this approach creates its own clustering problem which is called the secondary clustering. Additionally, [2] also indicate that it may not be able to traverse all slots in the hash table.

Another way to probe the hash table in the event of a collision is by using the Double Hashing strategy, which uses another hash function to come up with a value that determines the number of steps in probing. In their work, [1] makes use of prime numbers and map the result of the second hash function into a nonzero value to optimize the double hashing method. They also claimed that, by slightly changing the size of the hash table into a relative prime, the added hash functions could traverse all slots in the table. However, the method needs a rehash if the total of the two hash functions is greater than the table size (Fig. 4).

Accordingly, in [21], the probe sequence would be an iteration to test the slot $A[(i + f(j))$ mod $N]$ next, for $j = 1, 2, 3, \ldots$, where $f(j) = j * h'(k)$, if the first hash function h maps some key k to a slot $A[i]$, with $i = h(k)$, that contains a data. Also, the second hash function must have the following criteria: it must not be the same as the first hash function, it must be dependent on the search key as the first hash function, and it must not produce a zero value.

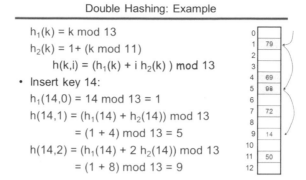

Double Hashing: Example

$h_1(k) = k \bmod 13$

$h_2(k) = 1 + (k \bmod 11)$

$h(k,i) = (h_1(k) + i\, h_2(k)) \bmod 13$

• Insert key 14:

$h_1(14,0) = 14 \bmod 13 = 1$

$h(14,1) = (h_1(14) + h_2(14)) \bmod 13$

$= (1 + 4) \bmod 13 = 5$

$h(14,2) = (h_1(14) + 2\, h_2(14)) \bmod 13$

$= (1 + 8) \bmod 13 = 9$

Fig. 4 Double hashing method Adapted from Presentation on B-trees, Heap trees, Hashing; In Slideshare. Retrieved from https://www.slideshare.net/hghhhhg/presentation-on-b-tress-heap-treeshashing

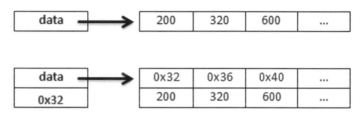

Fig. 5 Example of an array

Separate Chaining. Another known collision resolution is separate chaining which uses linked list data structure to dynamically allocate space for data with the same hashed key [5]. An array allocates space for all its elements as one chunk as they are in a contiguous memory location (Fig. 5).

On the other hand, a linked list allocates space for each element separately using pointers to connect all its nodes, and each node contains two fields; the one that holds the data and a pointer that holds the address to the next node [22]. So unlike arrays, a Linked list does not have a declared fix size [23]. The location of its elements can be anywhere in the memory and it can be dynamically grown to any extent as depending on the memory size and addressability of the system. Another difference between an array and linked list is the way they are accessed. Arrays can be randomly accessed through its indices, while the linked list must be traversed to get to the item's position. Arrays may waste some memory space since its size is fixed and the program running it may not utilize all the allotted size; however, linked list may have overhead memory spaces too, due to its pointer field (Fig. 6).

Reference [24] states that the number of memory accesses to acquire the value associated with the hashed key increases as the data to locate is toward the end of the linked list thus affecting the lookup performance. Reference [4] implied that collision resolutions cost additional memory accesses that may degrade lookup time that poses

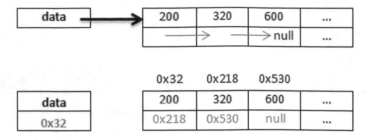

Fig. 6 Example of a linked list

a threat to time-critical systems, such as network packet processing segments. Subsequently, it is essential to keep the lookup operations fast and deterministic while not hurting the optimal load of the hash table. Reference [13] claimed that separate chaining is easy to implement; however, they also mentioned that the overhead cost of the pointers of the linked list is also a concern.

2.3 Vectors

Arrays and Linked List have their advantages and disadvantages over the other. Vectors are an array-like data structure which resides in contiguous memory [16], thus providing a faster access time [25]. Additionally, [4] in their research, used multiple vectors because of its random access property for faster lookup. Reference [15] indicated that another characteristic of a vector is that it can be dynamically grown and shrink, henceforth memory usage is optimized which means that space need not be allotted until it is to be used, unlike its array predecessor that once its size is declared, users can not change it.

3 The Proposed Hashing Algorithm Using 2D Vector on Its Separate Chaining Collision Resolution

The steps for the lookup procedure in the original Separate Chaining Collision Resolution is as follows:

1. Get the index of the hash table using the hash function ($i = H_1(x)$).
2. Move to the specified index where the head of the list is located. If NULL then Item is not in the hash table.
3. Otherwise, traverse the link list in the specified HT index.
4. Compare the item being searched for with each data in the nodes.
5. Repeat step 3 until item found or until end of the list is reached (Fig. 7).

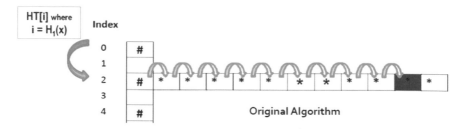

Fig. 7 Lookup procedure of the original separate chaining collision resolution

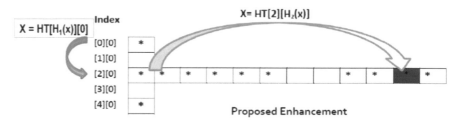

Fig. 8 The enhanced algorithm

The proposed algorithm employed two hash functions. The first hash function distributes the keys on the first index of the 2D vector, refer to as the first dimension or the hash table (HT). The same hash function will be used to search and look for items in the HT. The second hash function generates the keys for the chain, which is actually the second dimension of the vector, in case that the first hash function produced two or more same keys. The hash function in the chain must have a different methodology so that the collision encountered by the first hash function will not be repeated thus preventing the same collision problem in the chain. In case of collision in the chain, a rehash procedure is implemented only on that particular chain and not on the whole hash table. Although it is a multidimensional vector, its chain only allots space in case of collision, thus providing efficient use of memory; if there is no collision at a given hash table index, the chain does not grow. The purpose of another hash function in the chain allows an efficient lookup by computing the designated index of a given key. It accesses at once the data being searched for, eliminating the time consumed during a linear search (Fig. 8).

The steps for the lookup procedure in the proposed enhancement:

1. Get the index of the Hash Table where the item is located using the first hash function ($X = HT[H_1(x)][0]$).
2. If $HT[H_1(x)][0]$ is empty. Item is not in the HT.
3. Otherwise, it compares the item being searched for with the content. If items are the same; ITEM FOUND.
4. If not the same, locate item in the chain using the second hash function ($X = HT[H_1(x)][H_2(x)]$).

The code snippet below shows how the lookup procedure works using C++.

```
Code snippet in c ++:
    void SearchString(string strToSearch)
    {
    int i,j;
    i = hash1(strToSearch);
    j = 0;
    if (CheckIfOccupied(i,j)==true)
            {
            if (CURRENT_ITEM==strToSearch)//if it exist
    and they are equal, exit process
                    {cout << strToSearch << " Found in" <<
    i << ","<< j << "\n";}
        else
                    {j=hash2(strToSearch)+1;
                    if (CheckIfOccupied(i,j)==true)
                        {
                        if(CURRENT_ITEM ==strToSearch)//if
    it exist, exit process
                        {cout << CURRENT_ITEM << " = " <<
    strToSearch << " Found in" << i << ","<< j << "\n";}}
                        else
                        {cout << strToSearch << " Not
    Found empty slot in chain!\n";}
                        }}
        else
            {cout << strToSearch << " Not
    Found\n";} //Not in the table
    }
```

4   Conclusion

Combining the strengths of the array and linked list, a two-dimensional vector can be utilized to enhance the hashing algorithm on its separate chaining collision resolution to improve time spent during lookup procedure by only one or maximum of two memory accesses. Vectors reside in contiguous memory, thus it provides faster access time and its elements can be randomly accessed through its index. Hash functions can be used to generate the key using the value of the data that will point to its location just like a hash table does. Vector can be dynamically grown and shrink, henceforth memory usage is optimized which means that space need not be allotted until it is to be used.

5   Future Works

Future works will include comparison with other existing hash algorithm collision resolutions. Also, to find out the best combinations of hash function methodologies that may be utilized to optimize the collision resolution further.

References

1. Khorasani F, Belviranli ME, Gupta R, Bhuyan LN (2016) Stadium hashing: scalable and flexible hashing on GPUs. In: Parallel architectures and compilation techniques—conference proceedings, PACT, vol 2016–March, pp 63–74. https://doi.org/10.1109/pact.2015.13
2. Singh M, Garg D (2009) Choosing best hashing strategies and hash functions. In: 2009 IEEE international advance computing conference, IACC 2009, pp 50–55. https://doi.org/10.1109/pact.2015.13
3. He G, Du Y, Yu D (2016) Efficient hashing technique based on bloom filter for High-Speed Network. In: Proceedings—2016 8th international conference on intelligent human-machine systems and cybernetics, IHMSC 2016, vol 1, pp 58–63. https://doi.org/10.1109/ihmsc.2016.94
4. Li D, Li J, Du Z (2016) Deterministic and efficient hash table lookup using discriminated vectors. In: Proceedings of 2016 IEEE global communications conference, GLOBECOM 2016, 2016, pp 0–5. https://doi.org/10.1109/glocom.2016.7841732
5. Song T, Yang Y, Crowley P (2017) RwHash: rewritable hash table for fast network processing with dynamic membership updates. In: 2017 ACM/IEEE symposium architecture network communication system, pp 142–152. https://doi.org/10.1109/ancs.2017.28
6. Cantu M, Kim J, Finding hash collisions using MPI on HPC clusters, pp 1–6. https://doi.org/10.1109/lisat.2017.8001961
7. Kidon M, Dobai R (2017) Evolutionary design of hash functions for IP address hashing using genetic programming. In: Proceedings of 2017 IEEE congress on evolutionary computation, CEC 2017, pp 1720–1727. https://doi.org/10.1109/cec.2017.7969509
8. Liu D, Xu S, Liu H, Cui Z (2014) An empirical study on the performance of Hash Table. In: ICIS 2014
9. Wang D, Jiang Y, Song H, He F, Gu M, Sun J (2017) Verification of implementations of cryptographic hash functions. IEEE Access 5, no. c, 7816–7825. https://doi.org/10.1109/access.2017.2697918
10. Steinebach M, Klöckner P, Reimers N, Wienand D, Wolf P (2013) Robust hash algorithms for text. In: Lecture notes in computer science (including subseries lecture notes in artificial intelligence and lecture notes in bioinformatics), vol 8099 LNCS, pp 135–144. https://doi.org/10.1007/978-3-642-40779-6_11
11. Partow A Available hash function algorithm
12. Cormen TH, Leiserson CE, Rivest RL, Stein C (2001) Hash tables. In: Introduction to algorithms
13. Kumar S, Crowley P (2005) Segmented hash: an efficient hash table implementation for high-performance networking subsystems. In: 2005 symposium on architectures for networking and communications systems, ANCS 2005, pp 91–103. https://doi.org/10.1109/ancs.2005.4675269
14. Liu D, Xu S (2015) Comparison of hash table performance with open addressing and closed addressing : an empirical study 3(1), 60–68. https://doi.org/10.2991/ijndc.2015.3.1.7
15. Shaffer C (2012) Data structures and algorithm analysis, 2
16. Franek F (2003) Linked data structures, 132–158. https://doi.org/10.1017/cbo9780511584046.010
17. Bhullar RK (2016) A novel prime numbers based hashing technique for minimizing collisions. In: 2nd international conference on next generation computing technologies (NGCT-2016), 522–527. https://doi.org/10.1109/ngct.2016.7877471
18. Srinivasan J, Notes on hashing
19. Myers A (2014) CS 2112_ENGRD 2112 Fall 2014
20. Granville B, Del Tongo L (2008) Annotated reference with examples, p 101
21. CS240_ Data Structures & Algorithms I. (Online). Available: https://www.cpp.edu/~ftang/courses/CS240/lectures/hashing.htm. Accessed: 15 Sep 2017
22. Sullivan DG (2012) Fall 2012 alternative representation : a linked list, pp 1–19
23. Straub J (2008) C programming : data structures and algorithms

24. Cho C, Lee J, Ryoo J (2017) A collision-mitigation hashing scheme utilizing empty slots of cuckoo hash table, pp 514–517. https://doi.org/10.23919/icact.2017.7890143
25. Limon MR, Sharker R, Biswas S, Rahman MS (2014) Efficient de Bruijn graph construction for genome assembly using a hash table and auxiliary vector data structures. In: 2014 17th international conference on computer and information technology, ICCIT 2014, pp 121–126. https://doi.org/10.1109/iccitechn.2014.7073147

Truth Discovery Using the TrustChecker Algorithm on Online Quran Tafseer

Zaid J. Al-Araji, Sharifah Sakinah Syed Ahmad, Hayder A. Al-Lamy,
Mustafa W. Al-Salihi, S. A. Al-Shami, Hussain Mohammed
and Malik H. Al-Taweel

Abstract The World Wide Web (WWW) is an important source and repository of information. However, there are no guarantees regarding the correctness or trustworthiness of the information. Indeed, various websites frequently provide contradictory information on various matters, namely, various product stipulations for a similar product. Based on this study, the essential for research on the credibility of the online information is discussed. The primary goal of this paper is to test and propose an appropriate tool for constructing a system to help users in judging the validity of claims. Moreover, this paper focuses on the process to discover the 'true facts' from the vast quantity of contradictory information on Quran's Tafseer provided on different Internet websites. A framework was designed to validate the problem, through constructing an algorithm, called 'TrustChecker', which compares the interactions found between the websites and the information hosted on these websites. Accordingly, the website is 'trusted' if the site provides pieces of 'true' (correct) information, or the pieces of information are considered to be 'true' if supported by reliable websites.

Keywords Trustworthiness · Information · Claims · TrustChecker · Algorithm

1 Introduction

Traditional media channels and outlets, namely, online newspapers, streaming radio and television are rapidly being replaced by new easily accessible media via the

---

Z. J. Al-Araji (✉) · H. A. Al-Lamy · M. W. Al-Salihi · S. A. Al-Shami · M. H. Al-Taweel
Universiti Teknikal Malaysia Melaka (UTeM), 76100 Durian Tunggal, Melaka, Malaysia
e-mail: zaid.jassim4@gmail.com

H. Mohammed
Universiti Tun Hussein Onn Malaysia, Batu Pahat, Johor, Malaysia

S. S. Syed Ahmad
Department of Intelligent Computing & Analytics (ICA), Universiti Teknikal Malaysia Melaka
(UTeM), Hang Tuah Jaya, 76100 Durian Tunggal, Melaka, Malaysia
e-mail: sakinah@utem.edu.my

© Springer Nature Singapore Pte Ltd. 2019
V. Piuri et al. (eds.), *Intelligent and Interactive Computing*, Lecture Notes in Networks
and Systems 67, https://doi.org/10.1007/978-981-13-6031-2_39

Internet. As the primary source of information is on news events and other sources of the Interest, Internet websites and other forms of social media (i.e. Twitter, Facebook, etc.) are quickly influencing the social views of individuals and groups. Therefore, measuring social opinions is becoming increasingly difficult. For example, the younger generation spends considerable time on the Internet, and are easily influenced by what they viewed or are exposed to via the Internet. As an interactive environment, the Internet provides a reasonably straightforward mechanism for individuals to post comments, voice their political views, opinions and remarks about a variety of subjects on social and community forums, via blogs or chat rooms.

During the past few years, the use of social media websites has significantly increased and is continuing to grow at a remarkable rate. Many social media websites, such as MySpace, Facebook and LinkedIn, have become quite prominent and the 'norm' for many people as the preferred choice for communication. Indeed, the importance of these websites has resulted since the users spend majority of their time updating or customising information, including their personal profiles in order to interact with other Internet users and browse other member's online profiles [1]. By 2021, the social media visitors globally will exceed 3.2 billion as reported by statista.com.

In April 2018, Facebook was ranked the highest, as number one (as illustrated in Fig. 1) compared to other social media sites, with more than 2.234 billion users. Further, the number of Tencent QQ users ranked eighth overall was reported to have 783 million users and Twitter ranked twelfth, with 330 million users. The more recent '39th Statistical Report on Internet Development in China' released in December 2016 reported that about 53.2% of Chinese population are using the Internet.

Much of the information hosted on these websites and social media forums are searched daily, with often little to no control over what is being published. Regretfully, this can lead to misleading, poor quality and unreliable information. Nowadays, anyone and everyone has the opportunity to post whatever they like regarding their personal views, opinions and facts [2]. Given the vast amount of information currently hosted on the Internet, this undoubtedly led to credibility issues on what is read, and what is believed as factual and truthful.

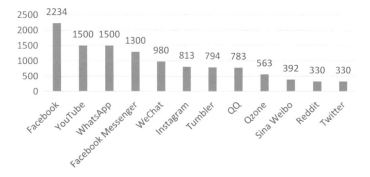

Fig. 1 Global social networks ranked by the number of users in 2018

The expression, 'trust', has been historically used in various civilisations, communities, groups and other disciplines, extending from computer science to sociology, philosophy and psychology. All disciplines consider trust based on different sets of criteria and are inevitably reliant upon its context. For instance, a sociologist tends to determine trust as being structural in nature. Whereas many psychologists have considered trust as the behaviour of a person against the information presented, the economists view trust as the methodology of optimal choice [3].

The ease of publishing news, events and the opinions of people has also affected the reliability and quality of information. Indeed, the news is becoming increasingly moderated, and verified by editors before publishing with little visible or transparent editorial oversight. When searching for news, people cannot readily recognise or differentiate between a trustworthy newspaper compared with another newspaper as being biased or non-factual.

Trustworthiness is also known as 'sentiment analysis' which is the procedure of finding user opinions on a specific topic. For instance, the topic could be an event, a product, service, movie, news, Quran Tafseer, etc. Furthermore, the information will often involve classification, aggregation and retrieval and is quickly becoming the basis for academic and non-academic research. Trustworthiness can further be observed in user or customer feedback regarding a product or analysis of a product and often influences the purchasing decision of potential customers. Whereas recovery is the procedure of gathering text from review websites [4]. It would be great to discover the means to assist the Internet users in verifying trustworthy or the claims made by news media from the blurred or hidden facts.

The goal, in this case, is to evaluate the accuracy of the claim. The claim is the passage or statement whose accuracy is unknown and could be the 'fact' (alternatively, 'correct claim') or the 'hoax' (alternatively, 'false claim'). Previous work in this area by researchers has called these claims 'facts' with unbiased terminology often used.

Based on the definition, in its elementary form, claims are relation tuples expressed or as a simple interaction used in the ordinary language. Alternatively, it can be described as a key attribute that recognises it from other sentences in which reasonably consistent truth values are linked with a claim that is verifiable.

The sources that reinforce the claims are the key elements in defining the accuracy of a demand which is the feature of the information sources that support the claim. The sources provide the facts for (or 'states') the demand. Trusted sources support correct claims or incorrect conflict claims, while untrusted sources reinforce incorrect claims or conflict correct claims. Also, a fact can be an artefact, passage or a document that reinforces or conflicts with a claim. Accordingly, 'Worthy' (reliable) facts boost the trustworthiness of the sources and the accuracy of claims.

There are many websites and algorithms, for example, CNNs Fact Check, FactCheck Blog and TruthFinder. These websites and algorithms are exactly entirely based on time-consuming, manual fact inspection assumed through writers, where the declarations made through politicians and other extraordinary objects are checked for precision.

This paper investigates the effectiveness of existing trustworthiness algorithms (TruthFinder and ClaimVerifier) in validating the consistency of the algorithms and constructing parameters that should be considered in constructing trustworthiness algorithms. Further, a new algorithm (TrustChecker) is proposed and compared with two trustworthiness algorithms using a real dataset.

2   Methodology

In this section, the methodology is described along with the data set and the algorithm that is adopted in this paper.

2.1   Data Collection

The real (actual) data set in this study contains the Quran's Tafseer which is available to view on many Internet websites. The dataset contains 100 websites of which contain 114 Surahs and 6236 ayats. For each website or claim, the name of the Surah, the number of the ayat and some keywords for example 'marriage, zakat, etc.' are used. Indeed, there are different Tafseer in some websites. For example, in first Surah 'Al Fatihah', in the third ayat, the website http://www.englishtafsir.com/ quotes: 'The word Rab which has been translated into the "Lord" stands for (a) Master and Owner, (b) Sustainer, Provider and Guardian, (c) Sovereign, Ruler, Administrator, and Organiser. Allah is the Lord of the Universe in all these senses'. Whereas, http://www.greattafsirs.com/ has quotes 'The Compassionate, the Merciful: that is to say, the One who possesses "mercy", which means to want what is good for those who deserve it'. Therefore, from these two (2) websites, the differences in Tafseer, this ayat and many others can be observed.

2.2   TrustChecker

Presently, trust frameworks function and perform on organised information and provide for precise information abstraction. Earlier researches on determining credibility [5, 6] have relied on the bipartite graph, containing the claim layer and the source layer (Fig. 2a). The bulges in a claim layer are connected to all source bulges that express a claim. Commonly, a claim mark relies on the mark of the source connected or related to a claim. Contrariwise, the source score depends on the score of the claim that the source is connected to. Therefore, the source contributes uniformly for all claims it expresses.

Two-layer architecture ignores both the content and context in which the source expresses the claim. Therefore, to overcome this constraint, a framework that includes

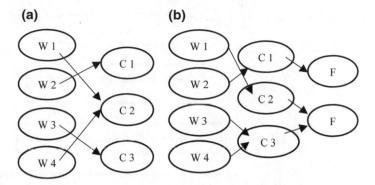

Fig. 2 Difference between two-layer and three-layer representations

content nodes as a middle layer is assumed. The framework is the three-tier graph consisting of the source, claim and the fact layers (see Fig. 2b). Every content node will represent a claim given by the source and connects to a claim node and source node. This allows the trust framework to explicitly capture the textual context in which the source gives the claim.

In this case, let $W = \{w_1, w_2, \ldots, w_M\}$ be a set of the M source website. Every website, w, is assigned a trustworthiness score $\tau(w) \in [0, 1]$. If the source is trusted; its trustworthiness score is $\tau(w) = 1$. Next, let $C = \{c_1, c_2, \ldots, c_N\}$ be N claims in a dataset. All the claims, c, have the overall confidence score, $\delta(c) \in [0, 1].\delta(c) = 1$ for the facts and 0 for the hoaxes.

A fact f is associated with parts of the claim, $\varepsilon(c_i) = \{f_1, f_2, \ldots, f_n\}$. A part of a claim, c, connects to a source website, $w(c) \in W$, to a claim, $c(f) \in C$, and has the confidence score of $\Psi(f) \in [0, 1]$, where (f) represents the level of confidence in the claim's trustworthiness. At the end, let $C(w_i) = \{c \in C | \exists e \in E : (c(f) = c) \wedge (w(f) = w_i)\}$ denote the claim for which the source w_i gives evidence.

Figure 3 shows the algorithm process depicted in this paper called 'TrustChecker'. At the beginning of the process, the information is collected which depends on the URL identified by the user. Next, the data is analysed to determine the claim before calculating the confidence, trustworthiness scores and the elapsed time (Table 1).

2.3 Computing Trust Scores

Accordingly, the trustworthiness of a website is the expected confidence in the facts that are presented on the website. For website w, its trustworthiness $\tau(w)$ is calculated in Eq. 1 via computing the average confidence of the facts provided via w as given below:

Fig. 3 Algorithm flow chart

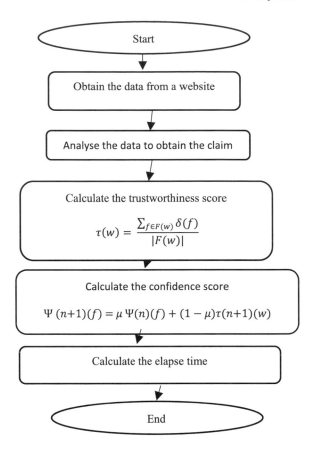

Table 1 Parameters and variables of TrustChecker

Name	Description
N	Number of facts
w	Website
M	Number of websites
f	Fact
$W(f)$	Set of facts provided via w
$t(w)$	Trustworthiness of the website
$\tau(w)$	Trustworthiness score of the w
$F(w)$	Set of the facts provided via w
$\delta(f)$	Confidence of the f
Ψ	Confidence score of the f

$$\tau(w) = \frac{\sum_{f \in F(w)} \delta(f)}{|F(w)|}, \tag{1}$$

where f is the fact, w is a website, $\delta(f)$ is the confidence of the fact and $F(w)$ is the set of facts provided via w.

Then, Eq. 2 re-estimates the source of trustworthiness based on the new claim confidence as follows:

$$\delta^{(n+1)}(f) = \sum_{f \in F(w)} \tau(f)^{(n)} + 1 \tag{2}$$

where f is the fact, $F(w)$ is the set of facts provided via w and $\delta(f)$ is the confidence of the fact.

Similarly, the confidence score of the fact is defined as shown in Eq. 3. The beneficial property is a confidence score of the fact f as the linear interpolation of previous and new estimates of the source trustworthiness are given as follows:

$$\Psi^{(n+1)}(f) = \mu \Psi^{(n)}(f) + (1 - \mu)\tau^{(n+1)}(w), \tag{3}$$

where μ denotes an interpolation parameter to control the bias of the previous knowledge of $\Psi(f)$ on future estimates. Once an algorithm converges, $\delta(f)$ provides an estimate of the confidence of claim c. Moreover, $\tau(w)$ shows the trustworthiness of the source w and $\Psi(f)$ gives the confidence score for a fact.

3 Result

In this section, the accuracy, efficient and elapsed time of TrustChecker on real data sets (marriage and zakat) are tested followed by comparing the results with two other algorithms (TruthFinder and ClaimVerifier).

3.1 Elapse Time

Table 2 displays the results from performing the elapse time tests. It is shown that the process of retrieving the data from the websites using TrustChecker has better elapse time compared to TruthFinder and ClaimVerifier.

In the initial test, the data is retrieved from the website, altafsir.com. The elapsed time for the process using TrustChecker is 19,245 ms, while the time using TruthFinder is 31,384 and 24,104 ms for ClaimVerifier. The difference between TrustChecker and TruthFinder is 12,139 ms, and the difference between TrustChecker and ClaimVerifier is 4859 ms.

Table 2 Elapse time for data set 1

Methods	Test 1	Test 2	Test 3	Average
TrustChecker	19,245	20,388	19,718	19,783
TruthFinder	31,384	31,958	31,684	31,675
ClaimVerifier	24,104	24,689	24,478	24,423

Fig. 4 Average elapse time for data set 1

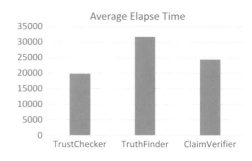

Table 3 Trustworthiness of data set 1

Methods	Test 1	Test 2	Test 3	Average
TrustChecker	0.971	0.975	0.974	0.973
TruthFinder	0.965	0.968	0.964	0.966
ClaimVerifier	0.971	0.970	0.970	0.970

In the second test, the data was retrieved from the website, quranwebsite.com. The elapsed time in milliseconds using TrustChecker is 20,388 ms, while the time using TruthFinder is 31,958 and 24,689 ms for ClaimVerifier. The difference between TrustChecker and TruthFinder is 11,570 ms, and the difference between TrustChecker and ClaimVerifier is 4301 ms.

In the third test, data was retrieved from the website, almizan.org. The elapsed time in milliseconds for the process using TrustChecker is 19,718 ms, while the time using TruthFinder is 31,684 and 24,478 ms for ClaimVerifier. The difference between TrustChecker and TruthFinder is 11,966 ms, and the difference between TrustChecker and ClaimVerifier is 4760 ms.

The average, (Fig. 4), time calculated based on the difference between TrustChecker and TruthFinder is 11,892 ms, and the difference between TrustChecker and ClaimVerifier is 4640 ms.

3.2 Accuracy

Table 3 shows the results from performing the accuracy tests. From the results, the process to assess the trustworthiness of the websites using TrustChecker displayed better accuracy compared to TruthFinder and ClaimVerifier.

Fig. 5 Average
trustworthiness of dataset 1

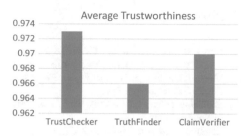

The initial test data was retrieved from the website, altafsir.com. The accuracy using TrustChecker is 0.973 while the accuracy using TruthFinder is 0.965 and using ClaimVerifier is 0.971. The difference between TrustChecker and TruthFinder is 0.006. However, TrustChecker and ClaimVerifier showed the same result.

In the second test, data was retrieved from the website, quranwebsite.com. The accuracy using TrustChecker is 0.975 while the accuracy using TruthFinder is 0.968 and using ClaimVerifier is 0.970. The difference between TrustChecker and TruthFinder is 0.007, and the difference between TrustChecker and ClaimVerifier is 0.005.

In the third test, data was retrieved from the website, almizan.org. The accuracy using TrustChecker is 0.974 while the accuracy using TruthFinder is 0.964 and using ClaimVerifier is 0.970. The difference between TrustChecker and TruthFinder is 0.01, and the difference between TrustChecker and ClaimVerifier is 0.004.

The average, (Fig. 5), calculated illustrates that the difference between TrustChecker and TruthFinder is 0.007, and the difference between TrustChecker and ClaimVerifier is 0.003.

4 Conclusion

In this paper, the TrustChecker algorithm has been shown to test the trustworthiness of several websites accurately. Moreover, to categorising the claims as either false or true, TrustChecker has the ability to allocate assurance notches to the claims, which gives the sources and confirmation based on their credibility. Further, the algorithm can provide a provisional summary of the confirmation with or in contradiction of the claims to assist the users decide their validity. In addition, a judgment based on the accuracy of a claim can be calculated according to the source's credibility in providing evidence for the claim. The experiments compared the proposed algorithm (TrustChecker) with TruthFinder and ClaimVerifier. The results indicated that TrustChecker has much higher accuracy in locating truthful or correct facts compared to the other two algorithms, and with less elapsed time.

References

1. Nepal S, Paris C, Pour PA et al (2015) Interaction-based recommendations for online communities. ACM Trans Internet Technol 15(2):1–21
2. Zhang Z, Gupta B (2016) Social media security and trustworthiness: Overview and new direction. Future Gener Comput Syst
3. Pattanaphanchai J (2014) Trustworthiness of web information evaluation framework. University of Southampton
4. Ortigosa A, Martín JM, Carro RM (2013) Sentiment analysis in Facebook and its application to e-learning. Comput Hum Behav J
5. Pasternack J, Roth D (2010) Knowing what to believe (when you already know something). In: Proceedings of the 23rd international conference on computational linguistics (COLING). Association for Computational Linguistics, pp 877–885
6. Yin X, Han J, Yu PS (2008) Truth discovery with multiple conflicting information providers on the Web. IEEE Trans Knowl Data Eng (TKDE) 20(6):796–808

Criteria Ranking Based on Volunteer Selection Using Fuzzy Delphi Method

Nurulhasanah Mazlan, Sharifah Sakinah Syed Ahmad
and Massila Kamalrudin

Abstract Volunteers are individuals who give their time, skills, and expertise in helping to bring measurable benefits to beneficiaries. These beneficiaries may include orphans, flood victims, the homeless, the poor, or people with disabilities. Word is spread through social media about the needs of beneficiaries. There would be a surge of volunteers who are willing to assist without expectation of payment or being rewarded. Therefore, it is important for an organization that deals in volunteer management to act fast and efficiently so that time and resource wastage can be prevented. However, it is difficult to find and recruit candidates suitable for volunteer organizations, as the volunteers may have too many criteria to be matched against the tasks being offered. Hence, the purpose of this study is to identify the important criteria for selecting the appropriate volunteers for specific tasks. This paper had identified seven aspects with 17 criteria based on the literature and from interviews conducted with experts from the non-government organizations (NGOs). The fuzzy Delphi method had been applied in choosing the essential criteria of volunteer selection. This method had been used to derive the value of fuzzy numbers as well as assigning ranks to the seven aspects along with their 17 criteria. The results show "Teamwork" and "Commitment" were top two main aspects.

Keywords Volunteer selection · Fuzzy Delphi method · Criteria volunteer

N. Mazlan (✉) · M. Kamalrudin
Faculty of Information and Communication Technology, Universiti Teknikal
Malaysia Melaka (UTeM), 76100 Durian Tunggal, Melaka, Malaysia
e-mail: nurulhasanah@student.utem.edu.my

M. Kamalrudin
e-mail: massila@utem.edu.my

S. S. Syed Ahmad
Department of Intelligent Computing & Analytics (ICA), Faculty of Information and
Communication Technology, Universiti Teknikal Malaysia Melaka (UTeM), 76100 Durian
Tunggal, Melaka, Malaysia
e-mail: sakinah@utem.edu.my

© Springer Nature Singapore Pte Ltd. 2019
V. Piuri et al. (eds.), *Intelligent and Interactive Computing*, Lecture Notes in Networks
and Systems 67, https://doi.org/10.1007/978-981-13-6031-2_41

1   Introduction

Voluntarism is an essential contribution to society. As such, volunteer selection is conducted as a means of finding a good match between the prospective volunteer and the opportunity on offer. It seeks to find a strong affiliation between persons who have the appropriate experience, skills, and interest and the opportunity that satisfies the needs and interests of the volunteer. However, matching volunteers and tasks could be difficult since selection is a question of whether a volunteer fits into the structure and is able to carry out the work involved. In the volunteer management process, there is a selection and recruitment process that will interview volunteers and conduct assessments to ensure that the organization's needs are being addressed effectively and efficiently. A needs assessment distinguishes differences between the knowledge, skills, and abilities or capabilities people possess and those they need to acquire to do their job effectively. As such, this is a complicated process since there would be too many criteria to be considered in the volunteering choices and the criteria needed for existing tasks. For example, volunteer management organizations would have to find volunteers with teaching skills to teach mathematics and English to secondary school students.

In this paper, we have identified a list of criteria that had been based on the past literature and interviews with experts in the selection of suitable volunteers and categorized the aspects according to these corresponding criteria. We have proposed a new methodology in this research, which is the multiple criteria decision-making approach that uses the fuzzy Delphi method (FDM). FDM is an integration of the fuzzy concept and the Delphi method that requires only a small sample of the survey to get objective and reasonable results. With this method, time and costs of collecting questionnaires can be reduced, and experts' opinions can be kept as they are without being twisted [1]. We use FDM to explore the important criteria that match suitable volunteers with existing tasks. FDM is, therefore, used to obtain the weight for critical factors of the model so that the decision-maker can select the most appropriate volunteer required for the tasks.

This paper has been divided into five sections. In Sect. 2, we elaborate on the existing literature that addresses the relevant criteria and methods for volunteering, while Sect. 3 touches on the details of the methodology used including choosing the criteria. The outcome of the criteria according to the perspective of volunteer organization and literature review is explained in Sect. 4, and finally Sect. 5 focuses on our conclusions and our plans for the future research.

2   Related Works

2.1   Volunteer Criteria and Task Criteria

Tasks from organizations managing volunteers are published in an unfamiliar solution group in the form of a public call for assistance. However, the suitability of a group in performing different tasks not only depends on the types of individuals involved but is also subjected to their location, skills, knowledge, expertise, and the availability of their time. Chen et al. [2] had explained the three primary criteria in selecting potential volunteers, which are time, skill, and location. They have also used questions such as quizzes to check on the similarity of values when comparing the volunteers' answers against the organizations. Cvetkoska et al. [3] had used knowledge, abilities, skills, and personality traits as part of the recruitment process, while other researchers such as Yu et al. [4] had used task criteria such as task number, the expertise needed, and the restricted time frame. As for recruiting volunteers for specific needs, they had used profiles such as location, skill, calendar, and social network. The Division of Industry and Community Network in Universiti Sains Malaysia [5] had suggested 18 criteria to evaluate volunteers and their experience in volunteerism.

Based on the literature surveyed, we have identified seven aspects and seventeen criteria for volunteer selection, which can be found in Sect. 4. Following that, we began conducting interviews with several voluntary organizations to determine the appropriate aspects and criteria required by the organizations. Among the organizations that we have conducted interview sessions with are Projek Iqra', Spot Community, Hidayah Centre Foundation, Mercy Malaysia, National Department for Culture and Arts, and a few other universities.

2.2 Volunteer Selection Methods

Based on past literature, even though the fuzzy Delphi method (FDM) has been widely used in questionnaires and evaluation analyses, it seems that this method had never been used in the selection of volunteers. There had been various methods used [6–8] in supplier selection, including FDM. Many selection methods have also been applied to volunteers such as analytic hierarchy process method (AHP) [3, 9] and fuzzy logic [10, 11]. These methods had not focused specifically on the criteria of volunteers, but rather on the general parameter of a disaster situation. Experts' suggestion more influences in the decision-making model. Delphi method provides an easy understanding of the group opinions through the twice provision of the questionnaire. We already use the Delphi method [12]. To deal with the fuzziness of human participants' decisions in traditional Delphi method, Ishikawa et al. [13] had posited the fuzzy set theory to improve time-consuming problems such as the convergence of experts' options. FDM was selected because people are often averse to rigorous assessments and for that reason, this study had employed the fuzzy linguistic scale as a way of overcoming this problem. The FDM has been used in various domains such as marine [14], an urban system [15], investment [16], human resources [17], maintenance strategies [18, 19], and ergonomic programs [20]. For that reason, we would like to use this method in a different domain, which is to provide the most important parameter for volunteer selection.

3 Methodology

The methodology used in this study is divided into eight phases, which is shown in Fig. 1. The FDM is used for evaluating and selecting the best criteria for a volunteer that uses a combination of quantitative and qualitative measurements. The step-by-step proposed methodology is as follows:

Step 1: Determine the aspects and criteria
We have categorized the aspects based on the criteria in the literature and the opinions given by experts. The panel of experts was chosen according to their experience in managing volunteers and was asked to answer the questionnaire.

Step 2: Construct the hierarchical structure
From these criteria, we have constructed a hierarchical structure of the aspect categories, which is further discussed in Sect. 4.

Step 3: Fuzzy Delphi method
The weaknesses of the traditional Delphi method include the consensus of low expert opinion, and customize individual opinion experts to achieve a consistent overall opinion and high implementing cost [1]. Delphi method provides a natural understanding of the group opinions through the twice provision of the questionnaire [15]. A pilot study on the fuzzy set modification of Delphi was conducted in 1985 [21]. The FDM was proposed by Murray et al. [21] to integrate the Delphi method and fuzzy theory to improve those disadvantages and was further elaborated on by Ishikawa et al. [13]. Noorderhaven [22] also suggested that applying the fuzzy Delphi method to group decision would solve the fuzziness of shared understanding among expert opinions.

Fig. 1 Conceptual model of the proposed methodology

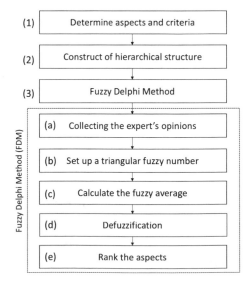

Table 1 Seven-point
linguistic scale

Linguistic variables	Fuzzy numbers
Extremely Unimportant (EU)	(0.0, 0.0, 0.1)
Not Very Important (NVI)	(0.0, 0.1, 0.3)
Less Important (LI)	(0.1, 0.3, 0.5)
Medium (M)	(0.3, 0.5, 0.7)
Important (I)	(0.5, 0.7, 0.9)
Very Important (VI)	(0.7, 0.9, 1.0)
Extremely Important (EI)	(0.9, 1.0, 1.0)

Fig. 2 Membership function
for the triangular fuzzy
number

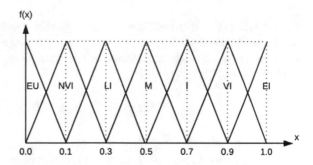

This method is based upon the group thinking of qualified experts that assures
the validity of the collected information. Fuzzy Delphi method not only saves survey
time and reduces the number of questionnaires but also considers the fuzziness that
confronts every survey process. In this way, the efficiency and quality of the questionnaires are improved [15, 23, 24]. The steps for conducting FDM are as follows:
Step (a): Collecting the expert's opinions
The experts' opinions are described by linguistic terms, which can be expressed
in triangular fuzzy numbers. To make the consensus of the experts consistent, we
have utilized the fuzzy Delphi method to adjust the fuzzy rating of every expert to
achieve the consensus condition. Each of the attributes that were given by the expert
is then assigned with a 1–10 evaluation score according to the linguistic description.
The experts use the fuzzy linguistic terms described in Table 1 to perform their
assessment of each alternative against each attribute [25]. The given weights are
"extremely unimportant (EU)," "not very important (NVI)," "less important (LI),"
"medium (M)," "important (I)," "very important (VI)," and "extremely important
(EI)". The linguistic scale variables and the linguistic weighting variables are then
converted into the triangular fuzzy numbers that match the linguistic variables of the
data as shown in Fig. 2 and Table 1. As a result of this, our data analysis had become
more reliable.
Step (b): Set up a triangular fuzzy number
Calculated and evaluated the triangular fuzzy number based on the experts [26, 27].
Step (c): Calculate the fuzzy average

This step involves summing up the decision-makers' given weights for each of the criteria and getting the mean of the fuzzy weight under consensus condition. Fuzzy number operations are then used to achieve the mean of weight (W_t) and to translate the linguistic terms into positive trapezoidal fuzzy numbers. Suppose that W_t is the linguistic weight given to aspects ($C_1, C_2,..., C_{k-s}$) and criteria ($C_{k-s+1},..., C_k$), as such, the aggregation can be formulated as follows:

$$D_{ij}^m = (1/m) \otimes \left(D_{ij}^1 \oplus D_{ij}^2 \oplus \cdots \oplus D_{ij}^m\right). \tag{1}$$

The symbol $\otimes$ is fuzzy multiplication, and the symbol $\oplus$ is a fuzzy addition. The D_{ij}^m
is the total of an average fuzzy number from the judgment of the decision-maker. Let D_{ij}^m indicated the fuzzy performance value regarding evaluator m toward strategy i under criterion j. It can be represented using a triangular membership function, which is shown in Eq. (2).

$$D_{ij}^m = \left(LD_{ij}^m, MD_{ij}^m, UD_{ij}^m\right). \tag{2}$$

The fuzzy weighting $\tilde{W}_t$ of j element is $\tilde{W}_t = (a_t, b_t, c_t), t = 1, 2, \ldots k$, formulated by Eq. (3), where L, M, and U represent the minimum and maximum values of the triangular fuzzy number.

$$LD_{ij} = \frac{\left(\sum_{ij}^m LD_{ij}^k\right)}{m}, MD_{ij} = \frac{\left(\sum_{k=1}^m MD_{ij}^k\right)}{m}, UD_{ij} = \frac{\left(\sum_{k=1}^m UD_{ij}^k\right)}{m}. \tag{3}$$

Step (d): Defuzzification
The result of each criterion fuzzy synthetic decision weighting is known as a fuzzy number. Subsequently, the non-fuzzy ranking method of fuzzy numbers should be used for each criterion. Converting fuzzy numbers into crisp real numbers is a technique called defuzzification that evaluates the best non-fuzzy performance (BNP) value. The defuzzification method is calculated by Eq. (4).

$$BNP_{ij} = \left[(UD_{ij} - LD_{ij}) + (MD_{ij} - LD_{ij})\right]/3 + LD_{ij} \forall i, j. \tag{4}$$

Step (e): Rank the aspects by descending order
From crisp value, we can arrange the results in descending ranking order. The symbol ">" means "superior to".

4 Results

After the questionnaires had been collected, the data were examined using content analysis and linguistic scale. Table 2 shows the hierarchy of the seven aspects along with the 17 criteria in a voluntary selection model that had been based on the literature and from interviews conducted with the experts.

Table 2 List of aspect and criteria

No.	Aspects	Criteria
1	Skills (A_1)	1.1 Years of working experience (C_{11}) 1.2 Ability to teach (C_{12}) 1.3 Job skill (C_{13})
2	Client Service (A_2)	2.1 Meeting people (C_{21}) 2.2 Social contact (C_{22})
3	Communication (A_3)	3.1 Written (C_{31}) 3.2 Oral (C_{32})
4	Workspace safety and well-being (A_4)	4.1 Contribution (C_{41}) 4.2 Impact (C_{42})
5	Learning (A_5)	5.1 Learn new skills (C_{51}) 5.2 Gaining experience (C_{52})
6	Teamwork (A_6)	6.1 Influence on other (C_{61}) 6.2 Facilitation of the discussion (C_{62}) 6.3 Open-mindedness (C_{63})
7	Commitment (A_7)	7.1 Reliability (C_{71}) 7.2 Positively (C_{72}) 7.3 Flexibility (C_{73})

Table 3 The result of the decision-makers

Aspect	D_1	D_2	D_3	D_4	D_5	D_6	D_7	D_8	D_9	D_{10}
A_1	M	I	M	EI	LI	M	M	I	NVI	I
A_2	I	I	M	M	NVI	NVI	I	M	LI	LI
A_3	EI	M	M	VI	M	M	I	M	M	M
A_4	I	I	LI	I	NVI	NVI	M	I	NVI	LI
A_5	M	M	LI	I	I	NVI	LI	LI	NVI	M
A_6	EI	M	EI	I	VI	EI	VI	VI	EI	M
A_7	I	I	M	VI	M	I	EI	VI	EI	M

A ten committee of decision-makers ($D_1, D_2, D_3, ..., D_{10}$) had utilized the weights to measure the level of each of the seven aspects: skills (A_1), client service (A_2), communication (A_3), workspace safety and well-being (A_4), learning (A_5), teamwork (A_6), and commitment (A_7). Table 3 illustrates the weight of the criteria that was obtained from the ten decision-makers.

Next, the expert opinions in the FDM questionnaires were converted into the positive triangular fuzzy numbers as described in Table 4. This step is used in Eq. (2). Then, the aggregate fuzzy weights for each criterion are considered by collecting linguistic judgments from ten decision-makers. By employing Eqs. (3) and (4), the results of the fuzzy mean are then obtained in Table 5.

Table 6 shows the fuzzy number values that are arranged in a descending ranking order: $A_6 > A_7 > A_3 > A_1 > A_2 > A_4 > A_5$.

Table 4 The average linguistic rating of aspects

Aspect	D_1	D_2	D_3	D_4	D_5	D_6	D_7	D_8	D_9	D_{10}
A_1	(0.3, 0.5, 0.7)	(0.5, 0.7, 0.9)	(0.3, 0.5, 0.7)	(0.9, 1.0, 1.0)	(0.1, 0.3, 0.5)	(0.3, 0.5, 0.7)	(0.3, 0.5, 0.7)	(0.5, 0.7, 0.9)	(0.0, 0.1, 0.3)	(0.5, 0.7, 0.9)
A_2	(0.5, 0.7, 0.9)	(0.5, 0.7, 0.9)	(0.3, 0.5, 0.7)	(0.3, 0.5, 0.7)	(0.0, 0.1, 0.3)	(0.0, 0.1, 0.3)	(0.5, 0.7, 0.9)	(0.3, 0.5, 0.7)	(0.1, 0.3, 0.5)	(0.1, 0.3, 0.5)
A_3	(0.9, 1.0, 1.0)	(0.3, 0.5, 0.7)	(0.3, 0.5, 0.7)	(0.7, 0.9, 1.0)	(0.3, 0.5, 0.7)	(0.3, 0.5, 0.7)	(0.5, 0.7, 0.9)	(0.3, 0.5, 0.7)	(0.3, 0.5, 0.7)	(0.3, 0.5, 0.7)
A_4	(0.5, 0.7, 0.9)	(0.5, 0.7, 0.9)	(0.1, 0.3, 0.5)	(0.5, 0.7, 0.9)	(0.0, 0.1, 0.3)	(0.0, 0.1, 0.3)	(0.3, 0.5, 0.7)	(0.5, 0.7, 0.9)	(0.0, 0.1, 0.3)	(0.1, 0.3, 0.5)
A_5	(0.3, 0.5, 0.7)	(0.3, 0.5, 0.7)	(0.1, 0.3, 0.5)	(0.5, 0.7, 0.9)	(0.5, 0.7, 0.9)	(0.0, 0.1, 0.3)	(0.1, 0.3, 0.5)	(0.1, 0.3, 0.5)	(0.0, 0.1, 0.3)	(0.3, 0.5, 0.7)
A_6	(0.9, 1.0, 1.0)	(0.3, 0.5, 0.7)	(0.9, 1.0, 1.0)	(0.5, 0.7, 0.9)	(0.7, 0.9, 1.0)	(0.9, 1.0, 1.0)	(0.7, 0.9, 1.0)	(0.7, 0.9, 1.0)	(0.9, 1.0, 1.0)	(0.3, 0.5, 0.7)
A_7	(0.5, 0.7, 0.9)	(0.5, 0.7, 0.9)	(0.3, 0.5, 0.7)	(0.7, 0.9, 1.0)	(0.3, 0.5, 0.7)	(0.5, 0.7, 0.9)	(0.9, 1.0, 1.0)	(0.7, 0.9, 1.0)	(0.9, 1.0, 1.0)	(0.3, 0.5, 0.7)

Table 5 The result aspects with fuzzy mean

Aspect	Fuzzy mean (Eq. 3)	Values (Eq. 4)
A_1	$W_1 = (0.37, 0.55, 0.73)$	0.550
A_2	$W_2 = (0.26, 0.44, 0.64)$	0.446
A_3	$W_3 = (0.42, 0.61, 0.78)$	0.603
A_4	$W_4 = (0.25, 0.42, 0.62)$	0.430
A_5	$W_5 = (0.22, 0.40, 0.60)$	0.406
A_6	$W_6 = (0.68, 0.84, 0.93)$	0.816
A_7	$W_7 = (0.56, 0.74, 0.88)$	0.726

Table 6 Ranking fuzzy aspect

Rank	Aspects		Values
1	A_6	Teamwork	0.816
2	A_7	Commitment	0.726
3	A_3	Communication	0.603
4	A_1	Skills	0.550
5	A_2	Client service	0.446
6	A_4	Workspace safety and well-being	0.430
7	A_5	Learning	0.406

Based on the ranking order, we can see that "teamwork" and "commitment" are the two aspects that had been rated highly by the experts of volunteer management. The criteria for teamwork, i.e., the ability to influence others (C_{61}), facilitating a discussion (C_{62}), and having an open mind (C_{63}) are regarded as the most crucial factors when assigning a volunteer to a general task. The other important aspect is a commitment, where the criteria include reliability (C_{71}), having a positive attitude (C_{72}) and being flexible to sudden changes (C_{73}).

5 Conclusion

Volunteerism is the heart and soul of the nonprofit sector. As such, it is important that the organization manage its volunteer resources wisely so that time and resource wastage can be avoided. However, to select the appropriate volunteers who fit the need tasks is difficult because volunteers have many criteria to be matched with the tasks offered. This study aims to identify the important criteria in the selection of suitable volunteers for the tasks being offered. For that reason, we have proposed a multi-criteria decision-making model that is based on seven aspects along with its 17 criteria as a way of conducting volunteer selection. The results from this study had indicated that "Teamwork" and "Commitment" are the two most important criteria in volunteer selection. Among the criteria listed under the teamwork aspect include the ability to influence others, being able to facilitate discussions, and having an open mind.

Meanwhile, the commitment aspect must include reliability, having a positive attitude, and being flexible to sudden changes. Therefore, this ranking is most important when assigning a general task to a selected suitable volunteer. The criteria are also task dependent, where a volunteer who is evaluated based on specified criteria will be matched against the task requirements. This method had been used to identify the criteria rank that relates the task to the volunteer. Further research on this method is now being conducted using the experience as an additional criterion in the evaluation of volunteer selection.

Acknowledgements The authors might want to acknowledge Universiti Teknikal Malaysia Melaka (UTeM) and the Ministry of Education Malaysia of the scholarship MyBrain15. This work was supported under grant PJP/2018/FTK (16A)/S01642. Dr. Ahmad Zaki A. Bakar and Dr. Farzad Tahriri for the help in this work. We also want to thank the International Conference on Intelligent and Interactive Computing (IIC) 2018 at Melaka, Malaysia.

References

1. Wu K (2010) Applying the fuzzy delphi method to analyze the evaluation indexes for service quality after railway re-opening—using the old mountain line railway as an example. Recent Res Syst Sci 474–479
2. Chen W, Cheng Y, Sandnes FE, Lee C (2011) Finding Suitable Candidates : the design of a mobile volunteering matching system. In: Human-computer interaction. towards mobile and intelligent interaction environments pp 21–29
3. Cvetkoska V, Gaber BS, Sekulovska M (2011) Recruitment and Selection of Student- Volunteers : a Multicriteria Methodology. Manag 139–146
4. Yu Z, Zhang D, Yang D, Chen G (2012) Selecting the best solvers: toward community based crowdsourcing for disaster management. In: 2012 IEEE Asia-Pacific Services Computing Conference. IEEE, Guilin, pp 271–277
5. Division of Industry and Community Network Universiti Sains Malaysia (2013) Volunteerism In Malaysia Fostering Civic Responsibility. Penerbit USM
6. Kuo RJ, Hong SY, Huang YC (2010) Integration of particle swarm optimization-based fuzzy neural network and artificial neural network for supplier selection. Appl Math Model 34:3976–3990. https://doi.org/10.1016/j.apm.2010.03.033
7. Tahriri F, Mousavi M, Hozhabri Haghighi S, Zawiah Md Dawal S (2014) The application of fuzzy Delphi and fuzzy inference system in supplier ranking and selection. J Ind Eng Int 10:66. https://doi.org/10.1007/s40092-014-0066-6
8. Cheng J-H, Tang C-H (2009) An application of fuzzy Delphi and fuzzy AHP for multi-criteria evaluation on bicycle industry supply chains. WSEAS Trans Syst Control 4:21–34
9. Hadi LA, Naim WM, Adnan NA et al (2017) GIS based multi-criteria decision making for flood vulnerability index assessment. J Telecommun Electron Comput Eng 9:7–11
10. Öztaysi B, Behret H, Kabak Ö et al (2013) Fuzzy inference systems for disaster response. In: Vitoriano B, Montero J, Ruan D (eds) Decision aid models for disaster management and emergencies. Atlantis Press, Paris, pp 17–44
11. Akyuz E, Ilbahar E, Cebi S, Celik M (2017) Maritime Environmental Disaster Management Using Intelligent Techniques. In: Kahraman C, Sari. Irem Uçal (eds) Intelligence Systems in Environmental Management: theory and applications. Springer International Publishing, Cham, pp 135–155

12. Mazlan N, Syed Ahmad SS, Kamalrudin M (2018) A crowdsourcing approach for volunteering system using delphi method. In: Zelinka I, Vasant P, Duy VH, Dao TT (eds) Innovative computing, optimization and its applications: modelling and simulations. Springer International Publishing, Cham, pp 237–253
13. Ishikawa A, Amagasa M, Shiga T et al (1993) The max-min Delphi method and fuzzy Delphi method via fuzzy integration. Fuzzy Sets Syst 55:241–253. https://doi.org/10.1016/0165-0114(93)90251-C
14. Etebarian A, Shirvani A, Soltani I, Moradi A (2013) The application of Fuzzy Delphi Method (FDM) and Fuzzy Analytic Hierarchy Process (FAHP) for evaluating marine casualties. Recent Adv Comput Sci Appl 54–69
15. Ho Y-F, Wang H-L (2008) Applying fuzzy Delphi method to select the variables of a sustainable urban system dynamics model. In: Proceedings of the 26th international conference of system pp 1–21
16. Mostafa K, Farnad N, Majid A (2010) Using Fuzzy-Delphi technique to determine the concession period in BOT projects. Proceedings - 2010 2nd IEEE international conference on information and financial engineering ICIFE 2010 442–446. https://doi.org/10.1109/icife.2010.5609396
17. Chang P-T, Huang L-C, Lin H-J (2000) The fuzzy Delphi method via fuzzy statistics and membership function fitting and an application to the human resources. Fuzzy Sets Syst 112:511–520. https://doi.org/10.1016/S0165-0114(98)00067-0
18. Jafari A, Jafarian M, Zareei A, Zaerpour F (2008) Using Fuzzy Delphi Method in Maintenance Strategy Selection Problem. J Uncertain Syst 2:289–298
19. Li C, Xu M, Guo S (2007) ELECTRE III based on ranking fuzzy numbers for deterministic and fuzzy maintenance strategy decision problems. In: Proceedings of the IEEE international conference on automation and logistics, ICAL 2007. pp 309–312
20. Ahmadi M, Zakerian SA, Salmanzadeh H, Mortezapour A (2017) Identification of the ergonomic interventions goals from the viewpoint of ergonomics experts of Iran using Fuzzy Delphi Method. Int J Occup Hyg 8:151–157
21. Murray TJ, Leo LP, van Gigvh JP (1985) A pilot study of fuzzy set modification of Delphi. Hum Syst Manag 5:76–80
22. Noorderhaven N (1995) Strategic decision making. Addison Wesley
23. Hsu Y-L, Lee C-H, Kreng VB (2010) The application of Fuzzy Delphi Method and Fuzzy AHP in lubricant regenerative technology selection. Expert Syst Appl 37:419–425. https://doi.org/10.1016/j.eswa.2009.05.068
24. Glumac B, Han Q, Smeets J, Schaefer W (2011) Brownfield redevelopment features: applying Fuzzy Delphi. J Eur Real Estate Res 4:145–159. https://doi.org/10.1108/17539261111157316
25. Habibi A, Jahantigh FF, Sarafrazi A (2015) Fuzzy delphi technique for forecasting and screening items. Asian J Res Bus Econ Manag 5:130–143. https://doi.org/10.1007/BF00027519s
26. Klir GJ, Yuan B (1995) Fuzzy sets and fuzzy logic: theory and applications. Prentice-Hall Inc
27. Laarhoven PJM va., Pedrycz W, van Laarhoven PJM, Pedrycz W (1983) A fuzzy extension of Saaty's priority theory. Fuzzy Sets Syst 11:229–241. https://doi.org/10.1016/S0165-0114(83)80082-7

The Effect of Two-Term Parameters on the Performance of Third-Order Neural Network on Medical Classification Problems

Nazri Mohd Nawi, Khairina Mohamad Roslan,
Shahrul Bariah Mustaffa Kamal and Noorhamreeza Abdul Hamid

Abstract Artificial neural networks (ANN) also known as backpropagation (BP) algorithm has recently been applied in many areas and applications. BP algorithm is known as an excellent classifier for nonlinear input and output numerical data. However, the popularity of BP comes with some weaknesses which are moderate in learning and easily getting stuck in local minima. A vigorous area of research and papers on enlightening the performance of BP algorithm are revised especially in terms of literature. Furthermore, the performance of BP algorithm also highly influenced by the order of learning and the parameters that been chosen. This paper presents an improvement of BP algorithm by adjusting the two-term parameters on the performance of third-order neural network methods. The effectiveness of the proposed method is proven through the methods of stimulation on medical classification issues. The results show that the proposed implementation significantly improves the learning speed of the general BP algorithm.

Keywords Backpropagation algorithm · Third-order method · Classification · Two term parameters

1 Introduction

Medical data or health information is organized by technicians in helping them to keep track medical records of their patients. They are responsible to make sure that their patient data is accurate, and all input data were kept into databases, where it can be analyzed for service quality and insurance reimbursement purposes. In addition, they ensure the confidentiality of sensitive medical information. It is important for the top management to do some decision making by using the data that been kept in database. The complicated nonlinear relationship consists of dependent and independent variable within medical data cause the doctor unable to identify all likely

N. M. Nawi (✉) · K. M. Roslan · S. B. M. Kamal · N. A. Hamid
Faculty of Computer Science and Information Technology,
Universiti Tun Husein Onn Malaysia, 86400 Parit Raja, Johor, Malaysia
e-mail: nazri@uthm.edu.my

© Springer Nature Singapore Pte Ltd. 2019
V. Piuri et al. (eds.), *Intelligent and Interactive Computing*, Lecture Notes in Networks and Systems 67, https://doi.org/10.1007/978-981-13-6031-2_20

interactions amongst predictor variables and the obtainability of numerous training algorithms which are impossible. Furthermore, there are not many statistical ways that were used by hospital to utilize the data and as a result the decision-making will become very difficult.

Recently, Artificial Neural Networks (ANN) had gained popularity among researchers and specialist in medical. ANNs are a branch of "Artificial Intelligence" where is a system roughly displayed founded with human brain. It consists of many titles which are, machine learning algorithm, parallel distributed processing, neuro-computing, natural intelligent systems, connectionism, and artificial neural network. ANNs. Furthermore, ANNs are a powerful technique to solve many real-world problems this is because of the ability to learn from experience in order to improve their performance and ability to adapt to the changes in the environment. Moreover, ANN is able to deal with incomplete information or noisy data and can be very effective especially in situations where it is not possible to define the rules or steps that lead to the solution of a problem. Backpropagation is the most used algorithm in artificial neural network which is utilized in updating weights using gradient descent [1].

There are many disadvantages using BP algorithm like easily converges in local minima, rather than global minimum, poor scale properties and temporal unsteadiness despite its status. Furthermore, the performances of BP algorithm are depending on the order of gradient descent (first-order method) learning as well as some parameter such as learning rate, momentum, target error, and hidden nodes. Therefore, researchers in ANN had proposed many improvements to increase the learning productivity of backpropagation algorithm with numerous developments [2–9]. Van Ooyen and Nienhuis [10] determined alterations in error function is for measuring global net performance.

Researchers also reviewed the impact in gain parameter to supervise the sharpness within the activation function (AF), which they discovered that the bigger the gain value, the larger the effect to upsurge learning performance [11, 12]. Instead of using global gain, a study by Nawi et al. [13] identified an adaptive gain within the AF can quicken training time which a revised value gain put modified for each node. The research was upgraded and projected, verified based on classification issues [13–15]. The result of improved proposed method shows an enhancement in backpropagation's performance.

Some researchers also focused their study on using optimized training algorithms. Haykin [16] proposed numerous enhanced information-driven training algorithm like mounted conjugate gradient (CG) algorithm and Levenberg–Marquardt helps to reduce local minima issues. Next, Kumar et al. [17] mentioned that improvement in performance of pulse radar detection is acquired by selecting the Bayesian standardizing of ANN. Grounded on Bishop [18] theory of numerical optimization, the book stated that second-order method like Fletcher and Powell [19] and the Fletcher and Reeves [20] and Hestenes and Stiefel [21] can improve further the efficiency of error minimization. Furthermore, those methods have been shown can deliver steady learning, sturdiness to fluctuations and enhance in convergence rates.

Later, Rivaie et al. [22] suggested an enhancement method to instruct backpropagation algorithm by implementing a new class of CG which its exploratory results show that intended CG approaches give an efficient and improve computational productivity and robustness of training process. Previous researchers claimed their improvement of BP using two parameters (learning rate and momentum) with second-order methods. Therefore, this paper presents the performance analysis of third-order implementation on BP together with adjusting two parameters (learning rate and momentum) on some medical datasets.

The following parts of this paper are arranged as follows. Section 2, the implementation of third-order method on BP algorithm is discussed. Section 3 presents the tests and recreation results on medical benchmark problems. The last part is the conclusion of the paper.

2 Third-Order Implementation on BP Algorithm

In this section, an improved method that enhanced the competency in implementing the gradient descent technique is acquired. Nawi et al. [23] proposed the method by altering the gain value consistently for each node and changing the direction of gradient search. Thus, from optimization side, finding a weight vector that reduces difference in errors $E(w)$ which is among original output to the desired output using training and testing data sets is the objective in learning process within artificial neural network.

$$\min_{w \in R^n} E(w) \tag{1}$$

Multilayer feed-forward NN consists of one output layer, one input layer, and many hidden layers. Each layer has sets of unit, node where it is fully linked to the preceding layers from direct related between not consecutive layers. Plus, every connection contains weight. In an input pattern, effort function is classified as:

$$E = \frac{1}{2} \sum_k \left(t_k - o_k^L \right)^2 \tag{2}$$

where o_k^L is an activation of kth node from layer L and w_{ij}^L is weight from the connection from the ith node in layer $L - 1$ to the jth node in layer L. Overall, the error on the training set is simply the sum, across patterns, of the pattern error E, whereas, the input net of jth node of layer L is referred as $\mathrm{net}_j^L = \left(w_j^L, o^{L-1} \right) = \sum_k w_{j,k}^L o_k^{L-1}$ and the activation of a node o_j^L is given by a function of its net input,

$$o_j^L = f\left(c_j^L \mathrm{net}_j^L \right) \tag{3}$$

f is function consist of restricted derivative, and c_j^L is real value referred as the gain of node. Note that for all nodes they used $c_j^L = 1$ and for activation function they used a typical logistic activation function. The updated weight expression weight update expression in Eq. (2) mix together with non-unit gain value is the result of separating the error term shown in Eq. (2) with from the w_{ij}^L as follows:

$$\frac{\partial E}{\partial w_{ij}^L} = \frac{\partial E}{\partial \text{net}^{L+1}} \cdot \frac{\partial \text{net}^{L+1}}{\partial o_j^L} \cdot \frac{\partial o_j^L}{\partial \text{net}_j^L} \cdot \frac{\partial \text{net}_j^L}{\partial w_{ij}^L}$$

$$= \left[-\delta_1^{L+1} \cdots -\delta_n^{L+1} \right] \cdot \begin{bmatrix} w_{1j}^{L+1} \\ \vdots \\ w_{nj}^{L+1} \end{bmatrix} \cdot f'(c_j^L \text{net}_j^L) c_j^L \cdot o_j^{L-1} \quad (4)$$

In precise, three factors consist of (1) are shown:

$$\delta_1^L = \left(\sum_k \delta_k^{L+1} w_{k,j}^{L+1} \right) f'(c_j^L \text{net}_j^L) \quad (5)$$

In Eq. (4), the δ_1^L is similar to a standard backpropagation [1] minus the gain value. Grouping of (1) and (2) produces the learning rule for weights:

$$\Delta w_{ij}^L = \eta \delta_j^L c_j^L o_j^{L-1}$$

$$= \eta \frac{\partial E}{\partial w_{ij}^L} \quad (6)$$

$$w_{\text{new}} = w_{\text{old}} + \Delta w_{ij}^L \quad (7)$$

where η is a "learning rate" while as the gradient vector or search direction is w_{ij}^L $d = \frac{\partial E}{\partial w_{ij}^L} = g$. Nawi et al. [23] proposed the calculation of the gradient of error $g^{(n)}$ at phase n is the function of gain $c_j^{L(n)}$ as:

$$d^{(n)} = -\frac{\partial E}{\partial w_{ij}^{L(n)}} \left(c_j^{L(n)} \right) = g^{(n)} \left(c_j^{L(n)} \right) \quad (8)$$

Gain value at stage n is designed with gradient of error w.r.t. to gain,

$$\frac{\partial E}{\partial c_j^L} = \left(\sum_k \delta_k^{L+1} w_{k,j}^{L+1} \right) f'(c_j^L \text{net}_j^L) \text{net}_j^L \quad (9)$$

Therefore, gradient descent rule of gain value is calculated as:

$$\Delta c_j^L = \eta \delta_j^L \frac{\text{net}_j^L}{c_j^L} \quad (10)$$

Last, every repetition of new gain value is reorganized by simple gradient-based method such as this formula,

$$c_j^{\text{new}} = c_j^{\text{old}} + \Delta c_j^L \tag{11}$$

There are numerous formula of the parameter $\beta_{(n+1)}$ however, the chosen of the formula for $\beta_{(n+1)}$ is problem dependent [23]. The paper consist of two most popular formula which are used Broyden–Fletcher–Goldfarb–Shanno (BFGS) algorithm and Davidon Fletcher Powell (DFP) [24, 25] where these formula are operated as an approximation to Newton's method where for error function $E(w)$ which aim to diminish the parameter vector w with respect, and the search direction d for Newton's method is create to resolve the system of equations

$$d = -[\nabla^2 E(w)]^{-1} \nabla E(w) \tag{12}$$

where $\nabla^2 E(w) = H$ is the Hessian matrix and $\nabla E(w) = g$. The method of converging one iteration for quadratic function is not complete without the needs of computation of the inverse using Hessian matric. This is a very difficult task for real-life applications. The BFGS algorithm allows constructing $\nabla^2 E(w)$ by using the only gradient information with the function of gain value $\nabla E(w) = g(c)$.

3 Experiments and Simulation Results

In this section, it demonstrates the recompenses of planned implementation which are using two specific benchmarks of difficulties and compared it using first-order, backpropagation (BP), third-order Halley with Broyden–Fletcher–Goldfarb–Shanno (BFGS) and third-order Davidon Fletcher Powell with Halley (DFP). The simulation testing was completed using Matlab R2010b software and later, performed on a CPU of Intel(R) 1017U, with 1.60 GHz processor. For comparison purpose, all algorithms were trained by using similar networks architecture and sigmoid activation function was functioned for all nodes where the initial weights are only small random values range [0, 1], and then it received input patterns for training in similar sequence initial random weights for the same problems. Every 100 trials, the means of epoch, CPU times, accurateness and the number of let-downs is noted. Any failure occurs when the network exceeds the maximum iteration limit. It means that each experiment is run up to 1500 epoch to reach the target error, otherwise, it is paused, and the run is recorded as failure. Convergence is achieved when the outputs of the network approve to error standard as compared to its desired outputs. The parameters assigned are examined and the results of best settings from simulations are shown in Table 1.

Table 1 Parameters settings

Variables	Value
Hidden nodes	5
Target error	0.001
Maximum epoch	1500
Trials total	100
Momentum	0.3, 0.5
Learning rate	0.3, 0.5

Table 2 Performance comparison for breast cancer datasets

	Learning rate = 0.3 and momentum = 0.3		
	BP	Halley-BFGS	Halley-DFP
Mean of epoch	457	51	**55**
CPU time (s)	12.57	2.92	**3.10**
Accuracy (%)	90.596	93.06	**93.44**
Failures	0	0	**0**
	Learning rate = 0.3 and momentum = 0.5		
	BP	Halley-BFGS	Halley-DFP
Mean of epoch	567	53	**57**
CPU time (s)	15.97	3.02	**3.17**
Accuracy (%)	91.568	93.23	**93.45**
Failures	0	0	**0**
	Learning rate = 0.5 and momentum = 0.5		
	BP	Halley-BFGS	Halley-DFP
Mean of epoch	497	53	**59**
CPU time (s)	14.34	3.04	**3.17**
Accuracy (%)	93.987	93.07	**93.36**
Failures	0	0	**0**

3.1 Breast Cancer Classification Problems

Breast cancer dataset is the first benchmark problem. It is one of three areas supplied by the Oncology Institute which repeatedly shown in the machine learning literature. Obtained from UCI Machine Learning Website, this data set includes 350 instances altogether. While for the testing example are 174 instances. The instances are labeled as 9 input attributes and 2 output attributes. The results of the testing were noted in Table 2.

Table 2 proved that third-order learning algorithms outperformed the conventional BP in term of accuracy and CPU times. Moreover, the adjustment of parameters such as learning rate and momentum had significantly improved further the performance

Table 3 Performance comparison for heart datasets

	Learning rate = 0.3 and momentum = 0.3		
	BP	Halley-BFGS	Halley-DFP
Mean of epoch	1289	134	**538**
CPU time (s)	30.34	8.16	**33.36**
Accuracy (%)	68.654	71.24	**73.38**
Failures	0	0	**0**
	Learning rate = 0.3 and momentum = 0.5		
	BP	Halley-BFGS	Halley-DFP
Mean of epoch	1387	136	**587**
CPU time (s)	32.33	8.40	**36.39**
Accuracy (%)	69.122	71.29	**73.81**
Failures	0	0	**0**
	Learning rate = 0.5 and momentum = 0.5		
	BP	Halley-BFGS	Halley-DFP
Mean of epoch	1399	120	**579**
CPU time (s)	35.344	7.00	**34.16**
Accuracy (%)	67.230	70.98	**74.07**
Failures	0	0	**0**

of the third-order algorithms. As summarized in Table 2, both third-order algorithms perform their best, however, Halley with DFP performed better than Halley with BFGS with higher accuracy even though it needs more epochs to converge. This is mainly because Halley with DFP need more time to calculate Hessian matrix in reaching for the target error.

3.2 Heart Classification Problem

This dataset consists of 37 attributes where 35 is input and the remaining 2 is output. The data sets were taken from 152 instances. Meanwhile, 76 instances were taken for testing. This dataset also was reserved from UCI Machine Learning Website where the dataset was already managed. As shown in Table 3, both third-order algorithms outperform BP in term of accuracy, number of epoch and CPU time. However, Halley with DFP performs better than Halley with BFGS with the highest average accuracy of 74.07% at parameter for both learning rate and momentum is 0.5. Even though, Halley with DFP performs its best it needs more time to calculate Hessian matrix in order to converge to global minima.

4 Conclusions

The limitation of BP algorithm has been improved in this research by implementing third-order method. Furthermore, the performance of the proposed network been further improved by adjusting two parameters which are learning rate and momentum. The performance of third-order neural network were differentiate with conventional BP algorithm. The three algorithms were confirmed by methods of simulation on two classification problems including breast cancer and heart classification problems. The simulation results show that Halley with DFP performs the best and an improved convergence rate than BP and Halley with BFGS.

Acknowledgements The authors would like to thank Universiti Tun Hussein Onn Malaysia (UTHM) and Ministry of Higher Education (MOHE) Malaysia for financially supporting this Research under IGSP grants note U420 and under Trans-disciplinary Research Grant Scheme (TRGS) vote no. T003.

References

1. Rumelhart DE, Hinton GE, Williams RJ (1986) Learning internal representations by error propagation. In: David ER, James LM, CPR Group (eds) Parallel distributed processing: explorations in the microstructure of cognition, vol 1. MIT Press, pp 318–362
2. Zhang SL, Chang TC (2016) A study of image classification of remote sensing based on back-propagation neural network with extended delta bar delta. Math Probl Eng 2015:10
3. Nawi NM, Rehman M, Khan A (2015) WS-BP: An efficient wolf search based back-propagation algorithm. In: International conference on mathematics, engineering and industrial applications 2014 (ICoMEIA 2014). AIP Publishing
4. Rehman MZ, Nawi NM (2011) The effect of adaptive momentum in improving the accuracy of gradient descent back propagation algorithm on classification problems. In: Mohamad Zain J, Wan Mohd WMB, El-Qawasmeh E (eds) Software engineering and computer systems: second international conference, ICSECS 2011, Kuantan, Pahang, Malaysia, Proceedings, Part I, 27–29 June 2011. Springer Berlin Heidelberg, Berlin, Heidelberg, pp 380–390
5. Nawi NM, Khan A, Rehman MZ (2013) A new back-propagation neural network optimized with cuckoo search algorithm. In: International conference on computational science and its applications, 2013. Springer Berlin Heidelberg
6. Liu Y, Jing W, Xu L (2016) Parallelizing backpropagation neural network using MapReduce and cascading model. Comput Intell Neurosci 2016:11
7. Chen Y et al (2016) Three-dimensional short-term prediction model of dissolved oxygen content based on PSO-BPANN algorithm coupled with Kriging interpolation. Math Probl Eng 2016:10
8. Cao J, Chen J, Li H (2014) An adaboost-backpropagation neural network for automated image sentiment classification. Sci World J 2014:9
9. Abdul Hamid N et al (2012) A review on improvement of back propagation algorithm. Glob J Technol 1
10. Van Ooyen A, Nienhuis B (1992) Improving the convergence of the back-propagation algorithm. Neural Netw 5(3):465–471
11. Maier HR, Dandy GC (1998) The effect of internal parameters and geometry on the performance of back-propagation neural networks: an empirical study. Environ Model Softw 13(2):193–209
12. Thimm G, Moerland P, Fiesler E (1996) The interchangeability of learning rate and gain in backpropagation neural networks. Neural Comput 8(2):451–460

13. Nawi NM, Ransing R, Hamid NA (2010) BPGD-AG: a new improvement of back-propagation neural network learning algorithms with adaptive gain. J Sci Technol 2(2)
14. Nawi NM et al (2010) An improved back propagation neural network algorithm on classification problems. In: Zhang Y et al (eds) Database theory and application, bio-science and bio-technology: international conferences, DTA and BSBT 2010, Held as part of the future generation information technology conference, FGIT 2010, Jeju Island, Korea, Proceedings, 13–15 Dec 2010. Springer Berlin Heidelberg, Berlin, Heidelberg, pp 177–188
15. Nawi NM et al (2011) Enhancing back propagation neural network algorithm with adaptive gain on classification problems. Networks 4(2)
16. Haykin S (1998) Neural networks: a comprehensive foundation. Prentice Hall PTR, p 842
17. Kumar P, Merchant SN, Desai UB (2004) Improving performance in pulse radar detection using Bayesian regularization for neural network training. Digit Sign Proc 14(5):438–448
18. Bishop CM (1995) Neural networks for pattern recognition. Oxford University Press, Inc., p 482
19. Fletcher R, Powell MJD (1963) A rapidly convergent descent method for minimization. Comput J 6(2):163–168
20. Fletcher R, Reeves CM (1964) Function minimization by conjugate gradients. Comput J 7(2):149–154
21. Hestenes M, Stiefel E (1952) Methods of conjugate gradients for solving linear systems. J Res National Bureau Stand 49(6):409–436
22. Rivaie M et al (2012) A new class of nonlinear conjugate gradient coefficients with global convergence properties. Appl Math Comput 218(22):11323–11332
23. Nawi NM, Hamid NA, Samsudin NA, Yunus MAM, Ab Aziz MF (2017) Second order learning algorithm for back propagation neural networks. Int J Adv Sci Eng Inf Technol 7(4):1162–1171
24. Sheperd AJ (1997) Second order methods for neural networks-fast and reliable training methods for multi-layer perceptrons. In: Taylor JG (ed). Springer, p 143
25. Byatt D, Coope ID, Price CJ (2004) Effect of limited precision on the BFGS quasi-Newton algorithm. ANZIAM J 45:283–295

Stock Market Forecasting Model Based on AR(1) with Adjusted Triangular Fuzzy Number Using Standard Deviation Approach for ASEAN Countries

Muhammad Shukri Che Lah, Nureize Arbaiy and Riswan Efendi

Abstract Traditional autoregressive (AR) time series models have been extensively applied to predict various stationary data sets based on single point data. However, real-world system involves uncertainty due to human behaviours and incomplete information. Since the single point data is not able to represent the nature of data, fuzzy approach is necessary to deal with such uncertainties in the analysis. This paper proposes AR(1) model building based on triangular fuzzy numbers. A procedural step for building triangular fuzzy number based on standard deviation approach is provided, to handle the existence of uncertain information and the biasness during data collection. The proposed model is applied to forecast buying–selling stock market prices by using real data sets from five ASEAN countries. The results from this study show that the proposed method with triangular fuzzy numbers exhibits smaller error. That is, the proposed method is able to achieve almost similar accuracy performance as obtained by the traditional autoregressive approach, yet it also solves the uncertainties issue in the analysis.

Keywords Left–right spread · Triangular fuzzy number · AR(1) · Standard deviation · Stock market

M. S. C. Lah (✉) · N. Arbaiy
Faculty of Computer Science and Information Technology, UTHM, Parit Raja, Malaysia
e-mail: shukrichelah89@gmail.com

N. Arbaiy
e-mail: nureize@uthm.edu.my

R. Efendi
Mathematics Department, UIN Suska Riau, Pekanbaru, Indonesia
e-mail: riswan.efendi@uin-suska.ac.id

© Springer Nature Singapore Pte Ltd. 2019
V. Piuri et al. (eds.), *Intelligent and Interactive Computing*, Lecture Notes in Networks and Systems 67, https://doi.org/10.1007/978-981-13-6031-2_22

1 Introduction

The stock market refers to public markets that issuing, buying and selling stocks that trade on a stock exchange [1]. Forecasting the stock market is aiming to develop approaches which successfully predict index values or stock prices at high profits using well-defined trading strategies. Economic analysis usually obtains the result based on the time series data. Autoregressive [2], Moving average [3], Multiple Kernel Learning [4] and Support Vector Machines [5] are an example of a widely used technique to analyze the behaviour of the stock market based on market history data. Among this, most of the models for the time series of stock prices have centred on autoregressive (AR) processes. The results obtained, assist people to select the best stock market. It makes forecasting model is importantly crucial to predict the stock market. The more accurate result implies the better model.

Stock market forecasts are constantly attracting researchers as it is extremely volatile and dynamic [6]. The most challenging task in stock prediction lies in the complexities of modelling market dynamics and uncertainty in stock market [7]. The existence of risk and uncertainty in economic brings a primary concern in this field [8]. Even if the traditional method can produce an impressive result, however, it is unable to cope with uncertainty. The traditional method is difficult to forecast accurately the stock market from the history of price with uncertain behaviour co-exist [5]. The existing models use single input for building univariate forecasting models. Since most of the data are obtained from secondary sources, it may contain validity, biasness, and representation issues which contributes to less accurate forecasting models [9]. Accordingly, Traders make predictions based on incomplete vague data, imperfect and uncertain [10]. Fuzzy logic provides a natural way to deal with subject observed in most financial time series models. Most of the fuzzy financial models that have been proposed to forecast stock market sectors, such as, fuzzy time series [10–14], fuzzy regression [15], fuzzy random [16] and fuzzy random auto-regression [17, 18]. However, the procedure to build a symmetric triangular fuzzy number (TFN) is not yet discussed clearly in the previous works.

In a conventional time series, the recorded values are represented by crisp numerical values, while fuzzy time series uses fuzzy numbers to represent such values. However, the procedure to construct TFN from crisp value is not discussed deeply. Most of the existing approach is focused on the forecasting model itself, while fuzzy data preparation is not thoroughly explained. Though data preparation, which involves fuzzy data transformation is important to obtain appropriate values for forecasting, i.e. the original meaning behind the crisp value is not abandoned. Motivated from the situation and previous studies [10–17, 19–23], this study proposes systematic steps to handle uncertainty in the development of model forecasts. Specifically, the triangular fuzzy number is used to represent the uncertainty in buying–selling data. The standard deviation approach is utilized to construct the TFN since it is able to measure the spread of scores within a set of data stock market using Fuzzy Autoregressive approaches. The concentration in fuzzy data preparation and model forecasting for both are very important to be investigated in improving the forecasting values and achieving the forecasting accuracy. The main contribution of this paper is to provide a better forecasting model for stock market using Fuzzy Autoregressive approach.

This paper is organized as follows: in Sect. 2, the fundamental theories in forecasting are described. The proposed method is presented in Sect. 3. In Sect. 4, the numerical experimental and analysis are explained. The discussion and a brief conclusion are explored in the final section.

2 Fundamental Theories

This section gives fundamental theories about AR, Fuzzy Number and Triangular Fuzzy Number in Sects. 2.1, 2.2, and 2.3, respectively.

2.1 Autoregressive

Based on [24], AR model predicts future behaviour based on past behaviour. It is used for forecasting when there are some correlations between values in time series and their lead and successful values. In the AR model, the value of result (Y axis) at some point t in time directly related to variable predicator (X axis). AR model is different with linear regression where Y depends on X and previous values of Y. An AR (p) model is an AR model where specific lagged values of y_t are used as predictor variable. The lag shows the outcome of a time period affecting the following periods. The value 'p' is called order.

$$y_t = c + \emptyset_1 y_{t-1} + \emptyset_2 y_{t-2} + \cdots + \emptyset_p y_{t-p} + e_t \tag{1}$$

where c is constant, e_t is white noise (error), and $y_{t-1}, y_{t-2}, \ldots, y_{t-p}$ are past series. The outcome variable in AR(1) is the process at some point in time, t is related only to line periods that are one period apart. Figure 1 shows an example of AR(1) graph.

From Eq. (1), the expected value of y_t for AR(1) is defined as:

$$y_t = c + \emptyset_1 y_{t-1} + e_t \tag{2}$$

where c is constant, e_t is white noise (error) and y_{t-1} is past series.

2.2 Fuzzy Number (FN)

Zadeh [25] defines the definition of fuzzy set as follows:

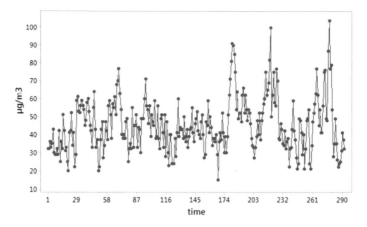

Fig. 1 Example of AR(1) model graph

Definition 1 Let U denote the universal set of the discourse. Then a *fuzzy set A* on U is defined in terms of the *membership function* m_A that assigns to each element of U a real value from the interval [0, 1]. A fuzzy set A in U can be written as a set of ordered pairs in the form $A = \{(x, m_A(x)) : x \in U\}$[1], where $m_A : U \rightarrow [0, 1]$.

The value $m_A(x)$, expresses the degree to which x verifies the characteristic property of A. Thus, the nearer is the value $m_A(x)$ to 1, the higher is the membership degree of x in A. The fuzzy number (FN) is a special form of fuzzy sets of real number sets on R. FNs plays an important role in fuzzy math, similar to the role played by ordinary numbers in classical mathematics.

Definition 2 A fuzzy set A on U with membership function $y = m(x)$ is said to be *normal*, if there exists x in U, such that $m(x) = 1$.

Definition 3 Let A be as in definition 2 and let x be a real number of the interval [0, 1]. Then the x-*cut* of A, denoted by A^x, is defined to be the set

$$A^x = \{y \in U : m(y) \geq x\}. \tag{3}$$

2.3 Triangular Fuzzy Number (TFN)

TFN is a fuzzy number represented with three points as follows: $A = (a, b, c)$.

Definition 4 Let a, b and c be real numbers with $(a < b < c)$. Then, the Triangular Fuzzy Number (TFN) $A = (a, b, c)$ is the FN with membership function:

$$y = m(x) - \begin{cases} \frac{x-a}{b-a}, & x \in [a, b] \\ \frac{c-x}{c-b}, & x \in [b, c] \\ 0, & x < a \, and \, x > c \end{cases} \tag{4}$$

From Eq. (4), we can define TFN as

$$TFN = y = [\alpha_l, c, \alpha_r] \tag{5}$$

From Eq. (5), if TFN is symmetry, $\alpha_2 - \alpha_1 = \alpha_3 - \alpha_2$, then y can be written as

$$y = [c, \alpha] \tag{6}$$

where a is spread of TFN and y is non-fuzzy number if $a = 0$.

3 Proposed Method: Forecasting Model Based on AR(1) with Standard Deviation Based TFN

This section discusses the systematic steps used for building forecasting model based on AR(1) with standard deviation based TFN (Δ_{σ^2}). The standard deviation approach is used to identify the spread value for building TFN. The proposed forecasting procedure serves as a guideline process in building forecasting model. Additionally, in this proposed procedure, times series model identification is provided to identify the model of Times Series before constructing TFN from crisp value. This step is important to ensure that only compliant dataset AR(1) model is selected. The flow of the proposed forecasting model based on AR(1) with Δ_{σ^2} is shown in Fig. 2.

The systematic steps for building AR(1) with Δ_{σ^2} is as follows:

Step 1.   Select n times series datasets in a form of input data as shown in Table 1.
Step 2.   Times Series model identification.

      i. Prepare datasets. Refer to Table 1.

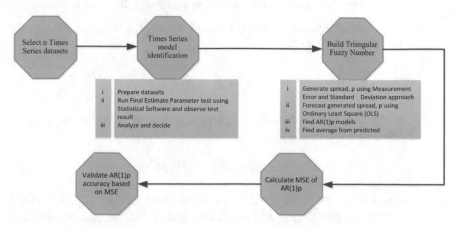

Fig. 2 Flow chart for forecasting model based on AR(1) with Δ_{σ^2}

Table 1 Format for input data

Data	1	2	...	n
Input	y_1	y_2	...	y_n

ii. Run Final Estimate Parameter test using Statistical Software to obtain Coefficient, Standard Error of the Coefficient (SE Coefficient), T-value and P-value.

iii. Check the result from estimate parameter to decide the AR(1) Times Series model. The p-value for each term tests the null hypothesis that the coefficient is equal to zero. The lowest p-value mean it is more meaningful to the model because changes in the predictor's value are related to the changes in the response variable.

Step 3. Build Triangular Fuzzy Number using Standard Deviation (SD).

i. Generate spread, s.

SD is a deviation that can be left or left from the centre. SD of the data is used as a spread because it leads to nature of SD.

$$\sigma = \sqrt{\frac{\sum_{i=1}^{n}(x_i - \bar{x})^2}{n}} \tag{7}$$

where σ, is standard deviation for population, x_i is each values of the data, $\bar{x}$ is the mean of x_i and n is the number of data points.

$$\tilde{y}_t^s = [y_t - s, y_t, y_t + s] \tag{8}$$

where $\tilde{y}_t^s$ is a fuzzy time series data at time, t with triangular fuzzy number data format. y_t is a time series data at time, $t (t = 1, 2, \ldots, n)$.

ii. Forecast generated spread, s using Ordinary Least Square (OLS).
iii. Find AR(1) models.

$$y_t = \beta_1 - \beta_2 y_{t-1} \tag{9}$$

where y_t is times series, β_1 is constant, β_2 is coefficient and y_{t-1} is past times series.

$$y_t^s = \left(\beta_1^L, \beta_1^R\right) - \left(\beta_2^L, \beta_2^R\right) y_{t-1} \tag{10}$$

where times series for standard deviation based models, $\left(\beta_1^L, \beta_1^R\right)$ is left and right constant, $\left(\beta_2^L, \beta_2^R\right)$ is left and right coefficient, y_{t-1} is past times

Table 2 Data format for average, $\bar{\bar{y}}_t^s$

y_t	y_1	y_2	...	y_n
$\bar{\bar{y}}_t^s$	$\bar{\bar{y}}_1^s$	$\bar{\bar{y}}_2^s$	...	$\bar{\bar{y}}_n^s$

series and $\bar{\bar{y}}_t^s$ is average predicted value. Table 2 shows the data format for average, $\left(\bar{\bar{y}}_t^s\right)$.

Equation 9 is the traditional model that can be build using OLS approach. While Eq. 10 presents AR(1) model with fuzzy data which is produced by SD approach.

iv. Find average for predicted values, $\bar{\bar{y}}_t^s$.

$$\bar{\bar{y}}_t^s = \frac{\left(\underline{\bar{y}}_t^s + s\right) + \left(\overline{\bar{y}}_t^s - s\right)}{2} \tag{11}$$

where $\underline{\bar{y}}_t^s$ is a left predicted value, $\overline{\bar{y}}_t^s$ is a right predicted value and spread of TFN uses SD, s of y_t. Table 2 shows the data format for average, $\left(\bar{\bar{y}}_t^s\right)$.

Step 4. Calculate the Mean Square of Error (MSE) of AR(1) with s-spread.
 After all stock market datasets have been tested, we need to compare the result of training and testing. To get total MSE for each y_t^s.

$$\text{MSE} = \frac{\sum_{i=1}^{n}\left(y_t - \bar{\bar{y}}_t^s\right)^2}{n} \tag{12}$$

where y_t is a time series data and $\bar{\bar{y}}_t^s$ is a predicted times series data at time, $t (t = 1, 2, \ldots, n)$ and n is sample size.

Step 5. Validate AR (1) accuracy based on MSE value.

The systematic steps presented here explains how to construct fuzzy number from crisp value, where the spread is derived from standard deviation. This step is important [16, 17] during data preparation since the crisp value which is observed as single points is insufficient to interpret such the financial or economic systems.

4   Numerical Experiment and Analysis

For the purpose of the experiment, stock market from five countries have been selected, namely, Malaysia (Kuala Lumpur Stock Exchange), Indonesia (Jakarta Stock Exchange), Singapore (Stock Exchange of Singapore), Philippines (Philippines Stock Exchange) and Thailand (Stock Exchange of Thailand). The data used was taken from 1 January to 26 March 2018.

Table 3 Datasets from five regions

Data	Input				
	Kuala Lumpur (57 data)	Philippines (58 data)	Thailand (58 data)	Indonesia (59 data)	Singapore (61 data)
1	1859.91	1328.30	1801.10	6200.17	3412.46
2	1865.22	1334.41	1794.21	6210.70	3421.39
...	...	...	...	...	...
56	1792.79	1795.45	1487.41	6353.74	3512.18
57	1782.70	1791.02	1477.91	6292.32	3489.45
58	–	1778.53	1477.18	6251.48	3501.16
59	–	–	–	6339.24	3464.28
60	–	–	–	–	3430.30
61	–	–	–	–	3402.92

Table 4 Kuala Lumpur stock market dataset

t	1	2	3	...	57
y_t	1782.70	1792.79	1803.45	...	1859.91

Table 5 Final estimation of parameters for Kuala Lumpur stock market

Stock market	Final estimates of parameters				
Kuala Lumpur	Type	Coefficient	SE coefficient	T	P
	AR(1)	0.9287	0.0659	14.09	0.00
	Constant	130.861	1.4330	91.33	0.00
	Mean	1835.88	20.10		

Step 1. Select five times series datasets as shown in Table 3.

Step 2. The following steps show times series model identification by using a Malaysia's stock market data as example.

    i. Prepare datasets as in Table 4.

    ii. Run Final Estimate Parameter test using Statistical Software and identify Coefficient, SE Coefficient, T-value and P-value as in Table 6.

    iii. From the result shown in Table 5, Kuala Lumpur stock market fulfills AR(1). This is based on the p-value that is 0.00.

Step 3. Build Triangular Fuzzy Number.

    i. Determine spread, s.
       The spread to build TFN are determined by using SD approach. The SD is calculated based on Eq. 7 and the value is 20.1057.

Table 6 The possibilities spread, s of TFN

n	y_t	$\overset{\leftarrow s}{y}_t$	$\overset{\rightarrow s}{y}_t$
y_1	1782.7	1762.594	1802.806
y_2	1792.79	1772.684	1812.896
y_3	1803.45	1783.344	1823.556
...	...	...	...
y_{57}	1859.91	1839.804	1880.016

Table 7 Predicted result for spread, s of TFN

n	$\tilde{y}_t$	$\overset{\leftarrow s}{\tilde{y}}_t$	$\overset{\rightarrow s}{\tilde{y}}_t$
y_1	–	–	–
y_2	1786.45	1766.34	1806.55
y_3	1795.82	1775.71	1815.93
...	...	...	...
y_{57}	1863.09	1842.98	1883.19

Using the SD value, the TFN is constructed as $\left(\overset{\leftarrow s}{y}_t, y_t, \overset{\rightarrow s}{y}_t\right)$ and is as shown in Table 6.

The spread to build TFN are generated using SD approach as shown in Table 7. Using the spread, TFN is written as $\left(\overset{\leftarrow s}{y}_t, y_t, \overset{\rightarrow s}{y}_t\right)$.

ii. OLS is used to obtain the forecast model. The predicted result using spread s as shown in Table 7.

iii. Find AR(1) models

$$y_t = 130.861 - 0.9287y_{t-1} \tag{13}$$

$$y_t^S = (129.427, 132.294) - 0.9287y_{t-1} \tag{14}$$

From Table 7, two equations are obtained which is Eqs. 13 and 14 that representing traditional and SD approach, respectively.

iv. Find single data point by taking the average $\overline{\tilde{y}}_t^S$ from predicted values based on Eq. (11). This is to transform the predicted values from TFN into a single point to measure the error. The forecasting result is shown in Table 8.

To get the average, $\overline{\tilde{y}}_t^S$ we will use Eq. 11. This is to transform the predicted values from TFN into a single point to measure the error.

Step 4. The result of MSE for Kuala Lumpur stock market is obtained based on Eq. 12 and presented in Table 9.

Table 8 Average predicted value for spread, s of TFN

y_t	y_1	y_2	$\cdots$	y_{57}
$\bar{\bar{y}}_t^s$	1786.45	1795.82	$\cdots$	1863.09

Table 9 MSE for Kuala Lumpur stock market

	Kuala Lumpur stock market MSE	
TFN	Training (46 data)	Testing (11 data)
yt	2.5424	[a]2.5188
y_t^s	[a]2.5342	2.5593

[a]Smallest MSE

Table 10 Summary of MSE

TFN	Stock Market MSE							
	Kuala Lumpur		Indonesia		Singapore		Thailand	
	Training (46 data)	Testing (11 data)	Training (48 data)	Testing (11 data)	Training (49 data)	Testing (12 data)	Training (47 data)	Testing (11 data)
yt	2.5424	[a]2.5188	5.6368	10.9316	4.4731	7.0399	[a]2.5847	2.6257
y_t^s	[a]2.5342	2.5593	[a]5.6367	[a]6.6593	[a]4.4727	[a]3.5830	2.7520	[a]2.0127

[a]Smallest MSE

Step 5.   Validate AR(1) p accuracy based on Mean Square Error (MSE).

In the early stage, five stock markets are evaluated. However, only four stock markets fulfill the requirement of AR(1) after AR(1) identification in Step 1. The accuracy of forecasting error can be verified by comparing MSE. Table 10 shows the summary of MSE.

In the validation step, although a result of MSE for training data is better, we consider the testing data results as MSE for testing forecast is made using its own data whilst training MSE uses previous data to forecast. By referring to Table 10, Training SD approach to Kuala Lumpur is better than traditional approach but when it comes to testing, the decision suggests otherwise. Additionally, other testings for Indonesia, Singapore and Thailand shows impressive result because the SD approach is better than traditional approaches.

5 Conclusions

This study presents a procedure to build an adjusted Triangular Fuzzy Number for AR(1) forecasting. The proposed procedure is important in handling uncertainty contained in the data used for forecasting. This procedure also suggests the technique to identify the spread of TFN clearly as compared with other approaches. The experiment result concludes that the proposed procedure is able to compete efficiently with the traditional approach, yet it solves the uncertainty issues in the input data.

Acknowledgements The author would like to extend his appreciation to the Ministry of Higher Education (MOHE) and Universiti Tun Hussein Onn Malaysia (UTHM). This research is supported by Tier 1 grant (Vot U895) and GPPS grant (Vot U975). The author thanks the anonymous viewers for their feedback. The author also would like to thank Universiti Tun Hussien Onn Malaysia for supporting this research.

References

1. What is the Stock Market? [Online]. Available https://corporatefinanceinstitute.com/resources/knowledge/trading-investing/stock-market/
2. Grunwald GK, Hyndman RJ, Tedesco LM (1996) A unified view of linear AR(1) models, pp 1–26
3. Lauren S, Harlili S (2015) Stock trend prediction using simple moving average supported by news classification. In: Proceeding—2014 International conference of advanced informatics: concept, theory and application, ICAICTA 2014, no 1, pp 135–139
4. Sirohi AK, Mahato PK, Attar V (2014) Multiple Kernel learning for stock price direction prediction. In: 2014 International conference on advances in engineering and technology research, ICAETR 2014, pp 2–5
5. Kaya M (2010) Stock price prediction using financial news articles. In: 2010 2nd IEEE international conference on information and financial engineering, ICIFE, pp 478–482
6. Schumaker RP, Chen H (2009) Textual analysis of stock market prediction using breaking financial news. ACM Trans Inf Syst 27(2):1–19
7. Sugumar R, Rengarajan A, Jayakumar C (2014) A technique to stock market prediction using fuzzy clustering. Comput Inf 33:992–1024
8. Barker KN (1980) Data collection techniques: observation. Am J Hosp Pharm 37(September):1235–1243
9. Efendi R, Samsudin NA, Arbaiy N, Deris MM (2017) Maximum-minimum temperature prediction using fuzzy random auto-regression time series model. In: 2017 5th International symposium on computational business intelligence, pp 57–60
10. Neenwi SL, Kabari G, Asagba P (2012) Nigerian stock market investment using a fuzzy strategy. J Inf Eng Appl 2(8):18–28
11. Chen MY, Chen BT (2015) A hybrid fuzzy time series model based on granular computing for stock price forecasting. Inf Sci (Ny) 294:227–241
12. Atsalakis GS, Protopapadakis EE, Valavanis KP (2016) Stock trend forecasting in turbulent market periods using neuro-fuzzy systems. Oper Res 16(2):245–269
13. Chang P-C, Wu J-L, Lin J-J (2016) A Takagi-Sugeno fuzzy model combined with a support vector regression for stock trading forecasting. Appl Soft Comput 38:831–842
14. Maciel L, Gomide F, Ballini R (2016) Evolving fuzzy-GARCH approach for financial volatility modeling and forecasting. Comput Econ 48(3):379–398

15. Kahraman C, Beskese A, Bozbura FT (2006) Fuzzy regression approaches and applications. In: Fuzzy Regression Model. Beyond Fuzzy Rule Base Model, vol 201, no 1, pp 589–615
16. Efendi R, Arbaiy N, Deris MM (2017) Indonesian-Malaysian stock market models using fuzzy random time series, pp 18–19
17. Efendi R, Arbaiy N, Deris MM (2018) A new procedure in stock market forecasting based on fuzzy random auto-regression time series model. Inf Sci (Ny) 441:113–132
18. Efendi R, Arbaiy N, Deris MM (2017) Estimation of confidence-interval for yearly electricity load consumption based on fuzzy random auto-regression model. In: Computational intelligence in information systems, vol 532
19. Singh P (2017) An efficient method for forecasting using fuzzy time series. In: Emerging research on applied fuzzy sets and intuitionistic fuzzy matrices, IGI Global, p 18
20. Efendi R, Ismail Z, Deris MM (2015) A new linguistic out-sample approach of fuzzy time series for daily forecasting of Malaysian electricity load demand. Appl Soft Comput J 28:422–430
21. Ismail Z, Efendi R, Deris MM (2015) Application of fuzzy time series approach in electric load forecasting. New Math Nat Comput 11(03):229–248
22. Efendi R, Deris MM, Ismail Z (2016) Implementation of fuzzy time series in forecasting of the non-stationary data. Int J Comput Intell Appl 15(02):1650009
23. Efendi R, Deris MM (2017) Prediction of Malaysian-Indonesian oil production and consumption using fuzzy time series model. Adv Data Sci Adapt Anal 9(1):1–17
24. Hyndman RJ, Athanasopoulos G (2014) Forecasting : principles and practice. [Online]. Available http://otexts.com/fpp/
25. Zadeh LA (1965) Fuzzy Sets. Inf Control 8:338–353

An Offshore Equipment Data Forecasting System

Ahmad Syazwan Sahdom, Alan Cheah Kah Hoe
and Jaspaljeet Singh Dhillon

Abstract In the oil and gas industry, various machineries and equipment are used to perform oil and gas extractions. The problem arises when there is unplanned maintenance on any equipment. Unplanned maintenance will result in unplanned deferments that disrupt business operations. Companies may have developed monitoring systems based on current and historical equipment statuses, but ideally there should be mechanisms to conduct or produce real-time forecasts on equipment conditions. In this paper, linear regression models were tested and deployed in a system developed to forecast flow rate of seawater lift pumps of an offshore platform. Apart from identifying and evaluating a suitable statistical model to derive the forecasts, this paper presents a tool that was developed using the selected model to automate real-time data extraction and execute the prediction process. The models were developed based on raw data that were accumulated from an oil and gas company over a period of 3 months. Of the 3 months' data, the first 2 months of data were used as the training data, and the last one month was used for testing the models. Data cleansing was performed on the dataset whereby unwanted values that could affect accuracy of the model or any other data with values not processable by the models were eliminated. Results indicated that Autoregressive (AR) model is suitable for a real-time prediction of an offshore equipment.

Keywords Forecasting system · Statistical models · Alarm alert system · Offshore equipment

A. S. Sahdom · A. C. K. Hoe · J. S. Dhillon (✉)
College of Computer Science and Information Technology, Universiti Tenaga Nasional,
Kajang, Malaysia
e-mail: jaspaljeet@uniten.edu.my

A. S. Sahdom
e-mail: syazwansahdom@gmail.com

A. C. K. Hoe
e-mail: alancheah@uniten.edu.my

© Springer Nature Singapore Pte Ltd. 2019
V. Piuri et al. (eds.), *Intelligent and Interactive Computing*, Lecture Notes in Networks and Systems 67, https://doi.org/10.1007/978-981-13-6031-2_25

1 Introduction

Extracting crude oil and gas is by no means an easy task. There are many challenges in terms of the Health and Safety Executive (HSE), as well as meeting production requirements since we deal with various machineries to successfully extract the raw resources. Hence, it is vital that we properly monitor every equipment used in the extraction process, and there are hundreds of equipments such as gas compressors, pumps, and pipelines to measure. Some of the oil and gas companies may have implemented means to measure the status of their equipment via a monitoring system (based on current and historical equipment statuses), but there are no means to conduct or produce any real-time forecasts on equipment conditions. The Internet of Things (IoT) technologies can be leveraged for remote monitoring purposes with observations taken at every few (e.g., 5–20) seconds interval. Thus, huge data can be accumulated over time. Millions of data points stored in databases are available for predictive analytics.

The data available within the huge data repository can be used for analytical purposes. By implementing proper statistical techniques, various previously unknown knowledge can be extracted from the available data. The goal of this paper is to use the data to develop statistical models which can extract knowledge from the data and perform predictions on real-time data. In this study, we have experimented with four models, in which three are statistical models and one is a neural-network-based model that will be able to predict the seawater lift pump flow rate (m^3/h) from the historical data acquired. Results obtained from these models were used to decide which one best fits the acquired data, in terms of both accuracy, and computing expenses. Computing expenses is crucial since we are dealing with a real-time prediction. A monitoring software is also developed to automate the data extraction, data cleansing, and forecasting. The software will also be the front-end to the whole system; it is what users will interact with.

2 Methodology

We have utilized an abundance of data collected from a seawater lift pump. Data are collected at every 20 s, 24 h per week, and 365 days per year. Hence, there are millions of historical data points available for us to test and analyze different models. For this study, we have used 3 months of data that consist of more than 350,000 rows of equipment measurements. The first 2 months of acquired data was used for training the models, and another one month for testing and validation. The time series data consist of two features which are the datetime and the value. The amount of data used for training is 200,000 data points that span up to 2 months. Figure 1 shows the data type used to train the model as well as to feed the models in real time after implementation.

Fig. 1 The time series data

1:17:50 PM	681.86
1:17:40 PM	688.08
1:17:30 PM	689.34
1:16:40 PM	Comm Fail
1:16:20 PM	682.96
1:16:10 PM	686.52
1:15:59 PM	681.89
1:15:49 PM	685.62
1:15:39 PM	684.81
1:15:29 PM	684.54
1:15:19 PM	685.37
1:15:09 PM	689.77
1:14:59 PM	682.45

2.1 Data Cleansing

To ensure proper model calibration and in turn, prediction accuracy, we have first performed data cleansing on the derived data. Since the forecasting to be done is real time in nature, we did not do an in-depth data cleansing. This is to ensure that the model can adapt to the irregularity of real-time data coming straight from the server [1]. From the PI Server, several invalid data due to transmission error were identified. Invalid data from the historical data were cleansed. Transmission error messages such as "Bad", "Bad data", "I/O error" and "Configure" were removed from the training dataset before it was fed into the models for training.

The data was then sorted in an appropriate sequence; oldest values on top and newest values at the bottom. This sequence is not necessarily the same for all models but we had to carefully sequence the data according to the model specification and how the model consumes data.

Before feeding the model, we had to transform the data into a "differenced" shape. The procedure is called "differencing". Suppose X3 is the most recent data, to difference the data we take the difference between value X3 and X2 and apply the value on X3. This process goes on until the end of the data. The differencing process ensures faster computations, as well as removing any property of the data that are showing nonstationarity. All computations within a model are made based on differenced real values. All predictions are generated on the differenced values, but generated predictions will not be in a differenced shape. Thus, we do not have to reverse difference the shape of the prediction values for plotting and our reading, we can readily use them.

2.2 Statistical Modeling

We have experimented with four types of statistical models to be used for forecasting: three of them are regression-based, and one of them is neural-network-based. Each model may work best for only certain types of data patterns; there is no model that fits all and every data. Different models may also yield different computing performance [2]. Some models may require higher computing load, and since we are using the models for real-time prediction, computing performance is very important. Models that can make accurate predictions but cannot cope with the real-time nature of the system cannot be used. Hence, the testing of the models is to measure not only the prediction accuracy of the models, but also the computing performance for each model to find the best to be used for the system. Regression-based models were developed using the StatsModels python library, and Neural Networks models were developed using Keras library and TensorFlow as the backend. The models experimented with in this paper are as follows:

- Seasonal Autoregressive Integrated Moving Average (SARIMA)
- Autoregressive Integrated Moving Average (ARIMA)
- Long Short-Term Memory Recurrent Neural Networks (LSTM RNN)
- Autoregressive (AR)

The aim of this study is to test each of the identified models with different combinations of their parameters (or hyperparameters). To achieve and exhaust all tests manually is a very tedious and time-consuming task. Thus, the tasks of testing the accuracy level of each parameters combination for statistical heavy models will be done within loops. A test environment has been setup in Python that loads the training data, and tests each combination while printing the results into a CSV file to be analyzed later. However, it is impossible to use loops to test the parameters combinations of the LSTM RNN model due to the complex nature of the model and the various hyper-parameters to tune. Thus, the testing of the LSTM RNN model is done manually, and it is possible that a perfect combination has not been found in this research project.

It is also worth highlighting that ARIMA and SARIMA (both are statistically heavy models) are very computationally expensive. To exhaust all tests will require a lot of computing power and time. Hence, an exhaustive test cannot be completed in this research, but tests within this research were completed with solid findings. An ARIMA and SARIMA based model is so computationally expensive that they are not fit for use in the case of real-time predictions.

All the models with their corresponding parameters are as follows:

1. Autoregressive Integrated Moving Average (ARIMA)

    a. p (the order of the non-seasonal autoregressive model)
    b. d (the number of non-seasonal differences)
    c. q (the order of non-seasonal moving average model)

2. Seasonal Autoregressive Integrated Moving Average (SARIMA)

    (a) p (the order of the non-seasonal autoregressive model)
    (b) d (the number of non-seasonal differences)
    (c) q (the order of non-seasonal moving average model)
    (d) P (the order of the seasonal autoregressive model)
    (e) D (the number of seasonal differences)
    (f) Q (the order of seasonal moving average model)

3. Long Short-Term Memory Recurrent Neural Network (LSTM RNN)

    (a) Number of epochs per training
    (b) Batch training size
    (c) Sequence length of data
    (d) Neurons
    (e) LSTM model shape (layers)

4. Autoregression (AR)

    (a) Max lag

For each model above (1, 2, 3, 4), it is required for us to tweak its parameters (e.g., a, b, c, d, e, f) to find out which combination of parameters will perform best in terms of accuracy, and in case of real-time predictions, we also had to consider the CPU processing time and performance of each model. To find out the best possible model with the best possible parameter combinations, a grid search must be performed. A grid search automates the testing of each parameter combination, in this case, a "for loop" was used to perform a grid search.

AR was tweaked first, since AR have less parameters to be compared to other models. The goal is to first develop a baseline model that will be able to at least perform some degree of predictions regardless of accuracy level, and continue working on the UI of the software system afterwards. This is to ensure that time is not wasted solely on developing (and testing) models, as models can still be developed and replace old models after the completion of a working prototype of the system. Models will be saved as a file which the Python interpreter along with libraries that supports them will be able to read.

2.3 Hyperparameters

The crucial steps are to identify and build a model based on the available data. In this stage, it is necessary to identify the hyperparameters of each model. The identification is only to find out which numerical combinations of the hyperparameters works best for a time series data.

SARIMA, ARIMA, and AR are based on regression. SARIMA is a model that employs the seasonal factor of the data into the model to perform predictions [3]. ARIMA, on the other hand, extends AR and considers the Moving Average of the data. AR is the simplest and takes in only the lag parameter; the length of training data to be considered for the model coefficients. As stated, hyperparameters are just numerical values that represent the significance of each properties (S, AR, I, MA) of the models.

A grid search is performed using loops programmatically. One can set a loop to start at $i = 0$ and end at $i = 100$, then fit the models with the data at every interval while incrementing the parameter values within each interval, and record the MSE result at each interval to find the best combinations of hyperparameters. For SARIMA and ARIMA, a nested loop was used since the models have more than one hyperparameter to be tuned. Within this research project, it is impossible to exhaustively perform a complete grid search on the hyperparameters of each model. Furthermore, the bigger the numerical values, the longer it takes to fit the data.

For the LSTM RNN model, tuning the parameters is highly based on estimation and a grid search was not performed. The most basic but important hyperparameters of LSTM RNN models are as below

1. Epoch—the number of iterations the model will undergo training on the same dataset
2. Batch size—the number of dataset batches to be fed into the model at each epoch, the lower the batch size, the slower the training. Lower batch size may or may not yield better model accuracy.
3. Sequence length—the length of the dataset shaped into sequences of a certain length. Using neural networks, it is necessary for us to reshape the dataset as such to initiate the supervised learning of the neural network.
4. Neurons—Neurons are the component within the network that is responsible to carry weights upon which the models will base their decisions. This parameter represents the number of neurons contained within a layer.

Training neural-networks-based models requires us to tune various other parameters such as the learning rate, the activation function to be used, what type of layer to be used, whether to make the learning "stateful" or not, as well as the structure of the network itself.

3 Results and Discussion

All four models mentioned above have been tested. The metric to measure accuracy for ARIMA and SARIMA is Akaike information criterion (AIC) and the metric for AR and LSTM is Mean Square Error (MSE). The goal is to find the best possible combination of the parameters with which the usage results in the lowest AIC or MSE depending on the model.

ARIMA model. A grid search has been performed on the ARIMA model to find the best parameter fit in terms of prediction accuracy. Parameters are a combination of three distinct numbers, p, d, and q. In a for loop, the grid search was performed in a range of 0–100, to find the best possible combination. For example, in the sequence of (p, d, q), the best possible combination could be (0, 10, 100), or (76, 54, 1). However, due to the increasing need for computing time of the model, the grid search was stopped at (0, 1, 50) with the lowest AIC of 31,886.6 to be compared to 60,945.23 for the first parameter test (0, 0, 1). It took approximately 2 h to perform the grid search on the parameter (0, 1, 50) alone, and the computing time will only grow as the parameter numbers increases, so it was stopped.

SARIMA model. A grid search was also performed on the SARIMA model. The case, however, is similar with ARIMA. The fitting process with SARIMA is even slower than ARIMA, resulting in a very slow grid search. The finding for SARIMA with the lowest AIC of 32,206.01 according to sequence of (p, d, q, P, D, Q) is (0, 0, 3, 2, 1, 3). The grid search was stopped to make room for the testing activities on other models. It can also be concluded that both ARIMA and SARIMA will not fit the scope of the project, which is to perform real-time predictions. The two models are too slow in performing computations.

LSTM Model. LSTM incorporates machine learning techniques, specifically the Recurrent Neural Network. This model consists of several parameters mentioned above. Several setups of the LSTM model have been tried and each of them yields different results. Some setups result in just an "imitation" of real data. Training an LSTM model takes time and computing power, however, models can generate predictions comparably faster than ARIMA or SARIMA [4]. One other concern regarding LSTM is to make it so that the model constantly learns as it predicts. That requires retraining the model with the most recent data as time goes by. The problem is, training an LSTM model takes time. Hence, until a better technique has been discovered, usage of the LSTM model will be discontinued.

AR Model. Using the stats models library there is only one main parameter to be tweaked which is the lag value. The lag value is the number of data we set the AR model to look back to in a given historical data just before the current value. That is because AR model assumes there are correlations between one value and another, and it uses previous values (or lag) to perform predictions on the next values. The AR model has been set to look back to up to 1800 values, and the model can predict fluctuations in the graph 60–70% of the time generating multi-step predictions up to 6 steps forward. However, sudden value increases that come with unstable graph readings cannot be predicted so far using the AR model. The AR model does a good deal of self-correction based on the last observed real values. More fine-tuning is still being done on the AR model, experimenting with other additional parameters that need to be learned and tuned such as the method, solver, and convergence tolerance. The AR model performs the fastest among the other models, with the ability to constantly update the model. So for the time being, the AR model is the most suitable for real-time prediction for its lightweight property and its ability to predict.

The results from the grid searches return a not-so-different MSE for each of the models, although some of them yield better accuracy than others. However, as mentioned before, the computing performance of each model is also considered when making the decision of choosing which model would be able to perform predictions on a real-time system. Results show that AR model triumphs over other models in terms of computing performance with a slight degradation of prediction accuracy

Table 1 AR model performance

(p) value	MSE (AR)
800	204.84
1200	127.60
1800	119.81
2800	110.49
3000	110.50

Table 2 ARIMA model performance

(p, d, q) value	MSE (AR)
3, 2, 0	205.845
5, 2, 0	165.663
6, 2, 0	150.419
4, 1, 0	118.997
5, 1, 0	117.315

from other regression models. LSTM RNN models were tested but yield the worst accuracy and are not able to perform forecasts that are as reliable as regression models. Hence, other models were discarded and would not be used in the system, and only the AR model will be used. Table 1 depicts the result of AR model for one equipment transmitter, the FICA7620 Flow Rate. It is apparent that the MSE will decrease with each increase of the (p) value. However, the (p) value has been increased beyond 2800 but yields a very slightly worse result than $(p) = 2800$.

Table 2 shows the accuracy of the ARIMA model on the same data with (p, d) tuned. The result shown in this table shows that the ARIMA model has a great deal of potential when all of its parameters are tuned but, in this application, it is too expensive to tweak the (p) parameter to a number more than 6 as this requires a significant amount of time to complete model computations. It is evident that in this case, the accuracy of AR outperforms the accuracy of ARIMA despite ARIMA being the more complex linear regression model. AR also is more suitable for real-time forecasting as the time to complete computing forecast results for each time step is significantly lower than the case with ARIMA.

4 Automating Forecasts

Based on the results obtained, the AR model was deployed in the development of a system to perform the remote forecasts. A software was written to extract data from the offshore server, to use the developed models to generate forecasts based on the extracted real-time data, and to display the results onto the GUI which users interact with. The system also handles the logic of alarm values, logging forecasts, and real-time values into CSV files, and send out emails in the occurrence of alarms.

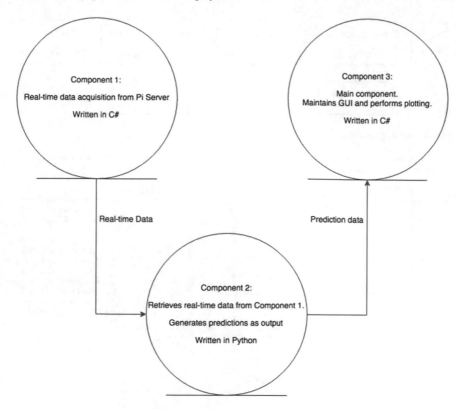

Fig. 2 Components of the offshore equipment data forecasting system

4.1 System Components

The system primarily consists of three components as shown in Fig. 2.

In Fig. 2, Component 1 will be responsible for extracting data from the PI Server. Component 2 will be responsible for reading the data from Component 1, performing data cleaning, and generating prediction values. Component 3 will then read all output from Component 2 and display them onto the GUI. All graph plotting, email alert sending, as well as logging will be handled by Component 3. Component 1 and 3 were written using C# and Component 2 was written using Python and compiled into a Windows Executable file using PyInstaller.

4.2 System Design

In Fig. 3, DataExtractor.exe is Component 1, Predict.exe is Component 2, and Real-TimePredictions.exe is Component 3. The system logs every alarm occurrence in a

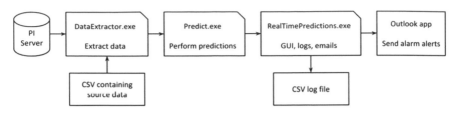

Fig. 3 The system flow

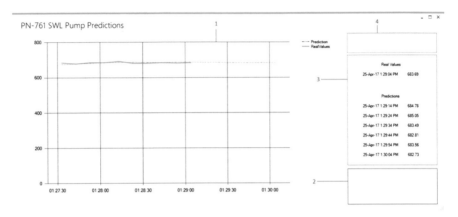

Fig. 4 User interface of the forecasting system

CSV file. Sending out alarm alert emails, will be carried out by using Outlook app through referencing Outlook objects in the C# project. Thus, for emails to work, it is advisable to let Outlook app run in the background.

Figure 4 shows a screenshot of the forecasting system. Essential elements are numbered and a brief explanation is given in Fig. 4

1. Tabs for selecting available transmitters to be forecasted.
2. Texts showing real-time values with dates (In the Most Recent Values groupbox).
3. Texts showing forecasted values (In the Forecasts groupbox).
4. Alarm threshold text and buttons that will pop up the user interface for users to change the threshold values.
5. Indicators (warnings) showing whether the predicted trend will go upwards, or downwards.
6. Text showing the confidence level of the predictions.
7. A graph to show real-time data.
8. All predictions as well as real values will be plotted in the graph in real time.
9. Start/Stop button.

5 System Limitation

The system meets all objectives as of now but has certain limitations. First, the system is not accessible from everywhere due to its nature of not being web-based. Although it was not a requirement, the initial idea is to make the system web-based, but due to the difficulty in plotting multiple values into one graph (2 lines, one for real-time values, one for predictions), it was then decided that the system will be a Windows Forms application. Second, the system only incorporates emails as a medium to send out alerts. Emails will generally be slower. One improvement that can be carried out regarding the alert feature is to implement other alert techniques such as Telegram. Telegram comes into the picture since it provides API that is ready to be used with the system. Also, the system is currently implementing the AR model as the prediction engine. When better models have been discovered and proven to work more effectively with this use case, they can replace the AR model as the main prediction engine.

6 Conclusion

Forecasting systems developed using suitable models have great potential in assisting engineers in mitigating situations, for which we cannot rely on human judgement alone. Automating the process of forecasting helps to make it convenient to remotely assess the health and performance of equipment. However, as any other system, the forecasting system will have its own limitations and drawbacks, and forecasts will not be 100% accurate. Using the system without human judgement and common sense will not make things better. Hence, the system is meant for assistance instead of heavy reliance.

AR model performs the fastest among the other models tested in this study, with the ability to constantly update the model. Hence, the AR model is suitable for real-time prediction due to its lightweight property and its ability to predict. In terms of system prediction accuracy, access methods, and alarm alert methods, plenty of improvements can still be made. Now that the prototype is ready, using efficient source code and other known techniques, we can further improve the system to make it reliable and ready for real-world.

References

1. Sfetsos A (2002) A novel approach for the forecasting of mean hourly wind speed time series. Renew Energy 27:163–174. https://doi.org/10.1016/S09601481(01)00193-8
2. Chang W-Y (2014) A literature review of wind forecasting methods. J Power Energy Eng 2:161–168. https://doi.org/10.4236/jpee.2014.24023

3. Subbaiah NKCHV (2016) SARIMA modelling and forecasting of seasonal rainfall patterns in India. Retrieved from https://tinyurl.com/y884qyco
4. Breiman L (2001) Statistical modeling: the two cultures. Retrieved from https://projecteuclid. org/download/pdf_1/euclid.ss/1009213726

Publication Model for Academic Talent Based on Apriori Technique

Zulaiha Ali Othman, Noraini Ismail, Roliana Ibrahim and Hamidah Jantan

Abstract Appropriate decisions-making for application or reversion of promotion is mainly based on essential information provided to the Human Resource Management (HRM). This is to ensure the continuity of excellence for academic staff in establishing and strengthening education towards the success of the universities. In addition, publishing is a core component in the evaluation of academic talent that is affected by the research, supervision, and conference at the local university in Malaysia. Therefore, this study aims to produce a model based on the main factors of academic talent using Association Rules. This study applies the Apriori Algorithm to identify a set of rules for the assessment of significant relevant talents for the promotion of academic staff in local University. The findings have successfully developed a model based on talent acquisition of knowledge related to the issuance and have been evaluated by comparing the guidelines for the promotion of academic experts. This knowledge helps to improve the quality of the evaluation process of academic talent management and future planning in the HRM.

Keywords HRM database · Association rule-apriori · Publication · Modeling

Z. A. Othman · N. Ismail (✉)
Faculty of Information Science and Technology, Universiti Kebangsaan Malaysia, Bangi, Selangor, Malaysia
e-mail: norainismail14@gmail.com

Z. A. Othman
e-mail: zao@ukm.edu.my

R. Ibrahim
Faculty of Computing, Universiti Teknologi Malaysia, UTM, Skudai, Johor, Malaysia
e-mail: roliana@utm.my

H. Jantan
Faculty of Computer and Mathematical Sciences, Universiti Teknologi MARA (UiTM), Dungun, Terengganu, Malaysia
e-mail: hamidahjtn@tganu.uitm.edu.my

© Springer Nature Singapore Pte Ltd. 2019
V. Piuri et al. (eds.), *Intelligent and Interactive Computing*, Lecture Notes in Networks and Systems 67, https://doi.org/10.1007/978-981-13-6031-2_2

1   Introduction

Human Resource (HR) is the most basic drives that determine the success of an organization's performance [1]. It owns a workforce that is competitive and has a talent that is very important in an organization to confront and overcome challenges [2]. Its aim is to ensure that the position given to a candidates is accordance with qualifications, talent, and achievements. However, the main problems or challenges faced in Human Resource Management (HRM) at a local university are recognizing, developing, and managing human resources academics efficiently and effectively [3]. HRM also needs to select the academician with a good qualifications, talents, and achievements for the right position in order to optimize the performance of a university.

The current challenges of HRM in a local university are talent management among employees to ensure that eligible employees are given a suitable position and provided in a timely manner [4]. Based on past research, several academic talent models have been proposed to tackle this problem [3, 5, 6]. Most of the results have shown that classification technique for J48 or C45 obtained the best model, but the model was very complicated to visualize and difficult to extract the meaningful pattern.

On the other hand, Association Rules technique is also one of the popular data mining methods that can be used to discover knowledge. However, less research has explored the important and interesting relationship toward the specific factors of academic aptitude such as demographic, publication, supervision, research, and conferences using Association Rules techniques. Therefore, this paper presents an academic publication model for academician using Apriori technique in Association Rules. The experiment was applied using HRM dataset collected from one of a local university in Malaysia.

This paper contains four sections. Section 2 discusses the related work on academic talent management and the techniques used. Section 3 discusses the preprocessing process of the data. Lastly, Sect. 4 presents the experiment result and research finding.

2   Materials and Method

In this study, a publication model for academic talent is created by applying an Apriori-based Association Rules algorithm to a set of data. Data mining model gets data from the mining structure and then analyzes the data by using data mining algorithms. The mining structure stores information that defines the data source. Then, the mining model saves information obtained from the processing of data statistics, such as the pattern found as an analysis result. This study has been divided the related work into 3 focus which is (1) academic talent management, (2) data mining for talent management, and (3) Association Rules using Apriori technique.

2.1 Academic Talent Management

Talent management is a very important process. It consists of the recognition of key talents in the organization, identifying people within organizations that make up the key talents, and pursuing development activities for talent groups to sustain and engage them as well as they are ready to move into a more important role [7]. In talent management, to identify existing talents is one of the top HRM challenges. There must be has a clear guideline and criteria for HRM to choose the right talent for the right position at the right time.

Furthermore, an appropriate decision-making needs to be made for application or reversion of staff promotion to establishing and strengthening education towards the success of the universities and the main decision is based on the essential information provided to the HRM [3, 8]. In local universities, there are four main competency assessments measured by HRM for an employee's academic staff which is talent in teaching, publishing, community service, and administration. However, publishing is a core component in the evaluation of academic talent that is affected by the research, supervision, and conference at the local university in Malaysia.

The main issues faced by the academic talent is the publication of an academician is influenced by the number of master and Ph.D. students that are under their supervision. One academic staff cannot supervise more than 7 Ph.D. students as the main supervisor at once. These issues are still in discussion among the higher education department to find the optimal value based on talent academic experience. Therefore, it is important to know the pattern of this relationship to enable the universities management to make an optimum supervisory planning and a new policy for academic staffs. This pattern can be found by using Data Mining technique in order to produce a model based on the main factors of academic talent.

2.2 Data Mining for Talent Management

Data mining is an area that has evolved as a result of increasing information and data. In industry, data mining is used to turn raw data into useful information and knowledge for their usability. Association rule, Association rule, text mining, sequential rule mining, classification, and cluster analysis are examples of data mining techniques. These techniques have been used by researchers in various domains such as customer and business analysis, finance, production management, education, health care, and hospital management [9].

In recent years, data mining becomes popular in HRM domain [5]. The research contribution refers to various HRM problems such as selecting employees, predicting employee turnover, career planning, performance pattern, and predicting evaluating employee performance in the function of performance management [4, 7]. As a result, data mining provides ample insights on how to generate advanced information and decision support within the HR domain. Based on a survey paper in [5], the researcher has been listed several methods and domains of HRM applications that use Data

Table 1 HRM application using data mining techniques

Data mining technique	Problem	N
Classification	Staffing	52
	Development	10
	Performance management	14
	Others	3
Association rules	Staffing	7
	Development	1
	Performance management	1
	Others	2
Clustering	Staffing	5
	Development	2
	Performance management	4
	Others	1

Mining technique to solve the problems. Table 1 shows the summary based on the research findings in [5].

However, based on the literature [5, 8, 10], there is less research applying data mining in HR academic field. In [11], the researchers have been applied Association Rules to elicit hidden relationship among the teacher's characteristics. The aim of the study is to help the HR in developing recruitment and policies to enhance the quality of research, teaching, and learning. While [3] use classification technique to predict talent performance based on past experience knowledge that discovered from the existing database. Then, [6] also use classification technique to predict the potential human talent for promotion recommendation. In addition, most previous studies also are more focused on classification technique to solve the problem in HRM [5]. Therefore, there are opportunities and space for additional research on HRM to use Apriori algorithm to find the pattern for promotion recommendation among the academician of local universities.

2.3 Association Rule Using Apriori Technique

In data mining, Association Rules are the most popular method on finding the co-occurring association among a collection in huge databases [12]. It aims to extract interesting associations, correlation, and a frequent pattern in databases using some interestingness measure [13]. Many algorithms have been proposed to find a frequent pattern, one of them is Apriori.

Apriori algorithm is a classical algorithm in data mining [14]. It is used for mining frequent itemsets and relevant Association Rules. It is devised to operate on a database containing a lot of transactions, for instance, items bought by customers in a store. It generates length $(k + 1)$ candidate itemsets based on length (k) frequent itemsets.

The frequency of itemsets is defined by counting their occurrence in transactions [13, 15]. Apriori algorithm is always regarded as an important Association Rules mining algorithm because the basic idea on finding all frequent itemsets in a given database is universal and easy to implement for any Association Rules mining problems.

According to the literature, besides Agrawal and Srikant themselves [14], many researchers have been applied Apriori algorithm to a variety of problems in a database system [16]. Later, it has become very useful in the field of retail trade and market basket analysis [17]. This algorithm also has been widely used in healthcare domain to find the relevant prevention factor for a different type of cancer using frequent pattern concept [18]. Other than that, several researchers have been applied this algorithm to solve a different agricultural problem such as to predict future crop, weather forecasting, pesticides and fertilizers to be used, revenue to be generated, and so on [19–21]. Lastly, [22] used past data to mining the frequent itemsets using Apriori algorithm to perform a comparative study of Association Rules algorithms for course recommender system in E-learning.

In conclusion, Apriori is one of the best algorithms for learning Association Rules and suitable for finding an interesting relationship between the itemset. Therefore, this algorithm can be used to identifying the association between the specific factors of academic aptitude such as demographic, publication, supervision, research, and conferences as required in this study.

2.4 Preprocessing Data

In general, the preprocessing data in this experiment involves five phases: (a) data collection, (b) data cleansing, (c) data integration, (d) data reduction, and (e) data transformation.

(a) **Data collection**

This research involves the use of the actual raw data sets taken from the database of human resources for academic staff at a local university in Malaysia from 1994 until 2010. The data consist of 15 raw datasets with 3220 data records and 1140 attributes. The data set contains data related to academic staff include Lecturer, Senior Lecturer, Associate Professor, and Professor. The focus is given to the talent publication for the experiments for the purpose of publication talent development model for academic staff at a local university.

(b) **Data cleansing**

Data cleaning is a process of correcting or removing inaccurate or missing values from dataset to improve the quality of the data. In this study, the data collected has irrelevant and repeated attribute and the data distribution for several attributes also are not consistent.

This study only focuses on five main factors that have an aptitude of five main criteria, namely demographic, publication, supervision, research, and conferences. Therefore, attributes that have no relevance or interest to be selected as an attribute

Table 2 Main criteria of academic talent

No.	Criteria	Attributes
1	Demographic	Gender, race, jobs status, current status, education, length of service, year be promoted to lecturer, year be promoted to senior lecturer, year be promoted to associate prof., year be promoted to prof., and actual post
2	Publication	Book publication, proceeding publication, journal publication, others publication, and total publication
3	Supervision	Total supervision
4	Research	Research grant 2006, research grant 2007, research grant 2008, research grant 2009, research grant 2010, and total of research grant
5	Conferences	Presenters, chairman, evaluator, participant, others, department conference, university conference, national conference, international conference, and total of conference

Table 3 Merging attribute

No.	Attributes	No. attributes merger	
		Before	After
1	Races	27	2
2	Job status	8	2
3	Current status	8	2
4	Education	31	3
5	Publication	20	4
6	Role in research	9	2

of this study were removed. In addition, the attribute that contains 80% value "0" and above also will be eliminated because the distribution of the data is not consistent. Only 35 attributes were selected from 1140 as listed in Table 2 and the rest will be removed from the dataset.

(c) **Data Integration**

Data integration involves combining data from different sources into meaningful and valuable information. The process of merging the data in this study is carried out through decisions or assessments by experts and analysis of the relevance of merged attributes and item in the attribute. In this study, the data were integrated with the relevance and suitable attribute. There are six attributes that were selected to undergo merged process which is Races, Job Status, Current Status, Education, Publishing, and Role in Research. The list of the integrated data shown is in Table 3.

The race consists of 27 types of people but the categories were combined into 2 categories for the academician which is Malays and non-Malays. For Job Status, the data were integrated from eight categories into two categories namely permanent and nonpermanent. Attribute Current Status also was integrated from eight categories into two categories which are active and inactive. Attribute for qualification in education was integrated into 3 categories, namely doctorate, master, and others from 31 categories. For publication, 20 publications had been divided into four main types which are published book, proceedings, journal, and others. While Role in Research attribute was integrated into nine types of roles in research into two roles which are researchers and head of research.

(d) **Data Reduction**

The overall picture of HRM dataset that contains 3220 data records shows only a part of the records hold value "1" and above, while the remaining records hold "0" value. Therefore, data reduction was conducted to reduce the size of data by removing the meaningless record using a formulation as below:

$$\text{Percent of reduction Academician record} = \frac{\text{Number of "0" value data}}{\text{Academician record}} \times 100 \quad (1)$$

The academician record that got a value below than 20% will be removed from the dataset. Only 1938 data records that get the values higher than 20% were taken from actual 3220 raw data record.

(e) **Data Transformation**

The last preprocessing step in this study is data transformation. Discretization of numerical data is one of the most influential data preprocessing tasks in data mining. It is considered as a data transformation mechanism because it transforms data from a large domain of numeric values to a subset of categorical values. This process resulting mining process becomes more efficient and the patterns that may appear easier to understand [23].

In this study, only several numeric attributes in HRM dataset were discretized using equal-depth partitioning. It divides the range into N intervals, each containing approximately same number of samples. To simplify the process of generating sets of items, the discretized values of every item in the attribute were changed to a symbol as shown in Table 4.

3 Result and Discussion

This experiment was conducted using Waikato Environment for Knowledge Analysis software to discover an Association Rules in HRM dataset and using Apriori algorithm to find the rules based on the data. Only 1938 dataset with 35 attributes was selected for this experiment.

In this study, the generation of rules was controlled by parameter tuning, such as minimum support and minimum confidence level. This experiment used four different minimum support values which are 0.05, 0.1, 0.15, and 0.2. While the threshold for minimum confidence level was set starting from 10% up to 90%. The support value

Table 4 Attribute transformation

No.	Attribute	Discrete	Symbol
1	Published book	0	A1
		1–2	A2
		≥ 3	A3
2	Published proceedings	0–1	B1
		2–6	B2
		≥ 7	B3
3	Published journal	0–1	C1
		2–6	C2
		≥ 7	C3
4	Others publications	0	D1
		1–3	D2
		≥ 4	D3
5	Number of publications	0–3	E1
		4–10	E2
		11–26	E3
		≥ 27	E4
6	Student supervision	0	F1
		1–34	F2
		≥ 35	F3
7	Actual post	A	G1
		B	G2
		C	G3
		D	G4

is an important parameter in Association Rules because the value will control the number and form of rule that will be generated [13].

While the highest confidence level shows best rules generated from the dataset. Conversely, if the rules generated were less than 30% confidence level, they were assumed as a weak and not a significant rule [15, 16]. However, if the experiment is set at a high confidence level, the rule generated will be less even though the rule produced is stronger. With a view to the needs of the study, the confidence is decreased to see the relevance of the results from a low confidence level.

The relationship between the "publication and research", "publication and supervision" and "publication and conferences" was tested to obtain the required rule. The production of talent model is based on the results of experiments on the HRM academic dataset that managed to find a set of Association rules. Based on the production Association rules of factors publication and demographics in this experiment, there are some demographic criteria that are useful but not stated clearly in the guidelines of promotion such as gender, race, level of education, level of positions, current status, number of years of service, and years of service before being promoted to an academic position. The HRM should be provided with adequate information for promotion academics to make sure it is at the proper level. Through this development talent model using the Apriori algorithm, HRM can make more accurate and effective result.

The conducted experiment involves the relationship between five main criteria for the publication talents factors which are demographic, publication, supervision, research, and conferences through four most important relationships of association which is "publication and demographic", "publication and research", "publication and supervision," and the last one is "publication and conferences." The strong relationship and interestingness among the selected attribute serve as the set of rules for the representation of knowledge based on experiment. Table 5a–d shows the sample of rules extracted for each relationship. The rules were sorted by the highest minimum confidence. Table 5a shows rules for Publication and Demographic, Table 5b shows rules for Publication and Supervision, Table 5c shows rules for Publication and Conference, and Table 5d shows rules for Publication and Research.

Figure 1 shows the production of talent model based on the extracted result of Association Rules using Apriori algorithm. The development of the model is based on the results of experimental relationships between "publication with research", "publication with supervision", "publication with conference" and "publication with demographics" talent attributes. The relevance association generated through this experiment are the basis of academic rank promotion selection criteria by generating the relevancy relationship frequency between the selected attributes. Based on the model below, all rules of publication association results were divided into four categories academician such as lecturer, senior lecturer, associate professor, and professor which are based on what components academic to be analyzed.

In overall, the results of experiments that have been conducted show that the relevance of all attributes is patterned with a uniform standard of academic positions. This can be proved by the production of interesting rules over frequency relationship occurs in academic transactions set item. However, this experiment found less interestingness Association Rules to observe the relationship between five main criteria because the dataset record containing almost 50% value zero (0). Only a few number of rules related were produced at minimum support 0.1 and 0.05.

4 Conclusion

Through this study, the production model of academic talent for publication talent factor by generating a set of Association Rules after undergoing several experiments showed the application of Association Rules by Apriori algorithm approach managed to generate a strong, interesting, and valid rules with the best accuracy. The data mining technique for Association Rules by Apriori algorithm has some distinct

Table 5 **a** Sample rules for publication and demographic. **b** Sample rules for publication and supervision. **c** Sample rules for publication and conference. **d** Sample rules for publication and research

No.	Rules	Min supp	Freq	Min conf
(a)				
1	IF other publication = 0 ACTUALPOST = A THEN year promoted to be B = 0 AND year promoted to be C = 0 AND year promoted to be D = 0	1	97	0.2
2	IF other publication = 0 OR published journal = 0–1 ACTUALPOST = A THEN races = M AND year promoted to be B = 0 AND year promoted to be C = 0 AND year promoted to be D = 0	0.89	163	0.15
3	IF education = Master OR other publication = 0 OR published journal = 0–1 races = M ACTUALPOST = A THEN year promoted to be B = 0 AND year promoted to be C = 0 AND year promoted to be D = 0	1	318	0.15
4	IF year promoted to be C = 0 AND other publication = 0 ACTUALPOST = B THEN year promoted to be D = 0	0.88	148	0.15
5	IF ACTUALPOST = A THEN races = M AND education = Master AND year promoted to be B = 0 AND year promoted to be C = 0 AND year promoted to be D = 0 AND published journal = 0–1 AND other publication = 0	0.42	1563	0.1
(b)				
1	IF total publication = 0–3 AND total supervision = 0 ACTUALPOST = A THEN published journal = 0–1 OR other publication = 0 OR published prosiding = 0–1	0.9	76	0.1
2	IF other publication = 0 AND total publication = 0–3 AND total supervision = 0 ACTUALPOST = A THEN published journal = 0–1	0.9	76	0.1
3	IF published book = 0 OR total publication = 0–3 OR published prosiding = 0–1 ACTUALPOST = A THEN total supervision = 0	0.8	155	0.1
4	IF published prosiding = 0–1 OR published journal = 0–1 OR other publication = 0 AND published book = 0 ACTUALPOST = A THEN total supervision = 0	0.8	155	0.1
5	IF published prosiding = 0–1 AND total publication = 0–3 ACTUALPOST = A THEN total supervision = 0	0.8	155	0.1

(continued)

Table 5 (continued)

No.	Rules	Min supp	Freq	Min conf
(c)				
1	IF published journal = 0–1 OR other publication = 0 OR published book = 0 ACTUALPOST = A THEN conference 5 = 0	0.9	28	0.15
2	IF other publication = 0 ACTUALPOST = A THEN conference 2 = 0	0.78	22	0.15
3	IF international conference = 0–2 OR university conference = 0 ACTUALPOST = A THEN other publication = 0	0.79	302	0.1
4	IF conference 1 = 0–3 ACTUALPOST = A THEN published journal = 0–1	0.79	302	0.1
5	IF published prosiding = 0–1 AND total publication = 0–3 ACTUALPOST = A THEN conference 5 = 0	0.9	257	0.1
(d)				
1	IF published book = 0 OR published journal = 0–1 OR other publication = 0 ACTUALPOST = A THEN G2006 = 0–74	0.91	154	0.15
2	IF total publication = 0-3 AND head of research = 0 ACTUALPOST = A THEN published journal = 0–1 OR other publication = 0	0.9	148	0.1
3	IF total publication = 0–3 AND head of research = 0 ACTUALPOST = A THEN published journal = 0–1 OR other publication = 0 OR G2006 = 0–74	0.96	803	0.1
4	IF published prosiding = 0–1 AND head of research = 0 ACTUALPOST = A THEN published journal = 0–1	0.9	148	0.1
5	IF published prosiding = 0–1 AND head of research = 0 ACTUALPOST = A THEN published journal = 0–1 OR G2006 = 0–74 OR G2007 = 0–107	0.9	1517	0.1

advantages in the production of interesting rule for not limiting the results that want to be found. In HRM, the results presented do not reflect accurately the real situation because the data collected is not complete. The use of computerized data that is complete after 2014 while the data used in this study is before 2010. However, this study managed to show a form of knowledge and design patterns to evaluate academic publications talent that led to the production model of academic talent.

The acquired knowledge representation model can be a source of reference for the improvement of management talent, especially on academics. It simultaneously addresses and resolves issues related to talent management in local universities, especially in identifying the factors and necessary talent criteria.

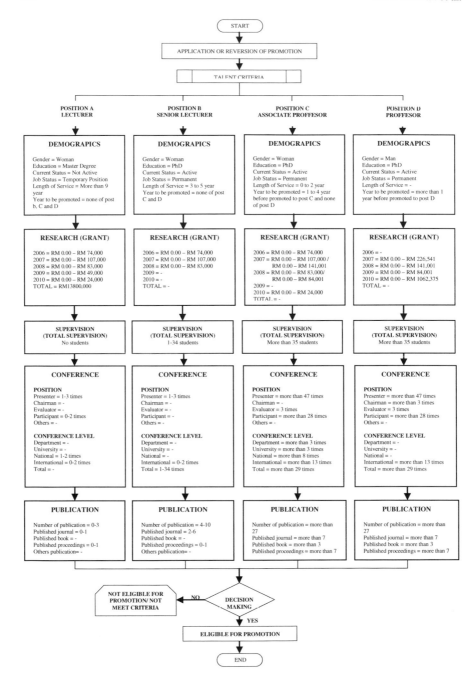

Fig. 1 Talent model based on publication

Acknowledgements The research was funded by Ministry of Higher Education research grant (Grant No. FRGS/1/2016/ICT02/UKM/02/8) and National University of Malaysia for supporting the students to do research.

References

1. Tafti MM, Mahmoudsalehi M, Amiri M (2017) Critical success factors, challenges and obstacles in talent management. Ind Commercial Training 49(1):15–21
2. Buller PF, McEvoy GM (2016) A model for implementing a sustainability strategy through HRM practices. Bus Soc Rev 121(4):465–495
3. Jantan H, Hamdan AR, Othman ZA (2010) Classification and prediction of academic talent using data mining techniques. In: International conference on knowledge-based and intelligent information and engineering systems, 2010, pp 491–500
4. Gallardo-Gallardo E, Thunnissen M, Scullion H (2017) A contextualized approach to talent management: advancing the field. Spec Issue Int J Hum Resour Manage, 1–4
5. Strohmeier S, Piazza F (2013) Domain driven data mining in human resource management: a review of current research. Expert Syst Appl 40(7):2410–2420
6. Jantan H, Hamdan AR, Othman ZA (2010) Human talent prediction in HRM using C4.5 classification algorithm. Int J Comput Sci Eng 2(8):2526–2534
7. Mishra SN, Lama DR (2016) A decision making model for human resource management in organizations using data mining and predictive analytics. Int J Comput Sci Inf Secur 14(5):217
8. Danping Z, Jin D (2011) The data mining of the human resources data warehouse in university based on association rule. J Comput 6(1):139–146
9. Jindal, R, Borah MD (2013) A survey on educational data mining and research trends. Int J Database Manag Syst 5(3):5
10. Kadappa V, Guggari S, Negi A (2016) Teacher recruitment data analytics using association rule mining in Indian context. In: International conference in data science and engineering (ICDSE), pp 1–7
11. Akhondzadeh-Noughabi E, Amin-Naseri MR, Albadvi A, Saeedi M (2016) Human resource performance evaluation from CRM perspective: a two-step association rule analysis. Int J Bus Perform Manag 17(1):89–102
12. Bhise S, Kale S (2017) Efficient algorithms to find frequent itemset using data mining. Int Res J Eng Technol (IRJET) 4(6)
13. Man M, Bakar WAWA, Abdullah Z, Jalil MA, Herawan T (2016) Mining association rules: a case study on benchmark dense data. Indonesian J Electr Eng Comput Sci 3(3):546–553
14. Agrawal R, Srikant R (1994) Fast algorithms for mining association rules. In: Proceeding 20th international conference on very large data bases (VLDB) 1215:487–499
15. Aziz AA, Ismail NH, Ahmad F (2013) Mining students' academic performance. J Theor Appl Inf Technol 53(3)
16. Kaur J, Madan N (2015) Association rule mining: a survey. Int J Hybrid Inf Technol 8(7):239–242
17. Gupta S, Mamtora R (2014) A survey on association rule mining in market basket analysis. Int J Inf Comput Technol 4(4):974–2239
18. Tomar D, Agarwal S (2013) A survey on data mining approaches for healthcare. Int J Bio-Sci Bio-Technol 5(5):241–266
19. Gandhi N, Armstrong LJ (2016) A review of the application of data mining techniques for decision making in agriculture. In: International conference in contemporary computing and informatics (IC3I), pp 1–6
20. Jantan H, Sa'dan SA, Baskaran AMF (2016) Artificial immune clonal selection based algorithm in academic talent selection. J Inf Math Sci 8(4):225–234

21. Othman ZA, Taghavi M, Aminudin N, Jantan H (2013) An R&D project leadership selection model for academician. Int J Digit Content Technol Appl 7(16):1
22. Aher SB, Lobo LMRJ (2013) Combination of machine learning algorithms for recommendation of courses in E-learning system based on historical data. Knowl-Based Syst 51:1–14
23. Othman ZA, Ismail N, Hamdan AR, Sammour MA (2016) Klang vally rainfall forecasting model using time series data mining technique. J Theor Appl Inf Technol 92:371–379

Electricity Price Forecasting Using Neural Network with Parameter Selection

Nik Nur Atira Nik Ibrahim, Intan Azmira Wan Abdul Razak,
Siti Syakirah Mohd Sidin and Zul Hasrizal Bohari

Abstract Price forecasting acts as an essential position in the current energy industry as to assist the independent generators in putting on a remarkable bidding system and scheming contracts, and helps with the selection of supply on the advance generation facility in the long term. These electricity prices are usually hard to predict as it always depends on the uncertainty factors which results in severe volatility or even spikes of price in the energy market. Therefore, determining the accuracy of electricity price forecasting had become an even more important task as there are often remains some crucial prices volatility in the electric power market. This approach focuses on the parameter selection (hidden neuron, learning rate, and momentum rate) and the selection of input data for three types of model. By using the appropriate parameters and inputs, the accuracy of the prediction can be improved. This approach is expected to provide market participants a better bidding strategy and will be used to boost profits in the energy markets using the artificial neural network.

Keywords Electricity price forecasting · Input feature · Neural network · Short-term forecast

N. N. A. N. Ibrahim (✉) · S. S. M. Sidin
Faculty of Electrical Engineering, Universiti Teknikal Malaysia Melaka (UTeM), Malacca,
Malaysia
e-mail: nikatira@yahoo.com.my

S. S. M. Sidin
e-mail: sitisyakirahsidin@gmail.com

I. A. W. A. Razak
Industrial Power Department of Faculty of Electrical Engineering/Center for Robotics and
Industrial Automation (CeRIA), Universiti Teknikal Malaysia Melaka (UTeM), Malacca,
Malaysia
e-mail: intan.azmira@utem.edu.my

Z. H. Bohari
Faculty of Engineering Technology, Universiti Teknikal Malaysia Melaka (UTeM), Malacca,
Malaysia
e-mail: zulhasrizal@utem.edu.my

© Springer Nature Singapore Pte Ltd. 2019
V. Piuri et al. (eds.), *Intelligent and Interactive Computing*, Lecture Notes in Networks
and Systems 67, https://doi.org/10.1007/978-981-13-6031-2_33

1 Introduction

The regulated electricity price market is traditionally practiced in the United States as it is not until 1992; the deregulated electricity market had been introduced along with the Energy Policy Act, which the restriction on the price that charged for general electricity had removed. Nowadays, the electricity market is shifting forward to a competitive structure and it is changing the conventional monopolistic market. The main reason of this revolution was the fact that the competition could bring the outcomes in improving the utilization of resources which leads to supply the consumer with a cheaper but yet more reliable energy supply [1].

In global, the deregulated electricity area had contributed to the expansion of a competing market organization where the contributors (producer, retailer, and trader) compete for the market share. However, electricity is a completely unique merchandise since it is not capable of being stored, and needs a consistent balance in the production and utilization [2]. Consequently, price forecasting acts as an essential position in the current energy industry as to assist the independent generators in putting on a remarkable bidding system and scheming contracts and help with the selection of investing on advanced generation facility in the long term [3].

2 Electricity Price Forecasting

Electricity price forecasting has gained many researchers interest as the expansion of the deregulated electricity market as it involved with competitive market between the participants; which including the generation and retail companies, the transmission network providers, and the market managers [4].

Unlike the power demand, the electricity price much complicated as it is influenced by other variables such as fuels, power plants, weathers. This will cause an existing of huge volatility and sudden spikes in the series of electricity price. These aspects make the price forecasting become much harder than the load forecasting.

The electricity price forecasting can be identified in a few classes; which are short term, medium term, and long term. The short-term electricity price forecasting has usually existed in the day-ahead market. In the day-ahead market, the hourly electricity blocks for the next day are traded where the participants will offer or bids the quantity or the price that will be sell or purchase [5].

The electricity price forecasting can be divided into two classes; which are hard computing and soft computing. The hard computing usually needs a huge data set, the accurate modeling of the system, and can cause a large computational strain [6]. The example of the hard computing can be General Autoregressive Conditional Heteroscedasticity (GARCH) [7], Autoregressive Integrated Moving Average (ARIMA) [6], and Wavelet Transform (WT) with ARIMA and GARCH.

Meanwhile, the soft computing adopts the learning process from historical data in order to predict the patterns such as Feedforward Neural Network [6], Recurrent Neural Network (RNN) [7], and Cascaded Neural Network (CNN) [8].

Recently, different kinds of forecasting approach primarily based on the artificial intelligence approach notably using the neural networks were proposed by the researchers [9, 10]. The artificial neuron process is motivated by models of the neural which recognized the pattern and then used the pattern to utilize and affect the formation of huge parallel networks and coaching those networks to solve specific problems.

3 Fundamentals of Neural Network

3.1 Neural Network

The ANN is an electronic clone based on the structure of neural in the brain. The artificial neuron process is motivated by models of the neural which recognized the pattern, and then used the pattern to utilize and affect the formation of huge parallel networks, and coaching those networks to solve specific problems [11, 12].

The ANN has the ability for adaption, where the memory corresponds to the activation maps of the neuron giving resistance to the noise. It capable starts the network with noisy data then recalls the right data. The ANN also is capable to derive from the complex information. It learns the rules by training from a set of sample and the apparently closed to the human derivation [13].

The ANN is essentially nonlinear circuits its output in linear or nonlinear mathematical functions its inputs. The input of the data can be the outputs or inputs of the other networks component. The ANN normally is having three layers. The first layer that is connected to the input variables is the input layer. The third layer that is connected to the output variables is the output layer. The hidden layer is in the between of input and output layer. The number of neuron in the input and output layer are the same as the number of input and output variable. The number of neuron in the hidden layer wills the one that will be varied.

The ANN models do not require ahead inference and the processing of information rely only on the input and output data [10]. The ANNs are not complex, but compelling and good methods for forecasting with the help of there are sufficient information for training, a suitable choice of the input and output data, a proper value of hidden neurons [14].

4 Experimental Results

4.1 Training Period

The electricity price forecasting will be started by selecting the suitable training period. The test market will be the Ontario electricity market, which the data is provided by the IESO data directory [15]. The historical data of actual demand and HOEP for the past 49 days will be used as the training period for a fair comparison with other previous works [16, 17].

4.2 Selection of Input

The suitable relevant input is determined by using the correlation relationship between the input and output data. The input will be chosen based on the highest correlation factor from the HOEP and actual demands. For the first and second models (NN1 and NN2), the input data are only considered the actual demand, while the third model (NN3) will consider both HOEP and actual demand as the input data.

From the correlation analysis, the NN1 and NN2 models will have the demand for $d-1$, $d-2$, and $d-7$ as the input for the neural network while the NN3 will have the demand $d-1$, $d-2$, and $d-7$ and HOEP $d-1$ as the input. The data are then normalized between $[-1, 1]$ to ensure that data is roughly uniformly distributed between the network inputs and the outputs [18].

$$x_n = \frac{x_j - \left[\frac{x_{max}+x_{min}}{2}\right]}{\left[\frac{x_{max}-x_{min}}{2}\right]} \tag{1}$$

4.3 MAPE Computation

The Mean Absolute Error Percentage (MAPE) of forecasted data will be calculated in order to compare the models with other types of methods. The following is the MAPE equations:

$$\text{MAPE} = \frac{100}{N} \times \sum_{t=1}^{N} \frac{\left|P_{\text{actual}_t} - P_{\text{forecast}_t}\right|}{P_{\text{actual}_t}} \tag{2}$$

where the notation of N is the number of hours, P_{actual_t} is the actual HOEP at hour t and P_{forecast_t} is the forecasted HOEP at hour t.

4.4 Neural Network Training and Testing

The Neural Network configuration for three models (NN1, NN2, and NN3) is using a single hidden layer with the tangent sigmoid activation function while the output layer with the pure linear activation function. In order to test the effect parameter to the neural network models, the parameter such as the hidden neuron, learning rate, and momentum rate is varied. Table 1 shows the parameter that tested on the NN1, NN2, and NN3.

For the first model, the learning rate and the momentum rate are fixed while the hidden neuron is varied (Table 2). The hidden neuron can cause the over-fitting or under-fitting. If there are too many neurons in the hidden layer it might cause the over-fitting while if the neuron is less as compared to the complexity of the data it might cause the under-fitting [19]. The average MAPE for the NN1 model is 24.46%. Next, the hidden neuron, learning rate, and momentum rate are varied for the NN2 model (Table 3).

The learning rate is the part of the error gradient, where the weights need to be modified. Bigger the learning rate can accelerate the convergence process; however, it may generate an oscillation across the minimal. The momentum rate resolves the fraction of the change in previous weights that need to be applied to calculate the new weights [1]. The average MAPE for the NN2 is 16.79%. Lastly, the NN3 model has varied all the three parameters, but with addition HOEP as the input data (Table 4). The average MAPE for the NN3 model is 15.70%.

Table 1 The parameter tested on the models

Model	Hidden neuron	Learning rate	Momentum rate
1	Varied	Fixed	Fixed
2	Varied	Varied	Varied
3	Varied	Varied	Varied

Table 2 The NN1 model with varied hidden neurons

Test week	Hidden neuron	MAPE (%)
1	3	19.30
2	3	18.18
3	3	21.31
4	9	30.79
5	4	22.94
6	6	34.36
Average		24.46

Table 3 The NN2 model with varied momentum rate and learning rate

Test week	Hidden neuron	Learning rate	Momentum rate	MAPE (%)
1	1	0.20	0.53	16.02
2	1	0.69	0.78	17.10
3	1	0.67	0.61	13.58
4	1	0.50	0.43	16.81
5	1	0.15	0.18	17.94
6	1	0.66	0.32	19.29
Average				16.79

Table 4 The hidden neuron, learning rate, and momentum rate for NN3 model

Test week	Hidden neuron	Learning rate	Momentum rate	MAPE (%)
1	1	0.98	0.29	15.04
2	1	0.65	0.49	16.65
3	1	0.8	0.89	14.34
4	1	1.0	0.1	15.25
5	1	0.65	0.33	15.83
6	1	0.95	0.73	17.06
Average				15.70

Table 5 The comparative MAPE for the proposed method with existing methods

Reference(s)	Method(s)	MAPE (%)
Proposed method	NN1	24.45
	NN2	16.79
	NN3	15.70
[20]	Heuristic	24.89
	IESO	20.21
	MLR	18.17
	NN	18.39
	Wavelet + NN	17.51

5 Comparative MAPE

The NN2 and NN3 have better MAPE compared to the NN1 model as it varied all the parameter. However, the NN3 has much accurate MAPE compared to the NN2 as the NN3 model has an additional input data. The NN2 and NN3 method have better MAPE compared to the author Aggarwal [20] which are 16.79 and 15.70%, respectively. By simply varied the parameter and input of the neural network, the effectiveness of the methods can be improved with the stand-alone method itself (Table 5; Fig. 1).

Fig. 1 Actual versus prediction of HOEP for week 1

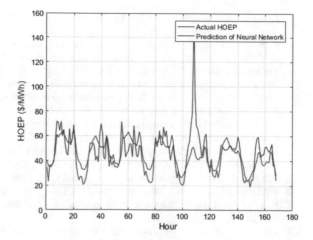

6   Conclusion

This paper proposed the artificial neural network in electricity price forecasting by varying the parameter and input data. By varying the parameter, the effectiveness of the neural network can be improved in addition to selecting the most correlated coefficient between the input and the target. For further improvement, other parameters such training algorithm can be varied to enhance the effectiveness of the neural network model.

Acknowledgements This study is supported in part by the Fundamental Research Grant Scheme (FRGS) provided by the Ministry of Higher Education Malaysia (FRGS/1/2017/TK04/FKE-CERIA/F00331). We also would like to dedicate our appreciation to Universiti Teknikal Malaysia Melaka (UTeM) for providing technical and moral support throughout conducting this study.

References

1. Catalao JPS, Mariano SJPS, Mendes VMF, Ferreira LAFM (2007) Short-term electricity prices forecasting in a competitive market: a neural network approach. Electr Power Syst Res 77(10):1297–1304
2. Weron R (2014) Electricity price forecasting: a review of the state-of-the-art with a look into the future. Int J Forecast 30(4):1030–1081
3. Voronin S, Partanen J, Kauranne T (2014) A hybrid electricity price forecasting model for the Nordic electricity spot market. Int Trans Electr Energy Syst 24(5):736–760
4. Szkuta TSDBR, Sanabria LA (2011) Electricity price forecasting using artificial neural networks. Int J Electr Power Energy Syst 33(3):550–555
5. Shayeghi H, Ghasemi A (2013) Day-ahead electricity prices forecasting by a modified CGSA technique and hybrid WT in LSSVM based scheme. Energy Convers Manag 74:482–491
6. Singh N, Mohanty SR (2015) A review of price forecasting problem and techniques in deregulated electricity markets. J Power Energy Eng 3:1–19

 7. Contreras J, Espinola R, Nogales FJ, Conejo AJ (2002) ARIMA models to predict next-day electricity prices. IEEE Power Eng Rev 22(9):57–57
 8. Garcia RC, Contreras J, Van Akkeren M, Batista J, Garcia C (2005) A GARCH forecasting model to predict day—ahead electricity prices. IEEE Trans Power Syst 20(2):867–874
 9. Osorio GJ, Matias JCO, Catalao JPS (2014) Electricity prices forecasting by a hybrid evolutionary-adaptive methodology. Energy Convers Manag 80:363–373
10. Sandhu HS, Fang L, Guan L (2016) Forecasting day-ahead price spikes for the Ontario electricity market. Electr Power Syst Res 141:450–459
11. Mohan A (2013) Mid term electrical load forecasting for State of Himachal Pradesh using different weather conditions via ANN Model 1(2):60–63
12. Hsu K, Gupta HV, Sorooshian S (1995) Artificial neural network modeling of the rainfall-runoff process that arise and based background and scope 31(10):2517–2530
13. Chaturvedi (2008) Artificial neural network and supervised learning. Springer, Berlin
14. Pousinho HMI, Mendes VMF, Catalao JPS (2012) Short-term electricity prices forecasting in a competitive market by a hybrid PSO-ANFIS approach. Int J Electr Power Energy Syst 39(1):29–35
15. IESO data directory. (Online). Available: http://www.ieso.ca/en/power-data/data-directory. Accessed 20 Jan 2018
16. Shrivastava NA, Panigrahi BK (2014) A hybrid wavelet-ELM based short term price forecasting for electricity markets. Int J Electr Power Energy Syst 55:41–50
17. Sharma V, Srinivasan D (2013) A hybrid intelligent model based on recurrent neural networks and excitable dynamics for price prediction in deregulated electricity market. Eng Appl Artif Intell 26(5–6):1562–1574
18. Rotich N (2012) Forecasting of wind speeds and directions with artificial neural networks, pp 1–59
19. Panchal FS, Panchal M (2014) Review on methods of selecting number of hidden nodes in artificial neural network. Int J Comput Sci Mob Comput 311(11):455–464
20. Aggarwal SKS, Saini LML, Kumar A (2008) Electricity price forecasting in Ontario electricity market using wavelet transform in artificial neural network based model. Int J Control Autom Syst 6(5):639–650

Part II
Applications of the Soft-Computing Techniques

Improved Ozone Pollution Prediction Using Extreme Learning Machine with Tribas Regularization Activation Function

Noraini Ismail and Zulaiha Ali Othman

Abstract Nowadays, increasing ozone (O_3) pollution is becoming a global issue. The increasing of these pollutions has a huge negative impact on human health and also to the ecosystem. In order to reduce the risk of high O_3 pollution, an accurate O_3 forecasting model should be developed, so that a preventive measure can be taken earlier. Therefore, this study proposed an accurate O_3 prediction model using improvement Extreme Learning Machine algorithm based on Regularization Activation Function (RAF-ELM). The experiment conducted by investigating RAF-ELM performance use four types of activation function, i.e., sigmoid, sine, tribas, and hardlim. In this study, RAF-ELM uses single hidden layer feedforward neural networks to predict the air quality index for O_3 pollutant based on meteorological variables (Temperature and Wind Speed) and other pollutants (NO_x, NO, NO_2, CO, PM_{10}, SO_2, CH_4, NMHC, and THC) in Malaysia using O_3 hourly time series data collected at Shah Alam station. It has 107,329 instances recorded from the year 1998 to 2010. The input weight and bias for hidden layers are randomly selected, whereas the best neurons' number of hidden layer is determined from 5 to 20. The number of neurons (11) with regularization (0.8) using tribas activation function showed the best model. The proposed model has obtained better accuracy performance (0.007999 MSE) and better processing time (2.699 s) compared with conventional MPE. It can be concluded that the proposed algorithm can be used as a good prediction technique for time series data.

Keywords Ozone · Extreme learning machine · Prediction · Neural networks · Data mining

N. Ismail (✉) · Z. A. Othman
Faculty of Information Science and Technology, Center for Artificial Intelligence and Technology, Universiti Kebangsaan Malaysia, Bangi, Selangor, Malaysia
e-mail: norainismail14@gmail.com

Z. A. Othman
e-mail: zao@ukm.edu.my

© Springer Nature Singapore Pte Ltd. 2019
V. Piuri et al. (eds.), *Intelligent and Interactive Computing*, Lecture Notes in Networks and Systems 67, https://doi.org/10.1007/978-981-13-6031-2_9

151

1 Introduction

Trioxygen also known as ozone is an inorganic molecule with O_3 chemical formula, which was discovered by Christian Friedrich Schönbein in the year 1840 [1]. It is a pale blue gas with a sharp strong odor. At normal temperatures and pressures, O_3 gas will form at the bottom of the troposphere layer where the location is closer to earth [2]. High concentrations of O_3 can harm both human health and also ecosystems. According to the World Health Organization, people who are exposed to O_3 for a long period of time are at the risk of inflammation and damage of lung cell lining that can cause death to the people with chronic disease [3].

Tropospheric O_3 is one of the greenhouse gases produced by human activity through pollutant emissions such as Nitrogen Oxide ($NO_x = NO + NO_2$), Carbon Monoxide (CO), Non-Methane Organic Compound (NMHC), and Organic Compounds (VOC), which react in the atmosphere with the presence of sunlight [4]. The major anthropogenic source that produces these pollutants is from vehicle smoke as well as industrial factories. O_3 is formed by a simple reaction by oxidation of CO with the presence of NO. While the Hydroxyl (OH), Hydroperoxy (HO_2), NO, and NO_2 chains act as catalysts in this reaction. In addition, organic compounds such as NMHC may also combine with the reaction chain when the carbonyl or ketone species are formed beside O_3 [5].

According to [6], industrial and densely populated areas have led to increased rates of O_3 due to heavy traffic. This is because vehicles and factories emit pollutants that help in shaping O_3 molecule. Especially if the area is hot and sunny throughout the year. Vehicles such as cars and light trucks using petrol are main sources of VOC, CO, and NO_x [7]. VOCs are produced from the exhaust by engine combustion, while emissions of NO_x and CO are produced during the combustion process and this only occurs in vehicles exhaust.

In addition, O_3 formation also depends on weather conditions and climate change. Variations in weather conditions play an important role in determining O_3 concentration. According to [5], the formation of O_3 is influenced by several meteorological factors such as temperature, humidity, wind speed, and wind direction. The correlation between O_3 and the meteorological parameters varies from region and also season. But the temperature is very closely related to the formation of O_3. According to [8], O_3 is more easily formed during hot and sunny days when the air stays because heat and sunlight provide energy for the formation of O_3.

Other than that, wind direction and speed also have a great impact on the formation of O_3 [2]. This is because the circulation of the wind can change the direction and accumulation of pollutant gases away from the pollution area and form O_3 in the new area. While the wind speed determines the conditions for the formation of O_3. For example, low wind speeds and high temperatures are excellent conditions for O_3 formation.

The increasing O_3 pollution has a huge impact on human health and ecosystem balance [9]. One of the methods that can be used in this problem is developing a more efficient forecasting model, so that a control measure can be structured in the future.

In order to predict O_3 concentration, the approach of machine learning techniques used is very important in providing accurate results [10]. There are various methods that can be applied, one of that is data mining method using Artificial Neural Network (ANN) techniques.

In forecasting, ANN serves as a supervised machine learning technique to predict future events, based on past data that have been analyzed using certain methods [11]. While the Extreme Learning Machine (ELM) algorithm is the enhancement of ANN techniques. This algorithm is an advanced learning method with the benefit of matrix theory in mathematical knowledge that use feedforward hidden layers as a learning process [12]. This algorithm has relatively fast learning techniques and capable of providing very small error [13]. This is because the algorithm trains the model by using input loads and bias weights selected at random. However, this technique has the potential to generate overfitting models and leads to less stablity when it comes to a certain condition.

2 Materials and Method

This study proposed an O_3 prediction model using ELM to predict air quality index of O_3 pollutant for the next hour based on meteorological variables (Temperature and Wind Speed) and other pollutants (NO_x, NO, NO_2, CO, PM_{10}, SO_2, CH_4, NMHC, and THC) in Malaysia using sample data from Shah Alam station. The background of the study area, data collection and preparation, and algorithm used are explained below.

2.1 Study Area

Shah Alam is the capital of the State of Selangor Darul Ehsan, Malaysia. It was inaugurated as a city in the year 2000. It is located in the highly populated residential areas of Kuala Lumpur's which is near to a recreational park, golf course, lake as well as industrial areas. It is also known as 'Industrial Town' because of the increasing number of multinational factories. This condition accelerates the rapid growth and increases the population in these areas. With the rapid development, the development of transport systems and the crowded population, several environmental pollution issues have arisen. One of them is the increase in O_3 concentration. Figure 1 shows the location of the sampling station selected for this study.

2.2 Data Collection and Preparation

Table 1 consists of the precursors, number of years, duration, missing data and total instances available for every selected variable. Overall, there is no low percentage of missing values for the selected dataset. The data were selected based on the

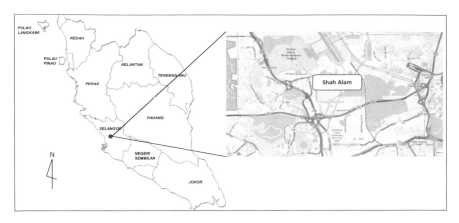

Fig. 1 Geographical map of the sampling station

Table 1 Period of data collection and percentage of missing values for each precursor

Attribute	Duration	Years	Instances	Missing data (%)	Total instances
CO	1997–2012	15	122,595	9.78	135,888
O_3	1997–2012	15	124,594	8.31	135,888
PM_{10}	1997–2012	15	130,026	4.31	135,884
SO_2	1997–2012	15	122,582	9.79	135,888
NO_x	1997–2012	15	123,382	9.20	135,888
NO	1997–2012	15	123,376	9.21	135,888
NO_2	1997–2012	15	123,375	9.21	135,888
CH_4	1997–2010	13	99,855	11.21	112,464
NmHC	1997–2010	13	99,861	11.21	112,464
THC	1997–2010	13	99,855	11.21	112,464
Temperature	1997–2012	15	126,625	4.76	132,960
Wind speed	1997–2012	15	133,974	2.48	137,376

completeness of the data and the only station that recorded NMHC concentration. The collected data includes the hourly average of O_3 concentrations, Oxide of Nitrogen (NO, NO_2, and NO_x), Carbon monoxides (CO) and NMHCs to represent the total composition of VOCs, Tetrahydrocannabinol (THC), methane (CH_4), sulfur dioxide (SO_2), particulate matter (PM_{10}), temperature, and wind speed as shown in Table 1. All parameter readings were recorded in ppm units. The collected data is an actual air quality data recorded at Shah Alam station. The data consist of 107,329 sets of records from February 1, 1998 to April 30, 2010 (12 years).

The collected data was prepared to turn the data set into a suitable form for further processing. This phase can improve the quality of data, thus increasing the algorithm performance. In this phase, the data was cleaned and transformed. This is because the collected data is not consistent and consists of empty tuples. Therefore, the missing values were filled with the nearer values in the neighbor which is the closest column in Euclidean distance. Even though ELM is powerful enough to deal with missing values, however, this method can help to improve the algorithm's learning process. On the other hand, the data was adjusted to the same date to deal with inconsistency and then was transformed into a suitable representation.

2.3 ELM Algorithm

ELM is a machine learning algorithm in ANN with a single hidden layer used for classification and also regression problem. This method was introduced by Huang [14] to overcome ANN's weaknesses. Particularly in the learning speed issues. This is because ELM does not train the input weight or bias like ANN. But randomly assigned and set it without iterative tuning to allow ELM to have a fast learning speed and generate a good generalization of performance (usually 10 times or better) [13]. Here is an ELM model for a different number of N samples (x_i, t_i), where $x_i = [x_{i1}, x_{i2}, \ldots, x_{in}]^T \in R^n$ and $t_i = [t_{i1}, t_{i2}, \ldots, t_{im}]^T \in R^m$, $(x_i, t_i) \in R^n \times R^m$ $(i = 1, 2, \ldots, N)$. The standard single-layer feedforward neural network with the sum of the hidden nodes of N and the activation function of $f(x)$ can be summarized as follows:

$$\sum_{i=1}^{\tilde{N}} \beta_i f_i(x_j) = \sum_{i=1}^{\tilde{N}} \beta_i f(w_i k \cdot x_j + b_i) = t_j, \; j = t_j, \; j = 1, \ldots, N \quad (1)$$

$$H \cdot \beta = T \quad (2)$$

where $w_i = [w_{i1}, w_{i2}, \ldots, w_{in}]^T$ is a weight vector that is connected with hidden node ith, input node, and threshold hidden node b^i. While $\beta_i = [\beta_{i1}, \beta_{i2}, \ldots, \beta_{im}]^T$ is a weight vector that is connected with hidden node ith and output node. Finally, $w_i \cdot x_j$ will represent the inner product w_i and x_j. Formula 1 can be summarized as Formula 2. In Formula 2, H is denoted as the output of hidden neural network matrix $(w_i \cdot x_j + b^i)$. Where ith column for H is a hidden output layer i according to $x_1, x_2, \ldots, x_N$. In ELM, input weight and bias are randomly assigned. While β is an output weight matrix and T is an output matrix [14].

$$H = \begin{bmatrix} f(a_1 \cdot x_1 + b_1) & \ldots & f(a_{\tilde{N}} \cdot x_1 + b_{\tilde{N}}) \\ \vdots & \cdots & \vdots \\ f(a_1 \cdot x_N + b_1) & \cdots & f(a_{\tilde{N}} \cdot x_N + b_{\tilde{N}}) \end{bmatrix}_{N \times \tilde{N}},$$

$$\beta = \begin{bmatrix} \beta_1^T \\ \vdots \\ \beta_{\tilde{N}}^T \end{bmatrix}_{\tilde{N} \times m}, \quad T = \begin{bmatrix} t_1^T \\ \vdots \\ t_N^T \end{bmatrix}_{N \times m} \quad (3)$$

ELM algorithm is very well known for its good generalization performance with extremely fast learning speed. However, ELM has several weaknesses as it tends to generate a large scale and overfitting models because it is implemented under the empirical risk minimization scheme. In addition, ELM also leads to a less stable in estimation when it comes to outliers. The improvements of conventional ELM have been done in this study by proposed a regularization factor into an activation function of ELM. The applied regularization is based on the principle of structural risk minimization according to statistical learning theory.

2.4   RAF-ELM Algorithm

In general, regularization is a technique used for an objective function to solve the optimization problem [15]. While in machine learning, regularization is known as a process of introducing additional information to solve the overfitting problem. This modification was inspired by kernel ELM to regulate the kernel function, which means the regularization was placed in a hidden layer to rank the neurons or placed in hidden weight known as L_1 and L_2 norms [16, 17]. However, we proposed the regularization to the activation function, known as RAF-ELM to make the activation function more balanced and resilient to random nonuniform distribution, thus can improve ELM performance. A formula of RAF-ELM with an added regularization component $R(f)$ to the activation function $f(x)$ can be written as follows:

$$\sum_{i=1}^{\tilde{N}} \beta_i \left[f_i(x_j) + R(f_i) \right] = T$$

$$R(f) = \lambda \sum_{k=1}^{N} \theta_k^2 \tag{4}$$

λ is a regularization coefficient parameter that is used to control the importance of regularization term as penalties on the complexity of activation function $f(x)$. It can be adjusted to help the algorithm find a good fit for a model. However, if the value of $R(f)$ is too low, the function might not do anything and if the value is too high, it may cause an underfit model and lose valuable information. The goal of this learning problem is to find the appropriate prediction results (labels) that minimize errors on all possible inputs and labels.

Normally, only a subset of input data and labels are available in learning problem and measured with some noise. Therefore, the estimated error is unmeasurable, and the best substitute available is the empirical error over the N available samples. Without limits on the complexity of the function space available, a model will be learned that suffers zero loss on the substitute empirical error. If measurements were made with noise, this model may suffer from overfitting and display poor expected error. Therefore, this function is introduced to give a penalty for exploring certain regions of the function space used to build the model, which can improve generalization.

2.5 Activation Function

On the other hand, activation function is a very important component in neural networks used for complex learning and mapping processes. The main purpose of the activation function is to change the input node to the output signal by specifying the output of the neural network such as "Yes" or "No". Then generated a value within a range of 0–1 or −1–1 and others (depending on the function). The activation function is essentially divided into two types which is linear and nonlinear.

Nonlinear activation functions are the most widely used functions. The absence of linear lines assists the function in adapting to multiple parameters and able to differentiate the output. The nonlinear activation functions are largely divided by their range or curve. There are four most popular nonlinear activation functions that have a good performance by calculating hidden matrix output neurons with activation function which is sigmoid [18, 19], sin [20], hardlim [21], and tribas [22]. The formula for each activation function is stated as follows:

Sigmoid Function:

$$f(x) = \frac{1}{\left(1 + \exp^{-H_{\text{init}}}\right)} \tag{5}$$

Sin Function:

$$f(x) = \sin(H_{\text{init}}) \tag{6}$$

Hardlim Function:

$$f(x) = \left\{ \begin{array}{ll} 1, & \text{jika } H_{\text{init}} \geq 0 \\ 0, & \text{jika tidak} \end{array} \right\} \tag{7}$$

Tribas Function:

$$f(x) = \left\{ \begin{array}{ll} 1 - \text{abs } H_{\text{init}}, & -1 \leq \text{jika } H_{\text{init}} \leq 1 \\ 0, & \text{jika tidak} \end{array} \right\} \tag{8}$$

2.6 Model Development for RAF-ELM Algorithm

In model development, data set separation for training and testing is an important part for evaluating data mining model. There are two competing concerns in splitting the data, which with less training data, the parameter estimates have greater variance. While with less testing data, the performance statistic will have greater variance [23]. However, if the dataset is large, it does not really matter whether to split the data into 80:20 or 90:10 [23]. Therefore, this experiment was carried out using 80% of the total sample data for training. While 20% of the rest is used for testing. The training phase is the process of learning algorithm with input data, while the testing phase is used to evaluate the ability of the model as a data prediction tool.

In this study, input weight and bias for hidden layers are randomly selected. Whereas the neuron of the hidden layer is tested from 5 to 20 to determine the best number of neurons for O_3 prediction model. While the activation function will be set as a sigmoid, sin, tribas, and hardlim. The model performance in this study is measured using mean square error (MSE) and processing time (sec). The proposed algorithm network and its variable number of hidden layer neuron for O_3 prediction were depicted in Fig. 2. While Fig. 3 shows RAF-ELM algorithm pseudocode step by step.

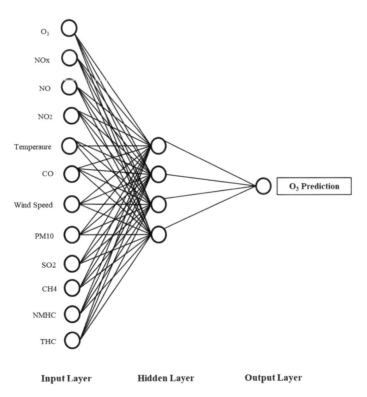

Fig. 2 Proposed algorithm network for the O_3 prediction model

RAF-ELM based regression algorithm-pseudocode

Input:	Samples of training dataset, regularization activation function [0.01-1], number of hidden neurons [5,20] and testing dataset.
Output:	Prediction of input data, T
Step 1)	For each hidden layer neuron, assign input, w_i from [-1,1], bias, b_i from [0,1] where $i = 1,...,N$.
Step 2)	Calculate the hidden output matrix $H_{init} = X.W^T + b$ using activation function $f(x)$ with regularization $R(f)$ added on activation function $f(x)$. While X is an input data.
Step 3)	Calculate the output weight β with value of hidden matrix output H.
Step 4)	Calculate the actual output $T = H.\beta$ of output hidden layer.

Fig. 3 Pseudocode of the RAF-ELM algorithm

3 Result and Discussion

The main objective of this experiment is to propose O_3 prediction model using RAF-ELM techniques. In this experiment, the variables for testing include the type of activation functions, the number of hidden neurons, and the regularization range of activation function. The experiment was set up by setting the number of hidden layer nodes to (5–20), the range of regularization to (0.01–1), the randomization range of the hidden weights to $(-1, 1)$, and the randomization range of threshold to $(0, 1)$. While activation functions are set to sigmoid, sine, hardlim, and tribas, respectively, to test the impact of different activation functions on the performance of ELM model with and without regularization factor.

Each activation function is used to generate a dataset. In each experiment, we select 80% of records for training and the remaining 20% of records for testing. The indicators for ELM performance testing include the Root-Mean-Square Error of Training (Training Error), and the Root-Mean-Square Error of Testing (Testing Error), Training Time, and Testing Time.

3.1 ELM Result for Each Activation Function with and Without Regularization Factor

Table 2 summarizes the optimal values of a hidden layer for each activation function. As shown in the result, we found that the choice of the activation function has a significant impact on the performance of ELM. In our experiments, the ELM models with tribas as activation function can achieve the best performance, while the models with other activation functions always obtain suboptimal results. In other words, the tribas function may perform better nonlinear transformations and obtain more meaningful feature mappings than the other functions for time series data.

Figure 4 depicted the difference between training and testing error and also the learning speed for each activation function. Based on the graft, we can see the big difference between tribas function and other activation functions. While hardlim, sigmoid, and sin functions do not show a big difference among themselves. Based on the past research in [22], the author conducts a similar experiment by comparing activation function and the result shows sigmoid function achieved the best performance in most cases, while tribas function always obtains a suboptimal result. However, the data set used is not explained in the paper. This shows unclear result because the proposed activation function is not well defined for which type of data.

While Table 3 summarizes the optimal values of hidden neurons for each activation function with regularization factor. Based on the result, we found that the regularization factor has a significant impact on the performance of ELM. In this experiment, the RAF-ELM model achieves better performance compared to the unmodified model in the previous experiment (Table 2) for all selected activation function. Also, tribas function performs better than others function in both conventional ELM and

Table 2 Summary values of the ELM optimal result for each activation function without regularization factor

Activation function	Training time	Training error	Testing time	Testing error
Tribas	6.622	0.006927047	0.005	0.006014417
Hardlim	8.019	0.017711792	0.006	0.015411197
Sigmoid	9.884	0.016083977	0.007	0.012582809
Sin	9.4	0.019099306	0.006	0.015276514

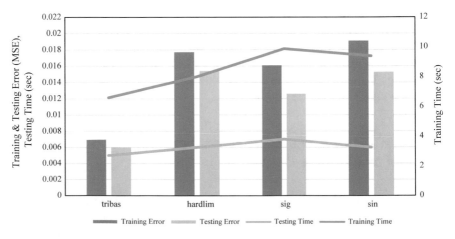

Fig. 4 Summary of the ELM optimal result for each activation function without regularization factor

Table 3 Summary values of the ELM optimal result for each activation function without regularization factor

Activation function	Training time	Training error	Testing time	Testing error
Tribas	5.343	0.006917	0	0.005979349
Hardlim	3.094	0.017758	0	0.015286471
Sigmoid	7.694	0.013791	0.005	0.010892465
Sin	10.203	0.013061	0.006	0.010056609

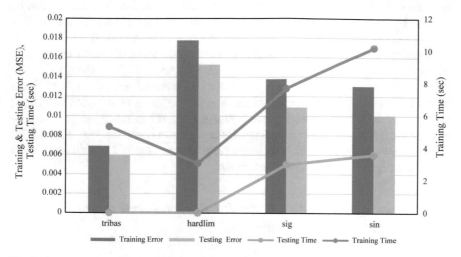

Fig. 5 Summary of the RAF-ELM optimal result for each activation function with regularization factor

RAF-ELM. RAF-ELM for tribas function achieves better performance with 0.00598 testing error and 0.00692 training error compared to tribas function for conventional ELM method with testing error 0.00601 and 0.00693 training error.

In addition, the learning speed for conventional ELM using tribas function shows the fastest speed in training time and testing time (Table 2). While RAF-ELM shows that hardlim function performs faster than tribas in training time with 2.249 s difference, but achieve the same learning speed in testing time with 0 s. However, the overall result shows that tribas function for ELM and RAF-ELM gives the most optimal result in any condition and more stable than any activation function as shown in Fig. 5.

3.2   Training and Testing Error for Each Neuron Based on Each Activation Function

In this phase, we conduct an experiment to see the best neuron for the ELM model with and without regularization factor. According to the review study in [14], the number of neurons has a significant impact on the performance of ELM. A suitable number of neurons should be identified because arbitrarily setting the number of neurons may affect ELM performance. Therefore, an automatic method must be applied by deducing the initial number of neurons using incremental constructive technique [12]. This experiment set the number of neurons to (5–20).

Based on the illustrated line graph in Fig. 6, the graph shows that too small number of neurons give a bad performance for ELM model, while ELM model starts to stablize when the number of neurons is 10 and above. On the other hand, the ELM model with 11 number of neurons achieves the best performance with 0.006917 training error and 0.005979 testing error (refer Table 4). Also, tribas activation function give the best performance for all optimal result for each number of hidden neurons. This result shows that tribas function is always stable in most cases. The next experiment involved a regulator factor against error rate.

3.3   Training and Testing Error for Each Neuron Based on Each Activation Function

Figure 7 illustrated RAF-ELM result based on the regulator values when the neuron is 11 and activation function is tribas. The result shows that the error rate stable

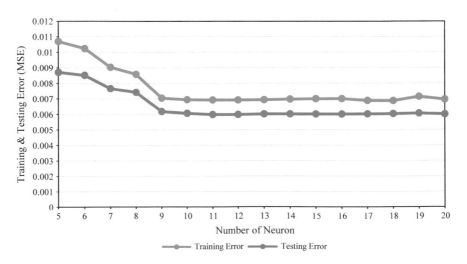

Fig. 6 Summary of the optimal result for each neuron

Table 4 Summary values of the optimal result for each neuron

Neuron	Training time	Training error	Testing time	Testing error	Function
5	2.787	0.010688	0.003	0.008706	Tribas
6	3.063	0.010238	0.015	0.008515	Tribas
7	3.531	0.009033	0	0.007654	Tribas
8	3.985	0.00858	0.016	0.007413	Tribas
9	4.453	0.007042	0	0.006179	Tribas
10	4.891	0.006938	0	0.006059	Tribas
11	5.343	0.006917	0	0.005979	Tribas
12	5.813	0.006917	0	0.005982	Tribas
13	6.266	0.006927	0.015	0.006014	Tribas
14	6.735	0.006966	0	0.006008	Tribas
15	7.172	0.00699	0	0.006008	Tribas
16	7.641	0.006998	0	0.006005	Tribas
17	8.094	0.006876	0.016	0.006013	Tribas
18	8.531	0.006867	0	0.006028	Tribas
19	8.985	0.007138	0	0.006058	Tribas
20	9.469	0.006951	0	0.006002	Tribas

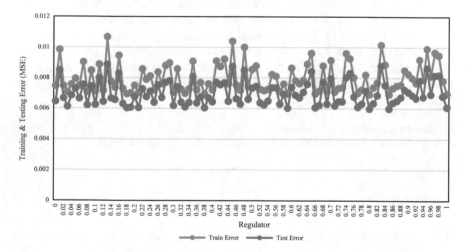

Fig. 7 Summary of the RAF-ELM result based on regulatory values

throughout the experiment where the values lie in the range (0.006914–0.010674) for training phase and (0.005979–0.008869) for the testing phase. While the best model generated when regulator values are 0.8 where the error values are 0.005979 for the testing phase.

Overall, this proposed method shows very promising performance for O_3 prediction. Similar research [24] has been published recently, where the author develops air pollution prediction using conventional ELM for Delhi, India case study area. The dataset used is a daily data from February to March 2016 that consist of only 37 instances. Based on the result, this model performs well, however, the experimental data do not support enough the conclusion of the research study. The parameter setting also was not declared in the paper. While this study used sufficient data and the data was carefully chosen to ensure that the generated model had enough learning from a complete dataset. We also provide complete information of parameter setting used for every experiment to facilitate researcher while referring to this paper.

4 Conclusion

The objective of this study is to propose an O_3 prediction model using RAF-ELM algorithm. This technique works with multivariables for nonlinear regression that can predict O_3 concentration in Malaysia precisely. The performance of RAF-ELM algorithm was compared with conventional ELM. The result shows that RAF-ELM improved the generalization performance and resulting in much faster learning speed for O_3 prediction with 0.005979 error rate and 0 learning speed for the testing phase.

In addition, we did a comprehensive study on the relationship of parameter setting and performance of ELM algorithm based on three testing variables includes the type of activation functions, the number of hidden neurons and the regularization range of activation function. Based on the result, we found that tribas activation function achieved the best performance in most cases. While the best number of hidden neurons is 11 and the best regulator values is 0.8.

Acknowledgements This research was supported by FRGS grant (FRGS/1/2016/ICT02/UKM/02/8), funded by the Ministry of Higher Education.

References

1. Rubin MB (2001) The history of ozone. The Schönbein period, 1839–1868. Bull Hist Chem 26(1):40–56
2. Austin E, Zanobetti A, Coull B, Schwartz J, Gold DR, Koutrakis P (2015) Ozone trends and their relationship to characteristic weather patterns. J Expo Sci Environ Epidemiol 25(5):535–542
3. WHO (1991) Guidelines for the assessment of herbal medicines. World Health Organization, Geneva
4. Othman ZA, Ismail N, Latif MT (2017) Association pattern of NO_2 and NMHC towards high ozone concentration in klang. In: 2017 6th international conference on electrical engineering and informatics (ICEEI), pp 1–6
5. Zhao W, Fan S, Guo H, Gao B, Sun J, Chen L (2016) Assessing the impact of local meteorological variables on surface ozone in Hong Kong during 2000–2015 using quantile and multiple line regression models. Atmos Environ 144:182–193

6. Ahamad F, Latif MT, Tang R, Juneng L, Dominick D, Juahir H (2014) Variation of surface ozone exceedance around Klang Valley, Malaysia. Atmos Res 139:116–127
7. Alvarez LJR, Lowe D, Badia A, Dorling SR, Lupascu A (2017) The source of ozone during an air pollution episode in the UK: a case study from the ICOZA campaign
8. Othman ZA, Ismail N, Latif MT (2017) Association rules of temperature towards high and low ozone in Putrajaya. In: 2017 6th international conference on electrical engineering and informatics (ICEEI), pp 1–5
9. Karlsson PE et al (2017) Past, present and future concentrations of ground-level ozone and potential impacts on ecosystems and human health in northern Europe. Sci Total Environ 576:22–35
10. Sajjadi S et al (2016) Extreme learning machine for prediction of heat load in district heating systems. Energy Build 122:222–227
11. Najah A, El-Shafie A, Karim OA, El-Shafie AH (2013) Application of artificial neural networks for water quality prediction. Neural Comput Appl 22(Suppl 1):187–201
12. Ding S, Zhao H, Zhang Y, Xu X, Nie R (2015) Extreme learning machine: algorithm, theory and applications. Artif Intell Rev 44(1):103–115
13. Deo RC, Şahin M (2015) Application of the extreme learning machine algorithm for the prediction of monthly Effective Drought Index in eastern Australia. Atmos Res 153:512–525
14. Huang G, Bin Huang G, Song S, You K (2015) Trends in extreme learning machines: a review. Neural Netw 61:32–48
15. Zaremba W, Sutskever I, Vinyals O (2014) Recurrent neural network regularization. Iclr, zenb. 2013, pp 1–8
16. Fakhr MW, Youssef ENS, El-Mahallawy MS (2015) L1-regularized least squares sparse extreme learning machine for classification. In: 2015 international conference on information and communication technology research, ICTRC 2015, pp 222–225
17. Martínez-Martínez JM, Escandell-Montero P, Soria-Olivas E, Martín-Guerrero JD, Magdalena-Benedito R, Gómez-Sanchis J (2011) Regularized extreme learning machine for regression problems. Neurocomputing 74(17):3716–3721
18. Izhab Z (1992) Effective reading is the way to ensure success of science students. Surat kepada Editor, New Straits Times, p 9
19. Dick T (2017) The effect of a missing at random missing data mechanism on a single layer artificial neural network with a sigmoidal activation function and the use of multiple imputation as a correction. West Virginia University
20. Yadav A, Sahu K (2017) Wind forecasting using artificial neural networks: a survey and taxonomy. Int J Res Sci 3:148–155
21. Venkitaraman A, Chatterjee S, Händel P (2018) Extreme learning machine for graph signal processing. arXiv Prepr. arXiv:1803.04193
22. Cao W, Gao J, Ming Z, Cai S (2017) Some tricks in parameter selection for extreme learning machine. In: IOP conference series: materials science and engineering, vol 261, p 1
23. Dobbin KK, Simon RM (2011) Optimally splitting cases for training and testing high dimensional classifiers. BMC Med Genomics 4:31
24. Bisht M, Seeja KR (2018) Air pollution prediction using extreme learning machine: a case study on Delhi (India), vol 79

Software Module Clustering Based on the Fuzzy Adaptive Teaching Learning Based Optimization Algorithm

Kamal Z. Zamli, Fakhrud Din, Nazirah Ramli and Bestoun S. Ahmed

Abstract Although showing competitive performances in many real-world optimization problems, Teaching Learning based Optimization Algorithm (TLBO) has been criticized for having poor control on exploration and exploitation. Addressing these issues, a new variant of TLBO called Adaptive Fuzzy Teaching Learning based Optimization (ATLBO) has been developed in the literature. This paper describes the adoption of Fuzzy Adaptive Fuzzy Teaching Learning based Optimization (ATLBO) for software module clustering problem. Comparative studies with the original Teaching Learning based Optimization (TLBO) and other Fuzzy TLBO variant demonstrate that ATLBO gives superior performance owing to its adaptive selection of search operators based on the need of the current search.

Keywords Search-based software engineering · Software module clustering · Adaptive teaching learning based optimization · Mamdani fuzzy

K. Z. Zamli (✉) · F. Din
Faculty of Computer Systems & Software Engineering, Universiti Malaysia Pahang, 26300 Gambang, Kuantan, Pahang, Malaysia
e-mail: kamalz@ump.edu.my

F. Din
Department of Computer Science & IT, Faculty of Information Technology, University of Malakand, Lower Dir, KPK, Pakistan
e-mail: fakhruddin@uom.edu.pk

N. Ramli
Department of Mathematics and Statistics, Faculty of Computer and Mathematical Sciences, Universiti Teknologi MARA Pahang, 2640 Jengka, Pahang, Malaysia
e-mail: nazirahr@pahang.uitm.edu.my

B. S. Ahmed
Faculty of Electrical Engineering, Department of Computer Science, Czech Technical University, Prague, Czech Republic
e-mail: albeybes@fel.cvut.cz

© Springer Nature Singapore Pte Ltd. 2019
V. Piuri et al. (eds.), *Intelligent and Interactive Computing*, Lecture Notes in Networks and Systems 67, https://doi.org/10.1007/978-981-13-6031-2_3

1 Introduction

Software module clustering can be used to assess and facilitate the software comprehension, evolution, and maintenance [1, 2]. In fact, evidence shows that modularized software could lead to better development and maintenance process [3]. Recently, researchers have formulated the software module clustering as an optimization problem within the area of search-based software engineering.

Many approaches have been adopted for addressing the problem of software module clustering by exploiting numerous meta-heuristic algorithms. Hill Climbing (HC) has been exploited by Mancoridis and Mitchell for this purpose (i.e., based on a tool called Bunch [4]). Similarly, HC has also been adopted by Mahdavi et al. [5]. Genetic Algorithm (GA) has also been used for software module clustering by Kumari and Srivinas [6]. Praditwong et al. [7] have pioneered the multi-objective approach for software module clustering by using Pareto optimality concept. More recently, Huang et al. [8] have proposed a multi-agent evolutionary algorithm.

Despite some useful progress, the adoption of newly developed meta-heuristic algorithm is deemed necessary. As the *No Free Lunch theorem* [9] suggests that it is not possible for a single optimization algorithm to solve all optimization problems or even variants of the same problem to optimality (e.g. [10–19]). For this reason, this paper explores the possibility of exploiting our proposed fuzzy Adaptive Teaching Learning based Optimization (ATLBO) [20] based on Mamdani fuzzy implementation for tackling the software module clustering problem. Specifically, we also compare the effectiveness of ATLBO against the original TLBO [21] and Fuzzy Adaptive TLBO (FATLBO) [22].

The organization of the paper is as follows. Section 2 presents the fundamentals of the software module clustering problem. Section 3 gives an overview of TLBO and its fuzzy adaptive variants. Section 4 presents our benchmark comparison between TLBO, ATLBO, and other TLBO variants. Section 5 presents our discussion. Finally, Sect. 6 presents the conclusion of the work.

2 Software Module Clustering Problem

Software module clustering problem can be defined as the problem of partitioning modules into clusters based on some predefined quality criterion. Typically, the quality criterion for software module clustering problem relates to two important concepts: coupling and cohesion. Coupling and cohesion are measurements where the former measures the dependency among module clusters, whereas the latter measures the internal strength of a module cluster. Thus, a good cluster distribution aids in functionality-cluster-module traceability provides easier navigation between subsystems and enhances source code comprehension.

To evaluate the cluster distribution, Module Dependency Graph (MDG) is used to model a software system [4]. Adding the weight of external edges entering into or

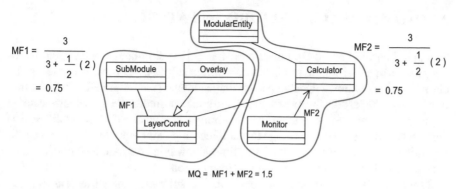

Fig. 1 Sample MQ calculation

leaving a cluster results in its coupling. This is also termed as *inter-edges*. Similarly, adding the weights of internal edges where the source and targets are part of a cluster partition results in cohesion. This is also termed as intra-edges. Combining coupling and cohesion [4], Modularization Quality (MQ) measure can be calculated by summing the ratio of inter-edges and intra-edges in each cluster. This ratio is called Modularization Factor (MF_k) for some cluster k. MF_k can be formally defined as follows:

$$MF_k = \begin{cases} 0 & \text{if } i = 0 \\ \frac{i}{i + \frac{1}{2}j} & \text{if } i > 0 \end{cases} \tag{1}$$

where i and j represent the weight of intra-edges and weight of inter-edges, respectively. The term $\frac{1}{2}j$ divides the penalty associated with inter-edges across the two clusters that are connected by that edge. The MQ can then be calculated as the sum of MF_k as follows:

$$MQ = \sum_{k=1}^{n} MF_k \tag{2}$$

where n denotes the number of clusters.

To illustrate the MQ calculation, Fig. 1 highlights a two-cluster modularization of a class diagram (i.e., referred to as MF1 and MF2). In this case, the class relationship is considered two ways when no navigation is specified (i.e., two relationships from source to destination and from destination to source). For the given class diagram in Fig. 1, the most minimum possible cluster is 1 while the maximum possible clusters are 6. The goal is to find the clusters from 2 to 5 that maximizes the MQ.

3   Overview of TLBO, FATLBO, and ATLBO

Teaching Learning based Optimization (TLBO) [21] simulates teaching and learning process in classroom to produce near optimal solutions. Basically, a teacher in a class is the most knowledgeable person than students (i.e., with better fitness function value). The responsibility of teacher is to enhance students' competency level by imparting his knowledge to them. As the competency levels of teachers differ, students can improve their knowledge through learning from other teachers as well (in any subsequent iterations). At the same time, knowledge exchange among students is also helpful for improving their competency levels.

Within TLBO, X represents a population of solutions known as learners. An individual learner X_i is one potential solution in this population. To be more specific, X_i is a vector containing D dimensions where each dimension represents a subject taken by students. The fitness function value denotes the grade of a student.

Searching process in TLBO is divided into two main phases; the teacher phase and the learner phase. In order to perform the search, TLBO undergoes both the phases sequentially one after the other per iteration (see Fig. 2a). The teacher phase involves invoking the global search process or exploration. At any point during the search in TLBO, the best individual X_i in X is always assigned to the teacher X_{teacher}. Based on the current mean value of the population X_{mean}, TLBO tries to improve the fitness function of each individual X_i by using X_{teacher} as given in Eq. 3.

$$X_i^{t+1} = X_i^t + r(X_{\text{teacher}} - T_{\text{F}} X_{\text{mean}}) \tag{3}$$

where X_i^t is updated to X_i^{t+1}, X_{teacher} represents the best individual in X, X_{mean} represents the mean of X, r is the random number in the interval [0, 1], and T_{F} denotes

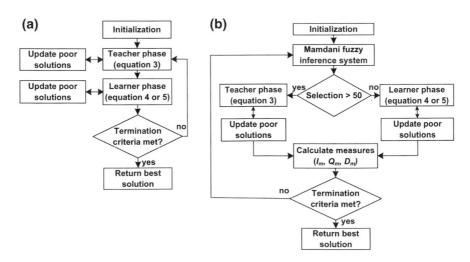

Fig. 2   a TLBO. **b** ATLBO

the teaching factor which can take two possible values 1 or 2, and it emphasizes the quality of teaching.

The learner phase simulates the local search operation or exploitation in TLBO. Specifically, the learner X_i^t is a current learner who increases his competency level through interaction with his random peer X_j^t (i.e., a potential new solution) within the population X. The learner's knowledge is updated if and only if his knowledge is poor than the peer. At any iteration i, if X_i^t is better than X_j^t, then X_j^t is moved toward X_i^t (refer to Eq. 4). Otherwise, X_i^t is moved toward X_j^t (refer to Eq. 5)

$$X_i^{t+1} = X_i^t + r\left(X_j^t - X_i^t\right) \tag{4}$$

$$X_i^{t+1} = X_i^t + r\left(X_i^t - X_j^t\right) \tag{5}$$

where X_i^t is updated to X_i^{t+1}, X_j^t is the random peer such that $i \neq j$, r is the random number from [0, 1].

Fuzzy Adaptive TLBO (FATLBO) proposed by Cheng and Prayogo [22] also adopted Mamdani-type fuzzy inference system to further strengthen the searching power of original TLBO. After monitoring the success rates of both the teacher phase and learner phase, FATLBO uses the crisp output (i.e., bar movement) of its Fuzzy Adaptive Teaching Learning Strategy (FATLS) to decide whether the teacher-learning process will be teacher-centric or learner-centric or both.

ATLBO [20] is the recent variant of TLBO that adaptively selects its global search (i.e., teacher phase) and local search (i.e., learner phase) operations by using a three inputs and one output Mamdani-type fuzzy inference system (see Fig. 2b). The inputs are called quality measure Q_m, the intensification measure I_m and the diversification measure D_m. Equations 6 through 8 correspond to the scaled values of these inputs.

$$Q_m = \left[\frac{X_{\text{current fitness}} - \text{min fitness}}{\text{max fitness} - \text{min fitness}}\right] \cdot 100 \tag{6}$$

$$I_m = \left[\frac{|X_{\text{best}} - X_{\text{current}}|}{D}\right] \cdot 100 \tag{7}$$

$$D_m = \left[\frac{\sum_{j=1}^{\text{population size}} |X_j - X_{\text{current}}|}{D}\right] \cdot 100 \tag{8}$$

The measure Q_m measures the quality of the current candidate solution. The intensification measure I_m measures whether the current solution is good as compared to the best solution. Finally, the diversification measure D_m measures the distance of current solution from the entire population. The single output of fuzzy inference system is "Selection." These inputs and output are transformed into linguistic variables which are fuzzified using trapezoidal membership functions. The rule base of the system is composed of four fuzzy linguistic rules with max-min fuzzy inference method. Interested readers are referred to [20] for further details of these membership functions and fuzzy rules. Finally, center of gravity (COG) is used as defuzzification method to obtain the single crisp output, the "selection." This value is then used to decide whether to launch global search or local search rather than executing both as in original TLBO. Figure 3 summarizes the main component of fuzzy inference system of ATLBO.

The pseudocode of the ATLBO-based strategy for software module clustering problem is shown in Fig. 4.

4 Benchmarking Case Studies

Our benchmarking experiments focus on characterizing the performances of ATLBO against the original TLBO and that of FATLBO. We note that evaluating the performance of ATLBO directly as compared to TLBO (i.e., even when a number of iterations are same) and FATLBO can also be unfair. With the same number of iteration, the fitness function evaluations for original TLBO will be twice as much of fitness function evaluations of ATLBO, because TLBO serially executes both exploration and exploitation steps per iteration. Furthermore, the number of iterations for both teacher and student phase is also nondeterministic for FATLBO. Hence, for the fair comparison, we opt to use the same number of fitness function evaluations.

Three case studies (refer to Fig. 5) have been selected for evaluating our approach. Case study 1 having 9 modules relates to the printer manager software. Case study 2

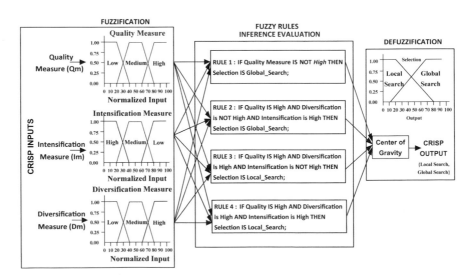

Fig. 3 Fuzzy inference system of ATLBO

Input:	The population $X = \{X_1, X_2, ..., X_n\}$	
Output:	Final best MQ Measure	
1	Set maximum partition = no. of modules/2	
2	Define the membership functions and fuzzy rules, Set $selection = 100$	
3	Initialize the random population of learners X and evaluate the mean X_{mean}, Select $X_{teacher}$	
4	While(Not maximum fitness function evaluations) do	
5	If $(selection > 50)$ then	//Teacher Phase
6	Select $X_{teacher}$ and calculate X_{mean}	
7	For(i = 1 to population size) do	
8	$T_F = round\left(1 + r(0,1)\right)$	
9	$X_i^{t+1} = X_i^t + r(X_{teacher} - T_F X_{mean})$	
10	If $(f(X_i^{t+1})$ is better than $f(X_i^t))$ then	
11	$X_i^t = X_i^{t+1}$	
12	End //For	
13	End //outer If	
14	Else	//Learner Phase
15	For($i = 1$ to population size) do	
16	Randomly select one learner X_j^t from the population X such that $i \neq j$	
17	If$(f(X_i^t)$ is better than $f(X_j^t))$ then	
18	$X_i^{t+1} = X_i^t + r\left(X_j^t - X_i^t\right)$	
19	Else	
20	$X_i^{t+1} = X_i^t + r\left(X_i^t - X_j^t\right)$	
21	End //For	
22	End //outer Else	
23	Compute $Q_m, I_m, and\ D_m$ and fuzzify them	
24	Defuzzify and set $selection = crisp\ output$	
25	Update the population X and return the best MQ	
26	End //While	

Fig. 4 ATLBO-based strategy for software module clustering

also having 9 modules relates to IOT controller while case study 3 having 6 modules deals with layer monitor. These case studies are actually students' class projects for undergraduate course. Although the chosen case studies are rather small, it is enough to demonstrate the usefulness of our approach.

In all our experiments, population size is set to 40, and the maximum fitness function evaluations are set to 5000. Our experimental setup is consisted of a PC having Windows 10 installed and with these hardware specifications: 16 GB RAM, CPU 2.9 GHz Intel Core i5 and a 512 MB of flash HDD. In all the experiments, we have executed ATLBO, TLBO, and FATLBO 20 times so as to have a fair statistical significance. The best MQ and the mean MQ for each case study are given in tabular form. Entries in bold font denote the best mean MQ.

5 Discussion

Table 1 summarizes the results. Referring to Table 1, we note that all TLBO variants can achieve the best MQ of 2.038, 2.011, and 1.500, respectively. ATLBO gives the

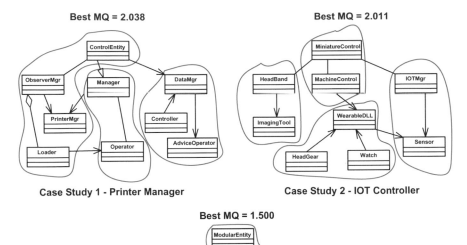

Fig. 5 Corresponding best clusters MQs for case studies 1 till 3

Table 1 Comparative MQ performance of ATLBO, TLBO, and FATLBO

	TLBO [21]		ATLBO [20]		FATLBO [22]	
Case studies	Best	Mean	Best	Mean	Best	Mean
Printer manager	2.038	2.004	2.038	**2.035**	2.038	2.012
IOT controller	2.011	1.992	2.011	**2.011**	2.011	1.992
Layer monitor	1.500	1.2860	1.500	**1.500**	1.500	**1.500**

best overall means for all three cases. For case study 3, ALTBO and FATLBO share the best performance. TLBO performs the poorest with no entry achieving the best mean.

Concerning time performance, we note that all algorithms have similar performances. In fact, from our observation, execution time of all the competing algorithms for a given experiment is equivalent.

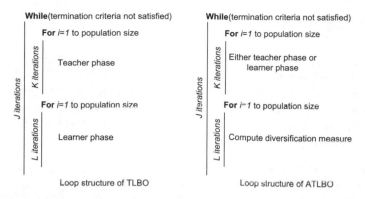

Fig. 6 Time complexity analysis of TLBO and ATLBO

Referring to Fig. 5, based on the best output from ATLBO, we cluster all the classes based on the best MQ that are generated. Ideally, these clusters can be considered as packages or sub-modules. It must be stressed to have a MQ measure that finds balance between coupling and cohesion but not to completely remove them. For instance, one can have only 1 cluster or n completely independent single module clusters to have zero coupling, but such an approach does not aid in functionality-cluster-module traceability.

Given that both ATLBO and FATLBO adopt Mamdani fuzzy as their improvement to TLBO, it is necessary to elaborate on their similarities and differences. The number of successful updates in each phase is called success rate in FATLBO. FATBLO uses this measure as the fuzzy input after its calculation for both teacher phase and student phase. Essentially, FATLBO relies on penalize and reward scheme to ensure the execution of either teacher or learner phase (or both) via fuzzy probabilistic bar movement. Performing phase has higher probability for re-selection in the next iteration. ATLBO, on the other hand, relies on the quality measure, diversification measure, and intensification measure to determine the best phase for selection. In this manner, ATLBO appears more sensitive on the current needs at specific instance of the searching process as compared to FATLBO.

Finally, for time complexity analysis, we need to consider the structures of ATLBO and TLBO. These structures are shown in Fig. 6.

If it is assumed that all other operations can be performed in $O(1)$ (i.e., constant time), the time complexity of FATLBO is $O(J \times (K + L)) \approx O(n^2)$ when J, K, and L approach large n. Similarly, the complexity of ATLBO is $O(J \times (K + L)) \approx O(n^2)$ when J, K, and L approach large n.

6 Conclusion

Summing up, this paper has demonstrated the effectiveness of ATLBO for software module clustering applications. ATLBO has outperformed FATLBO and the original TLBO in terms of generating the optimal MQ measure. As the scope of future work, we are exploring the adoption of ATLBO for other optimization problems particularly on wireless sensor network localization and software product line test suite generation.

Acknowledgements This work is supported by the Fundamental Research Grant from Ministry of Higher Education Malaysia (MOHE) under the title "A Reinforcement Learning Sine Cosine based Strategy for Combinatorial Test Suite Generation (grant no: RDU170103)." We are grateful to MOHE for this support.

References

1. Lucca GAD, Fasolino AR, Pace F, Tramontana P, Carlini UD (2002) Comprehending web applications by a clustering-based approach. In: Proceedings of the 10th international workshop on program comprehension. IEEE, Paris, France, pp 261–270
2. Jahnke JH (2004) Reverse engineering software architecture using rough clusters. In: Proceedings of the IEEE annual meeting of the fuzzy information processing. Alberta, Canada, pp 4–9
3. Sommerville I (2015) Software engineering, 10th edn. Pearson, Harlow
4. Mitchell BS, Mancoridis S (2006) On the automatic modularization of software systems using the bunch tool. IEEE Trans Softw Eng 32(3):193–208
5. Mahdavi K, Harman M, Hierons RM (2003) A multiple hill climbing approach to software module clustering. In: Proceedings of the international conference on software maintenance. Amsterdam, The Netherlands, pp 315–324
6. Kumari AC, Srinivas K (2016) Hyper-heuristic approach for multi-objective software module clustering. J Syst Softw 117:384–401
7. Praditwong K, Harman M, Yao X (2011) Software module clustering as a multi-objective search problem. IEEE Trans Softw Eng 37(2):264–282
8. Huang J, Liu J, Yao X (2017) A multi-agent evolutionary algorithm for software module clustering problems. Soft Comput 21(12):3415–3428
9. Wolpert DH, Macready WG (1997) No free lunch theorems for optimization. IEEE Trans Evol Comput 1(1):67–82
10. Alsewari ARA, Zamli KZ (2012) A harmony search based pairwise sampling strategy for combinatorial testing. Int J Phys Sci 7(7):1062–1072
11. Din F, Alsewari ARA, Zamli KZ (2017) A parameter free choice function based hyper-heuristic strategy for pairwise test generation. In: Proceedings of the IEEE international conference on software quality, reliability and security companion. IEEE, Prague, Czech Republic, pp 85–91
12. Din F, Zamli KZ (2018) Fuzzy adaptive teaching learning-based optimization strategy for GUI functional test cases generation. In: Proceedings of the 2018 7th international conference on software and computer applications. ACM, Kuantan, Malaysia, pp 92–96
13. Nasser AB, Zamli KZ, Alsewari ARA, Ahmed BS (2018) Hybrid flower pollination algorithm strategies for *t*-way test suite generation. PLoS ONE 13(5):e0195187
14. Othman RR, Zamli KZ (2011) ITTDG: integrated *t*-way test data generation strategy for interaction testing. Sci Res Essays 6(17):3638–3648

15. Younis MI, Zamli KZ, Isa NAM (2008) MIPOG-modification of the IPOG strategy for t-way software testing. In: Proceedings of the distributed frameworks and applications. IEEE, Penang, Malaysia, pp 1–6
16. Younis MI, Zamli KZ, Isa NAM (2008) Algebraic strategy to generate pairwise test set for prime number parameters and variables. In: Proceedings of the international symposium on information technology. IEEE, Kuala Lumpur, Malaysia, pp 1–4
17. Zamli KZ, Alkazemi BY, Kendall G (2016) A tabu search hyper-heuristic strategy for t-way test suite generation. Appl Soft Comput 44:57–74
18. Zamli KZ, Din F, Ahmed BS, Bures M (2018) A hybrid Q-learning sine-cosine-based strategy for addressing the combinatorial test suite minimization problem. PLoS ONE 13(5):e0195675
19. Zamli KZ, Din F, Kendall G, Ahmed BS (2017) An experimental study of hyper-heuristic selection and acceptance mechanism for combinatorial t-way test suite generation. Inf Sci 399:121–153
20. Zamli KZ, Din F, Baharom S, Ahmed BS (2017) Fuzzy adaptive teaching learning-based optimization strategy for the problem of generating mixed strength t-way test suites. Eng Appl Artif Intell 59:35–50
21. Rao RV, Savsani VJ, Vakharia DP (2011) Teaching-learning-based optimization: a novel method for constrained mechanical design optimization problems. Comput Aided Des 43(3):303–313
22. Cheng M-Y, Prayogo D (2017) A novel fuzzy adaptive teaching–learning-based optimization (FATLBO) for solving structural optimization problems. Eng Comput 33(1):55–69

Application of Bat Algorithm in Carbon Nanotubes Growing Process Parameters Optimization

M. I. M. Jarrah, A. S. M. Jaya, Mohd Asyadi Azam, Zakaria N. Alqattan, Mohd Razali Muhamad and Rosni Abdullah

Abstract Since their discovery, carbon nanotubes (CNTs) have become a subject of intense research for their potential use in various applications. Chemical vapor deposition (CVD) process is the most common method used to grow CNTs. However, the growing process suffers many difficulties in finding the optimal process parameters. Applying computational intelligence methods is a possible solution for optimization problems to reduce using conventional methods and experimental runs. In this work, a combination between bat algorithm (BA) and response surface methodology (RSM) is proposed to solve CNTs process parameters optimization problem. The study aims to maximize the CNTs yield percentage for mass production in two different datasets. BA search process is based on the objective function developed by RSM which represents the prediction mathematical model of growing process parameters. The optimized parameters from datasets are reaction temperature, reaction time, catalyst weight, and methane partial pressure. The algorithm search process was conducted with parameters tuning at different setting values to improve the algorithm's performance and CNTs yield value. Different evaluation metrics were

M. I. M. Jarrah (✉) · A. S. M. Jaya
Faculty of Information and Communication Technology, Universiti Teknikal
Malaysia Melaka, Hang Tuah Jaya, 76100 Durian Tunggal, Melaka, Malaysia
e-mail: aljarrahmuath@gmail.com

A. S. M. Jaya
e-mail: syukor@utem.edu.my

M. A. Azam · M. R. Muhamad
Faculty of Manufacturing Engineering, Universiti Teknikal Malaysia Melaka,
Hang Tuah Jaya, 76100 Durian Tunggal, Melaka, Malaysia
e-mail: asyadi@utem.edu.my

M. R. Muhamad
e-mail: muhdrazali@utem.edu.my

Z. N. Alqattan · R. Abdullah
National Advance IPv6 Center, School of Computer Sciences,
Universiti Sains Malaysia, 11800 Gelugor, Penang, Malaysia
e-mail: zakaria.alqattan@usm.my

R. Abdullah
e-mail: rosni@usm.my

© Springer Nature Singapore Pte Ltd. 2019
V. Piuri et al. (eds.), *Intelligent and Interactive Computing*, Lecture Notes in Networks
and Systems 67, https://doi.org/10.1007/978-981-13-6031-2_14

applied to compare the experimental results. The results have shown that BA has an efficient search performance and obtained better CNTs yield result than RSM in one of the datasets with 21% improvement of CNTs yield value. Besides, BA has shown a fast and stable convergence. Finally, the result was validated and found reliable to be used in real laboratory experiments.

Keywords Carbon nanotubes · Bat algorithm · Optimization · CVD

1 Introduction

In the past two decades, carbon nanotubes (CNTs) have been extensively investigated and considered as the most exciting discovered material due to their unique chemical, electrical, and physical properties. They have received a great deal of attention in industry and scientific community [1, 2]. CNTs are considered a promising material for various applications, such as novel electronics, field-effect transistors, energy storage, electrodes, biotechnology, sensors, fuel storage, and composites [3, 4].

Three main methods are used for CNTs growing, namely, chemical vapor deposition (CVD), arc discharge, and laser ablation methods. CVD is the most promising method due to its low cost and it can produce CNTs with high degree of structure in large scale [5].

In carbon nanotechnology, a high yield with high purity materials is required. Thus, synthesis process is one of the bottlenecks which must be optimized to achieve such desired nanomaterials [6]. CNTs synthesis process is influenced by several process parameters that affect the carbon yield and other characteristics. Examples of process parameters include reaction time, reaction temperature, gas pressure, gas flow rate, type of carbon precursor, type of active metal catalyst, etc. [5]. Generally, manufacturers rely on varying one process parameter while keeping others constant to optimize the manufacturing process parameters. This method has several issues. For example, it ignores the interaction effects between the process parameters, consumes long time, requires many experimental runs, and leads to inaccurate prediction of the optimal conditions [5, 7, 8].

In the last years, several studies have investigated and optimized process parameters for CNTs growing [9–11]. However, solving different optimization problems by adoption of artificial intelligence techniques is one of the most interesting trends especially metaheuristic algorithms [12–15]. Metaheuristic algorithms have been reported to have projected outstanding performance in solving several types of problems in different domains including numerical and complex optimization problems, engineering, and nanoprocess field [16–18]. However, carbon nanoproducts, i.e., CNTs growing parametric studies, widely use a statistical design of experiment (DOE), while the application of metaheuristic algorithms or its integration with DOE is rare in literature.

One of the recent proposed metaheuristic algorithms is bat-inspired algorithm (BA). Basically, it is better than other algorithms in its construction [19]. Due to its simplicity and reliability, BA is a new novel evolutionary algorithm that has been widely applied in various optimization problems [20, 21]. In this study, BA is adapted to solve the CNTs growing process parameters problem. This algorithm has been chosen because it is easy, has comprehensible structure, and able to tackle low-dimensional problems with reliable results [22]. Its effectiveness was demonstrated in solving engineering optimization problems [23–25]. At the moment, BA has not been applied in optimizing CNTs growing using CVD method.

In this study, BA is adapted with response surface methodology (RSM) and expected to find the optimal parameters combinations to improve the CNTs growing process in order to maximize CNTs yield%. The objective is to improve the conventional method and to propose efficient, productive, and better technique in the CVD process parameters optimization problem for CNTs growing, thus addressing the issues of laboratory experimental time and cost. Besides that, the capability of the adapted BA with this problem is expected to be used for another optimization problem in different domains.

2 Research Methodology

This section discusses the detailed methodology of the adapted BA for CNTs yield optimization including problem formulation, experimental data, and BA settings and parameters.

2.1 Problem Formulation

The objective function for this optimization problem is based on the mathematical prediction model developed by RSM. The RSM produces different objective functions based on the input dataset. Later on, the objective function will be used for maximizing the CNTs yield% by optimizing the process parameters using BA. There are five different CVD process parameters to be optimized which are reaction temperature, reaction time, catalyst weight, polymer weight, and methane partial pressure. For BA, the solution of Di dimension represents the process parameters of each data where $i \in$ process parameters. Therefore, the goal of applying BA is to find the optimal solution which maximizes the CNTs yield%.

2.2 Experimental Data

In the CVD method for CNTs growing, there are several process parameters that have been studied by researchers to maximize the CNTs yield%. However, this paper aims to study five parameters obtained from two different datasets denoted by case study 1 and case study 2. The first dataset was collected by [26] where the CNTs were grown using CVD. The waste of polypropylene (PP) bottles was considered as a hydrocarbon precursor with the mixture catalyst of Ni/Mo/MgO. This dataset investigates four process parameters, namely, reaction temperature, polymer weight, catalyst weight, and reaction time. Table 1 shows the collected dataset with 30 experimental runs.

The second dataset was collected by [27] where the CNTs were grown using CVD to measure carbon yield. The CNTs were synthesized over Co–Mo/Al$_2$O$_3$ catalyst by methane decomposition. This dataset investigates three process parameters, namely,

Table 1 Experimental data for case study 1

Run	Temperature (°C)	Polymer weight (g)	Catalyst weight (mg)	Time (min)
1	900	4	100	20
2	900	4	50	15
3	800	4	50	20
4	700	4	50	15
5	800	4	100	15
6	800	4	100	15
7	800	6	50	15
8	800	2	50	15
9	800	6	100	10
10	800	4	100	15
11	800	6	100	20
12	700	4	150	15
13	800	6	150	15
14	800	2	150	15
15	800	2	100	20
16	700	6	100	15
17	700	4	100	10
18	800	4	150	10
19	800	2	100	10
20	900	4	150	15
21	800	4	50	10
22	900	6	100	15
23	800	4	100	15
24	700	4	100	20
25	900	2	100	15
26	900	4	100	10
27	800	4	100	15
28	700	2	100	15
29	800	4	150	20
30	800	4	100	10

Table 2 Experimental data for case study 2

Run	Temperature (°C)	Partial pressure (atm)	Catalyst weight (g)
1	750	0.5	0.3
2	700	0.25	0.3
3	700	0.5	0.4
4	750	0.75	0.2
5	800	0.25	0.4
6	800	0.5	0.2
7	750	0.25	0.3
8	750	0.5	0.2
9	700	0.5	0.3
10	750	0.5	0.3
11	700	0.75	0.4
12	800	0.5	0.3
13	700	0.75	0.2
14	800	0.5	0.4
15	700	0.5	0.2
16	750	0.75	0.4
17	800	0.75	0.2
18	750	0.5	0.3
19	750	0.5	0.3
20	750	0.25	0.4
21	700	0.25	0.4
22	700	0.75	0.3
23	800	0.75	0.4
24	750	0.5	0.4
25	750	0.25	0.2
26	750	0.25	0.2
27	750	0.5	0.3
28	800	0.25	0.2
29	750	0.75	0.3
30	800	0.25	0.3
31	800	0.75	0.3
32	750	0.5	0.3

reaction temperature, methane partial pressure, and catalyst weight. Table 2 shows the collected dataset with 32 experimental runs.

In both datasets, RSM was applied to develop a prediction model and to optimize the process parameters in order to find the near optimal condition with best parameters

Initialize the bat population xi and velocity vi, where $i = (1, 2, ..., n)$
Find the best solution and its corresponding value $f(x)$
Define pulse frequency f_i at x_i
Initialize pulse rates r_i and the loudness A_i
while *(t <Max number of iterations)*
  Generate new solutions by adjusting frequency,
  and updating velocities and solutions/locations
if *(rand > r_i)*
  Select a solution among the best solutions
  Generate a local solution around the selected best solution
end if
Evaluate fitness(new solution)
if *(rand < A_i & $f(x_i) < f(x_*)$)*
  Generate a new solution randomly
  Evaluate fitness(new solution)
  Increase r_i and reduce A_i
end if
Evaluate the bats and find the current best x_*
end while

Fig. 1 The pseudocode of BA [28]

combination to maximize the CNTs yield. In this study, BA is expected to efficiently find the near maximum value of CNTs yield for both datasets.

2.3 Bat Algorithm

BA is a metaheuristic algorithm that is recently developed by [28]. It mimics the advanced capability of bats echolocation by emitting a very loud sound pulse and receiving its reflected echo after collided obstacle, prey, or any neighboring object. When bats are close to the target prey, the pulse emission rate is inversely proportional to the distance between them and dramatically increases by acceleration of about 200 pulses per second, and meanwhile, when bats fly to the prey, the loudness of emitted pulses will be decreased [25]. The pseudocode of BA is represented in Fig. 1.

3 Bat Algorithm for CNTs Yield Optimization Problem

3.1 Objective Functions for CNTs Yield%

In this study, the developed model of CNTs yield% from response surface methodology (RSM) for the process parameters which obtained from the original referred paper will be used as the objective function in BA for both datasets. The objective function of the CNTs yield% for the case study 1 using the waste of polypropylene (PP) bottles as a hydrocarbon precursor is represented in Eq. (1).

$$
\begin{aligned}
\text{Carbon Yield\%} = &-6748.77389 + 14.5483 \times x1 + 254.55528 \times x2 \\
&+ 4.80706 \times x3 + 66.57478 \times x4 - 0.00883101 \times x1^2 - 21.78832 \\
&\times x2^2 - 0.015475 \times x3^2 - 1.14291 \times x4^2 - 0.0075375 \times x1 \times x2 \\
&+ 0.000192 \times x1 \times x3 - 0.02847 \times x1 \times x4 + 0.39112 \times x2 \times x3 \\
&- 4.69312 \times x2 \times x4 - 0.13353 \times x3 \times x4
\end{aligned}
\tag{1}
$$

where $x1$, $x2$, $x3$, and $x4$ are reaction temperature, polymer weight, catalyst weight, and reaction time, respectively, while the objective function of the *case study 2* where the CNTs were grown by methane decomposition is represented in Eq. (2).

$$
\begin{aligned}
\text{Carbon Yield\%} = &-23707.88889 + 63.23889 \times y1 + 1396.77778 \\
&\times y2 - 1715.0 \times y3 - 0.041333 \times y1^2 - 554.66667 \times y2^2 - 250.00 \\
&\times y3^2 - 1.26667 \times y1 \times y2 + 1.633 * y1 \times y3 + 1143.33 \times y2 \times y3
\end{aligned}
\tag{2}
$$

where $y1$ represents the reaction temperature, $y2$ is the partial pressure of methane, and $y3$ is the catalyst weight.

3.2 Limitation Constraints

The limitation constraints are set in BA to define the search space in order to generate random values of the input parameters within the limits boundary range. In this study, the limitation constraints of the process parameters in *case study 1* are represented in Eqs. 3–6.

$$700 < \text{Reaction temperature} < 900 \tag{3}$$

$$2 < \text{polymer weight} < 6 \tag{4}$$

$$50 < \text{catalyst weight} < 150 \tag{5}$$

$$10 < \text{reaction time} < 20 \tag{6}$$

While the limitation constraints of the process parameters in *case study 2* are represented in Eqs. 7–9,

$$700 < \text{Reaction temperature} < 800 \tag{7}$$

$$0.25 < \text{Partial Pressure} < 0.75 \tag{8}$$

$$0.2 < \text{catalyst weight} < 0.4 \tag{9}$$

3.3   BA Default Parameters Settings

BA comprises five main parameters including population size, iterations number, initial emission rate (r_i^0), α constants which affect the loudness (A), and wavelength (γ) which affects the pulse rate (r). These parameters significantly influence the optimization result and algorithm search efficiency. Therefore, it is highly recommended to test the algorithm with tuning of these parameters in order to obtain the best optimization result. In this study, the default parameters settings for BA are set based on the optimal parameters setting used in [28–30] where they applied BA in cutting process parameters optimization. The values of r_i^0, α, and γ were tuned between 0.1 and 1.0 with increasing step by 0.1. The default value of α and γ was 0.8. Meanwhile, loudness (A) was fixed at $A = 0.7$, initial pulse rate (r) was fixed at $r = 0.5$, population size was ranged from 5 to 45 with default value of 25 bats, and iterations number was set to the values of 10, 15, 25, 50, 75, 100, 200, 300, 400, and 500 with default value of 100.

The tuning process was done by changing value of one tested parameter while fixing all other parameters. Each tuning case was repeated for 30 times of run, and then the results were recorded. The stopping condition in this study was considered as optimal value so far for the response yield. This condition was set first by testing BA algorithm using high number of iterations (5000) and population size of (100) with different tunings, while ignoring the time to get the best value among all algorithms.

3.4   Evaluation Metrics and Validation

The decision of performance measure of BA algorithm was evaluated by comparing eight terms of metric measurements, namely, *Best, Worst, Mean*, and *Success Rate percentage* (SR%) which refers to how many times the algorithm could obtain the target optimal value in all runs, standard deviation (Std.Dev), *Mean Execution Time (M_ET)* to exit the run, *Mean Required Iterations (M_RI)* to exit the run, and *Mean Function Evaluation Number (FEN)* which generally means how many times the objective function is called by the algorithm. Additionally, in order to validate the obtained optimal result, the optimal solution or parameters combination is passed to the objective function and the yield% result in the objective function should be similar to the result obtained by algorithm. All experiments were run using laptop with Intel (R) Core (TM) i3 CPU 2.13 GHz and 4 GB RAM.

4   Results, Discussion, and Validation

This section discusses the summary of the optimization result with performance enhancement progress for all tuned parameters in the parameters tuning process using BA. Tables 3 and 4 show the results for *case study 1* and *case study 2*, respectively. The best value of current testing parameter is reported in bold font.

Table 3 Performance enhancement of BA tuning process in *case study 1*

Parameter	Settings values[a]	Best	Worst	Mean	SR%	Std. Dev	M_ET	M_RI	FEN
r_i^0	**0.7**, 0.8, 0.8, 25, 100	534.47	370.48	441.95	0	40.7	0.07	100	2525
γ	0.7, **0.8**, 0.8, 25, 100	534.47	370.48	441.95	0	40.7	0.07	100	2525
α	0.7, 0.8, **1**, 25, 100	534.935	534.72	534.91	3.33	0.05	0.06	98	2497
Population size	0.7, 0.8, 1, **45**, 100	534.935	534.91	534.93	6.66	0.01	0.12	95	4361
Iterations	0.7, 0.8, 1, 45, **500**	534.935	534.89	534.93	16.66	0.01	0.56	474	21,380

[a]The sequence order of the values is r_i^0, γ, α, population size, and iterations number

Table 4 Performance enhancement of BA tuning process in *case study 2*

Parameter	Settings values	Best	Worst	Mean	SR%	Std. Dev	M_ET	M_RI	FEN
r_i^0	**0.8**, 0.8, 0.8, 25, 100	606.9180	564.118	586.789	0	8.768	0.059	100	2525
γ	0.8, **0.8**, 0.8, 25, 100	606.9180	564.118	586.789	0	8.768	0.059	100	2525
α	0.8, 0.8, **1**, 25, 100	606.9219	606.832	606.912	43.33	0.024	0.049	72.4	1834
Population size	0.8, 0.8, 1, **40**, 100	606.9219	606.893	606.919	60	0.007	0.063	60.2	2447
Iterations	0.8, 0.8, 1, 40, **100**	606.9219	606.893	606.919	60	0.007	0.063	60.2	2447

Fig. 2 Convergence behavior of BA in the optimal setting parameters for *case study 1*

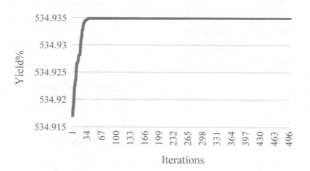

For the *case study 1*, as shown in Table 3, the maximum yield% of 534.9351% has been obtained at 0.7, 0.8, 1.0, 45, and 500 for r_i^0, γ, α, population size, and iterations number, respectively. Meanwhile, after tuning process, BA algorithms have obtained good result in terms of SR% with 16.66% with better performance in exploitation parts. The optimal solution which represents the best combinations of input parameters is at 806.72 °C, 5.97 g, 150 mg, and 10 min for temperature, polymer weight, catalyst weight, and time, respectively. Figure 2 shows the convergence behavior of BA in the optimal parameters tuning for CNTs yield% in *case study 1*. It is clear that BA has fast and stable convergence behavior. The maximum yield% was found at iteration number 466.

This optimum value shows better CNTs yield% than mentioned value in the referred original work where the yield% was optimized at 514% using RSM. In other words, an improvement of 21% of CNTs yield value at percentage of 4.07% was obtained by BA in this study.

To validate the optimal result from BA in this case study, the values of growing conditions that obtained maximum yield are transferred into the objective function or regression model in Eq. (1). The expected result from the model should be same as the obtained result by BA. By transferring the optimal values of growing conditions of BA into the regression model, the obtained value of predicted CNTs yield% is 534.935%. This value of transformation process is exactly the same as the obtained result of BA. This indicates that the same result of CNTs yield% = 534.935% could be obtained when the set of optimal growing conditions, estimated by BA, are used in the real experiment process using the waste of polypropylene (PP) bottles as a hydrocarbon precursor for CNTs growing using CVD.

For the *case study 2*, Table 4 shows that the maximum yield% of 606.9219% has been obtained when the algorithm's parameters were set at 0.8, 0.8, 1, 40, and 100 for r_i^0, γ, α, population size, and iterations number, respectively. Meanwhile, after tuning process, BA algorithms have obtained good result in terms of SR% with 60%. The optimal solution which represents the best combinations of input parameters at 761.43 °C, 0.75 atm, and 0.4 g for temperature, methane partial pressure, and catalyst weight, respectively.

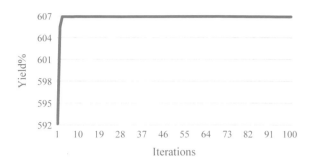

Fig. 3 Convergence behavior of BA in the optimal setting parameters for *case study 2*

Figure 3 shows fast and stable convergence behavior of BA in the optimal parameters tuning for CNTs yield% in *case study 2*. The maximum yield% was found at iteration number 68.

Comparing yield% value obtained by BA and RSM in the original work, it was found that the two values are same at ≈ 607, indicating that BA is also very efficient, besides RSM to maximize CNTs yield% in this case study. A validation process of the optimal result from BA in this case study was done by passing the values of growing conditions that obtained maximum yield into the objective function or regression model in Eq. (2). By transferring the optimal values of growing conditions of BA into the regression model, the obtained value of optimized CNTs yield% is 607%. This value of transformation process is exactly the same as the obtained result of BA. This indicates that the same result (CNTs yield% = 607%) could be obtained when the set of optimal growing conditions, estimated by BA, are used in the real experiment process using methane decomposition for CNTs growing using CVD.

In conclusion, BA is applicable algorithm for optimizing CNTs yield% and could be used as an effective optimization approach in other similar case studies.

5 Conclusion

This paper has studied CVD process parameters optimization for CNTs growing in two datasets. In this study, BA was proposed as a simple and efficient metaheuristic computational intelligence technique. The adapted BA with RSM in this study has demonstrated better result than RSM in maximizing CNTs yield%. In addition to its efficient performance with fast and stable convergence, the adapted BA with RSM could be proposed as an alternative technique of using conventional experimental method in the process parameter optimization problems, and therefore preventing repetitive laboratory experiments. Based on the result and analysis, parameters tuning of BA is highly recommended in order to improve the search process. Thus, BA is applicable and suggested to be also used in similar optimization problems. In future work, BA could be enhanced in order to improve its search performance to reduce function evaluation number (FEN) and mean required iterations (M_RI), and to increase success rate (SR%) and apply it in different case studies. Finally, it is vital to find a suitable optimization search algorithm for such complex engineering domain, i.e., CNTs growing process.

References

1. Abdel-Basset M, Hessin AN, Abdel-Fatah L (2018) A comprehensive study of cuckoo-inspired algorithms. Neural Comput Appl 29(2):345–361
2. Al-Betar MA et al (2018) Bat-inspired algorithms with natural selection mechanisms for global optimization. Neurocomputing 273(August):448–465
3. Allaedini G, Tasirin SM, Aminayi P (2016) Yield optimization of nanocarbons prepared via chemical vapor decomposition of carbon dioxide using response surface methodology. Diam Relat Mater 66:196–205
4. Alomari OA, Khader AT, Al-Betar MA, Abualigah LM (2017) MRMR BA: a hybrid gene selection algorithm for cancer classification. J Theor Appl Inf Technol 95(12):2610–2618
5. Alqattan ZN, Abdullah R (2015) A hybrid artificial bee colony algorithm for numerical function optimization. Int J Mod Phys C 26(10):1550109
6. Angulakshmi VS, Tamilarasan G, Karthikeyan S (2014) Optimization of CVD synthesis conditions for the synthesis of multiwalled carbon nanotubes using response surface methodology. J Environ Nanotechnol 3(2):81–91
7. Bajad GS, Tiwari SK, Vijayakumar RP (2015) Synthesis and characterization of CNTs using polypropylene waste as precursor. Mater Sci Eng B 194:68–77
8. Cai X et al (2018) Bat algorithm with triangle-flipping strategy for numerical optimization. Int J Mach Learn Cybernet 9(2):199–215
9. Chai SP, Lee KY, Ichikawa S, Mohamed AR (2011) Synthesis of carbon nanotubes by methane decomposition over $Co-Mo/Al_2O_3$: process study and optimization using response surface methodology. Appl Catal A Gen 396(1–2):52–58
10. Chakri A, Ragueb H, Yang X-S (2018) Bat algorithm and directional bat algorithm with case studies. Nat Inspired Algorithms Appl Optim 744:189–216
11. Fauzi NF, Jaya ASM, Jarrah MI, Akbar H (2017) Thin film roughness optimization in the TiN coatings using genetic algorithms. J Theor Appl Inf Technol 95(24):6690–6698
12. Jarrah MI et al (2016) Intelligence integration of particle swarm optimization and physical vapour deposition for tin grain size coating process parameters. J Theor Appl Inf Technol 84(3):355–369
13. Jarrah MI et al (2015) Modeling and optimization of physical vapour deposition coating process parameters for tin grain size using combined genetic algorithms with response surface methodology. J Theor Appl Inf Technol 77(2):235–252
14. Jaya ASM, AbdulKadir NA, Jarrah MI (2014) Modeling of TiN coating roughness using fuzzy logic approach. Sci Int (Lahore) 26(4):1563–1567
15. Jaya ASM, Jarrah MI, Muhamad MR (2015) Modeling of TiN coating grain size using RSM approach. Appl Mech Mater 754–755:738–742
16. Jayabarathi T, Raghunathan T, Gandomi AH (2018) The bat algorithm, variants and some practical engineering applications: a review. Stud Comput Intell 744:313–330
17. Khor CP, Jaafar M, Ramakrishnan S (2016) Optimization of conductive thin film epoxy composites properties using desirability optimization methodology. J Optim 2016:1–8
18. Koyee RD, Heisel U, Schmauder S, Eisseler R (2014) Experimental investigation and multi-objective optimization of turning duplex stainless steels. Int J Manuf Eng 2014:1–13
19. Lee IH, Lin JH, Wu CC (2015) A novel multi-objective optimization algorithm for social network. In: Multidisciplinary social networks research, vol 540. Springer, Berlin, pp 30–40
20. Liang H et al (2018) A hybrid bat algorithm for economic dispatch with random wind power. IEEE Trans Power Syst 8950:1–10 (May 2017)

21. Liu W-W et al (2012) Optimisation of reaction conditions for the synthesis of single-walled carbon nanotubes using response surface methodology. Can J Chem Eng 90:489–505
22. Mohammed IA et al (2017) Full factorial design approach to carbon nanotubes synthesis by CVD method in argon environment. S Afr J Chem Eng 24:17–42
23. Nayak BB, Mahapatra SS (2016) Optimization of WEDM process parameters using deep cryo-treated Inconel 718 as work material. Eng Sci Technol Int J 19(1):161–170
24. Norlina MS, Mazidah P, Sin NDM, Rusop M (2015) Application of metaheuristic algorithms in nano-process parameter optimization. In: 2015 IEEE congress on evolutionary computation, pp 2625–2630
25. Porro S et al (2007) Optimization of a thermal-CVD system for carbon nanotube growth. Physica E 37(1–2):16–20
26. Sahu RP et al (2018) Synthesis, characterization, and applications of carbon nanotubes functionalized with magnetic nanoparticles. In: Balasubramanian G (ed) Advances in nanomaterials. Springer International Publishing AG, pp 37–57
27. Tai NH et al (2012) Optimization of processing parameters of the chemical vapor deposition process for synthesizing high-quality single-walled carbon nanotube fluff and roving. Compos Sci Technol 72(15):1855–1862
28. Yang XS (2010) A new metaheuristic bat-inspired algorithm. In: González JR, Pelta DA, Cruz C, Terrazas G, (eds) Nature inspired cooperative strategies for optimization (NISCO 2010). Studies in computational intelligence. Springer, Berlin, pp 65–74
29. Yousefi AT et al (2015) Vectorial crystal growth of oriented vertically aligned carbon nanotubes using statistical analysis. Cryst Growth Des 15(7):3457–3463
30. Al Nuaimi ZNAM, Abdullah R (2017) Neural network training using hybrid particle-move artificial bee colony algorithm for pattern classification. J ICT 2(2):314–334

A New Variable Strength t-Way Strategy Based on the Cuckoo Search Algorithm

Abdullah B. Nasser and Kamal Z. Zamli

Abstract Considering systematic interaction of inputs, t-way testing is a sampling strategy that generates a subset of test cases from a pool of possible tests. Many t-way testing strategies appear in the literature to date ranging from general computational ones to metaheuristic-based. Owing to its performance, the metaheuristic-based t-way strategies have gained significant attention recently (e.g., Particle swarm optimization, genetic algorithm, ant colony algorithm, harmony search, and cuckoo search). Despite much progress, existing strategies have not sufficiently dealt with more than one interaction between input parameters (termed *variable strength* t-way). Complementing existing works, this paper proposes a new variable strength cuckoo search algorithm, called VCS. Experimental results have shown promising results as VCS can compete with many existing works.

Keywords Software testing · Variable strength interaction · Metaheuristic · Cuckoo search algorithm

1 Introduction

Software testing can be seen as the gatekeeper of quality for any software system (i.e., ensuring that the software meets the user's needs, business, and technical requirements). Ideally, exhaustive testing is desirable. Yet, exhaustive testing (i.e., involving all combinations of inputs values) is impractical and can lead to combinatorial explosion problem. In this respect, parallel testing can be an alternative approach; however, parallel testing is often resource hungry and expensive (e.g., in terms of the need to do load balancing). Alternatively, random testing selection also could be used; however, ensuring fair selection of test cases could be an issue.

A. B. Nasser · K. Z. Zamli (✉)
Faculty of Computer Systems and Software Engineering, Universiti Malaysia Pahang, 26300 Kuantan, Pahang, Malaysia
e-mail: kamalz@ump.edu.my

A. B. Nasser
e-mail: abdullahnasser83@gmail.com

© Springer Nature Singapore Pte Ltd. 2019
V. Piuri et al. (eds.), *Intelligent and Interactive Computing*, Lecture Notes in Networks and Systems 67, https://doi.org/10.1007/978-981-13-6031-2_17

Addressing the aforementioned issues, sampling strategy based on t-way testing (where t indicates the interaction strength) has been proposed to date. The usefulness of t-way based sampling strategies can be seen in many studies. For instance, pairwise (i.e., $t = 2$) can detect more than 70% of the bugs [1–3]. To have more chances of detecting bugs, one can generate the pairwise (i.e., two-way) test cases for all parameters and then compute three-way test cases for specific subset of parameters, called *variable strength* t-way testing.

Owing to its superior performance, the adoption of cuckoo search (CS) for t-way test suite generation appears to be an attractive possibility. In fact, many existing works prove that CS gives more efficient results than particle swarm optimization (PSO) and genetic algorithm (GA) as shown in [4–7]. In another study on scheduling optimization [8], CS performs better than GA and PSO. Additionally, CS is computationally efficient and lightweight relying only on one control parameter as well as provides aggressive exploration and exploitation through the adoption Lévy flight motion.

2   Overview of t-Way Testing

Nowadays, software development companies are now shifting their development efforts to software components based on the product lines approach. With product lines approach, a system can be seen as a decomposition of many components (or parameters) which interact with each other. As illustration, consider the Ms Word application in Fig. 1). Here, there are 12 parameters (i.e., Alignment values, Outline levels, Left and Right Indentation values, etc. as shown in Table 1); each parameter has different values. For instance, the first parameter has 7 values; the second parameter has 10 values; and the forth has 1 value. Hence, exhaustive testing of these 12 parameters values requires 5040 test cases (i.e., $7 \times 10 \times 1 \times 1 \times 3 \times 1 \times 2 \times 1 \times 1 \times 6 \times 1 \times 2$), which is hardly feasible in practice.

Considering two-way testing, the test case reduction can be significantly high, that is, from 5040 to 70 test cases. In this case, every two combinations will be considered at least one time in the final test suite. We can increase the interaction level to three-way, then the reduction is from 5040 to 430 test cases at $t = 3$. We observe that the difference between generated test cases sizes when $t = 2$ and $t = 3$ is very high. The fact that not all three-way combinations cause these faults warrants the possibility of computing variable strength interaction of two-way testing for all the 12 parameters and adding three-way interaction for selected parameters.

Mathematically, if symbolic values p, v, and t are used in place of parameters, values, and interaction strength, respectively, then CA $(N;\ t,\ v^p)$ refers to uniform covering array consisting of N rows of generated test cases from p parameters where each row has v values. If the number of parameters is nonuniform, the covering array representation takes the mixed coverage array notation of $MCA(N;\ t,\ v_1^{p1} v_2^{p2} v_3^{p3} \ldots, v_j^{pj})$. When a set of parameters is subjected to different

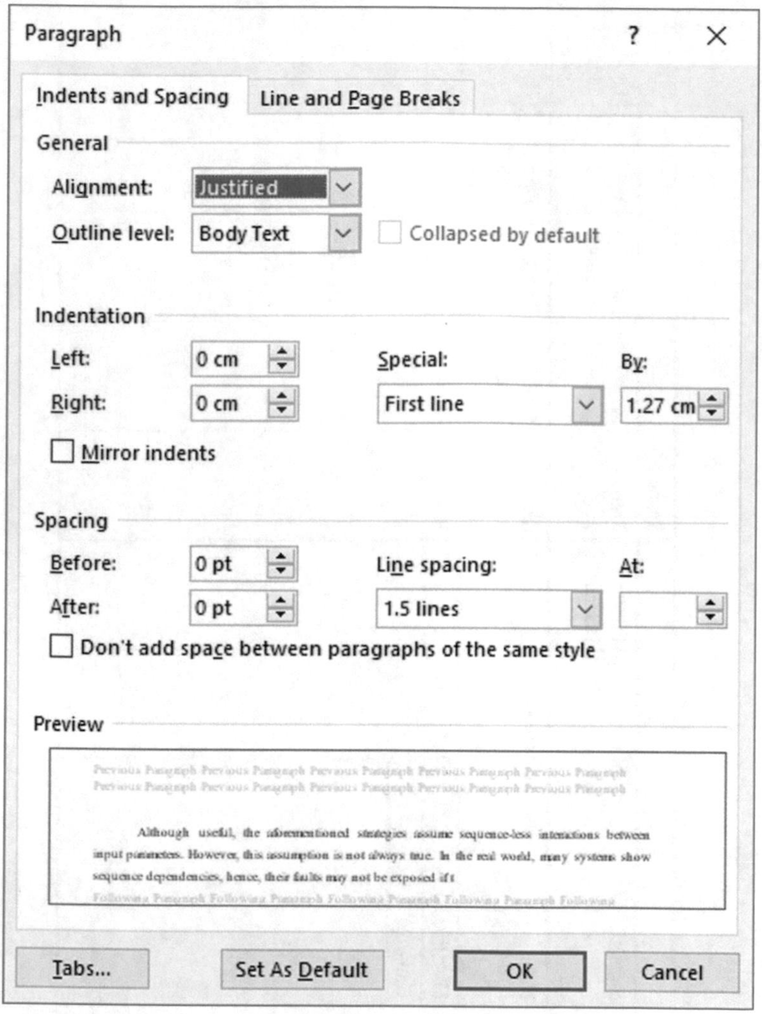

Fig. 1 Ms Word application

values of interaction strength (t), then the covering array is represented by variable or mixed strength covering array denoted as $\text{VCA}\left(\text{N}, v_1^{p1} v_2^{p2} v_3^{p3} \ldots . v_j^{pj} (\text{CA})\right)$ [9, 10].

Table 1 Ms Word application components and values

General		Indentation					Spacing				
Alignment	Outline level	Left	Right	Special	By	Mirror indents	Before	After	Line spacing	at	Don't add space
Left	Body test	One input value	One input value	Non	One input value	0	One input value	One input value	Single	One input value	0
Centered	OL_1			First line		1			1.5 line		1
Right	OL_2			Hanging					Double		
Justified	OL_3								At least		
Justify low	OL_4								Exactly		
Justify medium	OL_5								Multiple		
Justify high	OL_6										
	OL_7										
	OL_8										
	OL_9										

3 Related Works

Often, implementation of strategies for t-way test suite generation can be viewed as algebraic and computational approaches, respectively [10, 11]. Concerning algebraic approach, test sets are constructed using mathematical properties of covering arrays. As such, algebraic approach often adopts lightweight computations. Specifically, this approach exploits the mathematical properties of covering arrays in terms of arrangements of rows and columns [12–14]. Examples of algebraic-based strategies are orthogonal Latin squares (OLS) and TConfig. On the negative note, this approach is often limited to small configurations [9, 15].

Computational approaches exploit specific algorithm to ensure that all the interactions of combinations are fully covered in the greedy manner (i.e., covering as much interaction as possible within each test). This approach generates the test cases either using one-test-at-a-time (OTAT) (e.g., AETG [16], Jenny [17], TConfig [18], and WHITCH [19]) or one-parameter-at-a-time (OPAT) (e.g., in-parameter-order (IPO), IPOG [20], IPOG-D [11], and IPOF [21]).

Many metaheuristic algorithms have been adopted for generating t-way test suite owing to its performance as effective search algorithms (i.e., as the problem of t-way test suite generation is an optimization problem). Metaheuristic algorithm often starts with random population of solution. Then, the algorithm improves the current population by some random operator to generate a new solution. This new solution will replace the current population if and only if it has better objective value. The searching process continues until all combinations are covered. This is often done in greedy manner (i.e., maximizing the interaction coverage per test case).

To date, many metaheuristic-based t-way strategies have been developed in the scientific literature including SA [22], HS [10], GA [22, 23], PSO [9], ACA [23], FPA [24], hyper-heuristics (HHH) [25], fuzzy-based hyper-heuristics (FIS) [26], adaptive TLBO (ATLBO) [27], and sine cosine algorithm (SCA) [28]. Although useful, much of these algorithms have focused on general t-way strategy and lesser efforts are concerned on variable strength t-way. For this reason, our work contributes to this respect.

4 Cuckoo Search Algorithm

Cuckoo search (CS) is a new metaheuristic algorithm that mimics the behavior of brood parasitic of Guira cuckoos [29]. Essentially, CS has two search operators: local search operator which performs the local search around the current best and global search operator based on the Lévy flight motion. Local and global searches are controlled by probability (pa) to offer optimum balance local and global search.

In a nutshell, CS algorithm works as follows: the algorithm starts with a randomly generated nests (or populations). Then, the local and global search operators are repeatedly performed on the population. Ideally, the search operator generates a new

```
Objective function f(x), x = (x₁, ..., x_d) ;
Begin
1.     Initial a population of n host nests  x_i  (i = 1, 2, ..., n);
2.     while (t <MaxGeneration) or (stop criterion)
3.         Get a cuckoo (say i) randomly by Lévyflights;
4.         Evaluate its quality/fitness Fi;
5.         Choose a nest among n (say j) randomly;
6.         IF (Fi > Fj)
7.             Replace j by the new solution;
8.         End if
9.         Abandon a fraction (pa) of worse nests and
10.        build new ones at new locations
11.        Keep the best solutions
12.        Rank the solutions and find the current best;
13.    end while
14.    Postprocess results and visualization;
15.    End-Procedure
```

Fig. 2 CS algorithm

nest, using a random or Levy flight motion. If the new nest improves, the current nest is replaced accordingly. The algorithm may also remove the worse nest based on abandon probability (pa). Summing up, the complete pseudocode of CS is given Fig. 2.

5   Cuckoo Search Strategy for Variable Strength

VCS adopts two main steps for generating test suite. The steps are generating interaction elements and finding the optimal test cases, respectively. Figure 3 depicts the two steps using VCA $(N, 2, 4^{15}, \{CA (3, 4^3)\})$ as illustrated.

Generating interaction elements step starts by analyzing the system configuration into main using VCA $(N, 2, 4^{15})$ and sub-covering arrays CA $(3, 4^3)$. Based on the covering arrays configuration, interaction elements list is generated which represent all possible combinations of parameters values (Table 2).

The complete step of VCS is finding optimal test cases. In VCS, each nest represents one test case. VCS starts to generate population of nests randomly, and then updates the nests by performing a Levy flight. Based on nest weight, which represents the covered interaction elements by the nest, the best nest is selected and appended to the final test suite and subsequently removed from the interaction element list.

Generate Interaction Elements

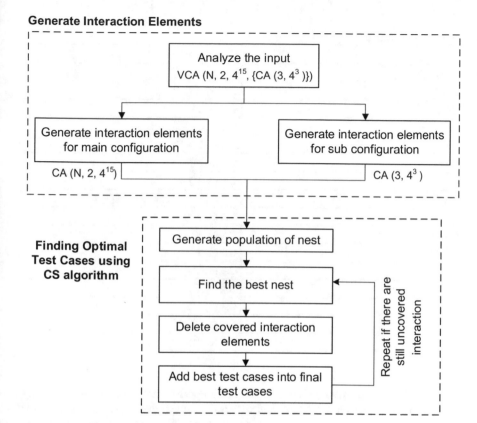

Fig. 3 The design of VCS strategy

6 Results and Evaluation

For experiment setup, the values of nest size $= 30$, Pa $= 0.25$, and iteration $= 500$ have been used based on [30]. The experiments have been performed on NetBeans 8.1.0 platform. The machine's specification used is HP, Intel Core i7 CPU, 16 GB RAM, Windows 10 professional, and 64-bit Operating System.

Table 3 depicts the VCS of interest consisting of VCA $(N, 2, 3^{15}, (C))$ along with the sub-arrays. Table 3 shows the comparison between VCS and existing strategies in terms of test suite size for VCA $(N, 2, 3^{15}, (C))$. Cell with bold text represents the best-obtained test suite size among the existing strategies. Cell marked "NA" represents the missing results which are not available in the literature.

Table 2 Array definition

	Covering arrays
CA-1	$\varnothing$
CA-2	CA $(3, 3^3)$
CA-3	CA $(3, 3^3)^2$
CA-4	CA $(3, 3^3)^3$
CA-5	CA $(3, 3^4)$
CA-6	CA $(3, 3^6)$
CA-7	CA $(3, 3^9)$
CA-8	CA $(3, 3^4)$, CA $(3, 3^5)$, CA$(3, 3^6)$
CA-9	CA $(3, 3^{15})$
CA-10	CA $(4, 3^4)$
CA-11	CA $(4, 3^5)$
CA-12	CA $(4, 3^7)$
CA-13	CA $(5, 3^5)$
CA-14	CA $(5, 3^7)$
CA-15	CA $(6, 3^6)$
CA-16	CA $(6, 3^7)$

At the first glance, we can observe that SA outperforms other strategies in many cases (7 out of 16 test cases), as marked with bold fonts; however, SA is a limited for small interaction strength. The table also shows that VCS and HSS strategies produce the best results in 10 out of 16 test cases and 9 out of 16 test cases, respectively. This early result demonstrates that VCS provides a viable solution for variable strength t-way test suite generation.

7 Conclusion

Many studies show that t-way strategies are very useful for reducing interaction faults among the system's components. However, some real-world applications' faults may not be exposed if a subset of the components is not considered by different interaction strengths in testing process.

Our work proposes VCS, a new cuckoo search-based t-way strategy that addresses the support for variable strength covering arrays. Experimental results show that VCS produces competitive results in most cases considered. As part of the future work, we are currently working on other experiments for improving and evaluating the performance of VCS for software product lines testing.

Table 3 Test suite size of VCS against existing strategies for VCA (N, 2, 3^{15}, (C))

Covering arrays	PSO	HSS	ACS	SA	TVG	PICT	Density	IPOG	ParaOrder	WHITCH	VCS
CA-1	19	20	19	**16**	22	35	21	21	33	31	18
CA-2	**27**	**27**	27	27	27	81	28	27	27	48	**27**
CA-3	**27**	**27**	**27**	**27**	30	729	28	30	33	59	**27**
CA-4	**27**	**27**	**27**	**27**	30	785	28	33	33	69	**27**
CA-5	30	**27**	**27**	**27**	35	105	32	39	27	59	**27**
CA-6	45	45	45	**34**	48	146	46	53	49	61	45
CA-7	57	60	57	**50**	62	177	60	65	62	94	55
CA-8	45	45	**40**	34	53	1376	46	51	44	114	44
CA-9	74	77	76	**67**	81	83	70	NA	82	132	72
CA-10	**81**	**81**	NA	NA	**81**	245	NA	**81**	NA	103	**81**
CA-11	97	**94**			103	301		122		118	**94**
CA-12	158	159			168	505		181		189	**155**
CA-13	**243**	**243**			**243**	730		**243**		261	**243**
CA-14	**441**	**441**			462	1356		581		481	**441**
CA-15	**729**	**729**			**729**	2187		**729**		745	**729**
CA-16	966	902			1028	3045		1196		1050	973

Acknowledgements This work is partially funded by the "Long Term National Plan for Science, Technology and Innovation (LT-NPSTI) Grant, King Abdul Aziz City for Science and Technology, Kingdom of Saudi Arabia (grant no: UIC141504)" and "the FRGS Grant—A Reinforcement Learning Sine Cosine based Strategy for Combinatorial Test Suite Generation (grant no: RDU170103)".

References

1. Othman RR, Zamli KZ (2011) ITTDG: Integrated t-way test data generation strategy for interaction testing. Sci Res Essays 6:3638–3648
2. Younis MI, Zamli KZ, Isa NM (2008) MIPOG-modification of the IPOG strategy for t-way software testing. In: Proceeding of the distributed frameworks and applications (DFmA)
3. Younis MI, Zamli KZ, Isa NAM (2008) Algebraic strategy to generate pairwise test set for prime number parameters and variables. In: International symposium on information technology, 2008. ITSim 2008, pp 1–4
4. Yang X-S, Deb S (2010) Engineering optimisation by cuckoo search. Int J Math Model Numer Optimisation 1:330–343
5. Yang XS (2010) Nature-inspired metaheuristic algorithms. Luniver press
6. Yang XS, Deb S, Karamanoglu M, He X (2012) Cuckoo search for business optimization applications. In: 2012 National conference on computing and communication systems (NCCCS), 2012, pp 1–5
7. Yildiz AR (2013) Cuckoo search algorithm for the selection of optimal machining parameters in milling operations. Int J Adv Manuf Technol 64:55–61
8. Burnwal S, Deb S (2013) Scheduling optimization of flexible manufacturing system using cuckoo search-based approach. Int J Adv Manuf Technol 64:951–959
9. Ahmed BS, Zamli KZ, Lim CP (2012) Constructing a t-way interaction test suite using the particle swarm optimization approach. Int J Innovative Comput Inf Control 8:431–452
10. Alsewari ARA, Zamli KZ (2012) Design and implementation of a harmony-search-based variable-strength t-way testing strategy with constraints support. Inf Softw Technol 54:553–568
11. Lei Y, Kacker R, Kuhn DR, Okun V, Lawrence J (2008) IPOG/IPOG-D: efficient test generation for multi-way combinatorial testing. Softw Test Verification Reliab 18:125–148
12. Hartman A, Raskin L (2004) Problems and algorithms for covering arrays. Discrete Math 284:149–156
13. Mandl R (1985) Orthogonal latin squares: an application of experiment design to compiler testing. Commun ACM 28:1054–1058
14. Bush KA (1952) Orthogonal arrays of index unity. Ann Math Stat 23:426–434
15. Williams AW (2000) Determination of test configurations for pair-wise interaction coverage. In: Ural H, Probert RL, von Bochmann G (eds) Testing of communicating systems: tools and techniques. IFIP TC6/WG6.1 13th International conference on testing of communicating systems (TestCom 2000), pp 59–74, Springer US, Boston, MA
16. Cohen DM, Dalal SR, Fredman ML, Patton GC (1997) The AETG system: an approach to testing based on combinatorial design. IEEE Trans Softw Eng 23:437–444
17. Jenkins B (2003, 5-Dec-2017). Jenny tool. Available http://www.burtleburtle.net/bob/math
18. Williams A (2014) TConfig download page [Online]. Available http://www.site.uottawa.ca/~awilliam/. Accessed 23 Dec 2014
19. Hartman A, Klinger T, Raskin L (2010) IBM intelligent test case handler. Discrete Math 284:149–156
20. Lei Y, Kacker R, Kuhn DR, Okun V, Lawrence J (2007) IPOG: a general strategy for t-way software testing. In: 14th Annual IEEE international conference and workshops on the engineering of computer-based systems (ECBS'07), pp 549–556
21. Forbes M, Lawrence J, Lei Y, Kacker RN, Kuhn DR (2008) Refining the in-parameter-order strategy for constructing covering arrays. J Res Nat Inst Stand Technol 113:287

22. Stardom J (2001) Metaheuristics and the search for covering and packing arrays. Simon Fraser University
23. Shiba T, Tsuchiya T, Kikuno T (2004) Using artificial life techniques to generate test cases for combinatorial testing. In: 28th Annual international computer software and applications conference, pp 72–77
24. Nasser AB, Zamli KZ, Alsewari AA, Ahmed BS (2018) Hybrid flower pollination algorithm strategies for t-way test suite generation. PLoS ONE 13:e0195187
25. Zamli KZ, Alkazemi BY, Kendall G (2016) A tabu search hyper-heuristic strategy for t-way test suite generation. Appl Soft Comput 44:57–74
26. Zamli KZ, Din F, Kendall G, Ahmed BS (2017) An experimental study of hyper-heuristic selection and acceptance mechanism for combinatorial t-way test suite generation. Inf Sci 399:121–153
27. Zamli KZ, Din F, Baharom S, Ahmed BS (2017) Fuzzy adaptive teaching learning-based optimization strategy for the problem of generating mixed strength t-way test suites. Eng Appl Artif Intell 59:35–50
28. Zamli KZ, Din F, Ahmed BS, Bures M (2018) A hybrid Q-learning sine-cosine-based strategy for addressing the combinatorial test suite minimization problem. PLoS ONE 13:e0195675
29. Yang X-S, Deb S (2009) Cuckoo search via lévy flights. In: World congress on nature and biologically inspired computing, pp 210–214
30. Nasser AB, Alsewari ARA, Zamli KZ (2015) Tuning of cuckoo search based strategy for t-way testing. In: International conference on electrical and electronic engineering, p 8948

A Divide-and-Conquer Strategy for Adaptive Neuro-Fuzzy Inference System Learning Using Metaheuristic Algorithm

Mohd Najib Mohd Salleh, Kashif Hussain and Noreen Talpur

Abstract Adaptive neuro-fuzzy inference system (ANFIS) has produced promising results in model approximation. The core of ANFIS computation lies in the training of its parameters. Metaheuristic algorithms have been successfully employed on ANFIS parameters training. Conventionally, a population individual in metaheuristic algorithm, considered as ANFIS model with candidate parameters, is evaluated for its fitness on complete training set. This makes ANFIS parameters training computationally expensive when dataset is large. This paper proposes divide-and-conquer strategy where each population individual is given a piece of dataset instead of complete dataset to train and evaluate ANFIS model fitness. The proposed ANFIS training approach is evaluated on accuracy on testing dataset, as well as, training computational complexity. Experiments on several classification problems reveal that the proposed methodology reduced training computational complexity up to 93%. Moreover, the proposed ANFIS training approach generated rules that achieved better accuracy on testing dataset as compared to conventional training approach.

Keywords Neuro-fuzzy systems · Fuzzy inference · Metaheuristic algorithms · Learning

1 Introduction

Computationally intelligent techniques such as neural network and fuzzy logic have been successfully used for modeling nonlinear problems in engineering, science, medical, business, etc. [1]. The fusion of neural network and fuzzy logic has resulted in highly accurate, adaptive, and flexible neuro-fuzzy system (NFS). An NFS uses learning ability of neural network and reasoning ability of fuzzy inference system to fuzzy IF-THEN rules that best map inputs with the desired output. The performance of any NFS depends on how the network learns its weights (parameters). Hence, learning algorithm employed on NFS is crucial to successfully modeling a problem [2]. There

M. N. M. Salleh (✉) · K. Hussain · N. Talpur
Universiti Tun Hussein Onn Malaysia, 86400 Batu Pahat, Johor, Malaysia
e-mail: najib@uthm.edu.my

© Springer Nature Singapore Pte Ltd. 2019
V. Piuri et al. (eds.), *Intelligent and Interactive Computing*, Lecture Notes in Networks and Systems 67, https://doi.org/10.1007/978-981-13-6031-2_18

has been extensive research pursuing high accuracy and efficient learning in NFSs. According to [3], an efficient NFS maintains (a) fast parameter learning for generating smallest possible error and (b) low network complexity for less computational cost.

Adaptive neuro-fuzzy inference system (ANFIS) [4] is based on Takagi—Sugeno–Kang (TSK) inference mechanism. It is the most successful among other NFSs, applied on almost all types of data modeling problems including regression, classification, approximation, as well as, control problems [2, 5]. However, ANFIS suffers from poor learning performance especially on problems where parameter optimization becomes highly multimodal and non-convex problem, due to gradient-based learning. Moreover, gradient-based learning algorithm in standard ANFIS is slow and prone to produce suboptimal parameter tuning. To cater this, researchers have proposed numerous alternatives to ANFIS learning using metaheuristic algorithms. This research also uses metaheuristic algorithm called particle swarm optimization (PSO) for ANFIS parameter training, but with a different approach than typically applied in literature.

The subsequent section discusses ANFIS and its standard learning algorithm. The alternative ANFIS learning algorithms proposed in literature have been highlighted in this section. Section 3 explains the proposed divide-and-conquer approach for the training of ANFIS parameters using PSO. The experimental environment is presented in Sect. 4. Section 5 discusses results, and the study is concluded in Sect. 6.

2 Adaptive Neuro-fuzzy Inference System

Jang in 1993 [4] introduced adaptive neuro-fuzzy inference system (ANFIS) which is a neural network-type structure with fuzzy logic embedded in first and fourth layers. The trainable parameters of ANFIS architecture are contained in first and fourth layers, and the remaining three layers perform inference operations to connect IF and THEN parts of a fuzzy rule. A fuzzy rule in ANFIS may be expressed as shown in Fig. 1, which has two parts: IF part which is also called premise part and the latter THEN part is known as consequent part. The k parameters in these two parts are trained by learning algorithm. This implies that if an ANFIS architecture has r rules, then $k \cdot r$ parameters will be trained, meaning that ANFIS training algorithm will have enormous computational cost if rules increase exponentially in a problem with large number of input variables.

The layer-wise illustration of ANFIS architecture is given in Fig. 2. The first layer of ANFIS takes crisp inputs and converts into degree of membership for the linguistic term defined. In Fig. 2, μ_{A1}, μ_{A2}, and μ_{B1}, μ_{B2} are membership functions for input

Fig. 1 IF-THEN rule in
ANFIS

$$\underbrace{\text{IF } x \text{ is } \mu_{A_1} \text{ AND } y \text{ is } \mu_{B_1}}_{\text{Premise Part}} \underbrace{\text{THEN } f = xp + yq + r}_{\text{Consequent Part}}$$

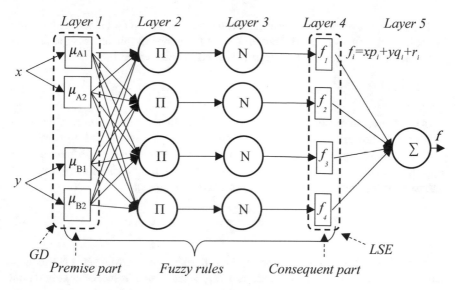

Fig. 2 ANFIS architecture with two inputs, four rules, and one output

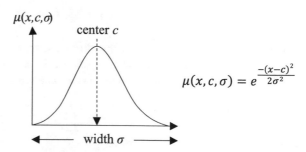

Fig. 3 Gaussian-type membership function

$$\mu(x, c, \sigma) = e^{\frac{-(x-c)^2}{2\sigma^2}}$$

variables x and y, respectively. Here, it is noteworthy to mention that the definition of membership function is user-defined. It is up to user who, based on experience or expert opinion, defines the shape of membership function which can be trapezoidal, triangular, Gaussian, generalized bell-shaped, or any other user-defined shape. Every node in this layer contains parameters updated during training process. For example, in case of Gaussian membership function (Fig. 3), the trainable parameters are center (c) and width (σ).

The output $O_{1,i}$ of every node in the first layer is membership degree, as expressed in (1) where μ_{Ai} and μ_{Bi} are membership functions:

$$O_{1,i} = \mu_{Ai}(x), \quad i = 1, 2$$
$$\text{or} \tag{1}$$
$$O_{1,i} = \mu_{Bi}(y), \quad i = 3, 4$$

The second layer simply performs product Π operation to calculate the firing strength w_i of a fuzzy rule. The output $O_{2,i}$ of the nodes in this layer is the product of combinations of membership degrees related to input variables.

$$O_{2,i} = w_i = \mu_{Ai}(x).\mu_{Bi}(y), \quad i = 1, 2, 3, 4 \tag{2}$$

In this layer, the rule base is generated with the help of different partitioning methods—mainly clustering and grid partitioning methods are used [6]. This research uses grid partitioning to generate m^n rules where m is number of membership functions per input and n is the number of input variables. The strength of each rule is normalized against all rules in the rule base in the third layer, using (3):

$$O_{3,i} = \bar{w}_i = \frac{w_i}{\sum_{j=1}^{n} w_j}, \quad i = n = 1, 2, 3, 4 \tag{3}$$

where $\bar{w}_i$ is the normalized firing strength of ith rule. The fourth layer is consequent part of fuzzy rules. The output of each rule $O_{4,i}$ is calculated using linear polynomial equation as expressed in (4):

$$O_{4,i} = \bar{w}_i f_i = (xp_i + yq_i + r_i), \quad i = 1, 2, 3, 4 \tag{4}$$

where p_i, q_i, and r_i are the consequent parameters associated with ith rule. These parameters are trained by learning algorithm. In the last layer, the function of the only node O_5 is to aggregate outputs of rules calculated in the previous layer (5):

$$O_5 = \sum_{i=1}^{n} O_{4i} \tag{5}$$

where n is the total number of rules (such as four rules in Fig. 1). It is obvious from (5) that the output of individual rule is important to the aggregated output of ANFIS.

2.1 Standard ANFIS Learning Method

To achieve minimum error, ANFIS learning algorithm updates trainable membership function parameters (such as c, σ) and consequent parameters (such as p, q, r). The standard learning algorithm uses a hybrid of gradient descent (GD) and least square estimator (LSE). For utilizing the two methods, ANFIS uses a two-pass learning algorithm. During the first pass, LSE tunes consequent parameters, while in the second pass, GD updates membership function parameters to further approximate the model with better accuracy.

Training of ANFIS parameters is often formulated as optimization problem. In this regard, according to [5], mainly three approaches related to ANFIS learning are found in literature: gradient-based learning, heuristic algorithms, and hybrid methods combining derivative and heuristic approaches. Since derivative-based learning approaches are prone to find suboptimal solutions [7], many researchers prefer metaheuristic approaches. The next section highlights some of the works proposing heuristic methods for the training of ANFIS parameters for solving classification problems.

2.2 Heuristic Approaches for ANFIS Learning

To counter suboptimal performance of derivative-based training methods in ANFIS training, various heuristic and metaheuristic algorithms have been utilized in literature. Premise and consequent parameters are often optimized by devising a combined search space and let metaheuristic algorithm to find optimal set of parameters that produce best results for ANFIS network.

The research by [8] employed ANFIS on identification of earthquake victims for disaster management operations in China. For optimizing the network parameters, the authors proposed novel differential biogeography-based optimization (DBBO) algorithm. When compared with other evolutionary algorithms, DBBO produced better classification results. Obo et al. [9] classified human gesture for human—computer interaction using ANFIS, where evolutionary algorithm was used to train parameters. Mine blast algorithm (MBA) was used on ANFIS parameters training for solving classification problems [10]. The research first improved MBA and then employed on both premise and consequent parameters. With an application in medical science, [11] proposed ant colony optimization (ACO) for improving the learning of ANFIS parameters. The research claimed to have improved ANFIS on mammogram classification problem by identifying breast cancer cases more accurately. Rini et al. [12] proposed modified ANFIS with linguistic hedge in membership functions. The research improved the network accuracy by optimizing all the trainable parameters using PSO. The authors contended to have achieved better accuracy on different classification problems.

Based on limited literature review, it is obvious that metaheuristic algorithms have been successfully applied to parameter training of ANFIS model. This implies that there is potential for further improvement in this area of research.

3 Proposed Method for ANFIS Learning

From the literature review mentioned above, it is clear that metaheuristic algorithms have been successfully used on ANFIS parameters optimization problem. In most of the works, each population individual in a metaheuristic algorithm is taken as ANFIS model, meaning that a population individual maintains a set of parameters (premise and consequent parameters). While evaluated on objective function (generally, error

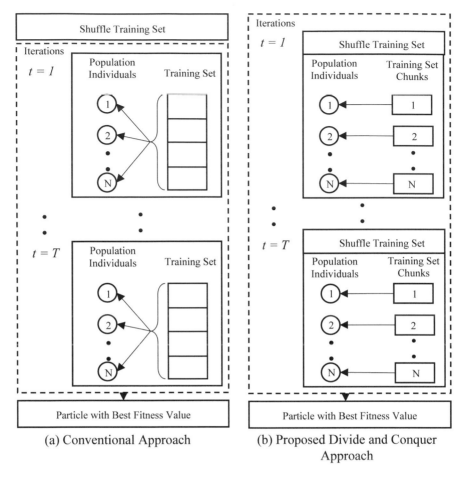

Fig. 4 Conventional and proposed approach for ANFIS parameters training using metaheuristics

measure), each individual is given all the training sets to evaluate model fitness. After training, the best-fit model is tested on testing set for validating model accuracy. This suggests that all the population individuals in a metaheuristic algorithm must try all M instances in a training set during parameter training process. It is therefore clear that if there are N population individuals in a metaheuristic algorithm, all will run complete training set for T times or iterations. This is well illustrated via Fig. 4a.

Opposite to the conventional approach proposed in existing literature, as shown in Fig. 4b, in each iteration t, the training set is shuffled and chunked into N parts so that each population individual is given a new bunch of training instances instead of whole training set. The idea behind shuffling instances, every time before breaking into pieces, is that every population individual should become fully aware of the data so that a best-fit model can be generated. Towards the end of iterations, the population individual with best objective function value is selected to test on testing instances.

It is noteworthy that the division of training set is proportional to population size. If there are N population individuals, N chunks of training data will be prepared.

As an example, we used PSO in this study. A comprehensive detail about PSO can be found in [13]. Algorithm 1 presents pseudocode of the PSO implementation for ANFIS training in this study. ANFIS training procedure starts with shuffling the classification dataset and partitioning into training and testing sets. Then, PSO initializes particles with random positions in search space. The search process is performed in iterations until stop criteria are reached (usually, maximum number of iterations). During iterations, the training dataset is first shuffled and then broken further into N chunks (equal to the swarm size) and each particle is assigned with a piece of training set. For example, if a training set has 100 instances and swarm size is 10, then each piece of training set will comprise of 100/10 instances. Each particle is then evaluated with the designated training set for its fitness value. The particle's personal best and swarm's global best positions are assigned. After the iterations, the particle with best solution is then taken as trained ANFIS model which is then tested on testing set for validation purpose. While testing, the trained ANFIS model is tested on the number of instances correctly identified.

4 Experimental Settings

Several experiments were performed to evaluate the proposed approach to train ANFIS parameters using metaheuristic algorithms. We solved classification problems by ANFIS using three classification datasets from the UCI (University of California) Machine Learning Repository [14]: Habermans, Iris Flower, Vertebral, and three from KEEL (knowledge extraction based on evolutionary learning) repository [15]: Banana, Phoneme, and Appendicitis. The datasets were divided into training and testing sets with the ratio of around 70:30 percent. Table 1 presents dataset details and further partitioning information.

Algorithm 1: PSO Steps for ANFIS Training
1: Partition dataset into training and testing sets
2: Initialize N particles with random positions
3: **Do**
4:     Shuffle and partition training set into N chunks
5:     Assign each particle with one chunk of training set
6:     Evaluate fitness of each particle on assigned training set
7:     Assign personal best and global best particles
8:     Update velocity and position of the particles
9: **Until** maximum iterations or stop criteria reached
10: Apply global best particle on testing set and evaluate results.

The objective function is mean squared error (MSE) for ANFIS training. After the model is trained, it is then evaluated on testing set for the number of instances correctly identified—hence, testing accuracy is computed as (6)

Table 1 Experimental settings

Datasets	Habermans (Attributes = 3, Instances = 306, Classes = 2)
	Iris Flower (4 Attributes, 150 Instances, 4 Classes)
	Vertebral (6 Attributes, 310 Instances, 3 Classes)
	Banana (2 Attributes, 5300 Instances, 2 Classes)
	Phoneme (5 Attributes, 5404 Instances, 2 Classes)
	Appendicitis (7 Attributes, 106 Instances, 2 Classes)
ANFIS Settings	Membership function shape = Gaussian
	Membership functions per input = 2
	Rule generation method = Grid partitioning
PSO Settings	Swarm Size = 15
	Cognition factor C_1 = Social factor C_2 = 2
	Inertia weight range = [0.4–0.9]
	Maximum iterations = 50

$$\text{Accuracy \%} = \frac{\text{Correct}_N}{\text{Instances}_T} \times 100 \qquad (6)$$

where Correct_N and Instances_T are the testing instances correctly identified and total number of testing instances, respectively. The major reason of the proposed methodology is reducing computational complexity of ANFIS training. We measure training computational complexity (CC_T) of the conventional and the proposed approach using (7)

$$CC_T = \text{Train}_P \times N \times \text{MaxItr} \qquad (7)$$

where Train_P, N, and MaxItr are training instances executed by each particle, swarm size, and maximum number of iterations of PSO algorithm.

5   Results

The proposed divide-and-conquer strategy for ANFIS training using PSO was implemented on six classification datasets. On the other hand, the conventional implementation of PSO on ANFIS training was also performed. The results are compared for accuracy and computational cost of ANFIS training. Because PSO is stochastic method, single run may not reflect the actual performance perspective; hence, the experiments on each dataset were run 10 times and mean of results is reported in this section. Table 2 shows performance results of ANFIS-PSO with proposed and conventional methodologies for comparison. According to the training MSE and testing accuracy, it is clear that there is no significant difference in results. However, from

Table 2 Performance results of ANFIS-PSO on classification problems

Problem	Proposed approach		Conventional approach	
	Training MSE	Testing accuracy (%)	Training MSE	Testing accuracy (%)
Habermans	0.0098	97	0.0100	98
Iris Flower	0.0047	100	0.0049	100
Vertebral	0.0108	96	0.0100	98
Banana	0.0139	94	0.0105	97
Phoneme	0.0122	94	0.0152	93
Appendicitis	0.0070	99	0.0050	100

Table 3 ANFIS parameters training computational complexity

Problem	Training computational complexity		Complexity reduced by (%)
	Proposed approach	Conventional approach	
Habermans	4200	64,200	93
Iris Flower	2100	31,500	93
Vertebral	4200	65,100	94
Banana	74,100	1,113,000	93
Phoneme	75,600	1,134,900	93
Appendicitis	1500	22,200	93

the perspective of computation complexity of training process given in Table 3, it is obvious that the proposed methodology reduced computational complexity significantly. Overall, around 93% of computation complexity of the training process was reduced.

6 Conclusions and Future Directions

A new approach for optimizing ANFIS parameters has been proposed in this study. Unlike traditional approach where population individuals in metaheuristic algorithms are given complete training set to evaluate model fitness, we proposed divide-and-conquer strategy which significantly reduced training computational complexity. In this paper, the training set is chunked into number of pieces equal to population size, meaning that each population individual is given a piece of training set instead of whole dataset. The proposed approach reduced training computational complexity up to 93% on most of the classification problems, whereas classification accuracy remained as good as conventional method. The proposed methodology can be handily implemented on more efficient population-based metaheuristic algorithms, which is the future consideration of this study.

Acknowledgements The authors would like to thank Universiti Tun Hussein Onn Malaysia (UTHM), Malaysia for supporting this research under Postgraduate Incentive Research Grant, Vote No. U560.

References

1. Subramanian K, Savitha R, Suresh S (2014) A complex-valued neuro-fuzzy inference system and its learning mechanism. Neurocomputing 123:110–120
2. Ghosh S, Biswas S, Sarkar D, Sarkar PP (2014) A novel Neuro-fuzzy classification technique for data mining. Egypt Inf J 15(3):129–147
3. Shihabudheen KV, Pillai GN (2018) Recent advances in neuro-fuzzy system: a survey. Knowl-Based Syst 152:136–162
4. Jang JS (1993) ANFIS: adaptive-network-based fuzzy inference system. IEEE Trans Syst Man Cybern 23(3):665–685
5. Karaboga D, Kaya E (2018) Adaptive network based fuzzy inference system (ANFIS) training approaches: a comprehensive survey. Artif Intell Rev 1–31
6. Kisi O, Shiri J, Karimi S, Adnan RM (2018) Three different adaptive neuro fuzzy computing techniques for forecasting long-period daily streamflows. In: Big data in engineering applications. Springer, Singapore, pp 303–321
7. Salleh MNM, Hussain K, Naseem R, Uddin J (2016) Optimization of ANFIS using artificial bee colony algorithm for classification of Malaysian SMEs. In: International conference on soft computing and data mining. Springer, Cham, pp 21–30
8. Zheng YJ, Ling HF, Chen SY, Xue JY (2015) A hybrid neuro-fuzzy network based on differential biogeography-based optimization for online population classification in earthquakes. IEEE Trans Fuzzy Syst 23(4):1070–1083
9. Obo T, Loo CK, Seera M, Kubota N (2016) Hybrid evolutionary neuro-fuzzy approach based on mutual adaptation for human gesture recognition. Appl Soft Comput 42:377–389
10. Salleh MNM, Hussain K (2016) Accelerated mine blast algorithm for ANFIS training for solving classification problems. Int J Softw Eng Appl 10(6):161–168
11. Thangavel K, Mohideen AK (2016) Mammogram classification using ANFIS with ant colony optimization based learning. In: Annual convention of the computer society of India. Springer, Singapore, pp 141–152
12. Rini DP, Shamsuddin SM, Yuhaniz SS (2016) Particle swarm optimization for ANFIS interpretability and accuracy. Soft Comput 20(1):251–262
13. Poli R, Kennedy J, Blackwell T (2007) Particle swarm optimization. Swarm Intell 1(1):33–57
14. Dua D, Taniskidou EK (2017) UCI machine learning repository. University of California, School of Information and Computer Science, Irvine, CA. http://archive.ics.uci.edu/ml
15. Alcalá-Fdez J, Fernandez A, Luengo J, Derrac J, García S, Sánchez L, Herrera F (2011) KEEL data-mining software tool: data set repository, integration of algorithms and experimental analysis framework. J Multiple-Valued Logic Soft Comput 17(2–3):255–287

A Simple Edges Extraction Method from Complex Digital Images (EEMI)

Abbas M. Al-Ghaili, Hairoladenan Kasim, Marini Othman and Zainuddin Hassan

Abstract This paper proposes a simple Edges Extraction Method from complex digital Images (EEMI). The proposed EEMI uses a simple image processing technique to detect edges of objects and regions inside complex scenarios of images. It highlights objects' edges by increasing their contrast levels and pixels' intensities using special masks. EEMI mainly uses two simple masks one of which is used to detect vertical edges while the other one detects horizontal edges. Results have revealed that EEMI is a robust edge detector with inclined and complex background images. EEMI has been found simple and has simple complexity that helps reduce the computational time existent with other competitive methods. Results have confirmed that EEMI's computation time could efficiently meet real-time requirements. EEMI has been compared to other competitive operators in terms of accuracy and computation time.

Keywords Edges' detection · Image processing · Pixel intensity · Adaptive thresholding

A. M. Al-Ghaili (✉) · M. Othman
Institute of Informatics and Computing in Energy (IICE), Universiti Tenaga
Nasional (UNITEN), 43000 Kajang, Selangor, Malaysia
e-mail: abbasghaili@yahoo.com; abbas@uniten.edu.my

M. Othman
e-mail: marini@uniten.edu.my

H. Kasim · M. Othman · Z. Hassan
College of Computer Science and Information Technology (CSIT), UNITEN,
43000 Kajang, Selangor, Malaysia
e-mail: hairol@uniten.edu.my

Z. Hassan
e-mail: zainuddin@uniten.edu.my

© Springer Nature Singapore Pte Ltd. 2019
V. Piuri et al. (eds.), *Intelligent and Interactive Computing*, Lecture Notes in Networks and Systems 67, https://doi.org/10.1007/978-981-13-6031-2_16

1   Introduction

Edge detection is a very essential task for many image processes. Usually, edges include sensitive and useful details that help perform additional processes that are required. Consequently, edge detection plays an important role in the system's accuracy and performance. Therefore, such a system could be affected in terms of its accuracy and efficiency. There are numerous applications from the field of digital image processing and computer vision that really need the help of edge detection and extraction process, e.g., medical images applications [1, 2], machine and robotics vision [3], video-based tracking and detection [4], intelligent transportation systems [5], security systems [6, 7], and so on.

There have been many edge detection methods proposed and reviewed in literature, e.g., region-based, vertical edge-based [8], text-based detection [4], and desired objects tracking algorithms.

Previous proposed edge detectors have contributed to image processing tasks in terms of several factors, e.g., accuracy of detected edges. However, there are drawbacks of other edge detectors need to be considered, e.g., low accuracy with complex background images, much computation time, edge detector's complexity, etc. This paper aims to enhance the edge detection performance in terms of computation time and code complexity.

Some edges' detectors are good for simple background images, and therefore this paper has proposed an Edges Extraction Method from Images (EEMI) to efficiently process and deal with complex background and inclined images.

One of the biggest drawbacks of edge detectors is the use of global thresholding during images' preprocessing. Such thresholding techniques fail to preserve important foreground details. Thus, the edge detection process will be affected and therefore accuracy does. In this paper, an adaptive thresholding technique [9, 10] has been used in order to keep most details inside images.

Different edges' detectors have used different mask-based techniques such as 3 × 3 [11]. However, those detectors might be affected in terms of computation time as revealed by obtained results and performances. EEMI has adopted a very simple mask which is 2 × 4 and 4 × 2 for both vertical and horizontal edges' detection, respectively. Thus, the code has become more straightforward and simple.

This paper is organized as follows: In Sect. 2, the proposed EEMI is explained in detail. Section 3, presents and discusses findings, whereas EEMI is evaluated in terms of accuracy, computation time, and robustness. Section 4 draws a conclusion.

2   The Proposed EEMI

2.1   Overview

The main steps of the proposed EEMI are unveiled as shown in Fig. 1.

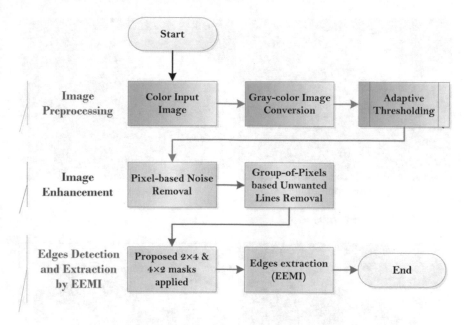

Fig. 1 The proposed EEMI flowchart

This figure shows three main steps which are image preprocessing, image enhancement, and edges' detection and extraction by EEMI.

2.2 Image Preprocessing

This subsection contains many processes which are color input image and gray-color image conversion. In the third process, adaptive thresholding will be applied to the gray-color image. They are explained in detail as follows:

Color Input Image. In this step, the color image is imported and loaded. In this step, the color image size is checked and verified. The maximum height and width are set to 1024 and 768 pixels, respectively.

Gray-color Image Conversion. The color image pre-imported will be converted to a gray-color image, as shown in Fig. 2.

Adaptive Thresholding (AT). This step is important because some details need to be carefully processed. An adequate thresholding technique must be used to guarantee whole details inside regions that are processed efficiently and sufficiently. The adaptive thresholding (AT) technique in [9, 10] has been applied to the grayscale input image, as shown in Fig. 2b, in order to constitute a binarized image. The researchers in [9] have proposed a real-time AT. AT has used an integral image to compute the mean of a local window inside image for each pixel. It is aimed to process pixel

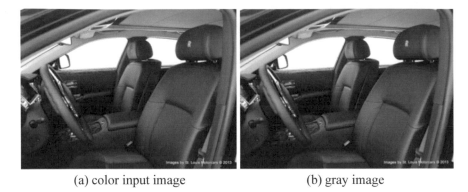

(a) color input image   (b) gray image

Fig. 2 Color image conversion

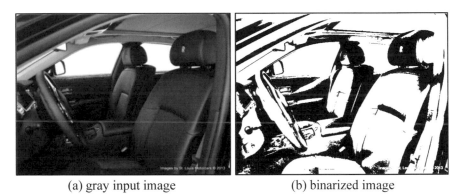

(a) gray input image   (b) binarized image

Fig. 3 Adaptive thresholding process

intensities for scale values between [0–255]. AT has been applied in Fig. 2b; the result is a binarized image as shown in Fig. 3b.

In Fig. 3b, it is noticeable that most details are preserved even though there are some low-contrast regions that their pixels' intensities are slightly different.

2.3 *Image Enhancement*

This section is dedicated to explain image enhancement processes. Discussed are two main enhancement processes using pixel-based and a group of pixels based on noise removal and unwanted lines removal processes, respectively. This process aims to reduce the computation time needed to process the image for further processes by reducing non-important details of regions inside the image.

Pixel-based Noise Removal. A special process is designed for this purpose in which every pixel in the binarized image shown in Fig. 3b is tested with its four neighbors and eight neighbors in two separate times to compare whether the current pixel belongs to its neighbor region or not. If NOT, it is considered as a background pixel which is not important, whereas foreground pixels are the Region-of-Interest (ROI).

Group-of-Pixels based Unwanted Lines Removal. For this process, a very efficient algorithm is reviewed in [12]. The authors have proposed an Unwanted Lines Elimination Algorithm (ULEA) to be used with unwanted lines and details inside images. In this paper, ULEA has been used to guarantee that unwanted details are eliminated from the binarized image. ULEA usually processes a group of pixels at once, so its processing time is short compared to other algorithms. It also processes diagonal lines with 45° and 135° angles to work with inclined images and to guarantee that regions in several directions are covered during processing.

After pixel-based noise removal and group-of-pixels based unwanted lines removal processes have been applied in Fig. 3b, the result is shown in Fig. 4b.

2.4 Edges' Detection and Extraction by EEMI

This process is dedicated to apply a very simple mask on the binarized image after enhancement has been implemented. The applied masks are aimed to detect objects' edges inside images. Each proposed mask uses a group of pixels to be processed at once, meaning masks are applied in a parallel processing technique to help reduce the computation time specifically with large image sizes and complex images with lots of details.

A 2 × 4 Mask-based Vertical Edges' Detection. The proposed mask implemented to detect vertical edges is shown in Fig. 5.

(a) binarized image   (b) enhanced binarized image

Fig. 4 Image enhancement process

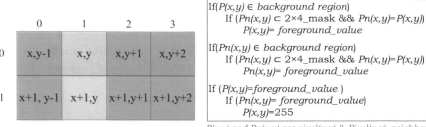

	0	1	2	3
0	x,y-1	x,y	x,y+1	x,y+2
1	x+1, y-1	x+1,y	x+1,y+1	x+1,y+2

If($P(x,y) \in background\ region$)
   If ($Pn(x,y) \subset 2\times4\_mask$ && $Pn(x,y)=P(x,y)$)
     $P(x,y)= foreground\_value$

If($Pn(x,y) \in background\ region$)
   If ($Pn(x,y) \subset 2\times4\_mask$ && $Pn(x,y)=P(x,y)$)
     $Pn(x,y)= foreground\_value$

If ($P(x,y)=foreground\_value$)
   If ($Pn(x,y)= foreground\_value$)
     $P(x,y)=255$

$P(x,y)$ and $Pn(x,y)$ are pixel(x,y) & Pixel(x,y)_neighbors

(a) the proposed 2×4 mask design   (b) the 2×4 mask pseudo-code

Fig. 5 Vertical edges' detection process using 2 × 4 mask

(a) binarized input image   (b) Vertical-edges detection and extraction

Fig. 6 Image enhancement process

EEMI-based Vertical Edges' Extraction. The proposed EEMI has a further process in order to finalize the shape of detected vertical edges to highlight edges of both sides of each object. It applies a special process to make the left edge with each object having 2-pixel width and the right edge of the object has 1-pixel width. This process highlights edges for every object in a distinctive way to ease the searching process of edges implemented by further processes depending on the application by which the searching process is applied. Therefore, the searching process can be done in a very short time. Once this process has been done, the result shows the vertical edges extracted and highlighted in Fig. 6.

3 Results and Discussion

The performance of EEMI is evaluated in terms of accuracy, robustness, and complexity. This work has been implemented and tested using C++ with the help of OpenCV Library running under Windows 7 OS, with an Intel Core i3 CPU 2.13 and a 4 GB RAM. Datasets are captured from real scenarios, whereas some samples are borrowed from the Internet.

3.1 Noise Removal-Based Subjective Evaluation of EEMI's Accuracy

This section presents and compares between two EEMI outputs using two different binarized images for the same input image in order to evaluate the ULEA-based noise removal performance to show its accuracy for detecting edges from different scenarios. The proposed block diagram of comparison process is shown in Fig. 7.

In this comparison, two different binarized images have been used. The first one is a binarized image before ULEA has been applied, whereas the second binarized image is a ULEA-based enhanced image. In other words, the first binarized image is used before the noise removal process has been applied using ULEA, while in the second binarized image, the noise has been removed using ULEA. In Fig. 8, the EEMI is subjectively evaluated based on eye's observation.

As noticeably shown in Fig. 8d, unwanted and non-important details are removed when compared to the binarized image shown in Fig. 8b, while the most important details are kept. Similarly, the vertical edge' extracted and shown in Fig. 8e are clear. One of the most important features to use ULEA to remove noise from binarized images is that most of non-important details are skipped from the vertical edge' images as shown in Fig. 8e compared to the image shown in Fig. 8c. These findings show good performance of EEMI in terms of accuracy.

3.2 Subjective Evaluation-Based Accuracy

Many details on foreground regions found in Fig. 8c are removed in Fig. 8e. It is clear that EEMI has reduced the number of pixels when the ULEA has been used.

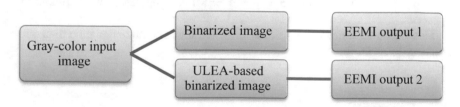

Fig. 7 Noise removal-based subjective evaluation

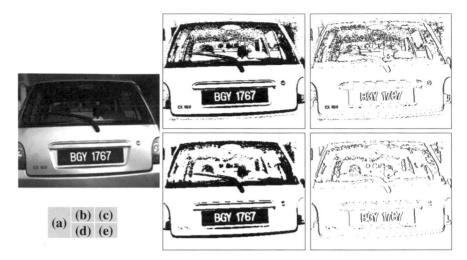

Fig. 8 EEMI subjective evaluation; **a** a gray input image, **b** a non-enhanced binarized image, **c** vertical edges' extraction of (**b**), **d** ULEA-based enhanced image, and **e** vertical edges of (**d**)

Image size	Operator		
	Computation time in ms		
	Sobel	VEDA [12]	Proposed EEMI
352 × 288	60	6.5	5.7
640 × 480	129.7	21.8	19.2
726 × 544	159.1	29	24.3

Table 1 EEMI's computation time compared to other competitive methods

3.3 Code Complexity and Computation Time

The proposed EEMI has a simple procedure that implements mainly parallel processing of pixels using an *IF* statement to reduce the *loop* function that has been used with 3 × 3 masks operators, e.g., in Sobel operator. In Table 1, a quantitative comparison is provided, whereas the proposed EEMI is compared to other competitive operators.

This table shows that the proposed EEMI has lower computation time than other competitive methods.

3.4 Robustness

The EEMI is evaluated in terms of robustness against complex background and blurry images. Provided are two examples one of which shows extracted horizontal edges

(a) (b)

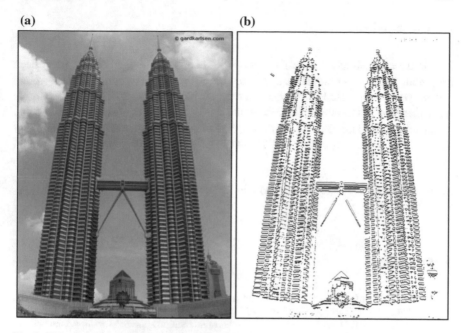

Fig. 9 Color input image (**a**) and its horizontal edges (**b**)

(a) (b)

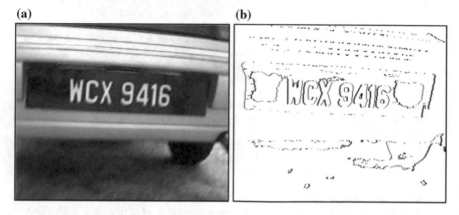

Fig. 10 Gray input image (**a**) and its vertical edges (**b**)

from complex background images including much details, while the other example shows extracted vertical edges from blurry and inclined images as shown in Figs. 9 and 10, respectively.

4 Conclusion

In this paper, a simple Edges Extraction Method from Images (EEMI) has been proposed. EEMI has confirmed that it accurately works with complex background images with much detail of objects and regions. The proposed EEMI uses a simple mask that does not require much processing time, while edges are being detected and extracted from images compared to 3×3 masks' operators. Despite its simplicity of code and computation time, EEMI works efficiently with complex background images.

Acknowledgements This research is funded by UNIIG-J510050781; which is supported by Universiti Tenaga Nasional.

References

1. Lin W-C, Wang J-W (2018) Edge detection in medical images with quasi high-pass filter based on local statistics. Biomed Signal Process Control 39:294–302
2. Gupta D, Anand RS (2017) A hybrid edge-based segmentation approach for ultrasound medical images. Biomed Signal Process Control 31:116–126
3. Luo L, Tang Y, Lu Q, Chen X, Zhang P, Zou X (2018) A vision methodology for harvesting robot to detect cutting points on peduncles of double overlapping grape clusters in a vineyard. Comput Ind 99:130–139
4. Min C, Jiqiang S, Lyu MR (2002) A new approach for video text detection. In: Proceedings. International conference on image processing, vol 1, pp I-117–I-120
5. Du S, Ibrahim M, Shehata M, Badawy W (2013) Automatic License Plate Recognition (ALPR): a state-of-the-art review. IEEE Trans Circuits Syst Video Technol 23:311–325
6. Liu Z, Chen H, Blondel W, Shen Z, Liu S (2018) Image security based on iterative random phase encoding in expanded fractional Fourier transform domains. Opt Lasers Eng 105:1–5
7. Kmieć M, Glowacz A (2015) Object detection in security applications using dominant edge directions. Pattern Recogn Lett 52:72–79
8. Al-Ghaili AM, Mashohor S, Ramli AR, Ismail A (2013) Vertical-edge-based car-license-plate detection method. IEEE Trans Veh Technol 62:26–38
9. Bradley D, Roth G (2007) Adaptive thresholding using the integral image. J Graph Tools 12:13–21
10. Shafait F, Keysers D, Breuel TM (2008) Efficient implementation of local adaptive thresholding techniques using integral images. In: Document recognition and retrieval XV, p 681510
11. Sobel I (1990) An isotropic 3×3 gradient operator. In: Freeman H (ed) Machine vision for three-dimensional scenes. Academic Press, New York, pp 376–379
12. Al-Ghaili AM, Mashohor S, Ismail A, Ramli AR (2008) A new vertical edge detection algorithm and its application. In: 2008 international conference on computer engineering & systems, pp 204–209

Face Recognition Using Texture Features and Multilayer Perceptron

Nanik Suciati, Chastine Fatichah, Wijayanti Nurul Khotimah,
Radityo Anggoro and Henning Titi Ciptaningtyas

Abstract Developing a robust system for face recognition is a challenge in the field of Computer Vision. Texture is one of low-level visual features that mostly used to represent a unique characteristic of an image in face recognition. In this study, a face recognition system is implemented using two well-known texture extraction methods, i.e., wavelet and Local Binary Pattern (LBP). The texture features of face image are then classified using multilayer Perceptron. The face dataset used in experiment consists of 2260 images from 113 people. The experiment is carried out by comparing the performance of the two texture extraction methods. Both methods achieve high accuracy in recognizing face images. The recognition rates achieved by the wavelet and LBP methods are 99.63 and 93.19%, respectively.

Keywords Face recognition · Local binary pattern · Wavelet · Multilayer perceptron

1 Introduction

In recent years, there have been many researches in developing face recognition system. These researches are motivated by potential applications of a robust face recognition system, such as controling an access using face image, room monitoring, and detection of unusual activities. Many face recognition systems are developed based on visual contents of a face image. In this case, the challenge is constructing image features which represent unique characteristics of an image, so that the features can identify image accurately. In feature-based recognition methods, feature extraction is an important step. In order to represent image features, many low-level features descriptors have been investigated. Those include color, shape, and texture descriptors. In this study, we focus on texture features which we consider as the important feature on a face image.

N. Suciati (✉) · C. Fatichah · W. N. Khotimah · R. Anggoro · H. T. Ciptaningtyas
Department of Informatics, Institut Teknologi Sepuluh Nopember, Surabaya, Indonesia
e-mail: nanik@if.its.ac.id

© Springer Nature Singapore Pte Ltd. 2019
V. Piuri et al. (eds.), *Intelligent and Interactive Computing*, Lecture Notes in Networks and Systems 67, https://doi.org/10.1007/978-981-13-6031-2_29

Among texture extraction methods, Local Binary Pattern (LBP) is one of the powerful local descriptors [1]. Developing a face recognition system by using LBP histogram is reported to be outperformed compared to other well-known methods such as Principal Component Analysis, Elastic Bunch Graph Matching, and Bayesian Intra/extra personal Classifier [2]. Previously, we have already investigated wavelet as a texture extraction method and implemented it into several applications with very good performance, such as batik classification [3], iris recognition [4], and mammogram classification [5]. The aim of this study is to implement the two texture extraction methods, i.e., wavelet and LBP, and investigate their performance in a face recognition system. We use multilayer Perceptron as the classification method. The comparison result of the two feature extraction methods will be used as consideration in our study on face recognition using video data.

2   The Face Recognition System

Algorithm of the face recognition system is shown in Fig. 1, which consists of training and testing steps. A classifier model for face recognition is developed in training step. First, a set of face images is pre-processed, and then, texture feature of each image is extracted using two kinds of texture extraction methods, i.e., wavelet and LBP features. The wavelet features and the LBP features are used to train face classifier model separately by using multilayer Perceptron method. Performance of each classifier model is evaluated in the testing step.

2.1   Pre-processing

Preprocessing used in the face recognition system is converting the face image from RGB color to gray-scale space. The aim of this process is to simplify the computation of the further processes with preserving all important image features.

2.2   Local Binary Pattern

Local Binary Pattern (LBP) is a method to analyze the texture feature of an image in the spatial domain by comparing pixel intensities in the center of a defined neighborhood window with its neighbors [1]. The area of the neighborhood window is defined by choosing a certain radius. LBP captures gradient information and is able to localize edge, dot, and other local image characteristics. The gradient information is often represented in the form of histogram, which describes the gradient distribution on the whole image.

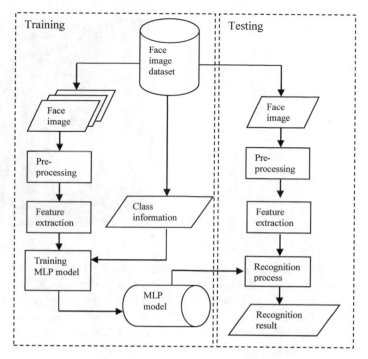

Fig. 1 The face recognition system

LBP value of a pixel at the center of neighborhood window is calculated by comparing its intensity and all other intensities inside the neighborhood window. Figure 2 shows the window defined using radius 1 ($R = 1$) with amount of neighbors 8 ($P = 8$), put on the top left part of a face image. LBP value of each pixel in the image is calculated from pixel at the top left corner until the bottom right corner, using Eqs. 1 and 2. I_C is the intensity of pixel at the window center; $I_{i,R}$ is all other pixel intensities in the neighborhood; $s(x)$ is the result of the threshold process.

$$\text{LBP}_{P,R} = \sum_{i=0}^{P} s(I_{i,R} - I_C) \times 2^{P-1-i} \tag{1}$$

$$s(x) = \begin{cases} 1, & x \geq 0 \\ 0, & x < 0 \end{cases} \tag{2}$$

Detail calculation of LBP value is illustrated in Fig. 3. For example, there is a part of an original image under 3×3 neighborhood window as shown in 3(a). The pixel intensity at the center of the window, which is 15, is set as a threshold. All other pixels in the neighborhood having an intensity greater or equal than the threshold is set to 1, and all other pixels with intensity smaller than the threshold is set to 0. The result of threshold process which is shown in 3(b) is then multiplied with predefined

Fig. 2 Neighborhood
window with $R = 1, P = 8$

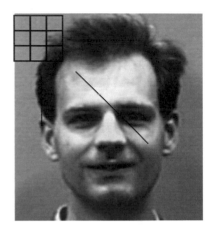

weight in 3(c). The result of element-by-element multiplication between those two
matrixes produces integer value for each pixel position as shown in 3(d). All integer
values are then summed up to obtain LBP of a pixel at the center position of the
original image 3(a). Calculating LBP of all pixel in image result an LBP image.
Instead of using the LBP matrix, some researches use LBP histogram representation
for further process. In histogram representation, the LBP value, which is equal to

Fig. 3 LBP calculation.
a Original image, **b** the result
after threshold process using
pixel at center as the
threshold, **c** the predefined
weight of 3×3
neighborhood window, **d** the
result after multiplying
(**b**) and (**c**), **e** LBP histogram

(a)

4	30	100
5	15	87
2	4	75

(b)

0	1	1
0		1
0	0	1

(c)

2^5	2^6	2^7
2^4		2^0
2^3	2^2	2^1

(d)

0	64	128
0		1
0	0	2

(e) LBP=1+2+64+128
        = 196

```
0  1  2     ....     196   ....  255
```

196, add the number of occurrence of LBP = 196 by 1, as shown in 3(e). Since the range of the LBP value is from 0 to 255, then the LBP histogram can be represented using a vector with size 1×256.

In this study, we represent LBP features of a face image using histogram with four different sizes, i.e., 1×256, 1×128, 1×64, and 1×32. Simplifying the LBP histogram from the original size 1×256 into three others size are calculated by summation each twos, fours, and eights neighboring LBP values, respectively. The visualization of LBP image used in this study is shown in Fig. 4.

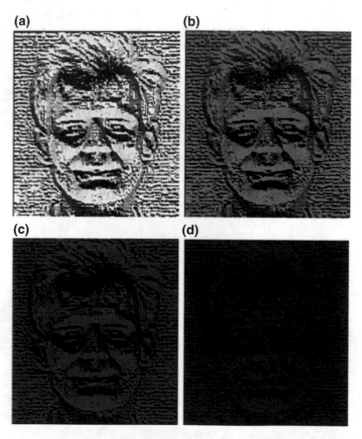

Fig. 4 LBP image using four different number of intensities **a** 256 intensities, **b** 128 intensities, **c** 64 intensities, **d** 32 intensities

2.3 Wavelet

A wavelet is a tool that can be used to represent multi-resolution image by decomposing an image into four sub-images, i.e., approximation, detail horizontal, detail vertical, and detail diagonal sub-images. The decomposition process can be performed repeatedly to the approximation sub-image to produce another smaller four sub-images. Illustration of two-level wavelet decompositions of a face image is shown in Fig. 5. The first decomposition to a gray-scale face image in (a) produces four sub-images in (b). The approximation, detail horizontal, detail vertical, and detail diagonal sub-images are shown in the left and right side of the first row, in the left and right side of the second row, respectively. Then, the second decomposition to the approximation sub-image in (b) produces another four sub-images in (c). For visualization purpose, the detail information in the horizontal, vertical, and diagonal sub-images are drawn after the enhancement process using histogram equalization.

Wavelet coefficients of all sub-images are used as texture features, which provides smoothness, coarseness, and regularity [6] characteristics of an image. Texture feature of an image is represented by using the mean and standard deviation of wavelet coefficients on each sub-images after decomposition. Mean and standard deviation, which denotes average and contrast of sub-image, are computed using Eqs. 3 and 4. In those equations, r is intensity, $p(r)$ is probability of pixels having r intensity, and L is the number of distinct intensities.

$$m = \sum_{i=0}^{L-1} r_i \, p(r_i) \tag{3}$$

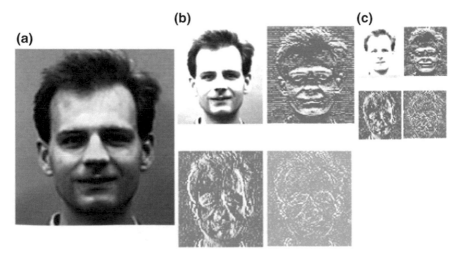

Fig. 5 Illustration of wavelet decomposition, **a** a gray-scale face image, **b** four sub-images after first decomposition, **c** four sub-images after second decomposition

$$\mu_2(r) = \sum_{i=0}^{L-1} (r_i - m)^2 p(r_i) \tag{4}$$

Computing mean and standard deviation using Eqs. 3 and 4 to the four sub-images will produce 16 descriptors in each decomposition level. In this study, we use texture descriptors with length 24, 32, and 40. Those descriptors are computed from wavelet coefficients after 3, 4, and 5 decomposition levels, respectively.

2.4 Multilayer Perceptron

Artificial Neural Network (ANN) is a model created for imitating the intelligence of the human brain which has problem-solving ability through learning process. In addition to find some patterns on data, ANN can also map some relationships between outputs and inputs.

Multilayer Perceptron (MLP) is one of Neural Network architecture which has at least one hidden layer to connect input and output layers. The input layer consists of some nodes with amount equal to the number of features in input data. The output layer consists of some nodes with amount equal to the number of classes. The learning algorithm applied to the MLP architecture adjust weights of all links connecting two layers. Each node, except all node in input layer, is activated using nonlinear activation function. The activation function, which is also referred to as threshold functions or squashing functions, maps the output of a node in possibly infinite domain to a prespecified range. Sigmoid (S-shaped) is one of the common activation functions with characteristics bounded, monotonic, non-decreasing, and nonlinear response within a prespecified range. The most common sigmoid function is logistic function as shown in Eq. 5, which provides an output value from 0 to 1.

$$f(x) = \frac{1}{1 + e^{-x}} \tag{5}$$

A supervised backpropagation learning algorithm is used to train MLP model. In training process, the backpropagation learning algorithm adjusts all weight values according to the predefined error function on the provided data. MLP has been a popular method to solve many complicated classification problems, i.e., a problem that is not linearly separable, such as character recognition [7], batik recognition [8], leaves recognition [9], and emotion recognition [10].

3   The Experiment

Experiments are performed to compare the performance of the two texture extraction methods, i.e., wavelet and LBP, in the face recognition system. Performance of face recognition is measured by using the recognition rate, which is calculated by the number of correct recognized images divided by the number of testing images. In experiment we use a part of Face94 dataset (cswww.essex.ac.uk). The dataset consists of 2260 face images from 113 distinct persons. There are 20 images with different expression for each person, since those 20 sequence images are captured when a person is speaking in front of camera. Size of each image is 200×180 pixels. The dataset is split as training and testing data with proportion of 80 and 20%, respectively. Using this proportion, 16 images of each person are used as training, and the rest (four images) are used as testing data, so that the total number of training and testing images is 1808 and 452.

Texture features of the training data are extracted using wavelet and LBP methods. Three sets of texture features are determined using the mean and standard deviation of wavelet coefficients on each sub-image produced by 3, 4, and 5 wavelet decomposition levels. By using 3, 4, and 5 wavelet decomposition levels, length of each texture feature vector is 24, 32, and 40, which produces three matrices of training data with size 1808×24, 1808×32, and 1808×40, respectively.

Four sets of texture features are computed using LBP histogram with length 32, 64, 128 and 256, which produces four matrices of training data with size 1808×32, 1808×64, 1808×128, and 1808×256, respectively. Each matrix of training data is used to train a multilayer perceptron classifier model, and then, performance of each MLP model is tested to recognize face image in the testing data.

The MLP architecture used for face recognition is a two-layer feedforward network. The first layer is a hidden layer with a sigmoid transfer function, while the second layer is an output layer with a softmax transfer function. The optimum number of neurons in the hidden layer is different depending on the number of neuron input. The number of neuron in output layer is fixed, i.e., 113, which is equal to the number of the different persons in the dataset. Illustration of the MLP architecture for neuron input with size 1×40 is shown in Fig. 6. In this case, the method used to extract the image features is wavelet with up to five decomposition levels. In the training phase, features vectors of 1808 training images are used to adjust weights of the MLP model. In the testing phase, 452 testing images are recognized to measure the performance of the built MLP model.

Performance of face recognition using three different sets of wavelet features is shown in Table 1. Each features set is composed by mean and standard deviation of wavelet coefficients on sub-images produced by wavelet decomposition. By using three decomposition levels, it produces 12 sub-images and 24 features per image. The recognition rate of the MLP classifier trained by using 24 features per image in the training data achieves 99.336%. Increasing the wavelet decomposition level and including the more numbers of wavelet features, also increase the recognition rate. The using of wavelet features with size 32 and 40 gives recognition rate 99.560 and

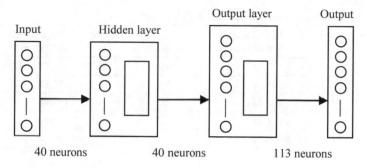

Fig. 6 Architecture of MLP classifier

Table 1 Performance of face recognition using wavelet features

Level of wavelet decomposition	Size of texture feature per image	Number of neuron in hidden layer	Recognition rate (%) of 452 testing
3	1 × 24	100	99.336
4	1 × 32	100	99.560
5	1 × 40	80	**100**
Recognition rate (%) of wavelet features			99.63

100%, respectively. The best recognition rate, i.e., 100% is obtained using features with size 40. There are two false recognitions in the second best recognition rate, which is obtained using features with size 32. These two false recognitions happened for face number 113. There are three false recognitions in the third best recognition rate, which is obtained using features with size 24. These three false recognitions happened for face number 99 and 113. Average of recognition rate of the three sets of wavelet features is 99.63%.

Performance of face recognition using four different sets of LBP features is shown in Table 2. Recognition rates of LBP features with length 256, 128, 64, and 32 are 85.62, 94.46, 97.12, and 95.57%, with number of false recognitions 65, 25, 13, and 20, respectively. The best recognition rate, i.e., 97.12% with 13 false recognitions (out of 452 total recognitions) is achieved using features with size 64. Two faces, which are difficult to recognize by using LBP features are face number 19 and 93. These two faces have the lowest recognition rate in all different sets of LBP features. Average of recognition rate of the four sets of LBP features is 93.19%.

All recognition rate written in Tables 1 and 2 is taken from the best recognition rate which is obtained from some experiments using different number of neurons in the hidden layer for each texture set. Comparison of the wavelet and LBP features to recognize face images shows that the wavelet features give better performance. The average recognition rate achieved by the wavelet method is 99.63%, while the LBP method is 93.19%. This result shows that the wavelet feature which is calculated using coefficients on multi-resolution can represent texture characteristics of face

Table 2 Performance of face recognition using LBP features

Size of histogram	Size of texture feature per image	Number of neuron in hidden layer	Recognition rate (%) of 452 testing
32	1 × 32	50	95.57
64	1 × 64	70	**97.12**
128	1 × 128	50	94.46
256	1 × 256	100	85.62
Recognition rate (%) of LBP features			93.19

image slightly better than the LBP feature. The two methods have different images that are difficult to recognize. Face number 99 and 113 are difficult to recognize by using the wavelet method. While face number 19 and 93 are difficult to recognize by using the LBP method. The best performances happened using feature length 40 in the wavelet method and using feature length 64 in the LBP method. In general, the two methods have good performance in recognizing face image with an average of recognition rate higher than 90%.

4 Conclusion

In this study, we compare the performance of two texture extraction methods in the face recognition system. These two methods are wavelet and LBP histogram. In general, both methods can achieve high accuracy in recognizing 452 face testing images. The wavelet method gives average of accuracy 99.63%. The LBP method obtains average of accuracy 93.19%. In the future, we will consider wavelet as the texture extracting method for face recognition system on video data.

Acknowledgements This study is supported by Institut Teknologi Sepuluh Nopember, Indonesia, as a part of Lab-based Education research on year 2018.

References

1. Ojala T, Pietikainen M, Harwood D (1996) A comparative study of texture measures with classification based on featured distributions. Pattern Recogn 29(1):51–59
2. Ahonen T, Hadid A, Pietikainen M (2004) Face recognition with local binary patterns. In: ECCV, pp 469–481
3. Suciati N, Herumurti D, Wijaya AY (2016) Feature extraction using gray-level co-occurrence matrix of wavelet coefficients and texture matching for batik motif recognition. In: 8th International conference on graphics and image processing
4. Suciati N, Anugrah AB, Fatichah C, Tjandrasa H, Arifin AZ, Purwitasari D, Navastara DA (2016) Feature extraction using statistical moments of wavelet transform for iris recognition. In:

International conference on information and communication technology and systems (ICTS), pp 193–198

5. Putra JA, Suciati N, Wijaya AY (2017) Ekstraksi Fitur Menggunakan Discrete Wavelet Transform dan Full Neighbour Local Binary Pattern untuk Klasifikasi Mammogram. Jurnal Ilmiah Teknologi Informasi Terapan 3(2)
6. Gonzalez RC, Woods RE (2002) Digital image processing, 3rd edn. Pearson Prentice Hall
7. Phangtriastu MR, Harefa J, Tanoto DF (2017) A comparison between neural network and support vector machine in optical character recognition. Procedia Comput Sci 116:351–357
8. Suciati N, Purwitasari D, AdlinaPratomo WA (2014) Batik motif classification using color-texture-based feature extraction and backpropagation neural network. In: IAAI international conference on advanced applied informatics, Kitakyushu, Japan, August 31–Sept 2014, pp 517–521
9. Şekeroğlu B, İnan Y (2016) A leaves recognition system using a neural network. Procedia Comput Sci 102:578–582
10. Rao KS, Saroj VK, Maity S, Koolagudi SG (2011) A recognition of emotions from video using neural network models. Expert Syst Appl 38(10):13181–13185

Recognition of Eye Movements Based on EEG Signals and the SAX Algorithm

Shanmuga Pillai Murutha Muthu, Sian Lun Lau and Chichang Jou

Abstract For patients with motor disabilities and difficulties to interact with computer and devices, brain–computers interface (BCI) may provide them with new ways to solve this problem. The patients may use a portable electroencephalography (EEG) device to instruct a computing device via eye movements. We propose the use of commercial off-the-shelf (COTS) EEG devices and pattern classification as a potential solution. In this paper, we investigate simple eye movement recognition using the symbolic aggregate approximation (SAX) algorithm and compare its suitability and performance against known classification algorithms such as support vector machine (SVM), k-nearest neighbour (KNN) and decision tree (DT). The SAX-based recognition performed better than the three classification algorithms. The SAX-based recognition was also able to achieve higher accuracy with only one single simple feature. These results showed that SAX could be a suitable and efficient technique to perform simple eye movement recognition using EEG signals.

Keywords Brain–computer interface · SAX · Eye movement

1 Introduction

In many everyday activities today, interactions with technology are made easier with modern human–machine interaction (HMI) approaches. For example, one can use facial recognition to authenticate his identity or give instruction using gestures to interact with a large display dashboard. When it comes to assistive technology for users with disabilities, HMI will play a crucial role to provide access to these users.

S. P. Murutha Muthu · S. L. Lau (✉)
Department of Computing and Information Systems, Sunway University, Bandar Sunway, Malaysia
e-mail: sianlunl@sunway.edu.my

S. P. Murutha Muthu
e-mail: 09015504@imail.sunway.edu.my

C. Jou
Department of Information Management, Tamkang University, New Taipei City, Taiwan
e-mail: cjou@mail.tku.edu.tw

© Springer Nature Singapore Pte Ltd. 2019
V. Piuri et al. (eds.), *Intelligent and Interactive Computing*, Lecture Notes in Networks and Systems 67, https://doi.org/10.1007/978-981-13-6031-2_38

Users who belong to this category include patients who have amyotrophic lateral sclerosis (ALS), cerebral palsy, or spinal cord injury. This group of users have very restricted movements and often will not be able to interact with a computer through conventional input approaches. One possibility is to use eye tracking as a computer input and interaction method [1, 2]. However, eye-gaze-based approach may face problem in determining a gaze from mere watch or gaze as an instruction [2]. Another option with potential is to utilise the brain–computer interface (BCI) to enable HMI. BCI is the approach that enables communication between a brain and a device. Suitable sensors are used to capture brain signals. These signals can be used with techniques, such as machine learning, to provide cues that may be interpreted as instructions.

Classification of eye movements using electroencephalography (EEG) signal may become an attractive approach if the signals produced can generate distinguishable signal patterns between different movement classes. However, one needs to investigate the usage of portable EEG devices to enable usage of this type of HMI in more situations and environments. This also means there will be a limited number of channels available, and the signal sample may be considered as lower quality as compared to the laboratory-grade EEG capturing devices. On top of this trade-off, one may also expect worst classification performance with fewer channels and reduced signal quality. We intend to apply suitable pattern classification techniques on commercial off-the-shelf (COTS) EEG devices and the algorithms should be producing good recognition accuracy even though there may be a limitation to the amount and quality of the data.

In this paper, we investigate the suitability of the symbolic aggregate approximation (SAX) algorithm [3] in detecting simple eye movements from EEG signals. We argue that SAX may be a more efficient approach as compared to classification algorithms. The paper will present an evaluation of the SAX algorithm used in detecting selected eye movement contexts.

2 Related Work

Brain–computer interface (BCI) is a combination of hardware and software systems that allow people with severe or partial disability to communicate with their environment. BCI goal is to improve quality of life, and its full potential has yet explored. Much research has so far focused on people with severe motor disabilities. For example, HMI is being investigated as techniques that use eye movements (EOG [4] or video-based eye gaze tracking [5]), body movement (limited limb movement, gestures head, facial expressions and so on.) [6] and brain signals (electroencephalography (EEG)) [7]. In this paper, the focus of the BCI technology will be based on EEG.

The first recorded electroencephalography effort is usually referred to the work of Hans Berger in discovering and measuring electrical activity from the surface of the human head [8]. This marked the beginning of EEG technology. EEG has been used in laboratory settings for various purposes, including neuroscience, medical, cognitive science, etc. The setup used in such settings are commonly complex and non-portable. It is only in the last decade, consumer-grade EEG devices have been developed and made available in the market [9].

The usage of EEG as a BCI input device can be an attractive option for patients with movement disabilities such as ALS or spinal cord injury. They can interact with computers and devices using BCI approach. For example, Carrino et al. proposed to use an Emotiv EPOC device to enable control of an electric wheelchair in a self-paced manner [10]. The investigation tested the EPOC using a motor imagery technique to perform wheelchair control. The best results obtained was not higher than 60%. Another example is the work of Vourvopoulos and Liarokapis. They have investigated both Emotiv EPOC and Neurosky devices (Mindset and Mindwave) to evaluate their suitability in navigation [11]. The Emotiv EPOC gave better performance, though the authors mentioned that latency is an issue. Navuluri et al. carried out an investigation to predict drivers' intentions while driving using the Emotiv EEG device [12]. EEG signal has also been applied to control a six-degree-of-freedom robotic arm [13], where an Emotiv EPOC was used for this purpose. The preliminary tests have verified that simple pick and place tasks can be performed by an operator after a relatively short learning period.

When it comes to algorithms for context recognition using EEG signal, classification algorithms have been utilised in previous work. Liu and Sourina applied Support Vector Machine (SVM) to detect emotions using EEG signal from an Emotiv EEG device [14]. In pattern recognition, the k-nearest neighbour algorithm (KNN) is a non-parametric method used for classification. In the work of Li et al., they used KNN to detect three levels of attention [15]. There are also other techniques used such as principal component analysis (PCA) [16], independent component analysis (ICA) [17] and multi-layer perceptron [18].

3 Using SAX for Eye Movement Recognition

In this paper, an investigation has been carried out to evaluate the feasibility of the SAX algorithm for eye movement context recognition based on EEG signals. Apart from SAX, three classification algorithms have been selected for comparison purposes —support vector machines (SVM), K-nearest neighbour (KNN) and decision tree (DT). The SAX algorithm converts the EEG signals from each channel into symbolic sequences. It can be seen as an extension of the piecewise aggregate approximation (PAA) technique. The latter divides a time series into equal parts. The arithmetic means of the signal points from each of these parts are then calculated. With the newly computed mean values, it forms new time series that may be better suited for pattern recognition. This is depicted in Fig. 1, where the coloured level lines are the PAA representations.

All evaluations were carried out using MATLAB and default parameters out of the box, except for the SAX algorithm. MATLAB was only used to generate the

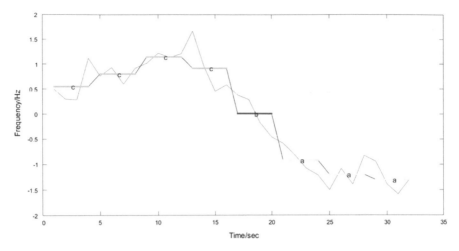

Fig. 1 Example representation of PAA and SAX from a time series

SAX string patterns, but the eye movement recognition was done using a custom code written in Java that performs simple string pattern matching.

The SAX algorithm extends PAA by converting the mean values into symbols such as '*a*', '*b*' and '*c*' as shown in Fig. 1. One can decide the number of symbols to be distributed so that the relevant ranges will be defined to separate the PAA representation levels into respective 'zones'. As depicted as light grey lines around 0.5 and −0.5 in Fig. 1, it divides the y-axis into three zones. Any mean levels belong to the higher zone will be converted to the symbol '*c*'. Similarly, the mean levels between −0.5 and 0.5 and below −0.5 will be converted to '*b*' and '*a*', respectively. With this conversion, a symbolic string series is the new representation of the time series. For the time series in Fig. 1, we then can say it is represented as '*ccccbaaa*' after applying the SAX algorithm.

The advantage the SAX algorithm brings is the possibility to apply string pattern matching techniques to detect patterns in time series. It also helps to reduce the dimensionality of the original time series data. More importantly, there exist different string pattern techniques that are efficient and accurate. By enabling these existing techniques for time series pattern matching, it may help to improve the accuracy for the latter.

3.1 Data Collection and Experiment Setup

For the data collection, EEG data were recorded from 15 subjects (11 males, 4 females; age range 19–31). The subjects had a normal or corrected-to-normal vision and had no history of neurological disorders. Each subject is requested to perform three types of movements—eye blink (*M1*), eyeball rotation (*M2*) as well as turning of eyeball to the left and right (*M3*). These movements have been selected as they are common movements used to instruct or interact with a computer. Each recording involved 10 repetitions of each movement, respectively. These three movements data are the first three sets of measurement data. Besides these three sets of measurements, a fourth set (*M4*) was collected from the subjects. This set contains all three movements.

The subjects were seated in a comfortable chair at a distance of approximately 70 cm from a 13 in. (LCD) panel in the lab. The recording environment was kept from any external disturbance so that no one gets distracted during the recording. The screen is at eye level position. The subjects were instructed to relax and stay as placid as much as possible. This is to avoid any possible effect on EEG signals with muscle artefact. The overall preparation process takes less than 5 min. The experiments were carried out during evening time where the participants were able to concentrate without any lack of sleep. The timing for the movements in each recording was predefined. Hence, it is possible to label the recorded signal based on these timing. The labelled data will be used for classification training and evaluation.

As shown in Figs. 2 and 3, there are regions of signals that displayed a spike or change in amplitude when an eye movement is carried out. If the signal patterns between non-movement and a movement are visually distinguishable, we would expect this pattern change behaviour can be made detectable using suitable machine learning techniques. Also, techniques such as SAX should also be able to distinguish signal patterns between non-movement and eye movements. The next subsection will describe the steps carried out to process and analyse the classification of EEG signal for eye movement recognition.

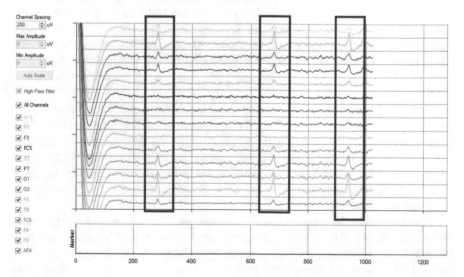

Fig. 2 Example EEG signal captured using Emotiv Testbench. Red boxes indicate eye blink movements

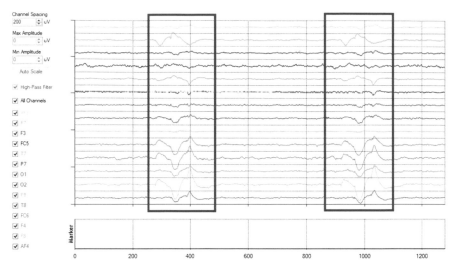

Fig. 3 Example EEG signal captured using Emotiv Testbench. Red boxes indicate eyeball rotation movements

3.2 Data Preprocessing and Classification

The raw EEG data was recorded using the tools provided by Emotiv EEG. The sampling rate of the raw data was 128 Hz for all channels. The first step of preprocessing was to perform independent component analysis (ICA) filtering of the signal to reduce noise in the EEG signals. Next, EEGLAB toolbox was used to perform the feature extraction. For the classification of the movements, simple statistical features (mean and standard deviation) have been selected. The raw EEG signals were downsampled to 128 Hz for all channels. For the feature extraction, the signals from each channel were segmented into windows of 0.25 s. No overlapping of the window was used.

The features generated will be more useful for the intended classification and pattern recognition, because not only the process reduces the dimensionality of the original data, but it may also highlight particular signal characteristics of the signal so that the patterns are more distinguishable among the different classes. For example, as shown in Figs. 4 and 5, the computed features for selected channels are more distinguishable than the rest and will be useful to produce models that can detect the expected movements.

The outcome of the feature extraction is used together with the labels at the corresponding timing of the windows as data for both training and testing. For the three classification algorithms, the classifier learner in MATLAB is selected to produce the desired models and perform the evaluation. As for SAX, the resulted time series of the features extracted will be converted into string sequences. For this paper, evaluation results for the string sequence length of four will be presented. Other lengths

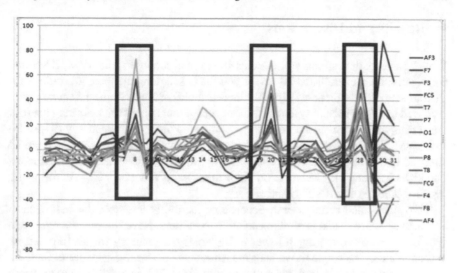

Fig. 4 Example mean feature extracted from EEG signal for the movement eye blink. Red boxes indicate eye blink movements

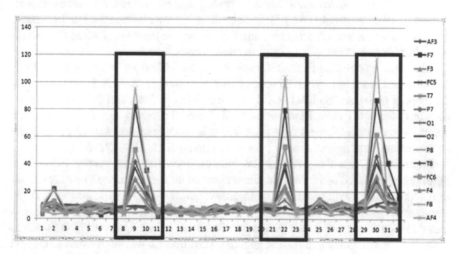

Fig. 5 Example standard deviation feature extracted from EEG signal for the movement eye blink. Red boxes indicate eye blink movements

have been also used for comparison and a sequence of four string characters gave the overall best outcome. From the training set, the string sequences that represent movements in each dataset were extracted. Evaluation for the classification is done by verifying the number of correctly matched movement string sequences with the test dataset.

4 Results and Discussions

Table 1 presents the summary of the classification accuracy for the SVM, KNN and DT algorithms. Generally, SVM performed the best among the three algorithms for all four sets of data (*M1–M4*) (highlighted in bold). Second highest accuracy was produced by DT consistently across all four sets of movements data. Among the four set of data, the movement eyeball rotation achieved the highest accuracy (91.29 %) with SVM. The fourth dataset that contains all three movements in one measurement obtained the lowest accuracy as compared to the single movement measurements. The highest among the classification algorithms was SVM with 71.17 %, following by DT (69.33 %) and KNN (68.35 %).

SVM performed better overall because of its ability to detect the best separation between classes and is not too sensitive to outliers. Nevertheless, the other two algorithms were not too much lower in the achieved accuracy for all four sets of measurements. The single movement measurements performed way better than the measurements that include all three movements (*M4*). This is mainly due to the more simplistic model for measurements (*M1–M3*). Each of these sets only contains patterns that different between a selected movement and no movement. The movements recorded in the fourth data set (*M4*) are three different eye movements, but there could be a similarity between them. This situation may cause the classification algorithms some trouble to successfully tell two or more movements apart.

The accuracy obtained from the evaluation of the SAX algorithm is found in Table 2. The evaluation was carried out under two different settings: SAX 1 used data from 10 subjects to extract top five SAX movement string patterns and then tested on the data from the exact same 10 subjects. SAX 2 used the same exact five patterns but tested them on measurement data from five subjects whose data was not part of the training data. The highest accuracy for each movement is highlighted in bold.

From Table 2, it is observed that generally eye movement recognition using SAX algorithm performed better than the three classification algorithms. The only exception was eyeball rotation, where the improvement was not significant (SAX using standard deviation, 92% compared to SVM 91.29%). Also, the evaluation for SAX-based recognition utilised only one single feature at a time. Comparing the settings SAX 1 to SAX 2 and which feature should be considered, we would want to discuss

Table 1 Evaluation of accuracy for all three classification techniques

Movement	Average accuracy (%)		
	SVM	KNN	DT
Eye blink	**86.87**	81.37	84.17
Eyeball rotation	**91.29**	87.39	88.70
Eyeball turn left and right	**83.65**	79.33	81.67
All three movements	**71.17**	68.35	69.33

Table 2 Evaluation of accuracy for SAX

| Movement | Average accuracy (%) | | | |
| | SAX 1 | | SAX 2 | |
	Mean	Standard deviation	Mean	Standard deviation
Eye blink	89.6	**92.6**	91.6	**93.6**
Eyeball rotation	**88.4**	83.2	86.8	**92.0**
Eyeball turn left and right	**91.2**	87.4	**94.8**	85.6

the performance of SAX 2 as it was testing the model from SAX 1 on a new set of data from additional subjects. The evaluation results for SAX 2 may be closer to real-world performance. For eye blink and eye rotation, the results from SAX applied on standard deviation feature only were higher than mean. For eyeball turn, the mean feature achieved the highest accuracy. From these results, the SAX technique produced relatively better accuracy using only one feature. It can be seen as more efficient in terms of computation.

As for the evaluation for all movements (*M4*) using SAX, the results are summarised in Table 3. The result using mean features and SAX performed similarly to the result from SVM. The accuracy achieved by SAX using only standard deviation is the highest among other two results. Using only standard deviation to create string patterns using SAX, the movements patterns for all three movements can still be detected and distinguished from each other.

To investigate whether one can also use only one single EEG channel for eye movement recognition, we analysed the recognition accuracy based on a single channel. The results are summarised in Table 4. It is observed that by selecting a particular

Table 3 Evaluation of accuracy using SAX algorithm for all movements

| Movement | Average accuracy (%) | |
	Mean	Standard deviation
All three movements	71.37	**90.29**

Table 4 Evaluation of accuracy using SAX algorithm for all movements

| Channel | Average accuracy (%) | |
	Mean	Standard deviation
F7	69.63	**78.14**
F8	50.37	**86.30**
FC5	94.81	**89.63**
FC6	73.70	**97.78**
T8	68.33	**99.63**

channel only, one can achieve higher recognition accuracy. Among all channels, the recognition obtained using only the channel T8 has almost 100% accuracy. While this may be too ideal, but the evaluation using individual channel showed that one can use only one or two channels, such as T8 and FC6, for the movement recognition to achieve good accuracy.

Overall, the movement recognition using SAX algorithm showed positive results that indicated its potential to be a suitable technique for movement recognition using EEG signal. As compared to typical classification algorithms such as SVM, KNN and DT, SAX algorithm allows accurate movement recognition with only one feature extracted from one or two channels of EEG signal obtained from a COTS EEG device such as the Emotiv EEG. This will allow better recognition efficiency since less data needs to be processed. It is also seen as attractive to be applied to resource-limited situations such as wearable and portable EEG for everyday activity scenario.

5 Conclusion

In this paper, we have presented an investigation of EEG-based eye movement recognition using three classification algorithms, namely, SVM, KNN and DT, and SAX. It is observed that the SAX-based approach achieved higher accuracy with fewer features and channels. This indicates a potential in using SAX for accurate eye movement recognition in a resource-limited scenario such as real-time eye movement recognition using a portable EEG device.

As future work, we wish to look into the possibility to implement the identified best settings as a prototype wearable system that allows accurate eye movement recognition using the Emotiv EEG device. This will allow us to verify and validate the results obtained in this investigation in a real-life scenario.

Acknowledgements The authors would like to thank all the participants involved in the data collection. Ethical approval has been given by the Sunway University Research Ethics Committee (SUREC 2016/001). Participants have provided their consent to be part of this study.

References

1. Porta M, Ravarelli A, Spagnoli G (2010) ceCursor, a contextual eye cursor for general pointing in windows environments. In: Proceedings of the 2010 symposium on eye-tracking research & applications, ETRA '10. ACM, New York, pp 331–337
2. Cantoni V, Porta M (2014) Eye tracking as a computer input and interaction method. In: Proceedings of the 15th international conference on computer systems and technologies, CompSysTech '14. ACM, New York, pp 1–12
3. Lin J, Keogh E, Wei L, Lonardi S (2007) Experiencing sax: a novel symbolic representation of time series. Data Min Knowl Discov 15(2):107–144
4. Barea R, Boquete L, Mazo M, Lopez E (2002) System for assisted mobility using eye movements based on electrooculography. IEEE Trans Neural Syst Rehabil Eng 10(4):209–218

5. Murata A (2006) Eyegaze input versus mouse: cursor control as a function of age. Int J Hum Comput Interact 21(1):1–14
6. Ju JS, Shin Y, Kim EY (2009) Intelligent wheelchair (IW) interface using face and mouth recognition. In: Proceedings of the 14th international conference on intelligent user interfaces, IUI '09. ACM, New York, NY, USA, pp 307–314
7. Iturrate I, Antelis JM, Kubler A, Minguez J (2009) A noninvasive brain-actuated wheelchair based on a p300 neurophysiological protocol and automated navigation. IEEE Trans Robot 25(3):614–627
8. Berger H (1929) Über das elektrenkephalogramm des menschen. Eur Arch Psychiatry Clin Neurosci 87(1):527–570
9. Lau SL, Ahmed A, Ruslan Z (2014) A review for unobtrusive COTS EEG-based assistive technology. In: Assistive technologies for physical and cognitive disabilities, pp 262–277
10. Carrino F, Dumoulin J, Mugellini E, Khaled OA, Ingold R (2012) A self-paced BCI system to control an electric wheelchair: evaluation of a commercial, low-cost EEG device. In: 2012 ISSNIP biosignals and biorobotics conference: biosignals and robotics for better and safer living (BRC), pp 1–6
11. Vourvopoulos A, Liarokapis F (2014) Evaluation of commercial braincomputer interfaces in real and virtual world environment: a pilot study. Comput Electr Eng 40(2):714–729
12. Navuluri K, Padia K, Gupta A, Nadeem T (2011) Poster: what's on your mind?: a mind-based driving alert system. In: Proceedings of the 9th international conference on mobile systems, applications, and services, MobiSys '11. ACM, New York, NY, USA, pp 415–416
13. Astaras A, Moustakas N, Athanasiou A, Gogoussis A (2013) Towards brain-computer interface control of a 6-degree-of-freedom robotic arm using dry EEG electrodes. Adv Hum-Comput Interact
14. Liu Y, Sourina O (2014) Real-time subject-dependent EEG-based emotion recognition algorithm. In: Transactions on computational science XIII lecture notes in computer science
15. Li Y, Li XW, Ratcliffe M, Liu l, Qi Y, Liu Q (2011) A real-time EEG-based BCI system for attention recognition in ubiquitous environment. In: Proceedings of 2011 international workshop on ubiquitous affective awareness and intelligent interaction, UAAII '11. ACM, New York, pp 33–40
16. Lin JS, Jiang ZYJ (2015) Implementing remote presence using quadcopter control by a non-invasive BCI device. Comput Sci Inf Technol pp 122–126
17. Bansal D, Mahajan R, Singh S, Rathee D, Roy S (2014) Real time acquisition and analysis of neural response for rehabilitative control. Int J Electr Robot Electron Commun Eng pp 697–701
18. Kaysa WA, Suprijanto, Widyotriatmo A (2013) Design of brain-computer interface platform for semi real-time commanding electrical wheelchair simulator movement. In: 2013 3rd international conference on instrumentation control and automation (ICA), pp 39–44

Effect of Supervised Region of Interest Against Edge Detection Method for Iris Localisation

Zuraini Othman, Azizi Abdullah, Fauziah Kasmin and Sharifah Sakinah Syed Ahmad

Abstract With the recent developments in information technology, health diagnosis based on iris analysis and biometrics has received considerable attention. For iris recognition, iris localisation, which is not an easy task, is an important phase. Moreover, for iris localisation, dealing with nonideal iris images could cause an incorrect location. Conventional methods for iris location involve multiple searches, which can be noisy and outdated. Such techniques could be inaccurate while describing pupillary boundaries and could lead to multiple errors while performing feature recognition and extraction. Hence, to address such issues, we propose a method for iris localisation of both ideal and nonideal iris images. In this research, the algorithm operates by determining all regions of interest (ROI) classifications through the use of a support vector machine (SVM) as well as the application of histograms that use grey levels as descriptors in all regions from those exhibiting growth. The valid region of interest (ROI) obtained from the probabilities graph of an SVM was obtained by examining the global minimum conditions determined using a second derivative model of the graph of functions. Moreover, this helped to eliminate the sensitive noises and decrease the calculations while reserving relevant information as far as possible. During the experiment, the comparison edge detection method was used with Canny and a multi-resolution local approach. The results demonstrated

Z. Othman · F. Kasmin · S. S. S. Ahmad (✉)
Department of Intelligent Computing and Analytics, Faculty of Information and Communication Technology, Universiti Teknikal Malaysia Melaka, Hang Tuah Jaya, 76100 Durian Tunggal, Melaka, Malaysia
e-mail: sakinah@utem.edu.my

Z. Othman
e-mail: zuraini@utem.edu.my

F. Kasmin
e-mail: fauziah@utem.edu.my

A. Abdullah
Faculty of Information Science and Technology, Center for Artificial Intelligence Technology, Universiti Kebangsaan Malaysia, Selangor Darul Ehsan, 43600 Bangi, Selangor Darul Ehsan, Malaysia
e-mail: azizia@ukm.edu.my

© Springer Nature Singapore Pte Ltd. 2019
V. Piuri et al. (eds.), *Intelligent and Interactive Computing*, Lecture Notes in Networks and Systems 67, https://doi.org/10.1007/978-981-13-6031-2_44

that the proposed ROI provided better results compared with those obtained without ROI.

Keywords Region of interest · Support vector machine · Region growing · Edge detection · Multi-resolution level

1 Introduction

For iris recognition systems, obtaining an accurate iris localisation from the eye image is the most important step. Information on edges acquired from all images does have a critical role in analyses of imaging properties. Research on effective imaging information has therefore emerged as a major field of research.

Edge detection lessens the amounts of data processed by filtering useless imaging information with each important image feature examined. The more common strategies in this field include the Sobel, Canny, Prewitt, and Log and Laplace techniques. Every method provides an ability to control less computation-intensive, more basic algorithmic methods. Nevertheless, each method presents particular weaknesses [1].

Practically, the Canny method, which was introduced by [2], provided excellent results and was extensively used. The technique subjects the images to processing with Gaussian filters, which smooths the high-frequency signals where edge pixel information resides and therefore minimises losses in edge details while restraining noise. With this strategy, high and low thresholds must be determined manually, necessitating more forward empirical examination. Numerous experiments will be required to determine possible appropriate thresholds. However, for real-time applications, lower and higher thresholds frequently change because of the different illuminations and natural scenes for each image. Established Canny operators lack a self-adaptive capability for obtaining satisfactory detection results in numerous cases [3].

Furthermore, there have been studies on edge-based and threshold-based segmentation. The Canny edge detector is considered to be an example of the edge-based method, while Otsu [4] thresholding is an example of threshold-based segmentation, which was performed by [5, 6]. Note that the Otsu method is based on grey histograms that were deduced using the least square method. It currently represents the most stable strategy for threshold segmentation of images and, statistically, the approach also obtains superior threshold values [7]. Similar to the studies done by [3, 7–11], to make Canny adaptively select higher threshold (H_t) and lower threshold (L_t) values, the Otsu method will be used.

Fixed partitioning strategy is fully utilised for analysing images locally and globally, with the aim of obtaining the localised spatial values of images. Through such methods, images are subjected to partitioning according to same size blocks even as local histograms are computed for each. The primary strength of this approach is that it obtains spatial distributions of imaged content such that additional histogram inputs are obtained [12]. In this proposed method, to produce edges for data of iris images, a comprehensive technique was used. Iris localisations are frequently utilised in determinations of pupil and iris boundaries. These are subject to modelling using ellipses, splines and circles. In an intuitive sense, more precise modelling can deliver superior recognition performances, which provides a basis for the use of commonly applied spline techniques in the field of iris recognition, for pupil and iris boundary estimations [13].

ROIs represent particular regions in scenes under examination. Selections of regions of interest are required to constrain the amounts of information to be processed, such that only those regions associated with the problem are to be designated for additional analyses. Numerous ROI definitions used in iris recognition currently exist, from rectangles that contain pupils, irises, and/or eyes, to irregular areas that contain no features other than iris textures. As eye image catalogues normally become increasingly larger in operation, speedy eye image processing is critical to success. ROIs are therefore considered in terms of rectangles that contain every obtainable iris texture, plus any intrinsically adjacent elements, i.e. eyelid, eyelashes and sclera features. Determining a suitable ROI remains problematic since, first, changes in camera-to-eye distances and relative positions can lead to changes in the ratio between the sizes of acquired irises and eye images, even in the very same iris.

Second, the brightness of eye images is typically distributed unevenly, due to uneven illuminations of the human subjects' facial features, although pupil dimensions will always vary between images. Neglecting ROI determinations for an inside eye image will inevitably bring forward the need to utilise brute-force search operations that identify limbic and pupillary limits in Hough space. The converse approach entails well-informed searches. It follows that ROI results represent nothing more than well-informed searches for pupils and those rectangles that are sufficiently large to contain every obtainable iris texture, plus every intrinsically adjacent eyelash, eyelid and sclera [14]. For recognition, the longest time was required for localising the iris boundaries. To address such issues, Li [15] suggested an algorithm that accurately detected the iris based on the region of interest (ROI) using a support vector machine (SVM) for training purposes, which reserved useful information as much as possible. Their method utilised the angular square region for detecting the iris' ROI.

In this study, we proposed an algorithm that is suitable for ideal and nonideal iris images rather than the algorithms that are merely suitable for ideal iris images [15]. For this purpose, the region growing with supervised learning using SVM will be used to obtain efficient and increase accuracy of iris localisation. In the training phase, information for the intensity value of the grey level histogram will be gathered. Next, the pattern of probabilities values that were obtained from the SVM for each region (R) from the region that grew provided valuable information on ROI classification based on the mathematical theories developed from the graph that was obtained. In this study, the dataset from CASIA V2 for ideal iris images and UBIRIS2 for nonideal iris images was used. Finally, the Canny edge detection [2] and adaptive multi-resolution level [9, 10] method were used for determining the proposed technique's effect on localising the actual iris boundaries.

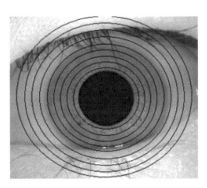

Fig. 1 Region growing in eye image [16]

The remaining paper has been organised in the following manner. In Sect. 2, the materials and methods used in the study have been briefly described. Section 3 illustrates the experiment and the results obtained using the proposed method while Sect. 4 discusses the conclusions obtained from this study.

2 Materials and Methods

In this study, we analysed the edge detection that was obtained using iris images for the supervised region growing for the iris' ROI [16]. First, we will discuss the training and testing phases for these supervised approaches. Then, to localise the iris image, the edge detection method will be used [9, 10].

2.1 Supervised Growing Approach for Region of Interest Detection in Iris Image

Previously, region growing has been discussed in Zucker [17]. There have been multiple different approaches, such as regional neighbour search, multiregional heuristics, function approximation and merging, split and merge, and regional interpretation and semantics, for region growing.

In this study, merging region growing will be used with the region being segmented from the pupil circle, which will then get completed at the end of the eye image. It has been represented as R (see Fig. 1).

From each region, the probabilities value will be represented as Pb(R) because this region's values are the probabilities (Pb) values that will be plotted for the respective value of R.

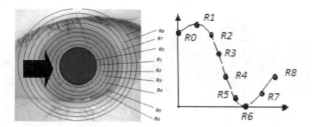

Fig. 2 Graph of SVM probability

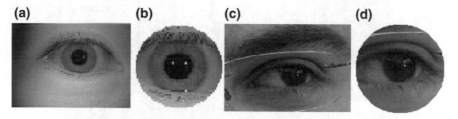

Fig. 3 An example of an original image and the ROI image. **a** Original ideal image from the CASIA V2. **b** ROI image extracted from (**a**). **c** Original nonideal image from UBIRIS2. **d** ROI image extracted from (**c**)

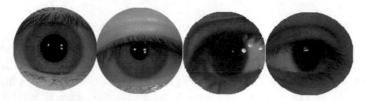

Fig. 4 Positive samples

Then, as shown in Fig. 2, each region R in region growing will be plotted as graphs. The value of $h = 1$ will be considered as a step size between each R. It starts from the smallest region R_0, R_1, R_2, …, R_n.

The ROI is an area that is graphically selected from a window that shows images containing the required information for a particular purpose. The ROI image that is selected is shown in Fig. 3.

Training Phase

To learn and classify the image ROIs, SVM classifiers [18] were used. For building models for training phases of every dataset, 50% positive samples and 50% negative samples data were used as shown in Figs. 4 and 5, respectively. The positive samples contained ROI images, whereas negative samples were circles that were reduced or excessively extracted from iris images.

Fig. 5 Negative samples

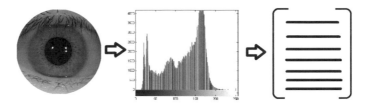

Fig. 6 Features descriptor

The grey level feature vector's histogram of positive and negative samples will be recorded. The positive samples will be labelled as 1 and negative samples as 0 (see Fig. 6).

Validating Phase

For the validation phase, similar to the training phase, data will be used and region growing will be observed. First, the pupil region will be extracted by determining the centre point and radius R of the pupil [19]. The feature descriptor, a histogram of grey levels, will be derived using the information from each circle. Note that each circle's area will be extracted from images and it will grow from pupils and finish at the end of the eye image (see Fig. 2).

From the pupil images, the iris images will be increased gradually. These images will be cropped in a circular shape, as shown in Fig. 7. Each cropped image represents

$$R_{i+1} = R_t + 5 \quad \text{with} \quad i = 0, 1, 2, \ldots, n \tag{1}$$

R_0 represents the pupil radius. From it, the region of R_1 until R_n will then be generated.

Note that each region R will provide the Pb value.

$$R_t = \text{Pb}(R_i) \quad \text{with} \quad i = 1, 2, \ldots, n \tag{2}$$

Subsequently, based on the Pb values obtained, a graph will be generated (refer Figs. 2). For this purpose, the validating data graph demonstrated the same pattern, i.e. a minimum value in the graph that contained the ROI image.

Fig. 7 Graph of SVM probability

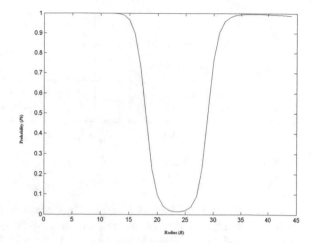

In the training and validating phase, the patterns were identified from the graph of SVM probability value (Pb). The condition from [16] will be considered such that the following equation will represent the Pb values that were obtained

$$Pb''(R_i) \cong Pb(R_{i+1}) - 2Pb(R_i) + Pb(R_{i-1}) \tag{3}$$

The localised iris was found at the minimum point of the graph. As shown in Fig. 7, the iris ROI is as per the radius R_{15}. Hence, ROI is the global minimum of the Pb values that were obtained and can be represented as

$$ROI = \mathop{\arg\min}_{R_0 < R_i < R_n} (Pb(R_{i+1}) - 2Pb(R_i) + Pb(R_{i-1}) > 0) \tag{4}$$

Testing Phase

Note that, at the beginning, the test images will go through similar processes as the validating phase. First, the pupil region will be detected and extracted, followed by region growing until the values of $Pb(R_i)$, $Pb(R_{i+1})$ and $Pb(R_{i-1})$ comply with conditions mentioned in Eq. (4) and that region selected is the ROI image.

2.2 A Statistically Based Multiple Resolution Levels Method for Canny Edge Detection

As images are retrieved, their content is transformed into grey levels. Then pre-processing phase will take part to smooth the images. Images are then subject to partitioning according to various levels. Canny edge detection is utilised in edge detections with support from Otsu techniques, although with modifications applied

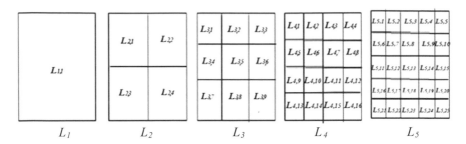

Fig. 8 Fixed image partitions

such that superior high threshold (H_t) and low threshold (L_t) values are obtained. In such cases and through the use of multiple image resolutions, global determinations of variance values for L_1 and localised spatial values in L_2 until L_5 are performed.

Let L denote $m \times m$ partitioned images, where all denote image partitions of level L_m wherein $m = 1, 2, \ldots, 5$. In all levels, partition L_{mj} provides for localised H_t and L_t values, wherein $j = 1, 2, \ldots, m^2$ (see Fig. 8).

Each non-partitioned, globally analysed image L_1, H_t and L_t are directly acquired [3, 10]

$$H_t = \text{Otsu Method [4]} \quad \text{and} \quad L_t = \frac{1}{2} H_t \tag{5}$$

The proposed strategy implements this approach for fixing partitioning in all L_m images, as performed in [10]. The modifications are involved with determining every spatial H_t and L_t value for all localised partitions. At all stages, the resultant minimum values of H_t are designated.

$$H_t = \frac{1}{n} \times \arg\min(H_t \in L_{mj}) \tag{6}$$

Then, for each L_{mj}, the corresponding L_t is

$$L_t = \frac{1}{2} H_t \text{ for each } H_t \in L_{mj} \tag{7}$$

Next, the minimum value of L_t

$$L_t = \frac{1}{n} \times \arg\min(L_t \in L_{mj}) \tag{8}$$

where $n = 1, 2, 3, \ldots, 10$.

Additionally, H_t and L_t values for providing precise edge images are designated.

2.3 Performance Measurement

In this step, the measurement used is the figure of merit (FOM) by Pratt [20]:

$$\text{FOM} = \frac{1}{\max(N_1, N_T)} \sum_{i=1}^{N_T} \frac{1}{1 + \alpha d_i^2} \tag{9}$$

wherein N_1 and N_T denote the numbers of ideal as well as actual edge points, whereas d_i denotes the Euclidean pixel distance of the detected ith edge, while α denotes the selected scaling constant $\alpha = \frac{1}{9}$, which is applied to penalised displaced edges. Higher FOM values from 0 to 1 denote better performances in the resulting image.

3 Experiment and Results

In this research, iris datasets of edge images from CASIA V2 [21] and UBIRIS2 [22] were utilised. For ground truth edge images, all were drawn manually as well as verified by a group of optometrists at Hospital Melaka. The optometrists select the best drawn iris and pupil boundaries from a set of edge boundaries provided, if they have suggested a new edge boundary out of the set given then the ground truth will be redrawn. With the aim of comparing experimental results, edge images derived from their supplied binary content were produced.

Figure 9 shows that the highest FOM result contained in between n equal to 1 and 10. Here, graph from CASIA and UBIRIS with using ROI show slightly higher for each level compared to CASIA and UBIRIS without using ROI. Table 1 displays the detailed results acquired through the use of Canny approaches that involve statistically based multiple resolution levels strategies for Canny edge detections. The findings shown were acquired using a statistically based multiple resolution levels strategy that outpaced the Canny edge approach. Through the use of the supervised ROI method, improvements are supported in every result when compared to non-ROI approaches.

In detail, graphs in Fig. 9 are generated based on the values obtained from Table 1 for 0020_005 image from CASIA and C518_S1_I6 image from UBIRIS. Table 2 displays the results obtained in three images from UBIRIS and CASIA, through which the use of ROI improved the results via the proposed techniques. The proposed results represented the best that can be derived using statistically based methods involving multiple levels of resolution, compared the Canny edge detection and Canny Otsu on global approach strategies.

Figure 10 shows one image obtained from CASIA and UBIRIS. Here, shows that for CASIA edge image generated without using ROI contained less noise by using statistically based methods involving multiple levels of resolution compared to Canny method and Canny Otsu approach. But for UBIRIS without using ROI, the important edge most has been completely eliminated by using the proposed approach.

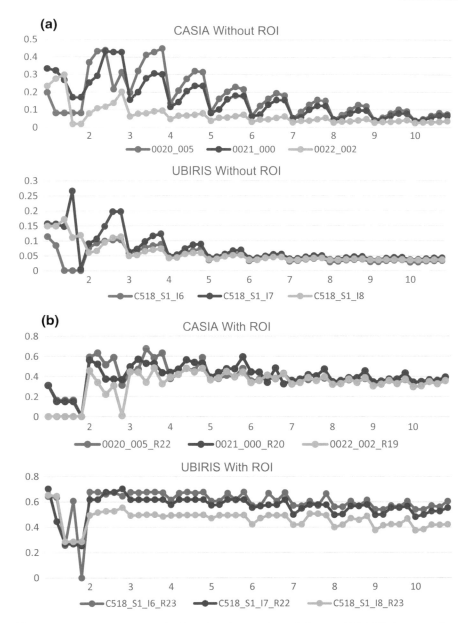

Fig. 9 Results of adaptive threshold on three images from CASIA and UBIRIS dataset: **a** data obtained without using ROI and **b** data obtained with using ROI

Table 1 FOM results on one image of CASIA and UBIRIS

n	Type	CASIA 0020_005		UBIRIS C518_S1_I6	
		With ROI	Without ROI	With ROI	Without ROI
	Canny	0.390916002	0.066298089	0.583171565	0.043396437
1	L_1	0.310703888	0.197900363	0.643892618	0.114915408
	L_2	0.162848167	0.080621408	0.635781512	0.085262241
	L_3	0.162848167	0.08046476	0.255702164	0.000552734
	L_4	0.162848167	0.080412403	0.607697327	0.000281797
	L_5	0	0.080432206	0	0.000220866
2	L_1	0.590492722	0.368690971	0.678452417	0.077130639
	L_2	0.631647927	0.433055849	0.677677775	0.092845375
	L_3	0.515659725	0.439558902	0.660873147	0.099638071
	L_4	0.587595764	0.216590308	0.674614289	0.102685278
	L_5	0.313358045	0.313649481	0.646180366	0.103197051
3	L_1	0.444872168	0.197075657	0.676241827	0.057174132
	L_2	0.471328418	0.321166353	0.673670911	0.066976026
	L_3	0.674809116	0.413629475	0.677677775	0.078343847
	L_4	0.586001992	0.429867737	0.678452417	0.084208523
	L_5	0.630973366	0.450219212	0.672654248	0.088455686
4	L_1	0.378493538	0.136885092	0.61536051	0.042554484
	L_2	0.421922022	0.209184882	0.668763741	0.050787879
	L_3	0.478985774	0.276137405	0.678452417	0.064499891
	L_4	0.457822518	0.320722961	0.669615601	0.066992315
	L_5	0.586001992	0.314764196	0.677677775	0.073754548
5	L_1	0.365072946	0.087400909	0.606355399	0.035189797
	L_2	0.378785608	0.162818437	0.606355399	0.040712952
	L_3	0.411268072	0.202907547	0.669615601	0.050143078
	L_4	0.395512801	0.230780642	0.61536051	0.055694594
	L_5	0.477914063	0.216421928	0.678452417	0.05934344
6	L_1	0.356143515	0.064840672	0.570462312	0.032328945
	L_2	0.367534415	0.122579058	0.571586286	0.035885532
	L_3	0.38135002	0.16307181	0.668763741	0.042367985
	L_4	0.377428234	0.195175377	0.606355399	0.046084023
	L_5	0.429370727	0.180856411	0.674755061	0.047858955

(continued)

Table 1 (continued)

		CASIA		UBIRIS	
		0020_005		C518_S1_I6	
n	Type	With ROI	Without ROI	With ROI	Without ROI
7	L_1	0.348463334	0.048914133	0.571690986	0.030889708
	L_2	0.359475479	0.087487021	0.568783911	0.033138413
	L_3	0.367539349	0.128319538	0.606355399	0.037618502
	L_4	0.367534415	0.156713283	0.570462312	0.039832941
	L_5	0.382278098	0.151232436	0.665382182	0.041981154
8	L_1	0.343699565	0.042834621	0.561042827	0.028143375
	L_2	0.353298677	0.068832233	0.561042827	0.030484072
	L_3	0.364692874	0.098100715	0.606355399	0.034553567
	L_4	0.359475479	0.126821949	0.568783911	0.036035244
	L_5	0.37758123	0.118649138	0.61536051	0.037862002
9	L_1	0.338136222	0.038870197	0.54044158	0.028654287
	L_2	0.349234802	0.054028926	0.540227375	0.02880948
	L_3	0.358345485	0.077475607	0.570462312	0.032388671
	L_4	0.353507775	0.10045225	0.562731088	0.034076058
	L_5	0.368478831	0.089801114	0.606355399	0.035461163
10	L_1	0.336700384	0.036167785	0.54044158	0.027432157
	L_2	0.344588209	0.046342891	0.54044158	0.027955911
	L_3	0.352698952	0.065250979	0.570462312	0.031346431
	L_4	0.349437308	0.081067805	0.561042827	0.032094413
	L_5	0.35933649	0.074164803	0.606355399	0.033138413

Furthermore, by using the same image and focusing on targeted area which is ROI the boundary of iris and pupil generated quiet similar to the ground truth image, and it is much better if compared to Canny and Canny Otsu approach.

4 Conclusion

In this study, we determine the effect of supervised ROI for iris localisation against the edge detection method. The modification of the Canny method was selected because edge images that were obtained provided the most complete edge boundaries, although more unwanted edges were detected. Thus, we proved that local spatial adaptive approach using the Otsu method enhanced the edges that were obtained by eliminating noises from the conventional Canny method. Findings show more accurate edge images as a result, since these contain foreground edge images, while background edge image content is ignored. Finally, the results that were obtained

Table 2 Comparison of FOM results on previous works—three images of CASIA and UBIRIS

	Image	Without ROI			With ROI		
		Canny	Canny Otsu	Proposed	Canny	Canny Otsu	Proposed
CASIA	0020_005	0.066298089	0.197900363	0.450219212	0.390916002	0.310703888	0.674809116
	0021_000	0.058520088	0.334119004	0.433342796	0.46195121	0.312255918	0.593886595
	0022_002	0.041099999	0.234548995	0.298025808	0.370326559	0.000524522	0.483207478
UBIRIS	C518_S1_I6	0.043396437	0.114915408	0.103197051	0.583171565	0.643892618	0.678452417
	C518_S1_I7	0.051454675	0.157669771	0.265920499	0.591415439	0.704348674	0.704348674
	C518_S1_I8	0.04957692	0.149392568	0.172161903	0.455844289	0.654467706	0.654467706

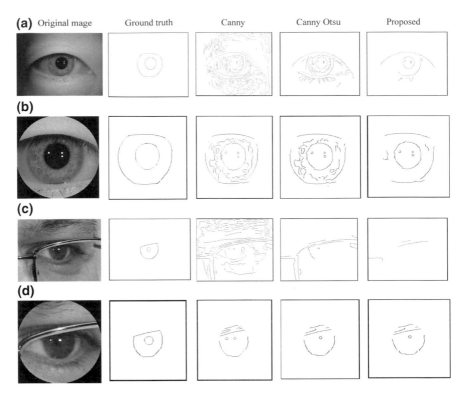

Fig. 10 Image obtained from CASIA and UBIRIS dataset: **a** CASIA image obtained without using ROI, **b** CASIA image obtained with using ROI, **c** UBIRIS image obtained without using ROI and **d** UBIRIS image obtained with using ROI

demonstrated that using ROI, all the results could outperform results obtained without using ROI for each method. The supervised ROI approach would help in eliminating unwanted noise from other areas; consequently, the localised iris could be readily used for the next stage. Therefore, in future studies, we need to consider the fitting method of segmentation, which will help complete the edge boundaries that were obtained.

Acknowledgements The authors express their deepest gratitude and thanks to Universiti Teknikal Malaysia Melaka (UTeM) in supporting this research PJP/2018/FTMK(2B)/S01629.

References

1. Wang M, Jin JS, Jing Y, Han X, Gao L, Xiao L (2016) The improved canny edge detection algorithm based on an anisotropic and genetic algorithm, vol 634, pp 115–124

2. Canny J (1986) A computational approach to edge detection. IEEE Trans Pattern Anal Mach Intell 8(6):679–698
3. Jie G, Ning L (2012) An improved adaptive threshold canny edge detection algorithm. In: 2012 International conference on computer science electronics engineering, pp 164–168
4. Otsu N (1979) A threshold selection method from gray-level histograms. IEEE Trans Syst Man Cybern 9(1):62–66
5. Al-Kubati AAM, Saif JAM, Taher MAA (2012) Evaluation of Canny and Otsu image segmentation. In: International conference on emerging trends in computer and electronics engineering, pp 23–25
6. Anitha R, Jyothi S (2014) Classifying penaeid prawns species using Canny and Otsu. Int J Adv Res Comput Sci. Manag Stud 2(11):35–42
7. Fang M, Yue G, Yu Q (2009) The study on an application of otsu method in canny operator. In: International symposium on information …, vol 2, no 4, pp 109–112
8. Zhao J, Yu H, Gu X, Wang S (2010) The edge detection of river model based on self-adaptive canny algorithm and connected domain segmentation. In: 2010 8th World congress on intelligent control and automation, vol 2, no 1, pp 1333–1336
9. Othman Z, Abdullah A (2017)An Adaptive threshold based on multiple resolution levels for canny edge detection. In: IRICT 2017 Recent trends information and communication technology, pp 316–323
10. Othman Z, Abdullah A, Prabuwono AS (2012) A statistical approach of multiple resolution levels for canny edge detection. In: Intelligent systems design and applications (ISDA), pp 837–841
11. Hui P, Ruifang Z, Shanmei L, Youxian W, Lanlan W (2011) Edge detection of growing citrus based on self-adaptive canny operator. In: 2011 International conference on computer distributed control and intelligent environmental monitoring, pp 342–345
12. Abdullah A, Veltkamp RC, Wiering MA (2009) Spatial pyramids and two-layer stacking SVM classifiers for image categorization: a comparative study. In: Proceeding of international joint conference on neural networks, pp 5–12
13. Zhao Y, Ti C, Huang X, Tokuta A, Yang R (2014) A performance comparison between circular and spline-based methods for iris segmentation. In: Proceeding—International conference on pattern recognition, pp 351–356
14. Celik ET (2016) Selecting a fuzzy region of interest in standard eye images. Adv Intell Syst Comput 357:793–804
15. Li Y, Li W, Ma Y (2012) Accurate iris location based on region of interest. In: Proceeding—2012 International conference biomedical engineering biotechnology, iCBEB 2012, pp 704–707
16. Othman Z, Abdullah A, Prabuwono AS (2018) Supervised growing approach for region of interest detection in Iris localisation. Adv Sci Lett 24(2), 1005–1011(7)
17. Zucker SW (1976) Region growing: childhood and adolescence. Comput Graph Image Process 5(3):382–399
18. Cortes C, Vapnik V (1995) Support-vector networks. Mach Learn 20(3):273–297
19. Business J (2010) An improved algorithm for iris location. In: 2010 International conference on computer, mechatronics, control and electronic engineering (CMCE), pp 219–221
20. Abdou IE, Pratt WK (1979) Quantitative design and evaluation of enhancement/thresholding edge detectors. Proc IEEE 67(5):753–763
21. CASIA-IrisV2. [Online]. Available http://biometrics.idealtest.org/
22. Santos R, Alexandre A (2010) The UBIRIS . v2 : a database of visible wavelength Iris images captured. Analysis 32(8), 1529–1535

Tracking Coastal Change Over Kuala Terengganu Using Landsat

Abd Rahman Mat Amin, Fathinul Najib Ahmad Sa'ad, Adida Muhammud, Wan Farahiyah Wan Kamaruddin, Mohd Rivaie Mohd Ali, Asmala Ahmad and Nor Haniza Bakhtiar Jemily

Abstract The main objective of this study is to track the coastal change over Kuala Terengganu coastal water from 1989 to 2018. In this study, dataset acquired from Landsat 5 and 8 have been used. The image was processed using Normalized Difference Water Index (NDWI) to differentiate land and water body. The result shows that, before the reclamation of the coastal for the extension of the airport runaway, accretion is a dominant process over this area. After the reclamation, the erosion occurs at the north and accretion at the south. During the period of 2006–2014, 0.0229 km^2 of coastal was lost due to the erosion. However, the reclamation and accretion processes contribute about 0.972 km^2 of new land over this area. Overall, for the period of 1989–2018, the reclamation and accretion contribute 1.337 km^2 of new land/coastal to this study area.

Keywords Landsat · NDWI · Coastal erosion · Reclamation · Coastal change

1 Introduction

Terengganu that located at the east coast of Peninsular Malaysia is the home of the longest coastline in Malaysia. The beach stretches 244 km from Kemaman at the

A. R. Mat Amin · F. N. Ahmad Sa'ad (✉) · A. Muhammud · W. F. Wan Kamaruddin
Faculty of Applied Science, Universiti Teknologi Mara, 23200 Bukit Besi,
Terengganu, Malaysia
e-mail: najib4496@yahoo.com

A. R. Mat Amin
e-mail: abdra401@tganu.uitm.edu.my

M. R. Mohd Ali
Faculty of Computer and Mathematical Sciences, Universiti Teknologi Mara, 21080 Kuala Terengganu, Terengganu, Malaysia

A. Ahmad
Faculty of Information and Communication Technology, Universiti Teknikal Malaysia Melaka (UTeM), 76100, Durian Tunggal, Melaka, Malaysia

N. H. Bakhtiar Jemily
Faculty of Mechanical Engineering, UniKL MSI, 09000 Kulim, Kedah, Malaysia

© Springer Nature Singapore Pte Ltd. 2019
V. Piuri et al. (eds.), *Intelligent and Interactive Computing*, Lecture Notes in Networks and Systems 67, https://doi.org/10.1007/978-981-13-6031-2_13

south to Besut at the north. These beaches are facing to the South China Sea and directly exposed to the high wave during monsoon season that already known as a main natural factor that contributes to the coastal change. In 2011, the worst erosion occurs over Tok Jembal beach that causes the collapse of coastal road, chalets and the building belong to Malaysia University of Terengganu. This worst erosion occurs just 4 years after the reclamation of the sea to extend the Kuala Terengganu Airport runaway [1]. Due to the rapid erosion rate over this area, it is considered interesting to track the coastal line change over this area.

To track the coastal change is a challenging issue. It requires long period of time in observation [2, 3]. The presence of remote sensing technologies such as aerial photography, satellite imagery and LIDAR could be utilized to overcome this problem. The satellite imagery such as acquired by the Landsat series that start monitoring our planet since July 1972 is the best choice for tracking the long-term change of our planet [4]. Their resolution and revisited time are enough to be used as a tool to monitor coastal change over the world. Instead of resolution and revisited time, the availability of this free data made is the best tool to track the coastal change. Many studies have been conducted to monitor/tracking coastal change using Landsat data. Tamassoki et al. [5] have used Landsat TM-5 to monitor coastal change over Bandar Abbas, Iran. Wang et al. [6] have utilized Landsat images to detect coastline change over Ningbo. In order to monitor reclamation status over China, Meng et al. [7] have used Landsat, SPOT, ZY-2 and ZY-3 dataset.

The main objective of this study is to track and map the coastal change in time domain over Kuala Terengganu coastal line using Landsat dataset for the period of 1989–2018. This study will be divided into three main periods. The period is before the reclamation of Kuala Terengganu Airport runaway. The second period is before and after the reclamation of the Kuala Terengganu airport runaway. The last period is after reclamation of the Kuala Terengganu airport runaway. The coastal changes' map will be constructed for these three periods.

2 Study Area

In this research, a beach stretches about 16 km from Mengabang Telipot beach at the north to the Teluk Ketapang beach at the south of Kuala Terengganu river mouth which is chosen as a study area. The main study area is Tok Jembal beach which is located in between of this beach. Tok Jembal beach stretches about 5 km from Sultan Mahmud Airport to Mengabang Gelam at the north. Residents in this area are diverse including fishermen and university students. Figure 1 shows the map of study area.

Fig. 1 The study area

3 Materials and Method

3.1 Satellite Data

Landsat images are well-known effective to extract coastline and water body. In this study, images acquired from various Landsat sensors over the study area are used. The images are obtained for the year 1989–2018. All images used are Level 1TP. Level 1TP means all images are radiometrically calibrated and orthorectified using ground control points and digital elevation model (DEM) to correct for relief displacement. These are the highest quality Level-1 product that is suitable for pixel-level time series analysis (https://landsat.usgs.gov/landsat-processing-details). The images used are carefully selected to ensure that no cloud covers along the coastal line. Image is downloaded in GeoTIFF format from the United States Geological Survey (USGS) Earth Explorer Website (http://earthexplorer.usgs.gov/). The Path/Row number for all acquisitions was 126/56 (Table 1).

Table 1 The Landsat images used	Acquisition date	Satellite	Path/row
	2 July 1989	Landsat 5 TM	126/56
	18 August 2006	Landsat 5 TM	126/56
	27 October 2014	Landsat 8 OLI	126/56
	21 February 2018	Landsat 8 OLI	126/56

3.2 Image Processing

Recently, many techniques and algorithms have been proposed to distinguish land and water body from remote sensing imagery [8–10]. However, in this study, the most established and robust algorithm, namely, NDWI, proposed by McFeeters et al. [8] was used. The algorithm was adopted from the format of normalized difference vegetation index (NDVI). Since the development of the algorithm, many studies have been conducted using this algorithm to detect the coastal line all around the world [7, 11]. The NDWI is defined as

$$NDWI = \frac{\left(\rho_{green} - \rho_{nir}\right)}{\left(\rho_{green} + \rho_{nir}\right)} \tag{1}$$

whereas

ρ_{green} reflectance of green wavelength.
ρ_{nir} reflectance of near-infrared wavelength.

In this study, ρ_{green} is the green band (band 2 for TM 5 and band 3 for OLI) and ρ_{nir} is the near-infrared band (band 4 for TM 5 and band 5 for OLI). This algorithm takes zero as the threshold to differentiate between water body and non-water body [12].

In order to achieve this goal, certain steps are conducted in this study. First, all images were radiometrically and geometrically corrected with the same projection coordination system (UTM Projection, WGS 84 reference system). The pixel size was resampled to 30 m. After that, the NDWI algorithm was applied to the images. The images were then masked and compared using 'change detection' in the ENVI 5.1 software. The difference between the two images (initial and final state) was obtained, and coastal map changes were then constructed. Due to the resolution of every pixel of the image, the pixel size is 30 m, so the area lost or increase can be calculated as

Area = number of pixels × 0.3 m × 0.3 m. Figure 2 shows the flowchart of this study.

Fig. 2 Image processing
flowchart

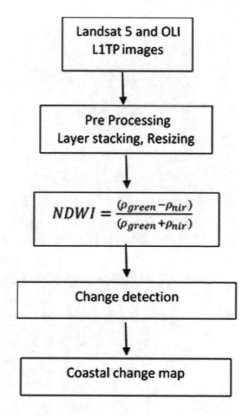

4 Results and Discussion

Figure 3 shows the true colour image (RGB) over the study area for 2 July 1989, 18 August 2008, 28 October 2014 and 21 February 2018. Images for July 1989 and 2006 are used to monitor the coastal change just before the construction of the Kuala Terengganu Airport. Images before July 1989 are not available for this scene.

4.1 Yearly Different

In order to track the coastal change over Kuala Terengganu coastal, the period of study is divided into three. The first period is before the construction of the Kuala Terengganu Airport in 2008. Due to limited suitable images, the period is just taken from 1989 to 2006. The second period is between 2006 and 2014. This period represents before and after the construction of the Kuala Terengganu Airport. The last is the whole period of the study that covers from 1989 to 2018.

(a)   (b)

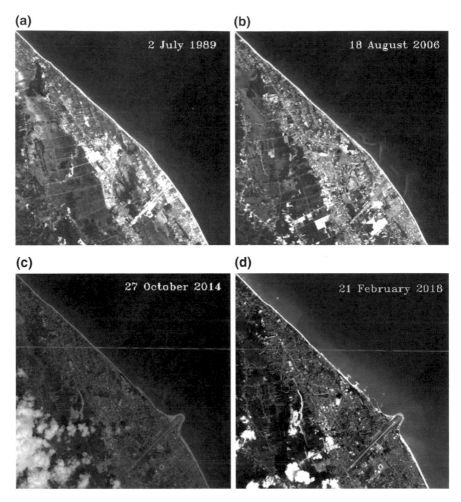

Fig. 3 RGB images over the study area for the year of **a** 2 July 1989, **b** 18 August 2006, **c** 27 October 2014 and **d** 21 February 2018

Figure 4a–c shows coastal change over the study period. Blue and red colours represent the increase and decrease in the coastal area. Figure 3a represents the difference in the coastal line before the construction of the Kuala Terengganu Airport. It covers the period of 1989–2006. Clearly seen that there is an increment in the coastal line during this period. The accretion area is about 0.972 km². There is only a small erosion area that covers about 0.229 km².

In order to track the coastal change before and after the construction of the Kuala Terengganu Airport, a difference between 2006 and 2014 map was constructed as shown in Fig. 4b. The worst erosion in this area happens in 2011 that causes the collapse of the road, chalet and the building belonged to University Malaysia Tereng-

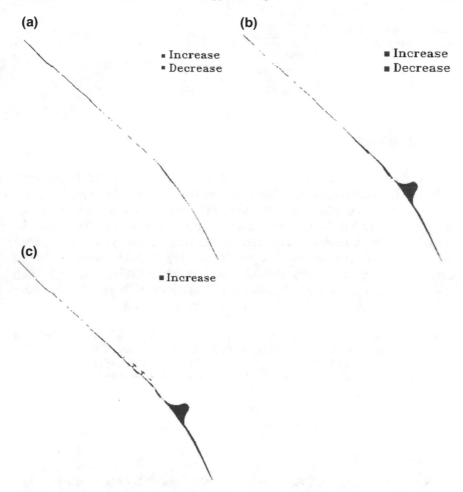

(a)

■ Increase
■ Decrease

(b)

■ Increase
■ Decrease

(c)

■ Increase

Fig. 4 Shoreline change **a** between 1989 and 2006, **b** between 2006 and 2014, **c** between 1989 and 2018

ganu. The affected coastline stretches about 4.2 km to the northwest of the airport as shown by the red colour in Fig. 4b. The erosion over this area was at the rate of 11 m annually [1]. Erosion at this area is classified under category 1 that is referring to an area suffering from coastal erosion where shore-based facilities are in imminent danger of loss or damage [13]. Clearly, as shown in figure, the huge erosion area at the northwest of the airport covers an area of 0.228 km² (Table 2).

Figure 4c shows the coastal line change from 1989 to 2018. As shown in figure, the coastline is increased almost along the study area except for some area that never changes. The increment in the coastline area is about 1.34 km². The main reason for this increment is the land reclamation done by the state government of Terengganu since 2011.

Table 2 Coastline change
from 1989 to 2018

Year different	Increase (km^2)	Decrease (km^2)
1989–2018	1.337	0
1989–2006	0.338	0.007
2006–2014	0.972	0.229

5   Conclusion

A study to track the coastal change over Kuala Terengganu area for the period of 1989–2018 has been conducted. The result shows that, before the extension of the Kuala Terengganu airport runaway in 2008, the accretion is dominating. However, since the construction of the Airport runaway, slight erosion occurs at the north of the airport. Overall, for the period of the study, the land along the study area has increased mainly due to the reclamation by the government. The result of this study could be used by the government agency such as Department of Irrigation and Drainage (DID) and Department of Survey and Mapping Malaysia (JUPEM). They can use this result to predict and identify the potential critical region over the study area.

Acknowledgements This project was funded by the Ministry of Higher Education (MOHE) Malaysia through the Fundamental Research Grant Scheme (FRGS) Project No. FRGS/1/2017/STG09/UITM/02/1. We would like to thank UiTM Terengganu management for their support and encouragement. Our appreciation also goes to the Landsat team for all the valuable data provided.

References

1. Ahmad MF, Subiyanto RY, Mamat M, Muslim AM, Husain ML (2014) Coastline changes in Vicinity of Runway Platform of Sultan Mahmud Airport, Kuala Terengganu: comparative analysis of one-line model versus satellite data. J Appl Sci 14(19):2234–2245
2. Addo KA, Walkden M, Mills JPT (2008) Detection, measurement and prediction of shoreline recession in Accra, Ghana. ISPRS J Photogramm Remote Sens 63(5):543–558
3. Mukhopadhyay A, Ghosh P, Chanda A, Ghosh A, Ghosh S, Das S, … Hazra S (2018) Threats to coastal communities of Mahanadi delta due to imminent consequences of erosion—present and near future. Sci Total Environ 637:717–729
4. Du Z, Li W, Zhou D, Tian L, Ling F, Wang H, … Sun B (2014) Analysis of Landsat-8 OLI imagery for land surface water mapping. Remote Sens Lett 5(7):672–681
5. Tamassoki E, Amiri H, Soleymani Z (2014) Monitoring of shoreline changes using remote sensing (case study: coastal city of Bandar Abbas). In: IOP conference series: earth and environmental science, vol 20, no. 1. IOP Publishing, p 012023
6. Wang X, Liu Y, Ling F, Liu Y, Fang F (2017) Spatio-Temporal change detection of Ningbo coastline using Landsat time-series images during 1976–2015. ISPRS Int J Geoinf 6(3):68
7. Meng W, Hu B, He M, Liu B, Mo X, Li H, … Zhang Y (2017) Temporal-spatial variations and driving factors analysis of coastal reclamation in China. Estuar Coast Shelf Sci 191:39–49
8. McFeeters SK (1996) The use of the Normalized Difference Water Index (NDWI) in the delineation of open water features. Int J Remote Sens 17(7):1425–1432

9. Rokni K, Ahmad A, Selamat A, Hazini S (2014) Water feature extraction and change detection using multitemporal Landsat imagery. Remote Sens 6(5):4173–4189
10. Yang Y, Liu Y, Zhou M, Zhang S, Zhan W, Sun C, Duan Y (2015) Landsat 8 OLI image based terrestrial water extraction from heterogeneous backgrounds using a reflectance homogenization approach. Remote Sens Environ 171:14–32
11. Xu N (2018) Detecting coastline change with all available Landsat data over 1986–2015: a case study for the State of Texas, USA. Atmosphere 9(3):107
12. Amer R, Kolker AS, Muscietta A (2017) Propensity for erosion and deposition in a deltaic wetland complex: Implications for river management and coastal restoration. Remote Sens Environ 199:39–50
13. Ann OC (1996) Coastal erosion management in Malaysia. In: Proceeding of the 13th annual seminar of Malaysian society of marine science, pp 1–11

Classification of Landsat 8 Satellite Data Using Unsupervised Methods

Afirah Taufik, Sharifah Sakinah Syed Ahmad and Ezzatul Farhain Azmi

Abstract The information from band extraction and calculation of indices which are used for classification imagery of Landsat 8 satellite data using unsupervised methods were studied. The visible and Near Infrared (NIR) bands of Landsat 8 satellite were used to derive Normalized Different Vegetation Index (NDVI) image. The Normalized Difference Water Index (NDWI) is a satellite-derived index from the NIR and Short Wave Infrared (SWIR) bands. Vegetation, non-vegetation, and water features classes were then analyzed by classification experiment of three unsupervised methods: ISODATA, K-means, and fuzzy c-means with guidance of ground truth information of the study area. The accuracy of the classified image is then assessed using a confusion matrix where classification accuracy and kappa coefficient are computed. The result shows that unsupervised methods classification is able to classify the Landsat 8 satellite data with FCM got a high accuracy compared to another two methods.

Keywords Accuracy assessment · Classification · Landsat 8 · Unsupervised

1 Introduction

The imagery satellite data provide the electromagnetic wavelength at different spectral and spatial resolutions. The visible and Near Infrared (NIR) bands of Landsat 8 satellite were used to derive Normalized Different Vegetation Index (NDVI) [1]

A. Taufik (✉) · S. S. Syed Ahmad
Department of Intelligent Computing & Analytics (ICA), Faculty of Information and Communication Technology, Universiti Teknikal Malaysia Melaka, Hang Tuah Jaya, 76100 Durian Tunggal, Melaka, Malaysia
e-mail: firafify@gmail.com

S. S. Syed Ahmad
e-mail: sakinah@utem.edu.my

E. F. Azmi
Faculty of Mechanical and Manufacturing Engineering Technology, Universiti Teknikal Malaysia Melaka, Hang Tuah Jaya, 76100 Durian Tunggal, Melaka, Malaysia

© Springer Nature Singapore Pte Ltd. 2019
V. Piuri et al. (eds.), *Intelligent and Interactive Computing*, Lecture Notes in Networks and Systems 67, https://doi.org/10.1007/978-981-13-6031-2_46

image [2]. The Normalized Difference Water Index (NDWI) [3] is a satellite-derived index from the NIR and Short Wave Infrared (SWIR) bands. Thus, this study aims to extract and calculate indices from multispectral images. The information from band extraction and calculation of indices which are used for classification imagery of Landsat 8 satellite data using unsupervised methods were studied. The unsupervised classification is about to find a natural group or clusters by assigning the imagery remote sensing data which is containing the multispectral bands and extracting the land cover information [4, 5]. This study includes the comparison study between three unsupervised clustering algorithms, namely, ISODATA, Fuzzy c-Means (FCM), and K-means toward classifying the Landsat 8 imagery remote sensing. The satellite imagery was then analyzed to represent three classes of interest: vegetation, water, and non-vegetation.

2 Related Works

The classification of the imagery using color K-means clustering algorithm and then the land cover classification were performed [6]. This study [7] proposes a new water body classification method using Top-of-Atmosphere (TOA) reflectance and Water Indices (WIs) of the Landsat 8 Operational Land Imager (OLI) sensor and its corresponding random forest classifiers. This research [8] was to compare Earth Observing-1 (EO-1) Hyperion hyperspectral data to Landsat 5 Thematic Mapper (TM) and Satellite (SPOT) 5 multispectral data for land cover classification in a dense urban landscape. By using Landsat Thematic Mapper (TM) dataset and classification scheme over Guangzhou City, China, they [9] tested two unsupervised and 13 supervised classification algorithms, including a number of machine learning algorithms. This study [10] is to compare two methods of image classification, i.e., Maximum Likelihood (ML), a supervised method, and ISODATA, an unsupervised method.

3 Materials and Methods

3.1 Data

The data used for this study was conducted in location of Kuala Krai, Kelantan. This area is positioned between longitudes 102° 12' to 102° 15' E and latitude 5° 25' to 5° 31' N. The satellite data were obtained from USGS Earth Explorer. The data were obtained on February 4, 2014. The interest study area was selected based on the Region of Interest (ROI) which contained seven bands with the 465 pixels for the rows and 406 pixels for the columns.

Fig. 1 RGB Kuala Krai

3.2 Preprocessing

The imagery of study area was executed in MATLAB. Preprocessing image was conceded for improving the color-infrared (CIR) composite using a decorrelation stretch as shown in Fig. 1. The whole study framework is shown in Fig. 2.

The image after preprocessing is adopted band 5, 4, and 3 which are allocated to red–green–blue (RGB) color, respectively. The NDVI and NDWI descriptions are shown below in (1) and (2):

$$NDVI = \frac{NIR - Red}{NIR + Red} \tag{1}$$

$$NDWI = \frac{Green - SWIR}{Greeen + SWIR} \tag{2}$$

3.3 Unsupervised Methods

Unsupervised classification or clustering is an effective method to extracting the land cover information. The process includes by assigning the pixels of imagery remote sensing data into thematic map information classes (e.g., water, forest). Compared to

Fig. 2 Framework of study

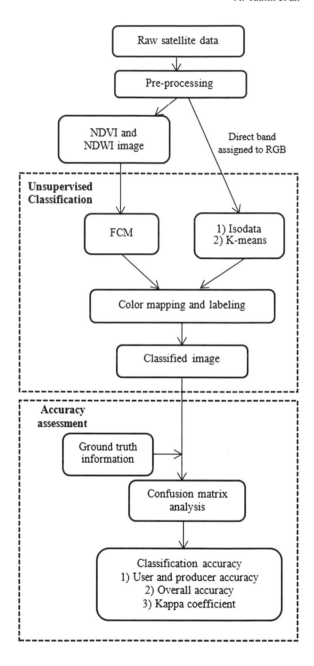

supervised classification, unsupervised classification requires the least input from the experimenter. Commonly, the unsupervised clustering does not need training data. The process is about to search for natural group of the spectral information, as tested in multispectral imagery data. In this study, the three of unsupervised clustering methods used are ISODATA, K-means, and FCM, which are used to identify the information class of interest. The FCM method is tested with the value information of two indices, which are NDVI and NDWI. On the other hand, the ISODATA and K-means method are tested with the direct band without convert into indices. The process of FCM classification that uses the NDVI and NDWI values image as input is executed in MATLAB. Otherwise, the ISODATA and K-means that are using direct data which are mapped into RGB are performed in ENVI. The results are recorded in the Result and Analysis section, and the three methods are compared. The three unsupervised methods are explained in the next section.

3.3.1 ISODATA Clustering

The iterative self-organizing data analysis technique (ISODATA) clustering method algorithm is about to compute the minimum distance of each data point into a certain cluster. The algorithm must specify the number of clusters and the maximum number of value which is the result of the clustering process itself. It classifies each point to the nearest cluster. After that, the mean vectors of a new cluster are computed based on entire pixels in one cluster. Process are repeats until the data point no longer moving [11].

3.3.2 K-Means Clustering

K-means clustering is determined to classify the land cover of imagery remote sensing using the reference of the ground truth. In this study the number of clusters is equal to three. This algorithm aims to minimize the objective function. The objective function can see in (3).

$$J = \sum_{j=1}^{k} \cdot \sum_{i=1}^{n} \left\| x_i^j - c_j \right\|^2 \tag{3}$$

3.3.3 FCM Clustering

The FCM works by allocating each data point with membership degree for each cluster center which is the distance between the data and the cluster center [12]. If the data are more near to cluster center, it becomes more membership to that cluster. The FCM algorithm is based on the minimization of the objective function which shown in (4).

$$J_m = \sum_{i=1}^{N} \cdot \sum_{j=1}^{C} u_{ij}^{m} \left\| x_i - c_j \right\|^2, \quad 1 \le m < \infty \tag{4}$$

3.4 Accuracy Assessment

The information of sampling truth map data was matched and compared with the labeled image to evaluate the classification accuracy. The user's accuracy and producer's accuracy are calculated to find the class accuracy. The overall accuracy (5) was calculated based on the confusion matrix that gained for user's accuracy and producer's accuracy. Here is the description of overall accuracy:

$$\text{Overall accuracy} = \frac{\text{Total number of correct classified}}{\text{Total number of pixels}} \times 100 \tag{5}$$

The additional measurement for classification accuracy is the kappa coefficient. The kappa coefficient (6) is used to measure the interest pixels area with the truth map data.

$$K = \frac{n \sum_{i=1}^{p} x_{ii} - \sum_{i=1}^{p} x_{io} x_{oi}}{n^2 - \sum_{i=1}^{p} x_{io} x_{oi}} \tag{6}$$

4 Results and Discussion

For this study, the classifications use the unsupervised method such as K-means, ISODATA, and FCM. The processing image is processed using the ENVI to obtain the Region of Interest (ROI). The selected pixels of Kuala Krai, Kelantan are continued to being classified using ENVI and MATLAB. The layers of the bands are extracting to use in the formula of NDVI and NDWI. The results for experimental of unsupervised classification are shown in Tables 1, 2 and 3. The results shown are based on the covariance matrix.

The results are based on user accuracy, producer accuracy, overall accuracy, and kappa coefficient. The classified image is compared with the truth map data of Kuala Krai, Kelantan that is shown in Fig. 3. In order to evaluate those unsupervised methods for classification, the selected study area (465 pixels × 406 pixels) of image is classified based on two input indices: NDWI and NDVI. The color mapping and labeling of three classes are presented. As shown in Fig. 4, the selected training pixels are grouped into three categories which are water, non-vegetation, and vegetation. Based on the results (Tables 1, 2 and 3), the total correctly classified pixels is very high rather than misclassified pixels for every class.

Table 1 ISODATA covariance matrix

	Non-vegetation	Vegetation	Water	Sum	User's accuracy (%)
Non-vegetation	4875	2	2	4879	99.9
Vegetation	26,925	116,691	17,946	161,562	72
Water	430	4688	17,231	22,349	77
Sum	32,230	121,381	35,179	188,790	
Producer's accuracy (%)	15	96	49		

Overall accuracy: 73.52%
Kappa coefficient: 0.37

Table 2 K-means covariance matrix

	Non-vegetation	Vegetation	Water	Sum	User's accuracy (%)
Non-vegetation	4139	103	637	4879	85
Vegetation	3511	138,722	19,329	161,562	86
Water	455	3189	18,705	22,349	84
Sum	8105	142,014	38,671	188,790	
Producer's accuracy (%)	51	98	48		

Overall accuracy: 85.6%
Kappa coefficient: 0.56

Table 3 FCM covariance matrix

	Non-vegetation	Vegetation	Water	Sum	User's accuracy (%)
Non-vegetation	4032	845	2	4879	83
Vegetation	3630	157,919	13	161,562	98
Water	2	2269	20,078	22,349	90
Sum	7664	161,033	20,093	188,790	
Producer's accuracy (%)	53	98	99		

Overall accuracy: 96.4%
Kappa coefficient: 0.86

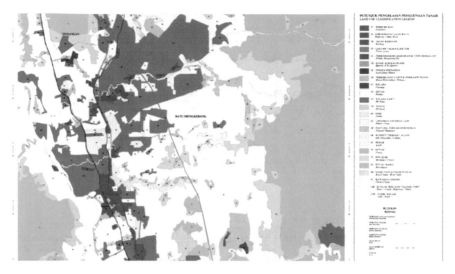

Fig. 3 Framework of study

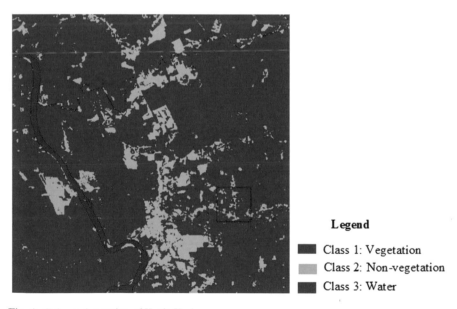

Legend

Class 1: Vegetation
Class 2: Non-vegetation
Class 3: Water

Fig. 4 Color and mapping of Kuala Krai

Based on the ISODATA classification, the class of vegetation showed that the misclassified pixels are 26,925 and 17,946 pixels to the non-vegetation and water, respectively. The non-vegetation class showed that the misclassified pixels to the other two classes are total 2 pixels each for vegetation and water. The class water showed that the misclassified pixels into the other two classes; non-vegetation and vegetation are 430 and 4688 pixels, respectively.

The user's accuracy and producer's accuracy were achieved from the addition of every row and column from the confusion matrix table. The total of correctly classified pixels of every row and column is divided with the sum of the total of training pixels. The overall accuracy of ISODATA classification is 73.52%, which is highly accurate. The other measurement is the kappa cocfficient which is 0.37 and the value is positive.

Based on the K-means classification, the class of vegetation showed that the misclassified pixels are 3511 and 19,329 pixels to the non-vegetation and water, respectively. The non-vegetation class showed that the misclassified pixels to the other two classes are total 103 and 637 pixels each for vegetation and water. The class water showed that the misclassified pixels into the other two classes: non-vegetation and vegetation are 455 and 3189 pixels, respectively.

The user's accuracy and producer's accuracy were achieved from the addition of every row and column from the confusion matrix table. The total of correctly classified pixels of every row and column is divided with the sum of the total of training pixels. The overall accuracy of ISODATA classification is 85.6%. The other measurement is the kappa coefficient which is 0.56 and the value is positive.

Based on the FCM classification, the class of vegetation showed that the misclassified pixels are 3630 and 13 pixels to the non-vegetation and water, respectively. The non-vegetation class showed that the misclassified pixels to the other two classes are total 845 and 2 pixels each for vegetation and water. The class water showed that the misclassified pixels into the other two classes; non-vegetation and vegetation are 2 and 2269 pixels, respectively.

The user's accuracy and producer's accuracy were achieved from the addition of every row and column from the confusion matrix table. The total of correctly classified pixels of every row and column is divided with the sum of the total of training pixels. The overall accuracy of ISODATA classification is 96.4%. The other measurement is the kappa coefficient which is 0.86 and the value is positive. The result was compared and the FCM classification showed that the FCM gained high accuracy, and it was better than the other two unsupervised methods that are used in this study, which are ISODATA and K-means.

5   Conclusion

As a conclusion, in the experiment of the land cover classification to compare between those three methods, the FCM is better than K-means and ISODATA. The experiment showed that FCM is better than the ISODATA and K-means classification and proved that the classification using indices is better rather than using image data from the direct band. The study shows that the unsupervised methods are excellent to use as the classification methods on this multispectral data.

Acknowledgements The authors would like to thank the Universiti Teknikal Malaysia Melaka for funding the study through PJP/2018/FTK(16A)/S01642. Besides, thank you to the Faculty of Information Technology and Communication for providing excellent research facilities.

References

1. Rouse JW, Haas RH, Schell JA, Deering DW (1973) Monitoring the vernal advancement and retrogradation (green wave effect) of natural vegetation
2. Taufik A, Sakinah S, Ahmad S, Ahmad A (2016) Classification of Landsat 8 satellite data using NDVI thresholds. J Telecommun Electron Comput Eng 8(4):37–40
3. Gao BC (1996) NDWI—A normalized difference water index for remote sensing of vegetation liquid water from space. Remote Sens Environ 58(3):257–266
4. Taufik A, Ahmad SSS, Khairuddin NFE (2017) Classification of Landsat 8 satellite data using fuzzy c-means. In: Proceedings of 2017 International Conference on Machine Learning Soft Computing—ICMLSC '17, January 2017, pp 58–62
5. Taufik A, Ahmad SSS (2016) Land cover classification of Landsat 8 satellite data based on fuzzy logic approach. In: IOP Conference Series Earth Environment Science, vol 37, p 012062
6. Usman B (2013) Satellite imagery land cover classification using K-means clustering algorithm computer vision for environmental information extraction. 63:18671–18675
7. Ko BC, Kim HH, Nam JY (2015) Classification of potential water bodies using landsat 8 OLI and a combination of two boosted random forest classifiers. Sensors (Switzerland) 15(6):13763–13777
8. Ferrato L-J, Forsythe KW (2013) Comparing hyperspectral and multispectral imagery for land classification of the Lower Don River, Toronto. J Geogr Geol 5(1):92–107
9. Li C, Wang J, Wang L, Hu L, Gong P (2014) Comparison of classification algorithms and training sample sizes in urban land classification with landsat thematic mapper imagery. Remote Sens 6(2):964–983
10. Ahmad A, Quegan S (2013) Comparative analysis of supervised and unsupervised classification on multispectral data. Appl Math Sci 7(74):3681–3694
11. Ball GH, Hall DJ (1965) ISODATA, a novel method of data analysis and pattern classification. Analysis no. AD699616, pp 1–79
12. Bezdek JC, Ehrlich R, Full W (1984) FCM: the fuzzy c-means clustering algorithm. Comput Geosci 10(2–3):191–203

Design of Automated Computer-Aided Classification of Brain Tumor Using Deep Learning

Nur Alisa Ali, A. R. Syafeeza, Liow Jia Geok, Y. C. Wong,
Norihan Abdul Hamid and A. S. Jaafar

Abstract In the recent years, health issues have inescapably become center of attention for many researchers. Brain tumor is now a leading cause of death among medically certified deaths. Brain image diagnosis is manually examined by the neurologist. It is time consuming and may lead to errors. The general idea of this research is to analyze the brain tumor based on magnetic resonance imaging (MRI) of medical images. The design of this system is aimed at detecting the brain tumor classifying the MRI samples. The system uses computer-based procedures to detect tumor blocks and classify the type of tumor to normal, benign, and malignant using tensor flow in MRI images of different patients. A promising method to perform the design is through a deep learning process. Deep learning is currently a well-known and superior method in the pattern recognition field. The performance measure for detection would be Equal Error Rate (EER), false acceptance rate (FAR), and false rejection rate (FRR). The higher percentage of accuracy of the biometric system depended on how much lower the ERR value would be. The samples are already available from a standard database, Multimodal Brain Tumor Image Segmentation Benchmark (BraTS). A comparison had been done between two different methods for classification of the brain tumor.

N. A. Ali · A. R. Syafeeza · L. J. Geok (✉) · Y. C. Wong · N. A. Hamid · A. S. Jaafar
Machine Learning and Signal Processing Research Group, Fakulti Kejuruteraan Elektronik Dan
Kejuruteraan Komputer, Centre for Telecommunication Research & Innovation (CeTRI),
Universiti Teknikal Malaysia Melaka, Durian Tunggal, Malaysia
e-mail: jiayu_0321@outlook.my

N. A. Ali
e-mail: alisa@utem.edu.my

A. R. Syafeeza
e-mail: syafeeza@utem.edu.my

Y. C. Wong
e-mail: ycwong@utem.edu.my

N. A. Hamid
e-mail: norihan.hamid@utem.edu.my

A. S. Jaafar
e-mail: shukur@utem.edu.my

© Springer Nature Singapore Pte Ltd. 2019
V. Piuri et al. (eds.), *Intelligent and Interactive Computing*, Lecture Notes in Networks
and Systems 67, https://doi.org/10.1007/978-981-13-6031-2_11

285

Keywords Brain tumor · Deep learning · Magnetic Resonance Imaging (MRI) · Convolutional Neural Network (CNN)

1 Introduction

The problem of the brain tumor in Malaysia is growing in recent years. It is now a leading cause of death among medically certified deaths. Brain image diagnosis is manually examined by the neurologist. It is time consuming and may lead to errors [1, 2]. Brain tumor occurrs when cells lose the ability to control their growth, divided too often, and without any order. Tumors can be classified as benign and malignant. Magnetic Resonance Imaging (MRI) is an advanced medical imaging technique used in hospitals to form the high quality of pictures of organs and structures inside the body. MRI assumes a significant part of brain tumor diagnosis and treatment arranging procedures. Fluid Attenuation Inversion Recovery (FLAIR) is commonly used in the diagnosis of brain tumor due to its higher contrast and resolution and describe the lesion precisely compare to other MRI modalities. Convolutional Neural Networks (CNN) is one of the deep learning classes. CNN is selected due to its robustness in handling image variabilities and noises contained inside the image. It consists of one or more convolutional layers, subsampling layers, and followed by one or more fully connected layers. Pattern recognition of CNN can be divided into three applications which are classification, recognition, and localization. In this research, CNN is used in the classification of the brain tumor. CNN has been used in several applications such as finger-vein recognition [3], face recognition [4], etc.

2 Methodology

There are four main processes, which are processing, designing, pretraining, and classification as shown in Fig. 1. For the designing stage, CNN architecture for AlexNet and VGG-16 are designed according to the research papers [5, 6]. The number of feature maps, number of CNN layers, and the size of filters are decided at this stage.

For the pretraining stage, the images of brain tumors are separated into testing and training files. The images in the training file will be used in the pretraining stage for the pretraining of CNN models. After that, the images in the testing file will be used to test the system and calculate the False Rejection Rate (FRR), False Acceptance Rate (FAR), Error Equal Rate (EER) and give the result in terms of accuracy (%).

Fig. 1 Flowchart of the methodology

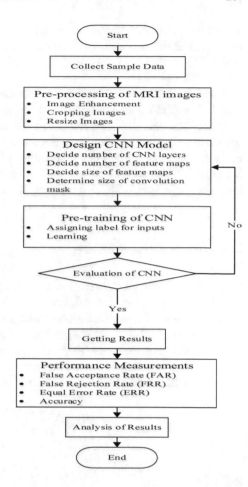

2.1 Clinic Samples

The Fluid Attenuated Inversion Recovery (FLAIR) dataset is downloaded from an online data source which is the Multimodel Brain Tumor Segmentation Challenge (BraTS 2015). The information is the MRI methodology with MHA arranged in 16 bits. The configuration of a dataset is then changed to Digital Imaging and Communications in Medicine (DICOM) design. The dataset was encoded in 8-bit DICOM. The parameter of the dataset is 240×240 of pixel resolution with 155 slices for each patient. The number of patients that are being analyzed is 141 patients in total. Images of each patient are extracted from 155 slices according to the number of images containing a tumor.

Multimodel Brain Tumor Segmentation Challenge (BraTS) Dataset. The dataset that used in this research is from the BraTS 2015 [7]. BraTS is an organization that held the gathering and workshop about the brain tumor. The data are in MHA format and three dimensional. The FLAIR MRI modality is used in this research. For this research, 7040 sample images of brain tumor are obtained, out of which 3952 samples are for malignant, 2783 samples for benign, and 305 samples for the normal brain.

Conversion Software. The original format of the image is MHA and three dimensional. The MHA format image is needed to convert to NII format by using ITK software. The data that used for this research is two dimensional. X Medcon software is used in order to convert the dimensional data. The selected slice is converted to the Digital Imaging and Communications in Medical (DICOM) format by using X Medcon. Since tensor flow can read the only image in JPG and PNG format, the image is converted to PNG format using X Medcon.

2.2 Image Preprocessing

Image preprocessing is utilized to enhance the pictures. The calculation of the picture standardization is utilized for this dataset. The informational index that given was in better condition. Besides, the input images are resized in the preprocessing. The MRI dataset is 8-bit significance pictures which are equivalent to 256 gray levels and the genuine arrangement of MRI framework has 16-bit significance pictures that are comparable to 65,535 gray levels. Equation 1 is used in preprocessing of the images. Grayscale changes of image are based on x, y (pixels) and N (number of bits).

$$I(x, y) = \frac{I(x, y)}{2^N - 1}, N = 8 \qquad (1)$$

2.3 Designing

Two CNN models are designed to classify brain tumors. The CNN models that proposed in this research are AlexNet and VGG-16. The architecture of CNN for VGG-16 Model 1 is designed according to [7]. Hereby, the number of feature maps is changed to 32 and 64 to compare the result of the model to find out the best architecture of the CNN model. The architecture of CNN for AlexNet Model 1 is designed according to [7]. The number of feature maps is changed to 32 and 64 to compare to the result of the model to find out the best architecture of the CNN model.

2.4 Pretraining

The CNN models were pretrained in this stage to learn from the feature maps. First, the images of brain tumor are separated into two files, which are the train and test part. 80% of the images (5633 samples of brain tumor) for train file and 20% (1402 samples of brain tumor) for test file. The best weight of the CNN model is saved into an hdf5 file

2.5 Classification

In the classification stage, a graphical user interface (GUI) is designed to show the result. By using the GUI, the user can load the CNN model that user wants to use and the images can be loaded from the test file to test the result of classification of the brain tumor. The brain tumor will classify according to the feature maps that CNN model learned before. The image of a brain tumor is classified into three classes; normal, malignant, and benign.

2.6 Performance Assessment Matric

The performance of the classification result is analyzed by the confusion matrix of the test file. The parameters that contribute to the performance is False Rejection Rate (FRR), False Acceptance Rate (FAR), Equal Error Rate (EER), and accuracy of the system. The equations are shown as follows:

$$FRR = \frac{FN}{(FN + FP)} \tag{2}$$

$$FAR = \frac{FP}{(FN + FP)} \tag{3}$$

$$EER = \frac{FRR + FAR}{2} \tag{4}$$

$$accuracy = \frac{TP + TN}{total} \tag{5}$$

where TP is true positive, FP is the false negative, TN is the true negative, FN is false negative, and total is the total number of images.

3 Results

3.1 Classification Process

The classification techniques used in this research are AlexNet CNN model and VGG-16 CNN model. Table 1 shows the results of classification using three different numbers of feature maps in both AlexNet and VGG-16 CNN models.

The result of the architecture of AlexNet CNN Model 1 is only 60.2% of accuracy. The best AlexNet model is the AlexNet model 3 which has 64 feature maps for each

Table 1 The results of CNN models

CNN model	Performance
AlexNet Model 1	FRR = 1.000 FAR = 0.000 EER = 0.500 Accuracy = 0.602 (60.2%)
AlexNet Model 2	FRR = 0.151 FAR = 0.849 EER = 0.500 Accuracy = 0.939 (93.9%)
AlexNet Model 3	FRR = 0.618 FAR = 0.381 EER = 0.500 Accuracy = 0.961 (96.1%)
VGG-16 Model 1	FRR = 0.480 FAR = 0.520 EER = 0.500 Accuracy = 0.982 (98.2%)
VGG-16 Model 2	FRR = 0.868 FAR = 0.132 EER = 0.500 Accuracy = 0.962 (96.2%)
VGG-16 Model 3	FRR = 0.981 FAR = 0.019 EER = 0.500 Accuracy = 0.851 (85.1%)

layer. The performance of VGG-16 architecture from [5] seemed to be the best model for classification of the brain tumor in which the accuracy was about 98%. A graphical user interface (GUI) is designed in order to show the result of the classification and ease user understanding the result of classification of the brain tumor.

4 Conclusion

Magnetic resonance imaging (MRI) is one of the latest technologies to analyze the brain tumors. The manual analysis of the classification of brain tumors is time consuming and could lead to the misdiagnosis. Thus, an automated computer system is very important in the analysis of brain tumors.

The accuracy of CNN models can be affected by some vectors, which are the number of feature maps, the number of filter layers, and the size of the sample data set. Overfitting occurs if the number of feature maps is too large and the accuracy decrease. However, the accuracy of CNN models is low if the size of the sample data set is too small for pretraining. In this research, the best CNN model is VGG-16 model 1, which has the highest accuracy of 98.2%. The classification in the earlier stage of this brain tumor is a serious issue in medical science. By enhancing the new imaging techniques, it helps the doctors to observe the occurrence and growth of tumor-affected regions at different stages so that they can provide a suitable diagnosis with these images scanning.

Acknowledgements The authors would like to thank Universiti Teknikal Malaysia Melaka (UTeM) and Ministry of Higher Education for supporting this research under PJP/2018/FKEKK (9D)/S01622.

References

1. Ostrom QT et al (2016) CBTRUS statistical report: primary brain and other central nervous systems tumors diagnosed in United States in 2009–2013. Neuro Oncol 18:v1–v75
2. Amiri S, Rekik I, Mahjoub MA (2016) Deep random forest-based learning transfer to SVM for brain tumor segmentation. In: 2nd international conference on advanced technologies for signal and image processing, ATSIP 2016
3. Itqan K et al (2016) User identification system based on finger-vein patterns using Convolutional Neural Network. ARPN J Eng Appl Sci 11(5):3316–3319
4. Syafeeza A et al (2015) Convolutional neural networks with fused layers applied to face recognition. Int J Comput Intell Appl 14(03):1550014
5. Makde V et al (2018) Deep neural network based classification of tumourous and non-tumorous medical images. In: Smart innovation, systems and technologies, pp 199–206
6. Shen H, Zhang J, Zheng W (2017) Efficient symmetry-driven fully convolutional network for multimodal brain tumor segmentation. In: ICIP (2017, to appear) Google Scholar, 2017
7. Menze B et al (2015) Brain tumor segmentation with deep learning, Multimodal brain tumor image segmentation (BRATS) challenge. In: MICCAI

Part III
System and Security Theories

Image Spam Detection on Instagram Using Convolutional Neural Network

Chastine Fatichah, Wildan F. Lazuardi, Dini A. Navastara, Nanik Suciati and Abdul Munif

Abstract Instagram is a social media to share moments in the form of photos and videos that is currently in great demand. However, the popularity of Instagram also widely used by certain people to spread spam for their personal interests such as advertising. Therefore, it requires a system for detecting spam on Instagram to obtain useful information and expected by users. The previous researches on image spam detection have been done to filter out inappropriate content on email using conventional classification methods. Recently, the Convolutional Neural Network (CNN) is a method that obtains higher accuracy than conventional classification methods for image classification problems without prior feature extraction process. We propose image spam detection using CNN on social media Instagram. The four architectures of CNN are used to compare the performance of each architecture, i.e., 3-layer, 5-layer, AlexNet, and VGG16. The performance of the system is evaluated by 8000 images taken from Instagram using web crawler. The results show that the highest accuracy achieves 0.842 by using a VGG16 architecture.

Keywords Image spam · Spam detection · Convolutional neural network · Social media · Instagram

C. Fatichah (✉) · W. F. Lazuardi · D. A. Navastara · N. Suciati · A. Munif
Department of Informatics, Institut Teknologi Sepuluh Nopember, Surabaya, Indonesia
e-mail: chastine@if.its.ac.id

W. F. Lazuardi
e-mail: wildanjawa@gmail.com

D. A. Navastara
e-mail: dini_navastara@if.its.ac.id

N. Suciati
e-mail: nanik@if.its.ac.id

A. Munif
e-mail: munif@if.its.ac.id

© Springer Nature Singapore Pte Ltd. 2019
V. Piuri et al. (eds.), *Intelligent and Interactive Computing*, Lecture Notes in Networks and Systems 67, https://doi.org/10.1007/978-981-13-6031-2_19

1 Introduction

Spam is the use of electronic devices to send irrelevant messages or content to a large number of recipients. The content of spam can be found in various electronic and internet media such as emails, short messages on mobile phones, search engines, websites, online games, and social media. Instagram is one of social media that is currently widely used by society. Due to having the unique features and user convenience as a means to share the moment in the form of photos and video. With a large number of active users on Instagram causes many entrepreneurs to use Instagram as one of the media to promote their products in the form of goods or services. This condition may cause the results of searching on Instagram to be irrelevant with that expected by the user.

Therefore, an automatic system for spam detection on Instagram to obtain useful information and expected by users is preferred. Spam on Instagram can be in the form of comments, direct messages, or post that not expected by the recipient or reader. In this research, the type of spam that will be detected on Instagram focuses on spam in the form of a post. We use image spam on Instagram data. In the future work, the research on spam detection is extended using a fusion of image and text from Instagram data. The previous researches on image spam detection have been conducted to filter out content in the email. The study literature of image spam detection methods have been presented in [1–3]. There are two categories of techniques for image spam detection, i.e., OCR-based technique and technique based on low-level image feature [1]. The previous research uses OCR-based techniques to extract the text embedded into images for email spam filtering [4, 5]. Combining the low-level feature and OCR-based for identifying the image spam has been done in research [6]. The literature [7] propose combining of text extraction and low-level image feature to categorize spam and ham. Due to the computational expensive of OCR-based, the research [8] propose spam image filtering based on extracted overlay text and color features. The literature [9] use edge-based features to compute the similarity score and classify the features using Support Vector Machine to detect spam image. To defeat OCR-based, [10] use low-level image by obscure detection approach. The research [11] present two approaches for detecting image spam. The first approach uses visual feature and Support Vector Machine as classification method. The second approach is near duplicate detection of image. Recently, the research [12] apply the Principle Component Analysis (PCA) and Support Vector Machine (SVM) for image spam detection. The Eigenspace of image spam is extracted as feature using PCA and SVM is used to classify the image into spam or not spam. The previous researches on image spam detection are generally applied in email system and use the conventional classification method such as Support Vector Machine. The literature [13] propose a method for spam detection on Instagram using the user profile feature and media feature. The Random Forest is used as classification method. The literature [14] study Recurrent Neural Network (RNN) and Convolutional Neural Network (CNN) for spam detection in social media however not yet implemented and evaluated the results. Recently, the classification method based on deep learning is popular method to solve the classification problem in large data with higher accuracy than conventional classification methods. Convolutional Neural Network (CNN) is one of deep learning method that developed from the artificial neural network that is widely used to detect and recognize objects in an image. We use CNN method to detect image spam on Instagram due to CNN obtain very high accuracy of image classification problem in previous research [15–17]. In addition, CNN method is a method that integrates the convolution process and the classification process, so we did not need to determine the feature extraction method. CNN belongs to a deep neural network because it has a large hierarchy of layers. CNN architecture generally consists of three types of layers: convolutional layer, subsampling layer, and fully connected layer. There are several architectures of CNN and we use four architectures of CNN to detect image spam, i.e., 3-layer, 5-layer, AlexNet, and VGG16. The performance of each architecture is evaluated by the accuracy and running time.

2 Research Method

The architecture of proposed method to detect image spam on Instagram using Convolutional Neural Network (CNN) is shown in Fig. 1. The proposed method consists of five phases, i.e., crawling data from Instagram, getting image data, manual labeling of training data, training process using CNN, and testing process using CNN model. The output of the system is labeling of image data that spam or normal. The first phase is crawling data from Instagram that conducted through a web crawler. Image data is taken from photos on each Instagram post. The second phase is labeling process of the image manually as training data. There are two categories of image labeling, i.e., spam and normal. The next phase performs the training process to build the CNN model. In this research, we compare the performance of four CNN architectures such as 3-layer, 5-layer, AlexNet, and VGG16.

2.1 The 3-Layer and 5-Layer Architectures

The 3-layer architecture has three convolution layers as well as two fully connected layers and each convolution layer ends with a pooling layer. While the 5-layer architecture has five convolution layers and two fully connected layers and each layer of convolution is also ends with a pooling layer.

2.2 The AlexNet and VGG16 Architectures

AlexNet is one of CNN architecture created by Alex Krizhevsky, Geoffrey Hinton, and Ilya Sutskever in the ImageNet Large Scale Visual Recognition Challenge

Fig. 1 Architecture of
image spam detection using
CNN method

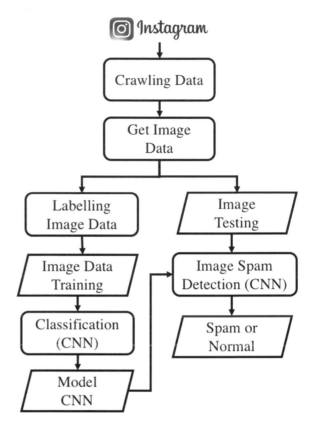

Fig. 1 Architecture of image spam detection using CNN method

(ILSVRC) competition in 2012 [15]. In the ILSVRC competition, AlexNet won the smallest error of 15.4%. This becomes a new breakthrough in visual recognition that deep learning able to deliver amazing accuracy results. The AlexNet architecture consists of five convolution layers and three fully connected layers. But in this study, we modify the number of neurons in each layer due to appropriate the existing hardware capabilities. VGG16 is also one of CNN architecture that gets runners-up position in the 2014 ILSVRC competition. This architecture is proposed by Simonyan and Zisserman. VGG16 consists of 16 convolution layers and three fully connected layers [16]. In this study, we also modify the number of neurons on each layer due to the hardware capabilities for testing. The last phase performs the testing process to determine label of the image includes spam or normal based on CNN model.

3 Experiments on Image Spam Detection

3.1 Dataset Description

This research uses 8000 image data taken from post of Instagram and is divided into 7600 as training data and 400 as testing data. We use one of our Instagram accounts for crawling the post from our Instagram wall. The testing data consist of two classes, i.e., spam and normal with 200 image for each class. The image example of spam image and normal image are shown in Fig. 2. All data used in the testing process is different from the data used during the training and validation process.

3.2 Experimental Results and Discussion

CNN architecture generally consists of three types of layers: convolutional layer, subsampling layer, and fully connected layer. The four architectures of CNN are used to compare the performance of each architecture, i.e., 3-layer, 5-layer, AlexNet, and VGG16. There are several parameters of CNN that are used in this research such as number of epoch, pool size, dropout rate, and optimizer algorithm. The parameter setting for the number of epoch is 50 with optimizer algorithm uses Adam and dropout rate is 0.5. In the CNN layers, we use 2×2 pool size in each layer of CNN. The first and second scenarios for testing process are designed to measure the performance of 3-layer and 3-layer architecture by comparing among several of the kernel dimensions. The kernel dimension has the same value on each layer.

The results of the first and second scenarios can be seen in Tables 1 and 2. The results of 3-layer and 5-layer architectures for image spam detection achieves the highest accuracy 0.832 with 5×5 kernel dimension and 0.730 with kernel dimension

(a) **(b)**

Segitiga Instan
Kode AS

Fig. 2 Image example of post on Instagram **a** spam image and **b** normal image

Table 1 The results on 3-layer architecture with varying kernel dimension

Kernel dimension	Accuracy	Precision	Recall	Running time (s)
2 × 2	0.697	0.820	0.658	9610
3 × 3	0.735	0.815	0.703	10,171
4 × 4	0.725	0.760	0.710	11,095
5 × 5	0.832	0.780	0.693	9415

Table 2 The results on 5-layer architecture with varying kernel dimension

Kernel dimension	Accuracy	Precision	Recall	Running time (s)
2 × 2	0.727	0.750	0.718	13,720
3 × 3	0.702	0.775	0.677	8981
4 × 4	0.725	0.760	0.710	9431
54333*	0.730	0.785	0.707	9500

*54333: Kernel dimension varying on each of layer in sequence with size 5 × 5, 4 × 4, 3 × 3, 3 × 3, and 3 × 3

Table 3 Comparing results on 3-layer, 5-layer, AlexNet, and VGG16 architectures

CNN architectures	No. of layer	No. of neurons	Accuracy	Precision	Recall	Running time (s)
3-layer	8	193	0.832	0.780	0.693	9415
5-layer	12	449	0.730	0.785	0.707	9500
AlexNet	11	2401	0.755	0.755	0.755	70,741
VGG16	17	4417	0.842	0.950	0.782	97,418

54333*, respectively. The results of 3-layer and 5-layer architectures for image spam detection achieves the highest accuracy 0.832 with 5 × 5 kernel dimension and 0.730 with kernel dimension 54333*, respectively.

The third scenario aims to compare the accuracy obtained by four CNN architectures, i.e., 3-layer, 5-layer, AlexNet, and VGG16. For the 3-layer and 5-layer architecture, we select the kernel dimension that has the highest accuracy value. The experimental results of the third scenario are shown in Table 3. The highest accuracy achieved by VGG16 architecture with 0.842, however, the running time is the worst compare to others architectures. Overall, the results show that the use of CNN method obtains good accuracy for all CNN architectures even though the training data contains varying images in each class. Additionally, the benefit of using CNN method is no need to perform the feature extraction process. In the conventional classification method, we should select the appropriate feature extraction method to obtain good classification results.

The image sample results of spam detection are shown in Table 4. The first column is the image sample, the second column is the label class, and the third column is the prediction results. There are 2 image samples of each correct and 1 image sample

Table 4 The image sample results of spam detection

Image sample	Label class	Prediction
	Normal	Normal
	Spam	Spam
	Spam	Normal

of incorrect prediction (misclassification) results. In the case of misclassification, the third image sample should be a spam image but the result is a normal image. This is due to the lack of similar sample images in the training data. In case of misclassification, there are some images which photo of a woman who uses hijab that should be a normal image but the result is spam image. We analyze that there are many images of hijab ads in the training data caused the prediction result is incorrect.

4 Conclusions

Image spam detection has an important role to obtain the relevant information expected by the user in post searching of social media such as Instagram. We present a system for detecting the image spam using Convolutional Neural Network (CNN) on Instagram. The performance of the system is evaluated by 8000 images taken from Instagram using a web crawler. The four architectures of CNN, i.e., 3-layer, 5-layer, AlexNet, and VGG16 are used to compare the performance of each architecture. The results show that the highest accuracy achieves 0.842 by using a VGG16 architecture however the running time of VGG16 is the worst compare to others architectures. The use of CNN method obtains good accuracy even though the training data contains varying images in each class. In addition, the advantage of the use of CNN method as the classification method is no need to perform the feature extraction process. In the future work, we extend the research on spam detection using a fusion of image and text from Instagram or other social media. Due to the information of spam post in social media can also be captured from text data.

Acknowledgements This research is funded by the Ministries of Research, Technology, and Higher Education Indonesia through PTUPT grant.

References

1. Biggio B, Fumera G, Pillai I, Roli F (2011) A survey and experimental evaluation of image spam filtering techniques. Pattern Recogn Lett 32(10):1436–1446. https://doi.org/10.1016/j.patrec.2011.03.022
2. Blanzieri E, Bryl A (2008) A survey of learning-based techniques of email spam filtering. Artif Intell Rev 29(1):63–92. https://doi.org/10.1007/s10462-009-9109-6
3. Ketari LM, Chandra M, Khanum MA (2012) A study of image spam filtering techniques. In: The fourth international conference on computational intelligence and communication networks, pp 245–250. https://doi.org/10.1109/cicn.2012.34
4. Fumera G, Pillai I, Roli F (2006) Spam filtering based on the analysis of text information embedded into images. J Mach Learn Res 7, 2699–2720. Retrieved from http://jmlr.csail.mit.edu/papers/volume7/fumera06a/fumera06a.pdf
5. Sathiya V (2011) Partial image spam E-mail detection using OCR. Int J Eng Trends Technol 1(1):55–59
6. Gupta A, Singhal C, Aggarwal S (2012) Identification of image spam by using low level & metadata features. Int J Netw Secur Appl (IJNSA) 4(2):163. https://doi.org/10.5121/ijnsa.2012.4213
7. Das M, Prasad V (2014) Analysis of an image spam in email based on content. Analysis 3(3):129–140
8. Aradhye HB, Myers GK, Herson JA (2005) Image analysis for efficient categorization of image-based spam E-mail. In: Proceedings of the international conference on document analysis and recognition, ICDAR, 2005(c), pp 914–918. https://doi.org/10.1109/icdar.2005.135
9. Nhung NP, Phuong TM (2007) An efficient method for filtering image-based spam E-mail. In: 2007 IEEE international conference on research, innovation and vision for the future, RIVF 2007, pp 96–102. https://doi.org/10.1109/rivf.2007.369141
10. Biggio B, Fumera G, Pillai I, Roli F (2007) Image spam filtering by content obscuring detection. In: Fourth 2005—second conference on email and anti-spam, CEAS, pp 3–7
11. Mehta B, Nangia S, Gupta M, Nejdl W (2008) Detecting image spam using visual features and near duplicate detection. In: Proceeding of the 17th international conference on World Wide Web—WWW '08, p 497. https://doi.org/10.1145/1367497.1367565
12. Annadatha A, Stamp M (2018) Image spam analysis and detection. J Comput Virol Hacking Tech 14(1):39–52. https://doi.org/10.1007/s11416-016-0287-x
13. Zhang W, Sun HM (2017) Instagram spam detection. In: Proceedings of IEEE Pacific rim international symposium on dependable computing, PRDC, pp 227–228. https://doi.org/10.1109/prdc.2017.43
14. Jain G, Agarwal B (2016) An overview of RNN and CNN techniques for spam detection in social media. Int J Adv Res Comput Sci Softw Eng 6(10), 126–132. Retrieved from https://ijarcsse.com/docs/papers/Volume_6/10_October2016/V6I10-0126.pdf
15. Krizhevsky A, Sutskever I, Geoffrey EH (2012) ImageNet classification with deep convolutional neural networks. In: Advances in neural information processing systems 25 (NIPS2012), pp 1–9. https://doi.org/10.1109/5.726791

16. Simonyan K, Zisserman A (2014) Very deep convolutional networks for large-scale image recognition, pp 1–14. https://doi.org/10.1016/j.infsof.2008.09.005
17. Wicaksono AY, Suciati N, Fatichah C, Uchimura K, Koutaki G (2017) Modified convolutional neural network architecture for Batik Motif image classification. 2(1), 26–33

A Review of Botnet Detection Approaches Based on DNS Traffic Analysis

Saif Al-Mashhadi, Mohammed Anbar, Shankar Karuppayah
and Ahmed K. Al-Ani

Abstract A botnet is a network of computing devices being commanded by an attacker, a daily Internet problem, causing extensive economic damage for organizations and individuals. With the avail of botnets, attackers can perform remote control on exploited machines, performing several malicious activities, since it enormously increases a botnet's survivability by evading detection, Domain Name System (DNS) nowadays is a favourable botnet communication channel. Fortunately, many strategies have been introduced and developed to undertake the issue of botnets based on DNS resolving; this review explores the various botnet detection techniques through providing a study for detection approached based on DNS traffic analysis. Some related topics, including technological background, life cycle, evasion, and detection techniques of botnets are introduced.

Keywords DNS-based botnets · Fast-flux · DGA · Botnets

1 Introduction

There is no doubt that the Internet becomes a core element of our day to day life, today its time about online presence, e-learning, social media presence, e-banking, and cloud computing [1, 2].

Unfortunately, the Internet is targeted by malicious attackers, one of the most sophisticated preferable attacker's tools is a botnet. A botnet is a network of computing devices that have been connected by malware [3], which is called the bot, controlled by a hostile and hidden commander called the botmaster, forming an army of compromised computing machines.

S. Al-Mashhadi (✉) · M. Anbar (✉) · S. Karuppayah · A. K. Al-Ani
National Advanced IPv6 Center, Universiti Sains Malaysia, 11800 Gelugor, Penang, Malaysia
e-mail: saifjawad@student.usm.my

M. Anbar
e-mail: anbar@usm.my

© Springer Nature Singapore Pte Ltd. 2019
V. Piuri et al. (eds.), *Intelligent and Interactive Computing*, Lecture Notes in Networks and Systems 67, https://doi.org/10.1007/978-981-13-6031-2_21

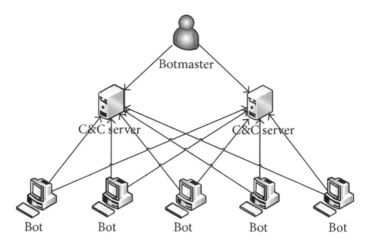

Fig. 1 Botnet structure

"Eggdrop" was the first botnet, evolved in 1993 by utilizing Internet Relaying channel IRC; IRC is a chat system that offers one-to-one and one-to-many instant messaging [4].

As shown in Fig. 1, the fundamental feature that botnets have is the power of its members (bots) to communicate with each other, or with the command and control servers (C&C) [5]. This way, the members of the botnet can obtain new commands. To establish this connection, botnet developers mainly have two options, use static IP addresses which are set on a code level within the malware, or the use of domain names, which are determined or generated via algorithms [6]. Instead of having names and numbers [7].

Moreover, since it enormously strengthens a botnet's survivability by evading detection [8], Domain Name System (DNS) nowadays is a convenient communication tool for the attackers.

Botmaster often relies on dynamic hosting strategies that are characterized by rapid changing of domain names and/or IP addresses as evading strategies. Dynamic DNS strategies include fast-flux [9] and domain-flux [8]. However, DNS provides evading strategies; it introduces some traits that can be employed for botnet detection [10]. Botnets are recognized one of the main threats to the security and privacy of Internet users, such threats are Distributed Denial of Service (DDoS), SPAMs, implementing ransomware [11], and phishing emails [12]. From analyzing peek attacks in Fig. 2 [13], we can notice that there is an increase in botnet DDoS attacks size from 24 Gbps in 2007 to 1.7 Tbps. in March 2018. As a result, many researchers are working to develop based-DNS mechanisms for detection of the botnets.

The rest of this review is organized as follows. Section 2 presents background on DNS, botnet evasion techniques and botnet life cycle. Section 3 presents existing work on identifying botnet malicious DNS traffic. Section 4 presents a Discussion. Finally, Sect. 5 concludes the paper.

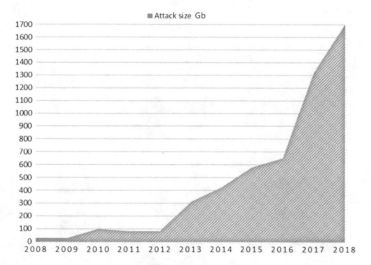

Fig. 2 Botnet DDoS attacks size for the last 10 years

2 Background

To construct further reviews and details, it is necessary to know some key terms about DNS and DNS-based botnet. This section discusses the key terms about DNS, based-DNS botnet, and botnet life cycle stages.

2.1 Domain Name System (DNS)

DNS is all about answering queries from any source, and mapping to IPs in a hierarchical way, it is a catalogue between human language and computer language. To query domain name IP address, a client called the DNS resolver, typically sends a DNS query to a local recursive DNS server (RDNS), or a known root DNS server if there is no local DNS server is available. The recursive server iteratively discovers Authoritative Name Servers (ANS) related to each zone; the name server will then answer with the stored record, or the name of another name server if the domain has been delegated. Using the glue record bundled with the DNS response, a DNS resolver will then be able to iteratively query the next name server until an address record is received [14].

The most common records are A-records and AAAA-records which contains IPv4 and IPv6 host addresses, and name server records which are used to delegate a subdomain to another name server. The typical procedure of domain name resolving is illustrated in Fig. 3.

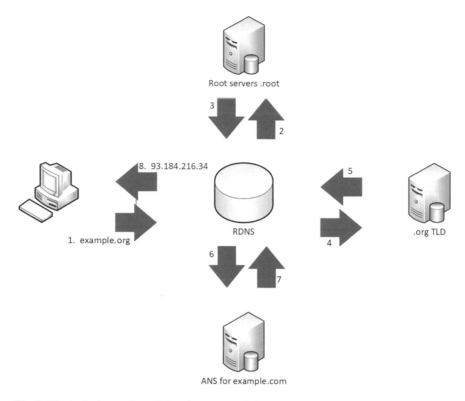

Fig. 3 The typical procedure of domain name resolving

A domain name is an order of labels separated by dots. For example, in the domain nav6.usm.my, .my is the TLD, usm is the second-level domain (SLD) of the .my TLD, while nav6 is a third-level domain (3LD) [15].

2.2 DNS-Based Botnet Evading Techniques

To avoid blacklisting utilizing DNS is one favourable approach for the attacker [15], more complex technology to change C&C servers IP addresses and domain name dynamically has been utilized by the botmasters, such as fast-flux [8] (both Single Flux and Double Flux) and Domain Generation Algorithm (DGA) [16], DGA categorized as domain fluxing.

Fast-Flux

The Honeynet project reported the fast-flux issue first in 2007 [17]. However, the first study of this phenomenon systematically was in 2008 [9]. Basically, "fast-flux"

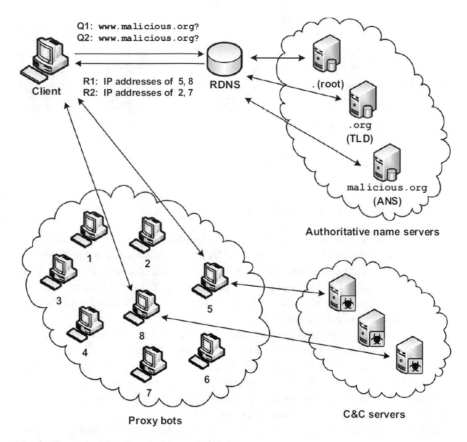

Fig. 4 The typical fast-flux domain resolving [9]

means that a domain name has many IP addresses assigned to it periodically. Fast-flux also used in different legitimate activities, for example, it can be utilised as a load balancer for the content delivery network (CDN) [8].

The attackers exploit fast-flux by combining round-robin IP addresses, with a short Time-To-Live (TTL) for the DNS Resource Record (RR), which lead to distribute the load for their botnet and extreme IP changes in their C&C server [17]. Moreover, the public IP address of the bot sometimes utilized as a proxy (flux-agent), the domain name of the C&C server is resolved to the flux-agent IP address; the real service server is hidden behind these flux agents. Double flux brings the fast-flux to a new level, the name server under the control of the botnet will resolve the domain inquiry for the C&C server. The name server's response will contain the IP addresses corresponded to the queried domain name. The resolved records correspond to the proxy bots that transmit information to and from bots and the C&C servers. The typical fast-flux domain resolving illustrated in Fig. 4.

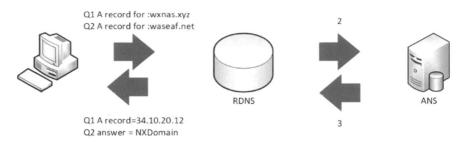

Fig. 5 A typical procedure for resolving a DGA

Domain Generation Algorithm (DGA)

DGA is carried out by dynamically generating pseudorandom domain. DGA has many unique characteristics. First, no need to hardcode the C&C domain, so it is difficult to predict the domain name that will be used as the C&C address [3]. Additionally, one can also use DGA as a failsafe channel to back up situations when the primary communication channels fail [18], for instance, Zeus worm used DGA as a safe backup communication mode [19].

By having a secret mutual seed algorithm between C&C servers and bot [20], botmaster usually registers one or few of these domains. To avoid blacklisting domain activated and utilized for a short period, For instance, Conficker can generate 50.000 domains per day, of which 500 are queried daily [22]. Reverse engineering such vast quantities of nonexistent algorithmically generated domain names is not sufficient, resource concentrated with low success rates [15].

If there is duplication with already active domain names, the botmaster will not be able to communicate with his bots, so the botmaster designs the code in a way such that the generated domains vary significantly from registered ones [22]. Crypotolocker is an example of ransomware that utilizes DGA, by adding time with domain generated, making malicious domain name filtering more difficult [11]. The typical procedure for resolving a DGA is illustrated in Fig. 5.

2.3   Botnet Life Cycle

For a targeted computer machine's (bot) to participate in a botnet, botnet life passes through four below different stages.

Initial infection and propagation phase

Since few will allow their devices to be a host of the attacker malicious activities, one of the primary concerns for the attackers is how to inject their malicious code into the victim computer machine; the machine could be a personal computer, smartphone or Internet of Thing devices.

To increase the chance of infection, the attackers will try each possible technique to inject his code; such efforts include the use of SPAMs, vulnerabilities caused by other software or botnets, file sharing, instant messaging and others. Some of these passive techniques might involve user intervention, in which case the attackers attempt tricks the user to download or install vulnerable software. Furthermore, the infection may be automated by scanning the networks for exploits known vulnerabilities, e.g., Mirai Botnet that has scanner process that actively seeks vulnerable machines to compromise [23], weak passwords or misconfiguration of specific services. After vulnerability is found the bot will try to connect to a remote server, then the remote server will try to install the software required to control the host. The connection with a remote server is established only after a DNS query by the bot [24], DNS lookup query method is the most common of almost all botnets [25].

Connection or Rallying Phase

In this phase, the bot attempts to locate and join the command and control server, in peer to peer botnet rallying is done by seeding, the initial list of peers is provided during the infection phase. The connection is made by providing the domain name of C&C server, or the proxy server that acts as a link between the bot and the actual C&C server. As stated before, the domain could be hardcoded, making the bot vulnerable for reverse engineering, or the domains generated automatically according to DGA. It is more fixable for the attackers to map a domain to their C&C server, let them easy to change the IP of the blocked C&C server. This phase occurs many times in the bot life cycle. Since the bots sending many discovery packets [26], bots will be exposed during this phase.

The Malicious or Attack phase

In this phase, the botmaster instructs bots to perform malicious activities; bots can perform several disruptive attacks [27] such as email SPAM, distributed denial of service attack, and distributing malicious software.

The maintenance and upgrading phase

In this phase, bots are in an idle state waiting for new commands from the botmaster, and these commands might include new target systems or a new update to the bots, the bot updates their behavior and performs new malicious activities based on the commands received.

The botmaster is trying to keep bot under its control as long as possible. As a result, bots are in constant update, either to bypass antivirus detection, improve propagation vectors [28] with new vulnerabilities and methods, or to fix any possible bug in the code (The botnet life cycle is illustrated in Fig. 6).

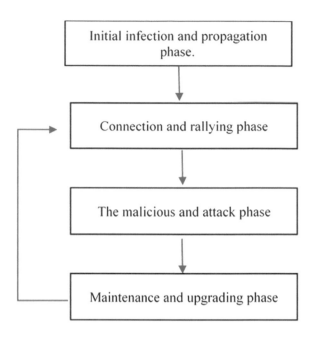

Fig. 6 Botnet life cycle

3 DNS-Based Botnet Detection Systems

There are two main proposed approaches for botnet detection based-DNS, the first is by placing honeypot on the network, the second by using Intrusion Detection Systems (IDS) [29]. Since the scope of this review on DNS-based detection; The survey will focus more on DNS-based approaches. The detection taxonomy is illustrated in Fig. 7.

3.1 Honeypots

Honeypots are widely deployed computer network machines, designed to be vulnerable with the aim to be infected in a short time; Honeynet is a network of honeypot [30]. The main goal of a honeypot is to analyze bots by collecting its information, measuring the technology used, attacker pattern, discover the C&C system, knowing attackers tools and motivations. Honeypots are running specialized software that can gather and monitor signature of bots for content-based detection and information of Botnet C&C.

Furthermore, honeypots can also help discover new attacks approaches and mechanism. Therefore, honeypots lead to more effective countermeasures [30], Gaonjur et al. [31] used honeypot for tracing botnets in the network by generating an early

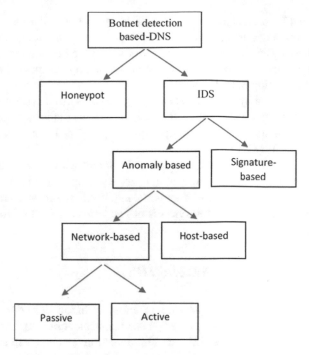

Fig. 7 Taxonomy of botnet detection based on DNS network traffic

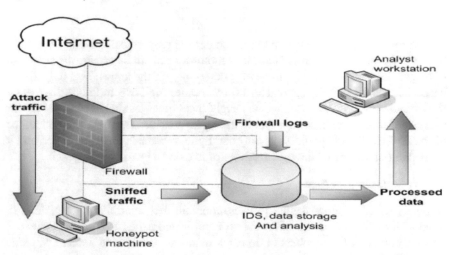

Fig. 8 A typical honeypot configuration [31]

report that could help to recognize the consequences of Botnets. Figure 8 shows a typical honeypot configuration [32].

Kumar et al. [33] used Honeynet as data feeders for a botnet detection framework, they analyses that data using different open source analyzing tool, botnet payload parser, which analysis payload on the network to get the bot command signatures. This study shows that it is too hard to deploy a single popular method to detect and mitigate all the existing botnet classes [34].

Honeydns proposed by Oberheide et al. [34] is a recognized system in Honeynet, that collect the DNS queries targeting unused (i.e., darknet) address spaces, by applying some basic statistics over the captured queries, it helped honeypots to limit the attackers from evading it [35].

A Honeynet is an excellent source to study botnet. However, it cannot detect all types of infections, such as bots that not propagate by scanning, and only could report the infected machine within its network [29]. Furthermore, attackers can utilize honeypot/Honeynet to attack other systems or machines outside the Honeynet [36].

3.2   Intrusion Detection System (IDS)

IDS group classified into two subgroups; anomaly-based detection system and signature-based [27]. Besides, anomaly-based subcategories to host-based and network-based detection system [29], current detection techniques in each category will be discussed.

Signature-based IDS

Signature-based Botnet detection technique detects only predefined bots by matching their signatures using IDS. Snort [37] monitor traffic to determine the signature of an existing bot, this lead to immediate determine for the known bots only. DNS-based blacklist (DNSBL) approach [38] by monitoring DNS traffic, it discovered the identities of the bots, based on the insight that botmaster should achieve "reconnaissance" lookups to determine their bots blacklist status. The main two limitations of the DNSBL-based approach, its only can figure reconnaissance botmaster, and it limited to bot advertised through SPAMs traffic using a heuristics approach.

Anomaly-based detection

Anomaly-based detection method utilizes numerous network traffic abnormalities in order to detect botnets. Some of the network traffic anomalies used are high network latency, domain TTL, numbers of domains requested per second domain patterns, high traffic volumes, and abnormal system behavior which may reveal the presence of bots in the network.

In other words, the expression "anomaly-based detection" alludes to the issue of finding an abnormal network traffic movement which differs from predicted behavior. In any case, this strategy meets the issue of identifying obscure botnets.

Anomaly-based can be classified under Host-based and Network based detection methods [12, 27, 29].

The host-based technique screens and examines the inner workings of a computing device aka "host level". One of the key strategies that emphasize observing DNS traffic at the host level is the EFFORT framework [38]. EFFORT brakes down five unique modules utilizing regulated machine learning algorithm in order to tell the difference between malicious or benign queried domain names; it is autonomous of network topology, communication protocol and fit for distinguishing encrypted protocols. In any case, EFFORT system is restricted to botnets that rely on DNS administrations for finding the addresses of their C&C servers. Host-based is generally an unadaptable approach. Hence, the monitoring apparatus must be installed in all network machines in order to be effective [27].

Oppositely, network-based procedures either passively or actively analyze system traffic [12, 27, 29].

Active monitoring depends on the capacity to embed test packets into the network, servers or application as opposed to measuring network reactions. The active DNS probing approach as proposed by Ma et al. [39] intends to vastly evaluate particular DNS question query characteristics based on DNS cache logs. This approach could identify the infected host in remote management networks, however, by transmitting packets, there is a high chance of being detected by attackers, and actively examining DNS could affect the privacy of the users. Furthermore, this approach is restrained to the Domain Name in cash entry, that cannot detect NxDomain request. In general, active monitoring can produce extra traffic generated from the test packet and probing [41].

Fluxor [42] was one of the first systems to detect and monitor botnet using fast-flux. The methodology of detection is by analyzing the observable characteristics of typical users. Uses active probing techniques to detect Fast-Flux domain, each suspect domain is monitored over a period, during which it is actively queried. FluxORs aim is not only to discover Fast-Flux domains but also to detect the number and identity of related proxy servers to prevent their reuse in a future Fast-flux service network [14]. However, Fluxor limited to mainly studying fast-flux domains advertised through SPAMs traffic [43].

Passive monitoring, on the other hand, utilizes some capturing devices otherwise known as "sensors" to review the passing traffic [44]. Subsequently, it does not increase the traffic on the inspected network. This procedure ordinarily requires quite a while to assess multiple stages or rounds of botnet correspondence and activity. In 2004, Weimer [45] was the first person to introduce a passive detection system [44].

NOTOS [46] is a dynamic domain name reputation system which analyzes DNS and secondary data given by Honeynet and malware analysis service. The reputation system inputs are characteristics and extracted from the list of IP addresses settled by the domain name, date of registration, known malware samples which contact a given domain name or IP address, and the number of blacklisted IP addresses of a given domain name. These features enable NOTOS to adapt to models of domains legitimacy and how malicious domains are operated, as well as calculate perfect reputation scores for new domains. NOTOS has accomplished high accuracy and low false-positive rates, and it is able to recognize the recently registered domains before being published to a public blacklist. Nevertheless, the reputation score requires various history for a given domain name to be assigned an accurate reputation score, and it is inaccurate against often changing C&C domains like the hybrid botnet architecture which utilizes numerous master C&C hubs to execute a command [47].

EXPOSURE [48] is a system which utilizes inactive DNS information to recognize domains expound with malicious activity. It holds an aggregate of 15 features disseminated over the four classes: time-based, DNS reply-based, TTL value-based, and domain name-based. Similarly, EXPOSURE uses an information gatherer module for recording the of the checked DNS traffic. It likewise has an element attribution component for relegating recorded areas in the database alongside its extracted features [41]. In any case, some present individual malware, which uses DGA techniques as their C&C methods, don't take after the above-grouped features inside the botnet. Such malware sends DGA queries in an unverifiable style, making it difficult to be detected by EXPOSURE [39].

Inversely to NOTOS, Mentor as presented by Kheir et al. [47], the system proposes to take away the legitimate domains from the botnet C&C blacklisted domains, which helps to diminish the false-positive proportion amid the identification procedure. In order to accomplish that, Mentor embeds a crawler to gather statistics concerning a suspicious domain name, for example, web content and domain properties, using its characteristics to manufacture a DNS pruning model. The findings of the Mentor system proved to be fruitful when tried over public blacklist, and it removed proper domain names having an insignificant false-positive rate.

Kopis [50] presented a novel traffic characteristic through observation of DNS traffic at root levels of top-level domain hierarchy, and it can precisely look through the malware domains by going over worldwide DNS query resolution patterns. In contrast to DNS reputation systems, i.e., NOTOS and EXPOSURE, Kopis was able to influence DNS administrators to examine malware domains freely (i.e., without the necessity of other networks data). Such a stage can be utilized to avert inappropriate use with squandering of assets. In addition to that, Kopis was able to check malware domains regardless of whether all IP notoriety data is inaccessible [49]. Pleiades [22] is effective in identifying recently administrated DGA domains by using a nonexistent domain (NXDomain) responses. Be that as it may, since the clustering strategy for Pleiades was reliant on structural and lexical features of domain names, it may be able to solely analyze DGA-based C&C queries. Moreover, one of the noteworthy issues of NXDomain based detection is if the host is tainted by an individually running malware which sends a couple of queries for DGA domains over an extended time period, the procedure will be pointless. It may be possible by comparing successful domain resolving entropy to a failed one [16], detect C&C addresses for domain fluxing botnet in the local network, the entropy calculated for the alphanumeric characters composing domain name, they used IPv4 offline only dataset captured from the Asia region, false-positive rate as low as 0.02%. However, their approach is limited to high entropy domain names (non-dictionaries words) with IPv4 domain resolving only. The drawback of botnet detection methods based on DNS traffic analysis is illustrated in Table 1.

Table 1 The drawback of botnet detection techniques based on DNS traffic analysis

Model/authors	Technique	Drawback
DNSBL [38]	Monitoring DNS traffic against an IP blacklist database, it discovered the identities of the bots	limited to bot advertised through SPAMs traffic using heuristics approach
NOTOS [46]	The dynamic DNS reputation system uses passive DNS query data to analyze the network feature of a domain name	It is inaccurate against frequently changing C&C, slow to a make reputation score
Mentor [47]	Removing the legitimate domains from the botnet C&C domains blacklist	Weak against a hybrid botnet
Yadav and Reddy [16]	Detect domain fluxes in DNS traffic	limited for high entropy domain names that included in the dictionary
Kopis [50]	Controlling DNS quires and replies from the root DNS	Limited to the local network as a real-time system
EFFORT [39]	Applied multi-module to correlate bot-related indications different client/network aspects	Vulnerable to different evasion techniques
Ma et al. [40]	Active probing on a big scale to assess DNS query features based on DNS cache activities.	High possibility of being detected by attackers and raises privacy concerns
EXPOSURE [48]	Passive DNS examination to detect malicious domain activities	Passive monitoring, weak against unclassified DGA
Fluxor [42]	Analyzing the observable features of average users	Limited to analyzing fast-flux domains advertised through email spams
Pleiades [21]	Detect newly DGA domain by utilizing nonexistent domain (NXDomain) responses	Could analyses DGA-based C&C queries only

4 Discussion

From studying Table 2 that based on a review of detection approaches, and as highlighted before that DNS is a critical part of modern botnet life cycle. We can find that many DNS features could lead to a botnet detection. Some features need a time period for profiling and classification, such as TTL, NXdomain and captured botnet group traffic activities. While early-stage detection is essential to prevent attacks generated from a botnet, the domain lexical entropy feature lead to a lightweight and real-time detection, detection threshold criteria is a critical element for such feature, which could lead to a false-positive rate. So, by improving it with other lightweight features, such as the registration date

Table 2 Summary of botnet detection approaches based on DNS traffic analysis

Author,Technique and Year	Detection technique					Detection criteria									
	Honeypot	IDS				Domain Time-To-live (TTL)	Domain entropy	Detection unknown bots	Encrypted Bot Detection	NxDomain	DNS blacklist logs analysis	Reputation score	Learning method	Time-Based	Light weight
		Anomaly-Based			Signature-based										
		Network-Based		Host-Based											
		Active	Passive												
DNSBL (2007) [41]					x						x				
Notos (2010) [47]	x		x					x	x			x	x		
Menton(2016) [48]			x								x	x	x		
Yadav(2012) [16]			x			x	x			x				x	
Kopis(2011) [50]			x					x	x						x
EFFORT (2012) [42]				x				x	x						
Ma et al. (2015) [43]		x						x							
EXPOSURE(2011) [49]			x			x	x	x	x				x	x	
Fluxor(2008) [36]		x							x				x		
Pleiades (2012) [21]			x				x		x	x					x

(most malicious domains are new registered), matching the request with domain and IP reputation database, could lead to a high positive rate approach. Supposed location for such system is the recursive and authoritative DNS server.

5 Conclusion

Host security levels can be improved by the installation of antivirus, system patches which clear the infected devices. However, increased use of IoT devices and smartphones limits host protection against botnet propagation. Modern botnets are undoubtedly one of the most severe threats to the Internet and its community. Despite extensive research in the detection of botnets, the conventional security firm reports have revealed that botnet attacks increase each year. This means that the readily available approaches have limited accuracy and efficiency. Thus, there is a critical need for novel approaches in botnet detection. This research shows that DNS traffic detection is one of the essential elements of the botnet communication, which represents a limitation in its lifespan. Hence, existing botnet detection techniques based on DNS traffic analysis need to be surveyed in order to further aid the research community in developing a better tool and method in mitigation of botnet threats.

Acknowledgements This research was supported by the Short Term Research Grant, Universiti Sains Malaysia (USM) No: 304/PNAV/6313332.

References

1. Stevanovic M, Revsbech K, Pedersen J, Sharp R, Jensen C, A collaborative approach to botnet protection
2. Al-Ani AK, Anbar M, Manickam S, Al-Ani A (2018) DAD-match: technique to prevent DoS attack on duplicate address detection process in IPv6 link-local network. J Commun 13(6):317–324
3. Abu Rajab M, Zarfoss J, Monrose F, Terzis A (2006) Proceedings of the 6th ACM SIGCOMM on internet measurement—IMC'06 (2006), p 41. https://doi.org/10.1145/1177080.1177086. URL http://portal.acm.org/citation.cfm?doid=1177080.1177086
4. Canavan J (2005) The evolution of malicious IRC bots. In: Virus bulletin conference, pp 104–114
5. Khattak S, Ramay NR, Khan KR, Syed AA, Khayam SA (2014) IEEE Commun Surv Tutorials 16(2):898. https://doi.org/10.1109/SURV.2013.091213.00134
6. Cant'on David (2015) Botnet detection through DNS-based approaches—CERTSI. https://www.certsi.es/en/blog/botnet-detection-dns
7. Satam P, Alipour H, Al-Nashif Y, Hariri S (2015) J Internet Serv Inf Secur (JISIS) 5(4):85
8. Yadav S, Reddy AKK, Reddy AN, Ranjan S (2010) Proceedings of the 10th annual conference on internet measurement—IMC '10, p 48. https://doi.org/10.1145/1879141.1879148
9. Holz T, Gorecki C, Rieck K, Freiling FC (2008) Measuring and detecting fast-flux service networks. In: NDSS
10. Stevanovic M, Pedersen JM, D'Alconzo A, Ruehrup S (2017) Int J Inf Secur 16(2):115. https://doi.org/10.1007/s10207-016-0331-3

11. Savage K, Symantec security response
12. Karim A, Salleh RB, Shiraz M, Shah SAA, Awan I, Anuar NB (2014) J Zhejiang Univ Sci C 15(11):943. https://doi.org/10.1631/jzus.C1300242
13. Morales C (2018) NETSCOUT Arbor confirms 1.7 Tbps DDoS Attack; The Terabit Attack Era is upon us https://asert.arbornetworks.com/netscout-arbor-confirms-1-7-tbps-ddos-attackterabit-attack-era-upon-us/. URL https://asert.arbornetworks.com/netscout-arbor-confirms-1-7-tbps-ddos-attack-terabit-attack-era-upon-us
14. Wielogorska M, O'Brien D (2017) DNS traffic analysis for botnet detection. In: 25th Irish conference on artificial intelligence and cognitive science, 7–8 Dec 2017, Dublin, Ireland
15. Ahluwalia A, Traore I, Ganame K, Agarwal N (2017) Lecture notes in computer science (including subseries Lecture Notes in Artificial Intelligence and Lecture Notes in Bioinformatics) 10618 LNCS, 19. https://doi.org/10.1007/978-3-319-69155-8_2
16. Yadav S, Reddy AL (2012) Lecture notes of the institute for computer sciences, Social informatics and telecommunications engineering 96 LNICST, 446. https://doi.org/10.1007/978-3-642-31909-9_26
17. William Salusky, R. Danford. The Honeynet project, know your enemy: fast-flux service networks [Internet]. https://www.honeynet.org/papers/ff. URL https://www.honeynet.org/papers/ff
18. Stone-Gross B, Cova M, Gilbert B, Kemmerer R, Kruegel C, Vigna G (2011) IEEE Secur Priv 9(1):64. https://doi.org/10.1109/MSP.2010.144
19. Luo X, Wang L, Xu Z, Yang J, Sun M, Wang J (2017) ACM international conference proceeding series, Part F1280, 47. https://doi.org/10.1145/3057109.3057112
20. Grill M, Nikolaev I, Valeros V, Rehak M (2015) Proceedings of the 2015 IFIP/IEEE international symposium on integrated network management, IM 2015. https://doi.org/10.1109/inm.2015.7140486
21. Antonakakis M, Perdisci R (2012) Proceedings of the 21st USENIX security symposium, p 16. URL https://www.usenix.org/system/files/conference/usenixsecurity12/sec12-final127.pdf
22. Truong DT, Cheng G (2016) Secur Commun Netw 9(14):2338. https://doi.org/10.1002/sec.1495
23. Kolias C, Kambourakis G, Stavrou A, Voas J (2017) Computer 50(7):80. https://doi.org/10.1109/MC.2017.201
24. Abdullah R, Abdollah M (2013) IJCSI Int J Comput Sci Issues 10(2):208
25. Manasrah AM, Hasan A, Abouabdalla OA, Ramadass S (2009) Detecting botnet activities based on abnormal DNS traffic. arXiv preprint arXiv:0911.0487
26. Liu L, Chen S, Yan G, Zhang Z (2008) Bottracer: execution-based bot-like malware detection. In: International conference on information security. Springer, Berlin, Heidelberg, pp 97–113
27. Silva SS, Silva RM, Pinto RC, Salles RM (2013) Comput Netw 57(2):378. https://doi.org/10.1016/j.comnet.2012.07.021
28. da Pedro Marques Luz P (2014) Botnet detection using passive DNS. Ph.D. thesis
29. Zeidanloo HR, Shooshtari MJZ, Amoli PV, Safari M, Zamani M (2010) Proceedings—2010 3rd IEEE international conference on computer science and information technology, ICCSIT 2010, vol 2(May), p 158. https://doi.org/10.1109/iccsit.2010.5563555
30. Dornseif M, Gärtner FC, Holz T (2004) Vulnerability assessment using honeypots. Praxis der Informationsverarbeitung und Kommunikation 27(4):195–201
31. Freiling FC, Holz T, Wicherski G (2005) Lecture notes in computer science (including subseries. Lecture Notes in Artificial Intelligence and Lecture Notes in Bioinformatics) 3679 LNCS, p 319. https://doi.org/10.1007/11555827_19
32. Gaonjur P, Bokhoree C (2006) School of Business Informatics, University of Technology, Mauritius (May)
33. Kumar S, Sehgal RK, Chamotra S (2016) Proceedings—2016 2nd international conference on computational intelligence and communication technology, CICT 2016, pp 6–13. https://doi.org/10.1109/cict.2016.12
34. Oberheide J, Karir M, Mao ZM (2007) Characterizing dark dns behavior. In: International conference on detection of intrusions and malware, and vulnerability assessment. Springer, Berlin, Heidelberg, pp 140–156

35. Aiello M, Mongelli M, Papaleo G (2015) Int J Commun Syst 28(14):1987. https://doi.org/10.1002/dac.2836
36. Bethencourt J, Franklin J, Vernon MK (2005) Mapping internet sensors with probe response attacks. In: USENIX security symposium, pp 193–208
37. Peng S, Yu (2009) Proceedings of the 3rd Hackers' workshop on computer and internet security (IITKHACK'09), 2(2), 2. https://doi.org/10.1145/1327452.1327492
38. Ramachandran A, Feamster N, Dagon D (2006) Proceedings of the 2nd conference on steps to reducing unwanted traffic on the internet, vol 2, 36, 8. https://doi.org/10.1109/tdsc.2008.35
39. Shin S, Xu Z, Gu G (2012) Proceedings—IEEE INFOCOM, pp 2846–2850. https://doi.org/10.1109/infcom.2012.6195713
40. Ma X, Zhang J, Li Z, Li J, Tao J, Guan X, Lui JC, Towsley D (2015) J Netw Comput Appl 47:72. https://doi.org/10.1016/j.jnca.2014.09.016
41. Alieyan K, Almomani A, Manasrah A, Kadhum MM (2017) Neural Comput Appl 28(7):1541. https://doi.org/10.1007/s00521-015-2128-0
42. Passerini E, Paleari R, Martignoni L, Bruschi D (2008) Fluxor: detecting and monitoring fast-flux service networks. In: International conference on detection of intrusions and malware, and vulnerability assessment. Springer, Berlin, Heidelberg, pp 186–206
43. Perdisci R, Corona I, Dagon D, Lee W (2009) Proceedings—Annual computer security applications conference, ACSAC, pp 311–320. https://doi.org/10.1109/acsac.2009.36
44. Zdrnja B, Brownlee N, Wessels D (2007) Passive monitoring of DNS anomalies. In: International conference on detection of intrusions and malware, and vulnerability assessment. Springer, Berlin, Heidelberg, pp 129–139
45. Weimer F (2005) Passive DNS replication. In: FIRST conference on computer security incident, p 98
46. Antonakakis M, Perdisci R, Dagon D, Lee W, Feamster N (2010) USENIX Security'10: proceedings of the 19th USENIX conference on security, pp 1–17. https://doi.org/10.1145/2584679. URL http://www.usenix.org/events/sec10/tech/fullpapers/Antonakakis.pdf
47. Kheir N, Tran F, Caron P, Deschamps N (2016) PositiveM, D.N.S. reputation, S.o. Benign, C.B.N. Cuppens-boulahia, F. Cuppens, pp 0–14
48. Bilge L, Kirda E, Kruegel C, Balduzzi M, Antipolis S (2011) Ndss pp 1–17. https://doi.org/10.1145/2584679
49. Li X, Wang J, Zhang X (2017) Future Internet 9(4), 1. https://doi.org/10.3390/fi9040055
50. Antonakakis M, Perdisci R, Lee W, Ii NV, Dagon D (2011) USENIX security symposium. 11, 1. DOI 2028067.2028094. URL https://dl.acm.org/citation.cfm?id=2028094

Features Selection for Intrusion Detection System Based on DNA Encoding

Omar Fitian Rashid, Zulaiha Ali Othman and Suhaila Zainudin

Abstract Intrusion detection systems detect attacks inside computers and networks, where the detection of the attacks must be in fast time and high rate. Various methods proposed achieved high detection rate, this was done either by improving the algorithm or hybridizing with another algorithm. However, they are suffering from the time, especially after the improvement of the algorithm and dealing with large traffic data. On the other hand, past researches have been successfully applied to the DNA sequences detection approaches for intrusion detection system; the achieved detection rate results were very low, on other hand, the processing time was fast. Also, feature selection used to reduce the computation and complexity lead to speed up the system. A new features selection method is proposed based on DNA encoding and on DNA keys positions. The current system has three phases, the first phase, is called pre-processing phase, which is used to extract the keys and their positions, the second phase is training phase; the main goal of this phase is to select features based on the key positions that gained from pre-processing phase, and the third phase is the testing phase, which classified the network traffic records as either normal or attack by using specific features. The performance is calculated based on the detection rate, false alarm rate, accuracy, and also on the time that include both encoding time and matching time. All these results are based on using two or three keys, and it is evaluated by using two datasets, namely, KDD Cup 99, and NSL-KDD. The achieved detection rate, false alarm rate, accuracy, encoding time, and matching time for all corrected KDD Cup records (311,029 records) by using two and three keys are equal to 96.97, 33.67, 91%, 325, 13 s, and 92.74, 7.41, 92.71%, 325 and 20 s, respectively. The results for detection rate, false alarm rate, accuracy, encoding time, and matching time for all NSL-KDD records (22,544 records) by using two and three keys are equal to 89.34, 28.94, 81.46%, 20, 1 s and 82.93, 11.40, 85.37%, 20 and 1 s,

O. F. Rashid (✉) · Z. A. Othman · S. Zainudin
Faculty of Information Science and Technology, University Kebangsaan Malaysia, Bangi, Malaysia
e-mail: Omaralrawi08@yahoo.com

Z. A. Othman
e-mail: zao@ukm.edu.my

S. Zainudin
e-mail: suhaila.zainudin@ukm.edu.my

© Springer Nature Singapore Pte Ltd. 2019
V. Piuri et al. (eds.), *Intelligent and Interactive Computing*, Lecture Notes in Networks and Systems 67, https://doi.org/10.1007/978-981-13-6031-2_23

323

respectively. The proposed system is evaluated and compared with previous systems and these comparisons are done based on encoding time and matching time. The outcomes showed that the detection results of the present system are faster than the previous ones.

Keywords Intrusion detection system · DNA encoding · Feature selection · KDD Cup 99 dataset · NSL-KDD dataset

1 Introduction

With the high usage and fast growth of the Internet, the network security has become an important issue; an Intrusion Detection System (IDS) is an effective approach that deals with network security problems. An IDS is a software application or a device that used to detect abnormal or suspicious activity in computer system or network. These systems can be classified either as misuse or anomaly; misuse detection system is done based on attacks patterns, while anomaly detection system is done based on user behaviour [1]. On the other hand, a feature selection is a process that is used to select relevant features and remove irrelevant or redundant features from the original dataset, which can be used in different areas such as statistical pattern recognition, machine learning, data mining, statistics [2] and IDS.

DNA is a genetic material in most creatures, DNA information stored as a code contains four chemical bases, these bases are called Adenine, Cytosine, Guanine and Thymine, and are referred as A, C, G and T. The problem is that it takes long time in order to check an entire DNA sequence, therefore, certain areas of DNA called Short Tandem Repeats are searched [3]; these STR include both keys and their positions, these STR detect DNA mutations.

2 Related Work

Existence of IDS is important for both computer security and network security. These systems are depend on the capability of their detections where they should be fast and with high detection rate; and so, IDS has many challenges such as time. Therefore, to deal with this challenge, a new feature selection method is proposed in order to reduce the computation and complexity [4], which helps to build a lightweight IDS with high detection rates and speed up the detection time. Various methods have been suggested for features selection in IDS.

Xian et al. [5] applied the improved information gain algorithm that depends on feature redundancy. Through this algorithm, 12 and 22 features are selected. A wrapper feature selection approach is proposed by using Bayesian classifier that applied for network IDS [6]. Eleven features are selected (feature no.: 3, 5, 6, 10, 17, 18, 32, 33, 35, 36 and 38). An investigation and evaluation of the performance of the combination models of both cuttlefish algorithm and decision tree methods that have been used to select feature for intrusion detection and its performance are carried out by [7]. A method is developed to improve the signature model by performing the Genetic Algorithm selection features based on customized features [8]. The following steps are used to remove the customized features: removing biased features, discretization using chi-merge, and remove features with string value. The Genetic Algorithm is applied on the customized features which improved the quality of the signature extracted and these are (feature no.: 5, 6, 13, 23, 24, 25, 26, 33, 36, 37 and 38). The CFA is modified to be used as a feature selection tool. Then, the DT classifier is applied as a measurement of the generated features. Empirical results indicated high DR especially when the number of produced features is equal or less than 20 features.

Improved version of the Binary Bat Algorithm system is proposed where two classifiers are used [9]. These classifiers are C4.5 and Support Vector Machine, the features selected are 19 based on Binary Bat Algorithm with SVM method, and 15 based on Binary Bat Algorithm with C4.5 classifier. An IDS for multiclass is presented to make improvements on the existing work in three perspectives [10]. First, pre-processed the traffic pattern and remove the redundant instances, next, adaption of a feature selection algorithm which minimize the classifier computational and complexity, and finally, used a neuro tree model for the classification engine, the extracted features by this method are 16. Al-Jarrah et al. [11] combined the filter and wrapper selection processes to establish hybrid feature selection method that was used as data classification for IDS. This approach is carried out by applying two main phases: (1) filter feature ranking and eliminating phase; and (2) wrapper feature selection using support vector machine and classification accuracy phase. The aims of this approach are to achieve both, the high accuracy of wrapper approaches and the efficiency of filter approaches. The six features extracted are (feature no.: 3, 5, 23, 32, 34 and 35).

A new feature selection method is proposed by [12] based on genetic with an innovation in fitness function that led to improving the performance of intrusion detection. A new feature selection model for anomaly IDS is proposed [13] based on features evaluation by using information gain by considering the information and the distance for each feature, the extracted features by this method are six (feature no.: 3, 4, 5, 6, 29 and 30). A features selection is proposed [14] by combining two techniques, and these techniques are consistency subset evaluation and DDoS characteristic features, the features selected are 17 (feature no.: 1, 2, 3, 4, 5, 6, 7, 8, 10, 14, 23, 29, 30, 32, 33 and 36). A features selection framework is presented by [15] by using various classifiers and using filter and wrapper methodologies.

The current paper is exhibiting a new features selection method that applied for IDS based on DNA encoding approach. This method is applied by using both KDD Cup 99 and NSL-KDD datasets in order to calculate the detection rate (DR), false alarm rate (FAR), accuracy, and time of the whole operation.

3 Proposed Approach

Selection features method for IDS based on DNA encoding is performed via three phases. These phases are pre-processing phase, training phase and testing phase.

The pre-processing phase is done before training phase, and its importance is to find the DNA position. The first step of pre-processing phase is converting 10% KDD Cup dataset records, which consist of 41 features, to DNA sequences. Teiresias algorithm is used to extract keys and their positions, then these positions are used for training phase. The pre-processing phase steps are shown in Fig. 1 and Algorithm (1).

Algorithm (1): Pre-Processing Phase
Input: Training dataset (10 % KDD Cup dataset)
Output: Keys and their positions
1. Encode all records of training dataset (all 41 features) based on random value with a specific size.
2. Calculate blocks numbers (block length is equal to the size of the key) for both attacks records and normal records separately.
3. Find normal blocks which is not like attacks blocks.
4. Calculate the repetition of these normal blocks.
5. Choose the highest repetition as the key, then find its location (position) in records.

In the training phase, build and apply a feature selection method on 10% KDD Cup dataset records with keys positions that gained from pre-processing phase to get the DNA locations. The training phase steps are shown in Fig. 2 and in Algorithm (2).

Algorithm (2): Training Phase
Input: Training dataset (10 % KDD Cup dataset), and keys positions
Output: DNA location
1. Find specific training dataset records features based on keys positions (features that contain positions values)
2. Encode all records of training dataset (for specific features).
3. Store the features numbers (DNA locations).

Fig. 1 The steps of the pre-proposing phase

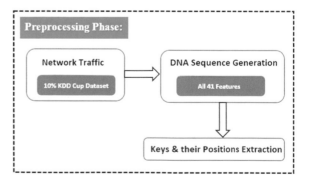

Fig. 2 Illustrates the
training phase steps

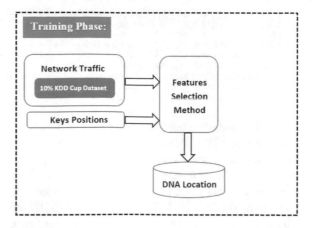

Fig. 3 Shows the testing
phase steps

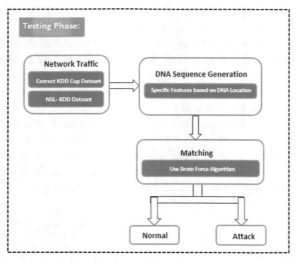

In the testing phase, the first step is used testing datasets that include both corrected KDD Cup 99 and NSL-KDD datasets as source information for testing phase;
convert specific features number of datasets records to DNA sequences based on
DNA location. Then, Brute Force algorithm is used for matching based on keys
that are extracted in the training phase. The testing phase steps are shown in Fig. 3
and in Algorithm (3). Finally, the experiment results are calculated based on DR,
FAR, accuracy and the time for the whole operation that includes both encoding and
matching time.

Algorithm (3): Testing Phase

Input: Testing dataset, Keys, and DNA locations
Output: Normal records number, and attack records number
  1. Encode all records of the testing dataset (for specific features based on DNA locations).
  2. Looking for the keys in these records after encoding.
  3. Calculate the normal records and attacks records based on matching or mismatching with these keys.
  4. Calculate the experiment results (DR, FAR, accuracy, and the time).

Testing datasets records are either normal or attacks, and the attacks can be classified into four attacks types. These types are Denial of Service (DoS), Probe, Remote to Local (R2L), and User to Root (U2R). Both datasets are widely used for evaluating the IDS performance and also included a large quantity of network traffic and a wide variety of attacks where the corrected KDD Cup 99 dataset consists of 311,029 records and NSL-KDD test dataset (KDDTest+) consists of 22,544 records. Therefore, in the current study, these datasets are used to evaluate the performance of the proposed system, these records are consisting of 41 features and these features can be either continuous or discrete.

The performance of the present system can be determined based on four measures. These measures are DR, FAR, accuracy and time. The DR result is the number of attacks record that were correctly detected over the total attacks record number. The FAR result is the number of normal records that were incorrectly detected as attacks over the total normal records number; the accuracy result is the number of correctly classified records (for both attacks and normal) over the total records number. Finally, the time is calculated for the whole operations which include both encoding time (convert dataset records features to DNA) and matching time. The formulae for calculating DR, FAR and accuracy are shown in Eqs. (1), (2), and (3), respectively [16], and the formulae for calculating time is shown in Eq. (4).

$$DR = \frac{TP}{TP + FN} \tag{1}$$

$$FAR = \frac{FP}{TN + FP} \tag{2}$$

$$Accuracy = \frac{TP + TN}{TP + TN + FN + FP} \tag{3}$$

$$Time = Encoding\ Time + Matching\ Time \tag{4}$$

4 Experiments and Results

Experimental environment: operating system is Microsoft Windows 10 Professional, CPU is Intel 2.50 GHz and memory is 4.00 GB. The data used in the experiment is all corrected KDD Cup dataset records which is equal to 311,029 records and all NSL-KDD dataset records which are equal to 22,544 records.

Table 1 DNA sequences for previous example

Feature number	DNA sequences	Feature number	DNA sequences
1	GCC	22	GCC
2	ACAG	23	AGA
3	CCGT	24	AGA
4	GCGT	25	GCCAGGGCCGCC
5	CGAGCCATA	26	GCCAGGGCCGCC
6	CGACAATAC	27	GCCAGGGCCGCC
7	GCC	28	GCCAGGGCCGCC
8	GCC	29	CGAAGGGCCGCC
9	GCC	30	GCCAGGGCCGCC
10	GCC	31	GCCAGGGCCGCC
11	GCC	32	AGAATAATA
12	GCC	33	AGAATACAA
13	GCC	34	CGAAGGGCCGCC
14	GCC	35	GCCAGGGCCCGA
15	GCC	36	GCCAGGGCCGCC
16	GCC	37	GCCAGGGCCGCC
17	GCC	38	GCCAGGGCCGCC
18	GCC	39	GCCAGGGCCGCC
19	GCC	40	GCCAGGGCCGCC
20	GCC	41	GCCAGGGCCGCC
21	GCC		

An example of how select the five features:

1. Convert all training dataset records to DNA Sequences, an example of one record is as follows:

   Network traffic record:
   0,udp,private,SF,105,146,0,0,0,0,0,0,0,0,0,0,0,0,0,0,0,0,0,2,2,0.00,0.00,0.00,0.00,
   1.00,0.00,0.00,255,254,1.00,0.01,0.00,0.00,0.00,0.00,0.00,0.00.

   The DNA sequences for previous network traffic record example are shown in Table 1.

2. Extract the keys and their positions, an example of the extracted keys and their position are as follows:

   GCCACAGCCGTGCGTCGAGCCATACGACAATACGCCGCCGCCGCC
   GCCGCCGCCGCCGCCGCCGCCGCCGCCGCCGCCGCCAGAAGAGCC
   AGGGCCGCCGCCAGGGCCGCCGCCAGGGCCGCCGCCAGGGCC
   GCCCGAAGGGCCGCCGCCAGGGCCGCCGCCAGGGCCGCC
   AGAATAATAAGAATACAACGAAGGGCCGCCGCCAGGGCCCGAGCC

AGGGCCGCCGCCAGGGCCGCCGCCAGGGCCGCCGCCAGGGCCGCC
GCCAGGGCCGCCGCCAGGGCCGCC.
where the first key is equal to CACAG and its position is 3, the second key is equal
to CGTGC and its position is 9 and the third key is equal to GCCAT and its position
is 19.

3. Find specific records features based on keys positions (features that contain positions values)

Key 1 is got from feature number 1 and 2, key 2 is got from feature number 3 and 4,
and key 3 is got from feature number 5. Therefore, the features numbers that have
been chosen are 1, 2, 3, 4 and 5. The features that have been selected by the current
proposed approach are 5 features and these features are shown in Table 2.

For specific features, the results of the DR, FAR and accuracy obtained by the
current proposed system are based on the number of extracted keys in pre-processing
phase (either two keys or three keys). The results that have been achieved by conducting the system on testing datasets are illustrated in Tables 3 and 4, respectively.

When corrected KDD Cup dataset is used, the achieved DR results for specific
features by using two keys and three keys are giving higher value than the DR result
for the whole features by 1.57, and 1.55%, respectively. For FAR, the obtained values
showed that the result for specific features by using two keys and three keys are giving
lower result than the result for all features by 0.21 and 0.04%, respectively. Finally,
for the accuracy, the achieved results for specific features by using two keys and
three keys are giving higher values than the result for the whole features by 1.22 and
1.27%, respectively.

Table 2 The five selected features used in the current system

Features	Description
Number	1, 2, 3, 4 and 5
Name	Duration, protocol type, Service, Flag and source bytes

Table 3 The obtained results by using current system on corrected KDD Cup dataset

Result	Two keys	Three keys
DR	96.97	92.74
FAR	33.67	7.41
Accuracy	91	92.71

Table 4 The obtained results by using the current system on NSL-KDD dataset

Result	Two keys	Three keys
DR	89.34	82.93
FAR	28.94	11.40
Accuracy	81.46	85.37

Fig. 4 Illustrates the DR
results achieved by applying
the current system

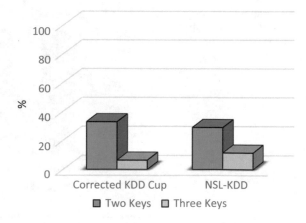

Fig. 5 Illustrates the FAR
results obtained from the
application of the current
system

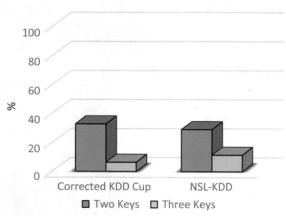

Also, when NSL-KDD dataset is used, the achieved DR results for specific features
by using two keys and three keys are giving higher values than the DR value for all
features by 6 and 5.85%, respectively. For FAR, the obtained results for specific
features by using two keys and three keys are giving higher values than the results
for all features by 3.29 and 0.91%, respectively. Finally, for accuracy, the obtained
values showed that the results for specific features by using two keys and three
keys are giving higher values than the results for all features by 1.99 and 2.94%,
respectively. The Detection Rate, False Alarm Rate and accuracy results for both
testing datasets are shown in Figs. 4, 5 and 6, respectively.

For the specific features, the encoding time and matching time results (in term
of seconds) obtained by the current proposed system based on two keys or three
keys conducted on all corrected KDD Cup dataset records and all NSL-KDD dataset
records are illustrated in Tables 5 and 6, respectively.

When corrected KDD Cup dataset is used, the matching time for specific features
by using two keys and three keys are decreased to 93 and 92.6%, respectively, than
the matching time for all features. However, for NSL-KDD dataset, the matching

Fig. 6 Illustrates the obtained accuracy results for both datasets

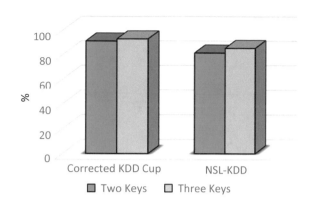

Table 5 Presented the calculated time by applying the current system on corrected KDD Cup dataset

Time (s)	Two keys	Three keys
Encoding time	325	325
Matching time	13	20

Table 6 Presented the calculated time by applying the current system on NSL-KDD dataset

Time (s)	Two keys	Three keys
Encoding time	20	20
Matching time	1	1

Fig. 7 The encoding and matching time results obtained from the application of the current system on corrected KDD Cup dataset

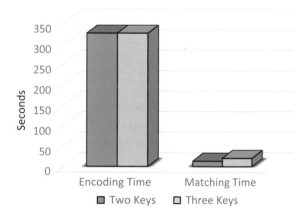

time for specific features by using two keys and three keys are decreased to 87.5% an 90%, respectively, than the matching time for all features. The results of the calculated time for testing datasets are illustrated in Figs. 7 and 8, respectively.

The obtained times of the executed system (in term of min) from the present system are compared with IDS based on least squares support vector machine classification method. These experiments are applied on KDD Cup 99 dataset by using modified mutual information-based feature selection (MMIFS) to select features for normal

Fig. 8 The encoding and matching time results obtained by applying the current system on NSL-KDD dataset

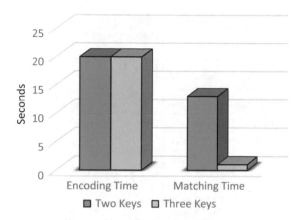

records and for all attack types. The features for normal records are equal to six features and these features are (5, 23, 3, 6, 35, 1), the features for DoS attack records are equal to 8 features and these features are (5, 23, 6, 2, 24, 41, 36, 3), the features for probe attack records are equal to 13 features and these features are (40, 5, 33, 23, 28, 3, 41, 35, 27, 32, 12, 24, 28), the features for R2L attack records are equal to 15 features and these features are (3, 13, 22, 23, 10, 5, 35, 24, 6, 33, 37, 32, 1, 37, 39), the features for U2R attack records are equal to 10 features and these features are (5, 1, 3, 24, 23, 2, 33, 6, 32, 4, 14, 21) mentioned by (Amiri et al. [17]). The encoding time (building time), the matching time and the features number for the current system and with the published one are shown in Table 7. It is clear, that the times obtained by the method of the present system are quite good. This system gives better times of detections than the previous system.

5 Conclusion

This paper proposed a feature selection method for IDS based on DNA encoding method and depends upon keys position (STR positions). This system consists of three phases, pre-processing phase, training phase and testing phase. The performance of the system is done by using both corrected KDD Cup dataset and NSL-KDD dataset. The obtained results are evaluated based on DR, FAR, accuracy, and time that includes both encoding time and matching time.

The obtained values of DR, FAR, and accuracy for all corrected 10% KDD Cup records (311,029 records) by using two keys and three keys are equal to 96.97, 33.67, 91%, and 92.74, 7.41% and 92.71%, respectively. The obtained values of DR, FAR and accuracy for all NSL-KDD records (22,544 records) by using two keys and three keys are equal to 89.34, 28.94, 81.46%, and 82.93, 11.40% and 85.37%, respectively.

The obtained results of encoding time and matching time for all corrected 10% KDD Cup records by using two keys and three keys are equal to 325, 13, 325 and

Table 7 Comparison between the proposed system and the published ones

Method	Class	Building time (min)	Matching time (min)
Amiri et al. [17]	Normal	25	11
Proposed system (2 Keys)		1.46	0.33
Proposed system (3 Keys)		1.46	0.33
Amiri et al. [17]	DoS	19	8
Proposed system (2 Keys)		4.56	0.18
Proposed system (3 Keys)		4.56	0.25
Amiri et al. [17]	Probe	35	13
Proposed system (2 Keys)		0.06	0.01
Proposed system (3 Keys)		0.06	0.01
Amiri et al. [17]	R2L	5	4
Proposed system (2 Keys)		0.08	0.01
Proposed system (3 Keys)		0.08	0.01
Amiri et al. [17]	U2R	23	10
Proposed system (2 Keys)		0.01	0.01
Proposed system (3 Keys)		0.01	0.01

20 s, respectively. The obtained results of encoding time and matching time for all NSL-KDD records by using two keys and three keys are equal to 20, 1 s, and 20, 1 s, respectively.

The proposed system performance is evaluated and compared with previous system and these comparisons are done based on encoding time and matching time, it showed that the current system results are faster than the previous one.

Acknowledgements This research was supported by FRGS grant (FRGS/1/2016/ICT02/UKM/02/8), funded by Ministry of Higher Education.

References

1. Mulay SA, Devale PR, Garje GV (2010) Intrusion detection system using support vector machine and decision tree. Int J Comput Appl 3(3):40–43
2. Liu H, Yu L (2005) Towards integrating feature selection algorithms for classification and clustering. IEEE Trans Knowl Data Eng 17(4):491–502
3. Soram R, Khomdram M (2010) Biometric DNA and ECDLP based personal authentication system: a superior posse of security. Int J Comput Sci Netw Secur 10(1):1–9
4. John GH, Kohavi R, Pfleger K (1994) Irrelevant features and the subset selection problem. In: Proceeding of the 11th international conference on machine learning, Morgan Kaufmann Publishers, pp 121–129

5. Xian J, Peiyu L, Wei G, Xuezhi C (2011) An algorithm application in intrusion forensics based on improved information gain. In: Web Society (SWS), 3rd symposium on date of conference, pp 100–104
6. Zhang F, Wang D (2013) An effective feature selection approach for network intrusion detection. In: 2013 IEEE eighth international conference on networking, architecture and storage, Xi'an, China, 17–19 July 2013
7. Eesa A, Orman Z, Brifcani A (2014) A novel feature-selection approach based on the cuttlefish optimization algorithm for intrusion detection systems. Expert Syst Appl 42(5):2670–2679
8. Othman ZA, Abu Bakar A, Etubal I (2010) Improving signature detection classification model using features selection based on customized features. In: 2010 10th International conference on intelligent systems design and applications, pp 1026–1031
9. Enache A, Sgarciu V (2015) A feature selection approach implemented with the binary bat algorithm applied for intrusion detection. In: 2015 38th International conference on telecommunications and signal processing (TSP), pp 11–15
10. Sindhu SS, Geetha S, Kannan A (2012) Decision tree based light weight intrusion detection using a wrapper approach. Expert Syst Appl 39(1):129–141
11. Al-Jarrah OY, Siddiqui A, Elsalamouny M, Yoo PD, Muhaidat S, Kim K (2014) Machine-learning-based feature selection techniques for large-scale network intrusion detection. In: 2014 IEEE 34th international conference on distributed computing systems workshops, pp 177–181
12. Gharaee H, Hosseinvand H (2016) A new feature selection IDS based on genetic algorithm and SVM. In: 2016 8th International symposium on telecommunications
13. Ullah I, Mahmoud QH (2017) A filter-based feature selection model for anomaly-based intrusion detection systems. In: 2017 IEEE international conference on big data
14. Yusof AR, Udzir NI, Selamat A, Hamdan H, Abdullah M (2017) Adaptive feature selection for denial of services (DoS) attack. In: 2017 IEEE conference on application, information and network security
15. Anwer HM, Farouk M, Abdel-Hamid M (2018) A framework for efficient network anomaly intrusion detection with features selection. In: 2018 9th International conference on information and communication systems
16. Wu S, Benzhaf W (2010) The use of computation intelligence in intrusion detection systems. Appl Soft Comput 10(1):1–35
17. Amiri F, Yousefi MR, Lucas C, Shakery A, Yazdni N (2011) Mutual information-based feature selection for intrusion detection systems. J Netw Comput Appl 34:1184–1199

Detection Techniques for DDoS Attacks in Cloud Environment: Review Paper

Sultan T. Alanazi, Mohammed Anbar, Shankar Karuppayah, Ahmed K. Al-Ani and Yousef K. Sanjalawe

Abstract Cloud computing security remains the goal of both cloud service providers and customers. With many of the security threats to the security of cloud computing, Distributed Denial of Service (DDoS) attacks is one of the most worrisome. The danger posed by the DDoS attacks are already known and continue to be the predominant security challenge in reaching an impervious and guaranteed safe cloud computing resources and service delivery. Many researchers have proposed many detection and defense techniques to protect cloud computing against DDoS attacks. In this paper, we present a review of many detection techniques that are useful in spotting DDoS attacks that are cloud-based and make a comparative analysis between them to find a suitable technique for spotting these cloud computing based DDoS attacks.

Keywords Cloud computing · Distributed denial of service (DDoS) · Detection techniques · Prevention technique

1 Introduction

Virtualization is that the main construct in Cloud Computing (CC), it could be a methodology that divides a physical Personal Computer (PC) into many machines that a partially or fully isolated and usually referred to as Virtual Machines (VM) or

S. T. Alanazi · M. Anbar (✉) · S. Karuppayah · A. K. Al-Ani · Y. K. Sanjalawe
National Advanced IPv6 Centre, Universiti Sains Malaysia, 11800 Gelugor, Penang, Malaysia
e-mail: anbar@nav6.usm.my

S. T. Alanazi
e-mail: sultan@nav6.usm.my

S. Karuppayah
e-mail: shankar@nav6.usm.my

A. K. Al-Ani
e-mail: ahmedkhallel91@nav6.usm.my

Y. K. Sanjalawe
e-mail: yousef@nav6.usm.my

© Springer Nature Singapore Pte Ltd. 2019
V. Piuri et al. (eds.), *Intelligent and Interactive Computing*, Lecture Notes in Networks and Systems 67, https://doi.org/10.1007/978-981-13-6031-2_34

337

guest machine [1]. Virtualization may be a key component that makes simpler the delivery of service to a group of dynamically assessable resources akin to storage, software, strength of processing, and varying resources for computing to the customers of CC that can be accessed on demand over the Internet. CC customers desire to get a browser and a web association to use CC resources [2]. CC is the distribution of varying computing facilities, servers, storage facilities, hosts of databases, computer networks, different software for analytics and a lot more over the web. CC technology is in continuous development and with various challenges relating to security [3]. Furthermore, CC is changing into the mandatory part of the web and it is necessary for cloud suppliers to make them stay obtainable. Due to its distributed nature, it has become very simple to attack. The protection of the cloud is compromised beneath the threat of DoS and DDoS attack [4]. Many ancient attacks influence the operation of CC, with these attacks influencing the integrity, confidentiality, and the handiness of Cloud resources and existing services provided at the network layer. The degree of the DDoS attack is associated with the lack of supply of the resources or services of one or multiple systems on the cloud [5]. Furthermore, DDoS attacks use many computers management them remotely known as zombies. Assaulter will attack with the assistance of zombies and targets the victim to make the information or resource unavailable. DDoS attacks server and cloud infrastructure-level attack [6]. We surveyed in this study, already known DDoS infiltrations aiming CC and dissimilar DDoS detection techniques.

The organization of this paper is thus: the second section deliberates CC, its architecture, its advantages, CC issues, and CC security issues. The third section gives an overview of DDoS attack, the types of DDoS attacks in CC. The fourth section comprises of related works in detections techniques of DDoS attacks in CC and preventing mechanisms. Section five presents discussions and finally, section six is the conclusion.

2   Cloud Computing: Overview

Today, Cloud Computing is quickly developing innovatively, which will remain generally utilized and propel more innovation in years to come. As it gives clients avenue to store, get, and control information with client's comfort paying little heed to time and area alongside, it gives the pool of assets, for example, system, stockpiling, and administrations on-request premise. Distributed computing offers clients' top of the line and versatile foundation at a reasonable cost. Virtualization is the way to opening distributed computing. Although virtualization has incredible advantages to the clients, the intricacy in its structure, acquaints inconspicuous and coercive dangers with the security of the information and to the framework foundation.

2.1 Cloud Computing Definition

Because of the Internet, CC has emerged, offering three key services—a platform, infrastructure, and software each as a service [7]. CC is a huge collection of virtual resources where it is easy to use and accessible, in addition to the possibility of increasing their ability to afford more and create resources for it [7]. Moreover, Availability of technological resources, storage of databases and applications through cloud platforms and price is determined against the required service. Many institutions have moved toward CC because of its advantages such as availability of services and access to resources, which reduce the cost to enterprises [8]. Besides, CC is defined as a range of services, resources, and applications that are available on cloud platforms that enable clients to access their services at all times and easily [9]. The National Institute of Standard and Technology (NIST) defined important characteristics of CC as self-service on request, resource pooling, quick flexibility, and measured service [10]. CC is a scheme that provides access to resources, applications, and databases when customers need them (e.g., storage of data, networking, provision of servers, and applications), which enables customers to access resources easily [11]. Service model can be classified in terms of the services they provide to Software-as-a-Service (SaaS), Platform-as-a-Service (PaaS), and Infrastructure-as-a-Service (IaaS), also this has been further divided as Private Cloud, Public Cloud, Community Cloud, or Hybrid Cloud [12]. CC boasts several attractive benefits for businesses and end users as shown in Table 1.

Table 1 Benefits of CC

Benefit	Explanation
Self-service delivery	CC enables customers to have the resources they need and can control these resources as needed
Resilience	One of the advantages added by CC is that it has provided institutions with the acquisition of infrastructure and the purchase of many expensive devices and provided services through the platforms of computers
Expenses	CC decrease cost of buying HW, SW, Servers, Networks, and other resources
Reliability	CC provides backups of data so that it can recover data in the event of disasters, damage, or breaches without harm to customers
Big data analytics	CC provides an opportunity for data scientists to analyze utilize large data in many different areas of life
Social networking	Social networks use cloud computing, including Facebook, Twitter, and others

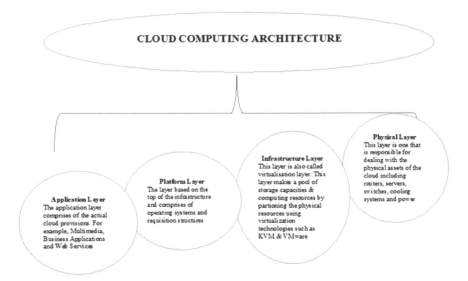

Fig. 1 Cloud computing architecture

2.2 Cloud Computing Architecture

CC structure is divided into different layers to include each layer or more of the cloud services. The architecture of CC can be further separated into four layers: physical, platform, application, and infrastructure layers. These are explained in Fig. 1.

2.3 Cloud Computing Issues

Several matters ought to be considered in cloud computing, these issues have been discussed in Table 2.

2.4 Cloud Computing Security Issues

One of the most challenging challenges for CC is security. One of the most important security problems facing CC is the distributed denial of service (DDoS) attack [13]. From cloud users' point of view, security is one of the main reasons for their use of cloud computing. This is because, the organizations assign the security function to other companies, in addition to the absence of security guarantees in agreements between service providers and service users.

Table 2 Cloud computing issues

Issue	Explanation
Security and privacy	Many organizations face security and privacy concerns when using CC
Downtime	Stopping the service of any organization is one of the most serious problems that can be faced
Interoperability and portability	CC should provide the ability to navigate to services anytime the organization needs it
Reliability and availability	The monitoring feature must be available to provide services and ensure availability throughout the day and there is no interruption in service
Performance and bandwidth cost	Many organizations need a high cost to buy hardware, but the issue in CC requires the cost of booking the service

This transfer of many data and applications between different clouds platforms leads to weak security process and poses threats. Table 3 identifies some of these threats.

3 DDoS Attack on CC

DDoS attack has become one of the most important security challenges in [14]. DDoS attack results in disabling services in CC. Figure 2 shows the DDoS attack on CC.

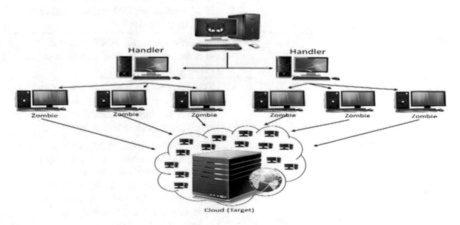

Fig. 2 DDoS attack on CC

Table 3 Cloud computing security issues

Threat	Explanation
Data breaches	Data penetration is a concern for cloud users, which can be from many sources, including human error or application malfunction
Anonymity, credentials, and access management	Unauthorized access to data and damage to end-user organizations and end users of CC
Application interfaces are unsafe (APIs)	CSA says CC security depends on APIs, so it must be protected
System and application vulnerability	Attackers exploit system vulnerabilities to break or control the system, steal data, or cause service disruption
Account hijacking	If attackers can gain unauthorized access to user data, they can control all transactions of the compromised customer and direct them to illegal sites that cause harm to them
Malicious insiders	Internal attackers are one of the most dangerous concerns in which they have access to data and resources, which increases the risk of attack
Persistent threats	Type of cyber-attack aimed at accessing infrastructure and stealing data
Loss of data	There are many reasons for the loss of data stored on CC. These causes may be as natural as earthquakes and volcanoes, resulting in loss of data forever if a copy of the data is not taken and kept in safe places
Insufficient due diligence	Providers of cloud services if they do not do the necessary precautions may cause multiple damages to themselves and to the users of the services
Cloud computing misuse	Fraudulent Accounts of CC and trial versions may cause attacks on their users, causing damage
Denial of service (DoS)	These attacks are used to prevent users from accessing their data and applications by consuming a large amount of system resources such as space and bandwidth
Common technological gaps	Cloud service providers are evolving through infrastructure and applications. Cloud technology categorized as service-as-a-service offering without hardware or software-based hardware change at times resulting in common technical weaknesses

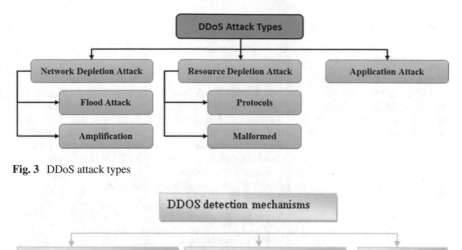

Fig. 3 DDoS attack types

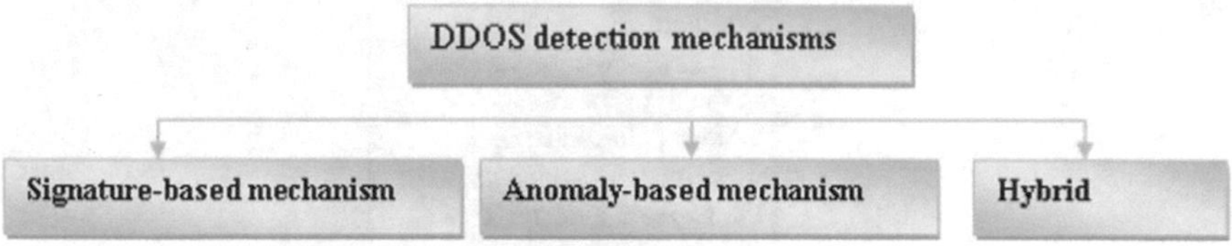

Fig. 4 DDoS-based detection mechanisms classification

3.1 Different Types of DDoS Attacks

There are various motivations that make the attacker perform DDoS attack such as intellectual challenge, cyber warfare, slow down the performance of the network, financial profit, the motives of revenge, and religious beliefs. DDoS attack on cloud computing has three main types as shown in Fig. 3 [15].

Three attacks explained in Table 4.

4 Classification of Detection Mechanisms to Secure Cloud Computing Against DDoS Attack

Several researchers have proposed different mechanisms to prevent DDoS attack on CC. Recent researches have shown the danger of DDoS attack on cloud computing. Research by Meng et al. [16] clarified that cloud blocking attacks adversely affect the victim server, network resources, and service providers. Further, research by Parwani et al. [17] shows that DDoS and the resulting disruption of customer services are leading to the collapse of the price of cloud services, which negatively affects service providers. In addition, Sri and Lakshmi [18] research clarified how a DDoS attack occurred on cellular networks and that these attacks caused service disruption (Fig. 4).

DDoS-Based Detection Mechanism
These mechanisms can only detect the attack without preventing it by using different methods such as monitoring, classification, and trackback. The most recently pro-

Table 4 Types of DDoS attacks

Attack		Explanation
Network depletion attack		
Huge traffic volume with the help of zombies (compromised machines) is triggered by an attacker to overwhelm the targeted network Prabadevi and Jeyanthi [39]	Most internetworking devices like routers have inbuilt broadcast feature, the attacker takes advantage of that feature to initiate the attack. The attacker broadcast packets to the internetworking device using the broadcast address. Internetworking devices further send those packets in range of the broadcast address; afterward that machine will send a reply to the targeted system. This will lead the targeted machine with malicious traffic [39]	Attacker attempts to consume all the targeted network bandwidth by flooding targeted network with malicious traffic which will eventually prevent the legitimate traffic from reaching the targeted network
Resource depletion attack		The attacker is keen to consume resources and weak server and memory processor capabilities
Protocol exploit attack	Malformed packet attack	
The consumption of a large amount of resources leading to the blocking of service [39]	Data packets and malware are combined, and an attacker sends them to the target server to disable the service [39]	
Application attack		The attacker gets vulnerability in the application protocol through which he can attack such as HTTPS, SMTP, FTP, VOIP, and other application protocols

posed mechanisms for detecting DDoS attack on cloud computing has been discussed below.

4.1 Signature-Based Mechanism

A set of attack rules and patterns are used in the databases, where the traffic is monitored for patterns and compared to the ability to detect the attack. This technique is one of the most important techniques for its accuracy in detecting attacks in the case of continuous updating of its databases. But it has a disadvantage that it cannot detect unknown attacks.

Research conducted by Lonea et al. [19] prevails VM-based IDS that features a graphical interface, organized MySQL information to observe signals of cloud fusion unit within the front-end of the Eucalyptus cloud design. They resolve to use the Barnyard tool to detect attacks, whereas a stamped based mostly SNORT was organized with predetermined DDoS rubrics to protect against DDoS attacks that are far-famed. The apprehended attack packets by Barnyard are kept in the binary unified file then spread employing a safe channel to central MySQL information on the front-end. DDoS incursions were mimicked using the Stacheldraht, a DDoS incursion instrument that produces an infrastructural resource reduction attack comprising of ICMP flooding, UDP flooding and transmission control protocol SYN. The key shortcoming of this method is the incapability to sight unidentified attacks—associate degree of intrinsic weakness of the signature based mostly method.

A study by Aghila and Karnwal [20] recommended the utilization of filter tree method to preserve in contrast to the application layer flooding. The five (5) components within the projected method were recreated to find and correct XML and hypertext transfer protocol-based DDoS incursions that happen in requisitions for resources victimization, Straightforward Object Access Protocol (SOAP), and electronic communication. The scientific discipline marking module uses the Versatile Settled Packet Marking (VSPM) theme to spot SOAP messages at the sting router; whereas scientific discipline traceback may be a logical file that keeps an inventory of a blacklist of scientific discipline addresses provided by the Cloud Defender. We conclude that the Cloud Defender sieves the intrusions by longing five totally dissimilar stages: detector filter, hop-count filter, scientific discipline frequency divergence filter. These ensure a genuine user scientific discipline filter and double signature filter. Primarily, four stages square measure is accustomed to find hypertext transfer protocol DDoS attacks whereas XML DDoS attack is detected within all five stages.

Furthermore, studies conducted by Negi et al. [21], anticipated the usage of attack detection styles supported by the VM profile optimization. When there was a TCP SYN flooding incursion resulting from the threshold for rule pattern at the start of the rule establishment stage, the rule-based recognition technique was used to match packets.

Furthermore, Gul and Hussain [22] proposed model for invasion discovery for handling massive movement of packets by examining as well as getting information on these packets. The proposed prototype uses multithreading methods in boosting the IDS output. NIDS detects as well as watch traffic within the network to examine for mischievous packets. When a mischievous packet is spotted, an alert is sent to an arbitrator observance system that notifies the CC management system. The NIDS prototype had three units, the capture and queuing unit, the processing/analysis unit, and reportage unit. This prototype used the .NET framework in the Windows operating system.

Further, research by Bakshi and Yogesh [23] used IDS in VMs that opposes DDoS incursions. IDS sensing element uses SNORT; a sign primarily built method which is positioned on the virtual interface of VMware virtual ESX machine to investigate each inward and outward traffic in real time. This form of protection is meant to deter DDoS incursions within the network/transport layer, that distinguishes the IP addresses that were utilized in these incursions by mechanically creating an admission management log to drop the complete packets from the proscribed IP addresses. When the incursions come from units characterized as zombie machines, this method will be designed to stop these types of traffic transfer the aimed application to a VM held in an alternative data center.

Furthermore, Lo et al. [24] proposed IDS primarily based mechanism scattered at intervals the cloud setting. Alerts are changed through the IDS nodes scattered all over the cloud setting once associate incursion is spotted. The IDS system includes four elements; the co-operative operation, the response and block, the threshold check and alert cluster, and attack discovery. A supportive agent is embedded in every IDS that is employed to see if it will admit or discard the alert directed by different IDS nodes. These associate methods will lessen incursions spotted and reported by different IDS nodes.

4.2 Anomaly-Based Detection

Anomaly-based includes a collection of patterns collected in a previously defined period; the primary objective of this technique is to predict future patterns [25], proposed a cloud DDoS incursion protection system referred to as DaMask. DaMask uses an extremely programmable network observation method that spots attack and reacts by employing a versatile management structure. DaMask is made up of 3 layers (i.e., network switches, network controllers, and network application) and 2 units (i.e., anomaly based mostly network incursion discovery component—DaMask-D, associated an incursion mitigation module—DaMask-M). Once associate incursions are spotted by DaMask-D, then a correspondingly an alert is sent. Based on this alert the packet info can then be sent to the DaMask-M module. When this packet is decided to be genuine, the packet is now sent to its proposed destination. DaMask-M accomplishes two roles: step choice and generation of the log. When getting an associate alert, DaMask-M chooses what step to use. A distinctive step is the packet dropping. Assessments exploitation with Amazon net service, public cloud, and a personal cloud running on Ubuntu 12. 10 with the UNB ISCX dataset instructed that Damask's work was almost like discovery structures supported constant time Bayesian network, however, DaMask supposedly encompasses a lesser procedure price.

Moreover, Marnerides et al. [26], offered associate irregularity discovery method, Ensemble Empirical Mode Decomposition (E-EMD), that might be accustomed use applied mathematics classification and breakdown of measured signals. E-EMD is enforced on the hypervisor level and roles by collectively looking at systems and network info from each VMs. The planned method was used because all supervised traffic display nonlinear as well as nonstationary characteristics.

In [27], it was proposed constant quantity method victimization the mathematician model to protect against the application layer DDoS incursions on cloud services, that use mischievous XML content contained during a SOAP. The traditional profile prototype is made from the dataset throughout the low-level formatting phase before triggering the proxy to pay attention to requests. Throughout discovery, they will be many stages within the first section, HTTP header review is going to be applied to forestall HTTP flooding. Additionally, SOAP starts to check and size outlier review. Within the succeeding section, processing of the XML content is done prior to checking if RSOAPAction were tricked as a result of consultations of earlier maps. Lastly, this SOAP attribute outlier detection (a filter method which estimates every attribute that corresponds to mathematical Structure) is applied. Thus, the ability to find requests representations ensuring from the new DDoS methods is the main weakness of this model, without the implementation of further options.

Furthermore, Shamsolmoali and Zareapoor [28], used a Cloud confidence DDoS Filtering, which is a filtering system based on statistics. The filtering is done on two (2) levels. First, the header field contained in the incoming packet is removed and compared to the TTL worth together with the hold worth within the IP to hop-count (IP2HC) table. Should there by a difference in the values, a packet is born that is characterized as fooled. The subsequent level is predicated by the Jensen–Shannon divergence that utilizes a hold on traditional profile hold on in an exceedingly information to match incoming packet header data. This helps to examine if there is data divergence.

Moreover, Choi et al. [29], used data mining method which used MapReduce structure to relieve application layer HTTP GET DDoS incursions. Meanwhile, batch job operations were quickened using the MapReduce model which is based on parallel processing. This anticipated model comprises of three units. They are the packet and log module analyses packet transmission and net server logs, and the pattern analysis module produces the incursion styles for DDoS detection. Structures that are examined embody computer hardware utilization, packet size, load, and data dispersal of the packet header. The discovery unit used a standard activity design to spot DDoS attacks. To judge the prototype, a map cut back formula were accustomed live rates amongst pattern rule with the recognition period of the projected system to external signatures. Outcomes obtained recommended that the projected methodology accomplishes higher than SNORT, because this establishes a new attack profile that is incorporated in a lesser period.

Furthermore, Alqahtani and Gamble [30] anticipated a way of spot DDoS incursions at the service level, tenant level, application level, and cloud level. The irregularity discovery is achieved locally at the service level by summering the stream of data with the use of a hash map. As the service level detection procedure is monitored, an increase in the flow of ingress will trigger an alert that is tipped off at the cloud interface. The source of this increase is pinpointed and tagged. The rate of the flow is determined using a conceptual information distance metric measured against an expected threshold. An activity is categorized an incursion when the measured rates are much more than the expected threshold. At the tenant level detection, indicators recognize possible incursions by merging hash maps gathered provided by local services. The application-level recognition compares DDoS incursions, flow rate, and performance declination of web services to spot dispersions resulting from DDoS incursions. The tenant and application levels outcomes are then referred to the cloud level for more confirmation. At this stage, key performance metrics would be utilized to watch both recognition and damages degree.

Additionally Zakarya [31], the research presents an entropy monitoring method that mainly utilizes the incursion packet dropping rule in spotting DDoS incursions in the CC situation and are subsequently applied to spot incursion movements supported by the dispersal magnitude relation. Anomaly recognition structure is organized on every edge router that later moves the stream to a router that is adjacent for validator checks immediately after DDoS is spotted. When established, DDoS packets square measure are born out of results from the cloud simulation.

Moreover, a mathematical model was used [32], and a variance matrix method to find flooding primarily founded DoS incursion against cloud services. Contained in this method initial part is a structure used in identifying traditional itinerary because of of the reference point, by charting apprehended previous traffic into an identical variance matrix. The next part is that the invasion discovery phase, where there is a measurement to detect the variance between the presently captured traffic and that obtained from the variance matrix obtained. The final part is that the bar phase that effectuates the earlier 2 stages.

Also, studies by Choi et al. [33], used MapReduce to categorize structures (i.e., CPU usage, packet size, protocol distribution, and information distribution of packet header) before using entropy figures to quantify the structures reliability.

Subsequently, [34], used Confidence-Base Filtering (CBF) technique, which uses correlation properties that might be setup in each attack and non-attack phases. Throughout the non-incursion amount, a minimal profile is made by removing associate in nursing characteristic from both the network and transport layers. The frequencies of the prevalence of those worth duos are removed and accustomed estimate their confidence worth. The correlation features abound among these 2 layers were accustomed to verify the validity of a packet throughout the flow of the traffic. Throughout the quantity of attack, characteristic worth pairs of inbound packets are collated and matched with the minimal profile to work out its confidence worth in genuine movement. Packet legitimacy is determined using the CBF score and the filtering criterion.

Moreover, Huang et al. [35] suggested a structure that utilizes multistage discovery and text-based turning check, which alleviate protocol request flooding incursions. The planned structure is created with 5 components, having: supply checking, counting, attack detection, Alan Turing check, and question generation. The supply scrutiny and investigation unit interrupt inbound packets. The inspected packet is tested by the Alan Turing check unit to verify if the packet is doubtful by responsive many text-based queries. The DDoS incursion discovery unit receives and keeps a log of the traffic performance of every Virtual Cluster (VC); this can be used as a profile to research every VC's instant traffic to visualize for attainable irregular traffic performances by malevolent packets. The text-based Alan Turing testing unit accepts transmitted cache of packets and randomly chooses an issue to be responded to by the requester. When a query is answered correctly, access to the destination is scarcely given. Question generation units intermittently update the collection of queries. The structure was enforced in UNIX kernel and user areas in keeping with necessities. Performance check steered a coffee reflection quantitative relation and high potency.

This method used the cosine resemblance method, while the best measure was utilized in the evaluation of self-resemblance. Throughout the analysis, the researchers utilize a preprocessor to isolate procedures from the windows security event log. When the self-resemblance is not genuine, a system alert is created, and the installed IDS inspects the outlier points and the information to the address of the supply. The supervisor is notified of this occurrence. An important attribute this method has is that it doesn't need a prolonged learning technique and the self-resemblance is often obtained in actual time.

4.3 Hybrid

Hybrid-based discovery method includes utilization of both signatures as well as anomaly-based technique. The method employs balancing options for each technique to attain a better discovery degree. Research by Lo et al. [24], suggests supportive interference discovery design sculpture with ECARGO to guard against cloud DDoS outbreaks. The planned design is created of 4 layers. The primary layer, a happening generator, gathers network packets and produces doubtful incursion events. The feature indicator works using a comparable style as that of SNORT, before it is then utilized to partition actions consistent with network protocols (e.g., ICMP and TCP). Applied math indicator utilizes knowledge packets from the feature indicator to work out the incursion event.

When the quantity of packets obtained at intervals for a definite period vary and is above the brink set, it is thought of associate in the nursing attack. Lastly, the fusion center uses mediators to achieve completely dissimilar roles, equivalent to knowledge preprocessing, space–time continuum fusion, and content fusion. Results obtained after the research indicated that planned methodology may be a feasible answer to notice DDoS and sluggish scanning incursions. Nevertheless, this becomes a constraint in distinguishing flash crowd traffic from malevolent incursions.

Moreover Modi et al. [36] developed an amalgam network primarily based on interruption to find cloud DDoS incursions. SNORT, associate degree open supply signature primarily based on discovery technique, which keeps guidelines of notable DDoS incursions styles, and theorem classifier, an applied mathematics categorizer which forecasts the likelihood of a network incident to a category either traditional or malevolent by high precision. Eucalyptus, an associate degree open supply cloud was utilized in this experimental system, wherever invasion discovery structure was put in every node controller with each port opened for verification purpose. In obtaining custom packets, Scapy was used whereas performance and feature were assessed for the exploitation of KDD '99 data set.

In another research by Cha and Kim [37], a 3-stage irregularity detection technique was designed. The primary phase handles observation, that utilizes a rule-based structure in preprocessing famed DDoS incursion styles. The next phase presents a lightweight irregularity recognition that forecasts the anticipated imminent load on every client interface mistreatment time-series modeling. The traffic capacity over the network is split to have huge and little capacities on the time axis, thereby theorem method is employed to research DDoS incursion candidate on the constellation. The final phase utilizes focused anomaly to spot the identified and unidentified DDoS incursion styles utilizing an uncontrolled learning rule (Table 5).

5 Discussion

CC is an important part of the development network based on the web and the availability of the cloud is the most precious. To make the availability of the cloud to the users at all time, it is necessary to have a mechanism to detect threats which hampers the accessibility of the cloud to its users. So, this review paper provided the techniques available for detection for DDoS attack which was effective in previous studies. Each discussed detection technique has strengths and corresponding limitations. Their strengths were aimed at filling the requirement of an exact constraint presented by the earlier procedure. Previously, a researcher undertakes the practicableness the procedure they make in contrasts to the ways of their predecessors. In reviewing a thought, a scientist should contemplate all the alternatives and items creating it up as well as their interrelationship [38]. Signature-based detection technique has an advantage of accuracy in detecting known incursions but has a disadvantage in that it cannot spot unidentified and zero-day incursions. The advantage the anomaly detection has is in the detection of patterns of behavior that is not in conformity with the known behavioral profile patterns this leads to detecting DDoS incursions. This anomaly has a disadvantage mainly when a misplaced or wrong value from the dataset is used, it will influence identification proficiency greatly. Hybrid detection technique has the advantage of efficiency as it combines multiple techniques to accurately classify rules but has disadvantage it requires high cost and be very complex. This review paper found that anomaly-based detection technique is the best suitable technique because it includes a lot of approaches especially technique related to big data and the reason for that availability of data give opportunity for early and accurate detection for DDoS techniques especially CC rely on big data and the developing of CC to frog computing and use in internet of things.

Table 5 DDoS-based detection mechanisms

Technique	Strengths	Limitations
Signature-based IDS	• Comparison is simpler • Precision in distinguishing identified attack signature having a little false positive rate • Offers low computational costs • Easy to track and stop an attack since log files are exhaustive • Offers the features of robustness, scalability, and flexibility • The incidence of DDoS incursion tags lets the system administrator control the sort of denial of service incursion that the user is facing	• Keeping an updated signature is demanding and expensive, if not unbearable, job • Unable to identify the new, novel attacks • It is very complex • Misrepresentation of signature style will give a high false negative rate • Huge traffic limits inspection of every packet causing unattended packets to pass through • Cannot track down intelligent intrusions • New attacks must be updated in the database
Anomaly-based IDS	• Ability to detect unknown attacks • Offers high detection accuracy • Gives a real opportunity to make the right decisions • System is self-learning. It gradually learns the network and builds a profile • The more it is used the higher the accuracy level • Higher the false alarm rate for unknown attacks • New threats are easily detectable without updating the database • Gives the opportunity to predict expected behavior • Effectively classifies unstructured network packets • Allow training and testing • Increase focus that leads to improving detecting process • High detection rate	• When malicious activities assume the features of normal traffic it is untraceable • Collected behavior and features determine the accuracy of detection • While building the profile, a network is left in an unmanaged state hence prone to attack • Genuine traffic is ranked as incursion traffic which can create a false positive of an attack • If the assumptions are wrong, they lead to an error • Has lesser flexibility • A problem occurs when over-training • Requires a lot of time at the training phase • Lost values are detrimental to accuracy
Hybrid detection	• Efficient as it combines multiple techniques to accurately classify rules	• Leads to high complexity and implementation cost

6   Conclusion

It is important to make sure that information within the cloud is free from all types of incursion. Safeguarding the cloud is difficult, however, inevitable. One of the various dreaded incursions in the cloud is that the DDoS attack. So, this paper described a comprehensive survey of detection and preventing techniques of DDoS incursions in CC. In this review paper, the researchers surveyed academic works on DDoS incursions against cloud services as well as the alleviation approaches published between 2009 and 2018. We clarified the classification of the diverse cloud DDoS incursion types of and the equivalent DDoS protection types. There are still challenges that require to be addressed in this area despite the amount of research done. Extra research is needed to find efficient discovery and protection ways to spot any type of DDoS incursion to make the variety of CC types safe from the threats of DDoS attacks.

Acknowledgements This research was supported by the Short-Term Research Grant, Universiti Sains Malaysia (USM) No: 304/PNAV/6313332.

References

1. Weng C, Guo M, Luo Y, Li M (2013) Hybrid CPU management for adapting to the diversity of virtual machines. IEEE Trans Comput 62(7):1332–1344
2. Jin H et al (2011) A VMM-based intrusion prevention system in cloud computing environment. J Supercomput, pp 1–19
3. Lonea AM, Popescu DE, Tianfield H (2013) Detecting DDoS attacks in cloud computing environment. Int J Comput Commun Control 8(1):70–78
4. Joshi B, Vijayan AS, Joshi BK (2012) Securing cloud computing environment against DDoS attacks. In: 2012 international conference on computer communication and informatics (ICCCI), pp 1–5
5. Yang SJ, Li YZ (2016) Design issues of enhanced DDoS protecting scheme under the cloud computing environment. In: Proceedings 2016 International Conference Networking and Network Application NaNA 2016, pp 178–183
6. Yang L, Zhang T, Song J, Wang JS, Chen P (2012) Defense of DDoS attack for cloud computing. In: CSAE 2012—Proceedings, 2012 IEEE international conference on computer science and automation engineering, vol 2, pp 626–629
7. Beale R (2011) Moving to the cloud; expect fog, no. February
8. Li J, Castiglione A, Dong C (2018) Special issue on security in cloud computing. J Netw Comput Appl 110(2017):97–98
9. Khorshed MT, Ali ABMS, Wasimi SA (2012) Classifying different denial-of-service attacks in cloud computing using rule-based learning. Secur Commun Netw 5(11):1235–1247
10. Mell P, Grance T (2011) The NIST definition of cloud computing, recommendations of the national institute of standards and technology. National Institute of Standards and Technology, p 7
11. Mell P, Grance T (2009) On-demand self-service. Nist 15:10–15
12. Tsai W-T, Sun X, Balasooriya J (2010) Service-oriented cloud computing architecture. In: 2010 seventh international conference on information technology: new generations, 2010, pp 684–689

13. Bhuyan MH, Kashyap HJ, Bhattacharyya DK, Kalita JK (2014) Detecting distributed denial of service attacks: methods, tools and future directions. Comput J Oxford J 57(4):537–556
14. Daffu P, Kaur A (2017) Mitigation of DDoS attacks in cloud computing. In: 2016 5th international conference on wireless networks and embedded systems, WECON 2016
15. Othman RAR (2000) Understanding the various types of denial of service attack no. URL:http://cybersecurity.my/data/content_files/13/72.pdf. 2 June 2018
16. Meng B, Andi W, Jian X, Fucai Z (2017) DDOS attack detection system based on analysis of users' behaviors for application layer. In: 2017 IEEE international conference computing science engineering, IEEE international conference embedded and ubiquitous computing, pp 596–599
17. Parwani D, Dutta A, Shukla PK, Tahiliyani M (2015) Various techniques of DDoS attacks detection and prevention at cloud: a survey. Orient J Comput Sci Technol 8(2):110–120
18. Sri KS, Lakshmi P (2017) DDoS attacks, detection parameters and mitigation in cloud environment, 3(01):1–4
19. Lonea AM, Popescu DE, Prostean O, Tianfield H (2013) Evaluation of experiments on detecting distributed denial of service (DDoS) attacks in eucalyptus private cloud. In: Advances in intelligent systems computing, vol 195 AISC, pp 367–379
20. Aghila G, Karnwal T (2012) A comber approach to protect cloud computing against XML DDoS and HTTP DDoS attack
21. Negi P, Mishra A, Gupta BB (2013) Enhanced CBF packet filtering method to detect DDoS attack in cloud computing environment. Arxiv, pp 2–6
22. Gul I, Hussain M (2011) Distributed cloud intrusion detection model. Int J Adv Sci Technol 34:71–82
23. Bakshi A, Yogesh B (2010) Securing cloud from DDOS attacks using intrusion detection system in virtual machine. In: 2nd international conference communication software networks, ICCSN 2010, pp 260–264
24. Lo CC, Huang CC, Ku J (2010) A cooperative intrusion detection system framework for cloud computing networks. In: Proceedings of international conference parallel processing workshop, pp 280–284
25. Wang B, Zheng Y, Lou W, Hou YT (2015) DDoS attack protection in the era of cloud computing and software-defined networking. Comput Netw 81:308–319
26. Marnerides AK, Spachos P, Chatzimisios P, Mauthe AU (2015) Malware detection in the cloud under ensemble empirical mode decomposition. In: 2015 International Conference Computing Network Communication, pp 82–88
27. Vissers T, Somasundaram TS, Pieters L, Govindarajan K, Hellinckx P (2014) DDoS defense system for web services in a cloud environment. Futur Gener Comput Syst 37:37–45
28. Shamsolmoali, P, Zareapoor M (2014) Statistical-based filtering system against DDOS attacks in cloud computing. In: Proceedings of the 2014 international conference on advances in computing, communications and informatics, ICACCI 2014, pp 1234–1239
29. Choi J, Choi C, Ko B, Kim P (2014) A method of DDoS attack detection using HTTP packet pattern and rule engine in cloud computing environment. Soft Comput 18(9):1697–1703
30. Alqahtani S, Gamble RF (2015) DDoS attacks in service clouds. In: Proceedings of annual Hawaii international conference system science, vol 2015–March, pp 5331–5340
31. Zakarya M (2013) DDoS verification and attack packet dropping algorithm in cloud computing. World Appl Sci J 23(11):1418–1424
32. Ismail MN, Aborujilah A, Musa S, Shahzad Aa (2013) Detecting flooding based DoS attack in cloud computing environment using covariance matrix approach. In: Proceedings of 7th international conference ubiquitous information management communication—ICUIMC '13, no. January, pp 1–6
33. Choi J, Choi C, Ko B, Choi D, Kim P (2013) Detecting web based DDoS attack using MapReduce operations in cloud computing environment. J Internet Serv Inf Secur 8111:28–37
34. Dou W, Chen Q, Chen J (2013) A confidence-based filtering method for DDoS attack defense in cloud environment. Futur Gener Comput Syst 29(7):1838–1850

35. Huang VSM, Huang R, Chiang M (2013) A DDoS mitigation system with multi-stage detection and text-based turing testing in cloud computing. In: Proceedings of 27th international conference advanced information networking application workshop. WAINA 2013, pp 655–662
36. Modi CN, Patel DR, Patel A, Muttukrishnan R (2012) Bayesian classifier and snort based network intrusion detection system in cloud computing. In: 2012 third international conference on computing communication networking technology, vol 39, pp 1–7
37. Cha B, Kim J (2011) Study of multistage anomaly detection for secured cloud computing resources in future internet. In: Proceedings of IEEE 9th international conference dependable, autonomic and secure computing DASC 2011, pp 1046–1050
38. Kaur P, Kumar M, Bhandari A (2017) A review of detection approaches for distributed denial of service attacks. Syst Sci Control Eng 5(1):301–320
39. Prabadevi B, Jeyanthi N (2014) Distributed denial of service attacks and its effects on cloud environment-a survey. In: 2014 international symposium networks, computing communication, pp 1–5

Network Traffic Classification for Attack Detection Using Big Data Tools: A Review

Zaid. J. Al-Araji, Sharifah Sakinah Syed Ahmad, Mustafa W. Al-Salihi, Hayder A. Al-Lamy, Mohammed Ahmed, Wisam Raad and Norhazwani Md Yunos

Abstract Network traffic classification is the foundation of many network research works. Network traffic classification is extensively required for some network management tasks, for example, prioritization, flow, diagnostic monitoring and traffic shaping/policing. Similar to network management tasks, many network engineering problems, like capacity planning, route provisioning and workload characterization and modelling, also benefit from accurate identification of the network traffic. The focus of this research is to classify traffic based on application type. This paper presents a review of different types of network classification methods and big data tools used to increase accuracy.

Keywords Network security · Classification methods · Network traffic · Machine learning

1 Introduction

Increased Internet technologies, in every aspect of our lives, make Internet security a major problem. Internet security is defined as the 'complex and complicated subject

Zaid. J. Al-Araji (✉) · S. S. Syed Ahmad · N. Md Yunos
Department of Intelligent Computing & Analytics (ICA), Universiti Teknikal Malaysia Melaka
(UTeM), Hang Tuah Jaya, 76100 Durian Tunggal, Melaka, Malaysia
e-mail: zaid.jassim4@gmail.com

S. S. Syed Ahmad
e-mail: sakinah@utem.edu.my

M. Ahmed
University of Mosul, Mosul, Iraq

W. Raad
Dijlah University College, Baghdad, Iraq

M. W. Al-Salihi · H. A. Al-Lamy
Faculty of Information and Communication Technology, Universiti Teknikal Malaysia Melaka,
Hang Tuah Jaya, 76100 Durian Tunggal, Melaka, Malaysia

© Springer Nature Singapore Pte Ltd. 2019
V. Piuri et al. (eds.), *Intelligent and Interactive Computing*, Lecture Notes in Networks
and Systems 67, https://doi.org/10.1007/978-981-13-6031-2_37

encompassing information assurance, computer security, comprehensive infrastructure protection, ubiquitous personal interactions and commercial integrity' [1].

Network traffic volume is increasing exponentially, because of the new multimedia applications and the advancements in Internet technology. The present invention relates to the method, system and computer program for performing network traffic classification. For example, network operators that handle network traffic between mobile phones and a web server classify network traffic to obtain information about the use of their network [2].

In this type of situation, application classification becomes extremely important for managing network quality of service (QoS) and security monitoring for various Internet service providers (ISP) and other governmental and private organizations. The accurate and efficient application classification is a key element of the networks' monitoring; and based on the classification results, the network administrators could design different policies to enhance network's security. However, it is a challenging task to classify application based on the traffic features due to a massive data involved in high-speed network. Although different methods for the traffic identification have been suggested, none has identified all kinds of Internet traffic. In this study, we will explain many network traffic classification methods and big data tools, in order to increase accuracy.

2 Attacks

Networks are subject to attacks from various malicious sources. Network traffic attacks can be divided into the following four categories: (Table 1).

3 Big Data

Big Data is a term used for massive datasets, large amounts of data available in a complex structure or in unstructured forms. These vast amounts of data are generated via social media networks, mobile devices, scientific instruments and sensor technology. One process of research into massive amounts of data reveals secret correlations and hidden patterns known as Big Data analytics. If data is beyond storage capacity and beyond processing power, that data is known as Big Data [6].

3.1 Big Data Tools

We try to describe current Big Data tools used for performance assessment and network traffic classification [7]. The following tools have been used by recent researchers.

Table 1 Types of attacks

Class of attack	Type of attack	Description
Probe	Ipsweep	Ipsweep is an observation sweep to define which nodes are listening on the network
	Mscan	Mscan uses both the DNS transfer or/and the 'brute force scanning' of the IP address to determine the machine and tests it for the vulnerability [3]
	Nmap	Nmap supports different kinds of port scan sùoptions, including ACK, FIN and SYN scanning, with UDP and TCP, and ICMP scanning [4]
	Resetscan	The resetscan sends a reset packet to the list of the IP address in the subnet to define the online machine. If there are no replies to a packet, that means it is still alive. If the routers or the gateway respond with 'the host is unreachable', that means it is not online
	Saint	The saint, or security administrator integrated networks tools, in simple mode, collects as much data about a remote host and its network as possible via examining NFS networks' service as a finger FTP, NIS, and statd, rexd, TFTF, and another services
	Satan	The Satan is the early predecessor of a Saint. Similar in purpose and design, a particular vulnerability which every tool checks for slightly differently [5]
Denial of service attacks	Teardrop	Teardrop generates IP fragments that overlap and affect target crashes or cause a reboot
	Land	Land attack uses IP spoofing combined with the opening of a TCP connection. In Land, IP addresses, destination and source are modified to become the same; and as a result, the kernel gets into the ACK war itself
	Slowloris	An attack does not complete a request to maintain a connection as open until a victim is not able to process the request from the legitimate client
User to root attacks	Yaga	Yaga adds an attacker to domain admin groups via hacking a registry. Attackers edit a victim registry so next time the system services crash on a victim; attackers added to a Domain's Admin groups

(continued)

Table 1 (continued)

Class of attack	Type of attack	Description
	Loadmodule	Loadmodule is user to root attack versus a SunOS 4.1 system which uses xnews windows systems
	Perl	Perl is user to root attack which exploits bugs in many Perl implementations. The suidperl is one of the Perl types which support saved set-users-ID and set-groups-ID script
	Ps	Ps attacks get benefits of race conditions in a model of the 'Ps' distributed with the Solaris 2.50. It allows the attackers to implement tyrannical codes with the root privileges
	CaseSen	The CaseSen is user to root attacks which exploits cases sensitivity of NT objects directories. The attacker's FTP three attack files of victims: editwavs.exe, psxss.exe, soundedt.exe
	Xterm	The Xterm exploits the buffer overflow in Xaw libraries distributed with the Redhat Linux and allows the attackers to execute tyrannical instructions with the root privileges
Remote to local attacks	http_tunnel	Attacker gains local access to a machine under attack and sets up and configures the HTTP clients to query web servers periodically which the attacker had set up at many remote hosts
	Ppmacro	Ppmacro is a remote to local attacks that uses trojans in PowerPoints macros to read secret files. This type of attacks is based on specific scenario. The victim users commonly receive PowerPoint template from external sources by email attachments
	Imap	This type of attack exploits the buffers overflow in Imap servers of the Redhat Linux, which allow the remote attacker to implement tyrannical instruction with the root privilege
	FTP write	This attack is a remote to local user attacks which take advantage of mutual anonymous FTP misconfigurations. An anonymous FTP root directory and its subdirectory must not be owned via an FTP account or be in groups as FTP accounts
	Xsnoop	In this attack, the attackers watch keystrokes via unprotected X servers to try and gain data which may be used to gain the local access of a victim's systems

Hadoop

Hadoop is open-source platform, established to manage, process and collect large datasets. Organizations usually depend heavily on proprietary hardware, along with expensive and huge system to process and store data. However, Hadoop minimizes the requires of organizations to exploit in expanding or replacing hardware by enabling distributed parallel processing of large amount of data over existing server. Implementing Hadoop enables organizations to harness the ability to process and store data efficiently and faster via adding hardware outwardly limits. This is one benefit, as there is the requirement for organization to collect data from new source daily. The ability to invoke organizations to find and keep valuable data that was once considered worthless [6].

Hive

Hive is open-source data repository infrastructure, run on the top of the Hadoop. It suggests high-level program languages that abstract implementations of the MapReduce job to grant user-friendly interfaces to the Hadoop. The command is expressed in the form of an SQL-like query. The language itself is called HiveQL [8].

Pig

Pig is high-level distributed program model, which builds on the top of the Hadoop. The main variation between Pig and Hive is the aim. Hive appropriates for the database's user, but Pig targets the experienced programmer who is not used to writing declarative SQL query [9].

Spark

Spark [10] suggests distributed information processing solutions for data-heavy applications with a variation that data to the process is stored in the memory. The spark proves to be up to hundred (100) times faster more than the Hadoop for a particular task.

Shark

Shark [7] is the sub-project of the Spark which implements Hive on the top of the Spark. It is therefore fully suitable with the Hive.

4   Classification Methods

Many techniques are used for Internet traffic classification (Fig. 1) and will be briefly described in this section.

4.1   Port-Based Method

In this method, the application type can be identified using port number registered in an Internet-Assigned Numbers Authority (IANA) list. Port-based method is con-

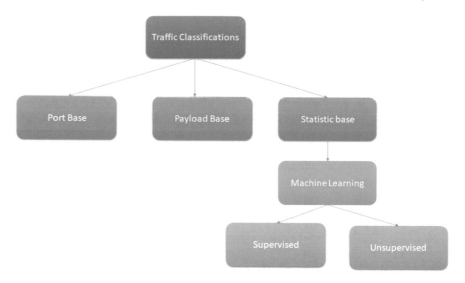

Fig. 1 Traffic classifications methods

sidered as simple and easy to implement classification of applications for online and real-time traffics. Moore and Papagiannaki [11] describe the technology and the user's skills that have adapted and developed. As we know, port number is no longer trustful for use in the classification. For this reason, Internet applications need to be designed to use another standard dynamic port selection, port number, or other protocols. Moore and Papagiannaki showed that changes in the design lead to low the accuracy classifications (50–70%) of the network traffics while using the port-based classifications from an IANA list [8]. Roughan et al. [12] proposed classification network traffic over the use of the statistical application's signature. This signature is derived from a manner in which an application is used, creating sets of rules for the classification based on IP address or port number, prepared to be more insensitive to a specific applications layer protocol. The authors limit their results with a low error rate (5–8%). Meanwhile, Haffner et al. [13] suggest a method using the application-level data taken from the packet to match the usual application's signature.

Payload based

In this scheme, the application type can be identified using the contents of the packet payload, by extracting the signature from the payload [14]. The results show that problems of the port-based method are solved given the high accuracy in recent Internet traffic, but fail in encryption communication. Furthermore, they take more time to extract signatures, need high computation resources and privacy laws, and may not let the administrators check the payload [15].

Statistic based

In this scheme, the application type is defined using statistical properties, like inter-arrival time, flow duration and packets length, to classify traffic. All problems mentioned above are solved, and high accuracy is gained depending on the number of features [16].

Machine learning It is a subset of algorithms that are developed in disciplines of the artificial intelligence. Those algorithms used various characteristics to learn the sets of rules to recognize various types [17]. The two types of machine learning methods are supervised and unsupervised [18].

Unsupervised Method

In this technique, the input does not require labelled data. It can naturally group datasets, known as clusters. However, these clusters need to be labelled by an expert. The earliest work done using the unsupervised technique used an expectation maximization algorithm for the IP traffic classifications based on applications [19]. The disadvantages of unsupervised learning include the cluster does not map 1:1 to the application. Ideally, the numbers of the clusters formed are equalled to the number of applications to be mapped; however, this is not the main case. The number of the cluster is usually bigger than a number of applications, i.e. one application may control over a number of the clusters or the applications may be spread but do not control clusters [20]. Zander et al. [21] used AutoClass (the Bayesian clustering algorithm based on an EM) in the method of traffic classifications. One of the problems in this classification method is that many applications, like Web traffic, Telnet and FTP, overlap together or have the wide ranges of features, which made them hard to classify. Meanwhile, average classifications accuracy of the techniques, over all those specific applications, is 86.5%. These diversities of features result in the false positive rate of up to 40%, depending on applications.

Supervised Methods

Supervised methods create knowledge based on the structures that can help classify new instances of various classes. Input instances are provided during a learning process that is pre-classified into the output. For example, the processes depend on those generalized instances [22]. A dataset provided for the training is labelled and (in this stage) time which it takes to process a sampled flow does not matter. More numbers of the attributes or characteristics will require time to process them and will be an accuracy of the classifier. Various algorithms have various sets of the rules that were developed from provided datasets and the performances [23]. The most popular 'machine learning algorithms' that are applied to the traffic classification are neural network, C4.5 Decisions Tree, Support Vector Machines (SVM), naïve Bayes kernel estimation, Bayesian network K-NN and naïve Bayes. The successful results were gained from SVM and C4.5 [24]. Li and Moore [25] suggest online that a near real-time classifications systems, using a packet-header-based attitude characteristic and the C4.5, produced overall average classification accuracy of 99.834%.

Tabatabaei et al. [26] suggested seven as a required number of the packets needed for the online traffic classifications, compared between SVM and k-NN techniques. Three classification techniques are compared using complete traffic flows and first seven packets resulted in a higher average accuracy of 84.9%, occurring with an SVM fuzzy one-against-all, using just a first seven packets.

5 Conclusions

In this paper, several literature researches concerned with this area, i.e. port analysis, signature-based methods and the ML method, were discussed. Performance, in terms of the accuracy and trust in resulting classification of the traffic, was illustrated. Several machine learning classifiers were evaluated; the results showed that C4.5 decision tree was the most stable supervised machine learning classifier, and was able to provide 82.7% accuracy from a training dataset captured by WIRESHARK and 88.9% from a dataset capture by Snort. Port analysis and signature-based methods were discussed; the results showed an average accuracy rate of 50–80% from the training dataset. For future work, we recommend a plan to examine machine learning classifiers with different datasets in addition to different numbers of features, in order to choose the optimum algorithm for different situations.

References

1. Katzan H (2012), Essentials of cybersecurity. In: Southeastern INFORMS conference, Myrtle Beach
2. Chen M, Kotay S, Robinson J (2018) Network traffic classification. U.S. Patent No. 9,948,605
3. CERT Incident Note (1998) (Online Available): http://www.cert.org/incident_notes/IN-98.02.html
4. NMAP Homepage. http://www.insecure.org/nmap/index.html
5. COAST FTP Site, (Online Available): ftp://coast.cs.purdue.edu/pub/tools/unix/satan/
6. Hadoop, (Online Avaliable). http://hadoop.apache.org
7. Engle C, Lupher A, Xin R, Zaharia M, Franklin MJ, Shenker S, Stoica I (2012) Shark: fast data analysis using coarse-grained distributed memory. In: Proceedings of the 2012 ACM SIGMOD international conference on management of data, ser. SIGMOD '12
8. IANA Port Numbers, (Online Available): http://www.iana.org/assignments/service-names-port-numbers/service-names-port-numbers.xhtml. Accessed 15 Jan 2014
9. Thusoo A, Sarma JS, Jain N, Shao Z, Chakka P, Anthony S, Liu H, Wyckoff P, Murthy R (2009) Hive: a warehousing solution over a mapreduce framework. Proc VLDB Endow 2(2):1626–1629
10. Zaharia M, Chowdhury M, Franklin MJ, Shenker S, Stoica I (2010) Spark: cluster computing with working sets. In: Proceedings of the 2nd USENIX conference on hot topics in cloud computing, ser. HotCloud'10
11. Moore AW, Papagiannaki K (2005) Toward the accurate identification of network applications. In: Proceedings of the 6th passive active measurement (PAM) Workshop, vol 3431, pp 41–54
12. Roughan M, Sen S, Spatscheck O, Duffield N (2004) Class-of-service mapping for QoS: a statistical signature-based approach to IP traffic classification. In: Proceedings of ACM SIGCOMM internet measurement workshop, Sicily, pp 135–148
13. Haffner P, Sen S, Spatscheck O, Wang D (2005) ACAS: automated construction of application signatures. In: Proceedings of the 2005 ACM SIGCOMM workshop on mining network data, Philadelphia, pp 197–202
14. Finamore A, Mellia M, Rossi MMaD (2010) Kiss: stochastic packet inspection classifier for udp traffic. IEEE/ACM Trans Networking 18(5):1505–1515
15. Kim J, Hwang J, Kim K (2016) High-performance internet traffic classification using a Markov Model and Kullback-Leibler divergence. Mobile Information Systems

16. Peng L, Yang B, Chen Y, Chen Z (2016) Effectiveness of statistical features for early stage internet traffic identification. Int J Parallel Prog 44(1):181–197

17. Mula-Valls O (2011) A practical retraining mechanism for network traffic classification in operational environments. Master Thesis in Computer Architecture, Networks and Systems, Universitat Politecnica de Catalunya

18. Jain AK, Mao J, Mohiuddin K (1996) Artificial neural networks: a tutorial. IEEE Comput 29(3):31–44

19. Namdev N, Agrawal S, Silkari S (2015) Recent advancement in machine learning based internet traffic classification. Procedia Comput Sci 60:784–791

20. Zhang J, Xiang Y, Zhou W, Wang Y (2013) Unsupervised traffic classification using flow statistical properties and IP packet payload. J Comput Syst Sci 79(5):573–585

21. Zander S, Nguyen T, Armitage G (2005) Automated traffic classification and application identification using machine learning. In: 30th annual IEEE conference on local computer networks, Sydney, pp 250–257

22. Taylor FR (2016) Evaluation of supervised machine learning for classifying video traffic. Doctoral dissertation. Nova Southeastern University. Retrieved from NSUWorks, College of Engineering and Computing

23. Bin H, Shen Y (2012) Machine learning based network traffic classification: a survey. J Inf Comput Sci 9(11):3161–3170

24. Huang N-F, Jai G-Y, Chao H-C, Tzang Y-J, Chang H-Y (2013) Application traffic classification at the early stage by characterizing application rounds. Inf Sci 232:130–142

25. Li W, Moore AW (2007) A machine learning approach for efficient traffic classification. In: 15th international symposium on modeling, analysis, and simulation of computer and telecommunication systems, 2007. MASCOTS '07

26. Tabatabaei TS, Karray F, Kamel M (2012) Early internet traffic recognition based on machine learning methods. In: IEEE Canadian conference on electrical and computer engineering (CCECE), Montreal

Propose a New Approach for Securing DHCPv6 Server in IPv6 Link-Local Network

Ayman Al-Ani, Mohammed Anbar, Rosni Abdullah, Ahmed K. Al-Ani
and Saif Al-Mashhadi

Abstract IPv6 is becoming more and more entrenched, especially as the shortage of IPv4 address became obvious recently. Internet Protocol version 6 (IPv6) uses Dynamic Host Configuration Protocol for IPv6 (DHCPv6) for assigning the IPv6 address to hosts and provides the host with network configuration parameters. The DHCPv6 protocol may use to reveal host information and inject fake information into a host. Thus, the authentication and privacy of the DHCPv6 messages are a vital security component in the IPv6 network. This paper proposes a new security approach to provide privacy and security for the DHCPv6 message. The proposed approach utilized a hybrid cryptosystem and Message Authentication Code (MAC) algorithm to provide privacy and authentication. The paper also presents the expected results for the proposed approach and the future works.

Keywords IPv6 network · DHCPv6 server · Rogue DHCPv6 server attack

1 Introduction

The number of internet nodes is significantly increasing [1], making limitations in the Internet Protocol version 4 (IPv4) addressing space limitation critical and serious, and strong worldwide problem. Internet Protocol version 6 (IPv6) is the next generation of the Internet Protocol that intents to replace IPv4. The researches show that IPv6 is getting high used by the host [2, 3]. Currently, hosts access Google over IPv6 exceed

A. Al-Ani · M. Anbar (✉) · R. Abdullah · A. K. Al-Ani · S. Al-Mashhadi
National Advance IPv6 Centre (NAv6), Universiti Sains Malaysia,
11800 Gelugor, Penang, Malaysia
e-mail: anbar@nav6.usm.my

A. Al-Ani
e-mail: ayman@nav6.usm.my

R. Abdullah
e-mail: rosni@usm.my

A. K. Al-Ani
e-mail: ahmedkhallel91@nav6.usm.my

© Springer Nature Singapore Pte Ltd. 2019
V. Piuri et al. (eds.), *Intelligent and Interactive Computing*, Lecture Notes in Networks and Systems 67, https://doi.org/10.1007/978-981-13-6031-2_42

21% based on the Google statistic [4]. Although IPv6 came to improve the network security and quality of service, IPv6 faced some of the security challenges such as Daniel of Service (DoS) and Man-In-The-Middle (MITM) attacks [5].

IPv6 network uses either DHCPv6 protocol or Stateless Autoconfiguration (SLAAC) mechanism to configure global host's addresses [6, 7]. The DHCPv6 gives the network administrator more control over the network than SLAAC mechanism [1]. Further, DHCPv6 is used to distribute other network configuration parameters to host in IPv6 network link-local network [8, 9]. The international organization of Internet Engineering Task Force provides more than 30 documents for the services [10], such as Domain Name System (DNS) server address and Network Time Protocol (NTP) server address [11, 12], which can be provided by the DHCPv6. Thus, DHCPv6 is widely used in IPv6 network. The DHCPv6 can be exploited by the attacker to get critical information related to the IPv6 hosts, and it also can be used to insert rogue services to the hosts. Thus, provide the privacy and authentication for the DHCPv6 message is crucial.

This study aims to highlight the main security vulnerabilities of the DHCPv6 message in IPv6 Network link-local network and propose a new security approach for securing the DHCPv6 process. The rest of this paper is organized as follows: Sect. 2 explains the process of DHCPv6 and security issues of DHCPv6. Section 3 discusses related works. Section 4 introduces the design of the proposed approach, and Sect. 5 provides the expected result. Finally, Sect. 6 presents the conclusion and future work.

2   Background

2.1   Dynamic Host Configuration Protocol

DHCPv6 is used by the IPv6 network to assign IPv6 addresses and distribute other network configuration information such as Domain Name System (DNS) server address and Network Time Protocol (NTP) server addresses [13, 14]. DHCPv6 has two operation modes: stateless and stateful. The DHCPv6 stateful mode is used for assigning IPv6 and distributing other network configuration information. The DHCPv6 stateless mode is used for distributing the network configuration information and assigning the IP address to hosts when the DHCPv6 is used with the SLAAC mechanism.

When the host joins the IPv6 network during DHCPv6 stateful mode, the host should multicast DHCPv6 solicit message to all DHCPv6 servers, which are located on the link-local network. The server will then respond through DHCPv6 advertise message with a related configuration information to the host. Next, the host will send DHCPv6 request message to confirm the configuration. Finally, the server should send DHCPv6 reply message to confirm the configuration [15]. Figure 1 illustrates the basic process of the DHCPv6 server during the stateful mode.

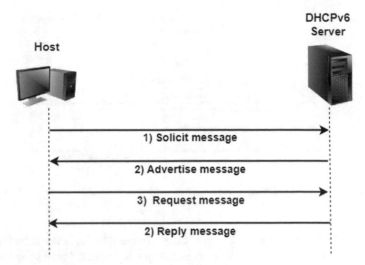

Fig. 1 DHCPv6 processes

Since the DHCPv6 server is the one that sends these configurations information to all the hosts located on the same network, therefore, DHCPv6 is considered one of the main components in the IPv6 network, and its protection is needed.

2.2 Rogue DHCPv6 Server Attack

As mentioned above, upon receiving the DHCPv6 server messages, the host will configure its IP address and other network configuration without any verification about whether these messages come from a legitimate or fake server [16]. Therefore, any attacker placed on the same network can pretend to be a legitimate server by sending fake messages to the hosts. When the host sends a solicit message asking the server to reply. In case, there is an attacker on the network will respond back via an advertise message containing a wrong configuration information. Since the host does not have any mechanism to verify the source of this message, it will accept the message and configure its IP address and other network configuration with a wrong configuration. Hence, the host is under attack such as DoS or redirect the user to rogue servers [17–19] (Fig. 2).

2.3 Privacy Issues

The DHCPv6 messages are transmitted in clear text. Thus, the DHCPv6 message reveals some information about the host such as device vendor, host operating system.

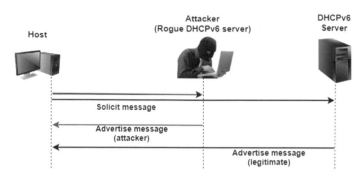

Fig. 2 DHCPv6 security challenge

This information can be exploited by the attacker to monitor the host and find security vulnerabilities of the host. RFC 7824 presents privacy considerations for DHCPv6 and the possibility of exploiting this information [20]. However, currently, DHCPv6 messages are transmitted in clear text and do not be protected from passive attacks, such as pervasive monitoring [12].

3 Related Work

Several approaches that are proposed to secure DHCPv6 server messages between host and server. The main current existing approaches are discussed in the following text.

RFC 3315 [21] defines two mechanisms to secure DHCPv6 server messages, namely, delayed authentication protocol to authenticate host solicit message and reconfigure key authentication. Both mechanisms do not provide privacy for DHCPv6 messages. Further, the two mechanisms required a shard key to be distributed between the host and server. The first mechanism does not provide any key distribution mechanism. Therefore, the shared keys are usually distributed manually which conflicted with the goal of minimizing the configuration data needed at each host. Besides, it is difficult to be deployed in a wide-scale network and mobility nodes [22]. The second mechanism is used for protection against misconfiguration of host caused by a reconfigure message send by a rogue DHCPv6 server. The shared key is sent as plaintext to the host in the initial exchange of DHCPv6 message for future use. Due to the fact that the key is transmitted as plaintext, an attacker can hijack the initial messages that contain the key and used the key for authenticating a malicious reconfiguration message [16].

Moreover, there are some approaches such as [18, 23] that attempted to enforce authentication without privacy by using an RSA signature algorithm. These approaches require to distribute the certificate to each DHCPv6 host manually, thus making these approaches difficult to be deployed in a wide-scale network and mobility nodes. In addition, this mechanism does not provide privacy for the DHCPv6 messages.

The study [18] proposed an approach to provide privacy and authentication for DHCPv6. This approach used certification and digital signature to provide authentication. In addition, it used an asymmetric encryption mechanism to encrypt the DHCPv6 message. Although this approach can provide privacy to DHCPv6 messages, it has some disadvantages: first, it cannot work with the security mechanism required to monitor DHCPv6 message such as Source Address Validation Improvement (SAVI) mechanism; second, it uses digital signature algorithm for providing authentication and Asymmetric Key Cryptography (AKC) for providing privacy. These mechanisms required high computation process compared with the Message Authentication Code (MAC) and hybrid cryptosystem. Additionally, this mechanism required preconfiguring the host with certification which should be conducted manually.

4   Design of Proposed Approach

As explained earlier, the main problem with the current authentication approaches, it does not provide a mechanism to distribute the key between the host and DHCPv6 server. Additionally, the current approach that intends to provide privacy to the DHCPv6 messages cannot work with monitor application such as SAVI mechanism. The core goal of this study is to propose an efficient approach to secure DHCPv6 in IPv6 network by (i) providing the authentication and privacy for the DHCPv6 message (ii) introducing a mechanism to distribute the key among hosts without human intervention, and (iii) the new mechanism should be applicable with SAVI mechanism.

To introduce an efficient approach that is able to provide privacy and allow the monitoring system applicable to the proposed approach, the proposed approach utilizes hybrid cryptosystem and MAC algorithm. The Hybrid cryptosystem is used to encrypt DHCPv6 options for providing privacy and authentication. Hybrid cryptosystem uses two separate cryptosystems: Symmetric Key Cryptography (SKC) algorithm and AKC algorithm. SKC algorithm uses to encrypt DHCPv6 options messages between hosts and the DHCPv6 server by using the same key which is called shared key. Further, AKC algorithm is used to distribute shared key among hosts and DHCPv6 server. AKC algorithm has the public and private keys, the public key will be used by the host to encrypt the shared key and send to the DHCPv6 server, the DHCPv6 server decrypts the shared key by using the public key. The private key will be deployed manually at the DHCPv6 server. On the other hand, to distribute the public key to the host, the proposed approach utilizes the RA message, since the RA message is needed for informing the host with the network configuration information such as stateless or stateful modes. Therefore, the proposed approach utilizes the RA message to distribute the public key to the host; the proposed approach considers the RA message is fully secure. Thus, the host does not require to preconfigure with the

Fig. 3 DPK option format

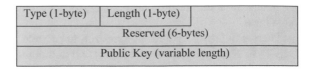

Type (1-byte)	Length (1-byte)	
Reserved (6-bytes)		
Public Key (variable length)		

Fig. 4 Distributed shared key host (DSKC) option format

Option Type (2-bytes)	Option_Len (2-bytes)
Encrypted Shared Key (8 bytes)	
Encrypted DHCPv6 Options (variable length)	

public key. Moreover, to allow the proposed approach works with SAVI mechanism, two of DHCPv6 options are not encrypted as will illustrate in Sect. 4.2. Moreover, the MAC algorithm is used to provide the authentication for the unencrypted field and the DHCPv6 server options; the MAC algorithm is considered faster than a digital signature algorithm.

4.1 The Proposed Option DHCPv6 Options

Several options have been introduced by the proposed approach to permitting the host and DHCPv6 server to provide privacy and authentication for the exchange messages. These options have been listed below:

- DHCPv6 Public Key (DPK) Option

DPK option introduces to allow the RA message to convey the public key to the host. Later, the host is going to use the public key for encrypting the shared key and other DHCPv6 options except are rapid commit and IA address options. Besides, DPK option follows the original format of RFC 8200 RA option [24]. The DPK option should be appended to each RA messages that are sent out by the router. DPK option consists of four main fields as shown in Fig. 3.

- Distributed Shared Key Host (DSKC) Option

DSKC option is designed to convey encrypted shared key and encrypt DHCPv6 options [25], for providing privacy. Figure 4 shows the ESK option format.

- Distributed Shared Key Server (DSKS) Option

DSKS option is designed to convey key ID for future communication, timestamp for preventing reply attack, MAC to provide authentication for unencrypted fields such as (timestamp and transaction ID), unencrypted DHCPv6 options (rapid commit and IA address), and encrypted DHCPv6 option field to provide privacy to other DHCPv6 options as shown in Fig. 5.

Fig. 5 Distributed shared key server (DSKS) option format

Option Type (2-bytes)	Option_Len (2-bytes)
Key ID (8 bytes)	
Timestamp (8 bytes)	
MAC (16 bytes)	
Encrypted DHCPv6 Option (variable length)	

Fig. 6 Encrypted DHCPv6 host (EDC) option format

Option Type (2-bytes)	Option_Len (2-bytes)
Key ID (8 bytes)	
Timestamp (8 bytes)	
MAC (16 bytes)	
Encrypted DHCPv6 Options (variable length)	

Fig. 7 Encrypted DHCPv6 server (EDS) option format

Option Type (2-bytes)	Option_Len (2-bytes)
Timestamp (8 bytes)	
MAC (16 bytes)	
Encrypted DHCPv6 Options (variable length)	

- Encrypted DHCPv6 Host (EDC) Option

EDC option is designed to convey key ID and the encrypted DHCPv6 host options as shown in Fig. 6.

- Encrypted DHCPv6 Server (EDS) Option

EDS option is designed to convey timestamp and encrypted DHCPv6 option as shown in Fig. 7.

4.2 Unencrypted DHCPv6 Options

The proposed approach aims to allow monitoring tools such as SAVI mechanism work with the proposed approach. The SAVI mechanism is used to keep track of all the IP addresses that have been assigned by the DHCPv6 to each device by snooping on the DHCPv6 message. Encrypted entire DHCPv6 messages prevent SAVI mechanism to work [26]. To allow SAVI mechanism to work with the proposed approach, the proposed approach does not encrypt the two DHCPv6 options which are used by SAVI mechanism. The unencrypted DHCPv6 options are rapid commit option and IA address option.

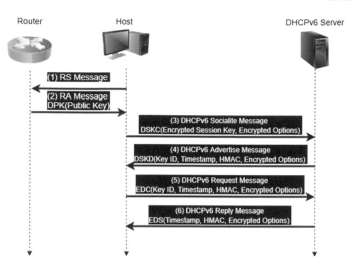

Fig. 8 Proposed approach message exchange

4.3 Proposed Approach Procedure

The DHCPv6 server should generate public and private keys, the private key should be saved at the DHCPv6 server and the public key should be deployed to the router. This operation is performed manually. Figure 8 shows the operation of the proposed approach, which illustrates the messages that are exchanged between a router, a host, and the DHCPv6 server.

When a new host joins the network, the host exchanges message with the router by multicast RS message, and the router replies by an RA message as a response to NS message with DPK option to convey the public key to the host. The host sends DHCPv6 solicit message contain the unencrypted options and DSKC option which consist of encrypted shared key field and encrypted DHCPv6 options field. Then, the ASC algorithm and the public key will be used to encrypt the shared key and DHCPv6 Option except for (rapid commit and IA address) options for hiding the shared key and provide privacy for the DHCPv6 Options. The shared key will be used later for MAC and SKC algorithms by the host and DHCPv6 server for future communication. The DHCPv6 solicit message header has a transaction ID to prevent a reply attack. Figure 9 shows the DHCPv6 solicit message with DSKC option.

After receiving the DHCPv6 solicit message, the DHCPv6 server decrypts the encrypted shared key and encrypted DHCPv6 option that is located at DSKC option by using ASC algorithm via the private key. Next, the DHCPv6 server should generate a key ID for shared key, to use it later to bind the key ID with the host shared key. Key ID and shared key need to be saved in a cache file at the DHCPv6 server. Subsequently, the DHCPv6 server will reply to the host by sending a DHCPv6 advertise message with DSKS option which consists of key ID, timestamp, MAC, and encrypted DHCPv6 options. All the options in DHCPv6 advertise message must

Fig. 9 DHCPv6 message
with DSKC option

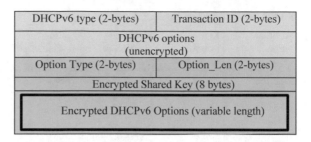

DHCPv6 type (2-bytes)	Transaction ID (2-bytes)
DHCPv6 options (unencrypted)	
Option Type (2-bytes)	Option_Len (2-bytes)
Encrypted Shared Key (8 bytes)	
Encrypted DHCPv6 Options (variable length)	

be encrypted by using SKC algorithm with a shared key. Whereas the unencrypted fields such as (timestamp, key ID and transaction ID) and unencrypted DHCPv6 options (rapid commit and IA address) will be digested by using MAC algorithm with a shared key for providing authentication. The digest value will be inserted into MAC field in DSKS option. In addition, the transaction ID of the DHCPv6 advertise message should match the received DHCPv6 solicit message to inform the host that DHCPv6 advertise message responses to its DHCPv6 solicited message.

After the host receives the DHCPv6 advertise message, the host should verify first the MAC by using the shared key to check if the DHCPv6 server message comes from a legitimate DHCPv6 server. Then, the host verifies the timestamp and transaction ID to prevent reply attack. Next, the host needs to save the key ID and using it with each DHCPv6 message that sends by the host. By doing so, the host and the DHCPv6 server share the shared key and key ID. The DHCPv6 server and the host should use the shared key for the MAC and SKA algorithms for next exchange messages. Then, the host decrypts the encrypted options in the DSKS option and processes the advertise message as per RFC3315 [15]. The host generates a DHCPv6 request message, with EDC option which consists of key ID, timestamp, MAC, and encrypted DHCPv6 option. The DHCPv6 options should be encrypted by using SKA algorithms and shared key to provide privacy. The MAC as well is used to provide the authentication for unencrypted options and field by using the shared key.

After receiving the DHCPv6 request message, the DHCPv6 server verifies the MAC and timestamp, and then decrypts the encrypted field of EDS option. The DHCPv6 server replies to host by sending a DHCPv6 reply message with EDC option which consists of timestamp, MAC, and encrypted DHCPv6 options fields.

After receiving the DHCPv6 reply message, the host verifies the MAC, transaction ID, and timestamp. Then, the host decrypts the encrypted DHCPv6 options.

Whenever the verification fails in any operation, the message should be discarded which indicates that the message is unauthenticated. By doing so, the proposed approach can prevent rogue DHCPv6 server and provide the privacy for DHCPv6 messages exchange between DHCPv6 servers and hosts.

Table 1 Approaches compression

Approach name	Key technology based on	Defense capability	Pre-configuration host	Compatible with SAVI mechanism	High computational cost
Delayed authentication protocol [15]	HMAC	Authentication integrity	Yes	Yes	Low
Reconfigure key authentication [15]	HMAC	Authentication integrity	Yes	Yes	Low
[23, 27]	Digital signature	Authentication integrity	Yes	Yes	Medium
Secure DHCPv6 [18]	Digital signature asymmetric cryptography	Integrity authentication privacy	Yes	No	Very high
The new proposed approach	MAC hybrid cryptosystem	Integrity authentication privacy	No	Yes	High

5 Expected Results

The proposed approach will compare with other approaches in term of defense capability (i.e., Integrity, privacy, and authentication), key technology based on, defense capability, deployment difficulty, compatible with SAVI mechanism, computing cost. Table 1 shows the comparison of these between other approaches and the proposed approach of this paper.

- Defense capability: Most of the approaches provide authentication and integrity. Only the proposed approach and secure DHCPv6 are able to provide integrity, authentication, and privacy.
- Preconfigure the host: All the existing approaches need a pre-configuration which are difficult to deploy in wide-scale network and mobility nodes; however, the proposed approach does not require for pre-configuration since it is used RA message to distribute the public key to the hosts.
- Compatible with SAVI mechanism: Most of the current approaches can work with SAVI mechanism, but failed to provide privacy. The only approach that is able to provide privacy is the Secure DHCPv6 approach, but it cannot work with SAVI mechanism due to encrypts all the DHCPv6 options which are not able to be monitored by SAVI mechanism. The proposed approach is able to work with SAVI mechanism and providing the privacy since the proposed approach does not encrypt the DHCPv6 options that are used by SAVI mechanism.

- High computational cost: the proposed approach is considered as having lower computation cost compared with the Secure DHCPv6 approach, as the proposed approach using hyped cryptosystem which has less computation cost compared with asymmetric cryptosystem. In addition, the proposed approach has a higher computation cost compared with other approaches because the proposed approach provides privacy which requires to encrypt the DHCPv6 messages.

6 Conclusion and Future Work

The paper presents the security vulnerabilities of the DHCPv6 message and makes an overview of the proposed works that have been done to secure the network against these attacks. Further, the paper shows that most of the current approaches required to be distributed the key manually to host and DHCPv6 server. The proposed approach attends to use a router to distribute the keys to the hosts without human intervention. Further, most of the proposed approach does not provide privacy for the DHCPv6 message. In addition, the proposed approach uses MAC and hybrid cryptosystem to provide privacy which is considered faster than a digital signature and asymmetric key cryptosystem that is used by the secure DHCPv6 approach. Moreover, the proposed approach is compatible with SAVI mechanism and can provide authentication and previous for the DHCPv6 messages. Selecting the algorithm for encryption and authentication will be the future work, in addition, to implement the proposed approach and test with existing approaches in terms of security and speed.

References

1. Groat S, Dunlop M, Urbanksi W et al (2012) Using an IPv6 moving target defense to protect the Smart Grid. In: Innovative smart grid technologies (ISGT), 2012 IEEE PES. IEEE, pp 1–7
2. Al-Ani AK, Anbar M, Manickam S et al (2017) Proposed DAD-match security technique based on hash function to secure duplicate address detection in IPv6 Link-local Network. In: Proceedings of the 2017 international conference on information technology. ACM, pp 175–179
3. Al-Ani AK, Anbar M, Manickam S et al (2017) Proposed DAD-match mechanism for securing duplicate address detection process in IPv6 link-local network based on symmetric-key algorithm. In: International conference on computational science and technology. Springer, pp 108–118
4. Google IPv6 (2018) IPv6 – Google
5. Elejla OE, Anbar M, Belaton B (2016) Icmpv6-based dos and ddos attacks and defense mechanisms: review. IETE Tech Rev 1–18
6. Ruiz JMV, Cardenas CS, Tapia JLM (2017) Implementation and testing of IPv6 transition mechanisms. In: 2017 IEEE 9th Latin-American conference on communications (LATIN-COM), IEEE, pp 1–6
7. Yousheng G, Lingyun Y, Lijing H (2017) Addressing scheme based on three-dimensional space over 6LoWPAN for internet of things. In: 2017 13th IEEE international conference on electronic measurement & instruments (ICEMI), IEEE, pp 59–64

8. Tirkkonen L (2016) Utilising configuration management node data for network infrastructure management
9. Tripathi N, Hubballi N (2017) Detecting stealth DHCP starvation attack using machine learning approach. J Comput Virol Hacking Tech 1–12
10. Wei D, Kerr J, Huth T, Mihajlovski V (2018) Dynamic Host Configuration Protocol for IPv6 (DHCPv6). https://www.iana.org/assignments/dhcpv6-parameters/dhcpv6-parameters.xhtml. Accessed 18 Jun 2018
11. Brzozowski J, Van de Velde G (2017) Unique IPv6 Prefix per Host
12. Horley E (2014) IPv6 and DHCP. In: Practical IPv6 for Windows Administrators. Springer, pp 191–207
13. Kaltio J (2016) IPv6 in soho environment: a study of basic functionality
14. Kharche MPS, Jawandhiya PM (2016) A case study of IPv4 and IPv6. In: National conference "CONVERGENCE. p 6
15. Droms R, Bound J, Lemon T et al (2003) Dynamic host configuration protocol for IPv6 (DHCPv6) (RFC 3315)
16. Shen S, Lee X, Sun Z et al Enhance IPv6 dynamic host configuration with cryptographically generated addresses. ieeexplore.ieee.org
17. Gont F, Liu W, Van de Velde G (2015) DHCPv6-Shield: protecting against rogue DHCPv6 servers
18. Li L, Ren G, Liu Y, Wu J (2018) Secure DHCPv6 mechanism for DHCPv6 security and privacy protection. Tsinghua Sci Technol 23:13–21. https://doi.org/10.26599/TST.2018.9010020
19. Alangar V, Swaminathan A (2013) IPv6 security: issue of anonymity. Int J Eng Comput Sci 2:2486Å93
20. Krishnan S, Mrugalski T, Jiang S (2016) Privacy Considerations for DHCPv6
21. Droms R, Arbaugh W (2001) Authentication for DHCP messages
22. Su Z, Ma H, Zhang X et al Secure DHCPv6 that uses RSA authentication integrated with self-certified address. ieeexplore.ieee.org
23. Su Z, Ma H, Zhang X, Zhang B (2011) Secure DHCPv6 that uses RSA authentication integrated with self-certified address. In: 2011 Third International Workshop on Cyberspace safety and security (CSS), IEEE, pp 39–44
24. Hinden R (2017) Internet protocol, version 6 (IPv6) specification
25. Troan O, Droms R (2003) IPv6 prefix options for dynamic host configuration protocol (DHCP) version 6
26. Bi J, Wu J, Yao G, Baker F (2015) Source address validation improvement (SAVI) solution for DHCP. RFC 7513, https://doi.org/10.17487/rfc7513, May 2015, http://www.rfc-editor.org/info/rfc7513
27. Jiang S, Shen S (2012) Secure DHCPv6 Using CGAs. Work Prog

A Proposed Framework: Enhanced Automated Duplication Algorithm for Flash Application

Aqilah Rostam and Rosni Abdullah

Abstract Duplicator machine is typically used to improve the data transfer over a shorter time. The process of duplication is done by copying each data bit from the master device to the targeted device including unused memory region. However, to duplicate a 64 GB embedded MultiMedia Card (eMMC) memory is usually very time-consuming which is between 2 and 7 h; plus the speed specification claimed by the vendor is not the same when it is tested during benchmarking process. Therefore, this paper describes a proposed technique to improve eMMC duplication process by adopting the data storage and data transmission concepts such as reducing the data size using compression and sorting technique, where the algorithm needs to be optimized from the sequential level (data structure, data flow, etc.) which includes the significant impact of data type and data size during the duplication process. This study also extends these techniques by using heterogeneous collection of dataset or digital data to serve the true purpose of flash storage.

Keywords Duplication · Data transfer · Sorting · Compression

1 Introduction

Digital data has created convenience to the user every day [1]. Transfer or duplication of digital data is important as data has been an important asset to many industries. Consequently, backup process of digital data has become a very important and crucial process. To speed up the process of transferring data, the duplicator machine has been used by some industries. Moreover, content from the flash application such as USB drive, SD storage, and embedded MultiMedia Card (eMMC) and other flash application can be copied using the duplicator machine. Since flash memory content

A. Rostam (✉)
School of Computer Science, Universiti Sains Malaysia, Gelugor, Penang, Malaysia
e-mail: aqilahrose@gmail.com

R. Abdullah
Sophic Automation Sdn Bhd, Bayan Lepas, Malaysia
e-mail: rosni@cs.usm.my; director@nav6.usm.my

© Springer Nature Singapore Pte Ltd. 2019   377
V. Piuri et al. (eds.), *Intelligent and Interactive Computing*, Lecture Notes in Networks and Systems 67, https://doi.org/10.1007/978-981-13-6031-2_1

can only be reproduced through duplication, flash duplicators make a great accessory to any office or workshop. General duplication process is done by copying each data bit from the master device to the slave including the unused memory region. In addition, to duplicate a 64 GB eMMC memory typically takes up between two to seven hours. Even though specifications provided by vendors claim to have higher average data duplication speed, the duplicators are lower than what it claims in real experience during product analysis and benchmarking process.

Moreover, most of the studies in the literature discuss more on the architecture and limitation of flash such as Flash Translation Layer (FTL) especially on the mapping scheme [2–4] NAND flash chip on how to reduce energy or power consumption during erase operation [5, 6] and how to lengthen the flash storage duration or lifetime [7–9] rather on the way data is transfer and stored during the duplication process. Moreover, FTL has limitation due to "erase before write" characteristic, because no matter how little data is written, any write access to the device requires erasing a whole flash erase unit and worsen the flash performance besides various techniques of FTL are only described in patents but not in technical articles [10, 11]. Other issues such as the quality of the data being stored, delay perceived by the user during data duplication from source to the target destination with a big amount of data have created transmission problem [12]. For example, in case of difficult to send high data volume from source to target destination, the same problem is faced during the duplication process. To reduce data transmission time for a large multimedia data, normally, the size of data is reduced by using data compression.

Common performance measures for an algorithm are speed, memory usage, and power consumption, and other measures really depend on the input size or on the arrangement of data (e.g., sorting) [13]. Therefore, to design a faster duplication algorithm using eMMC as the memory storage, the algorithm needs to be optimized from the sequential level (data structure, data flow, etc.). This triggered us to investigate the correlation between data size, data type, and how it is sort with duplication performance.

2   Background Study

Data transmission is a process of transferring data from a source destination to the targeted destination whether is in analog or digital format and in the form of bits. Data transmission can be divided into serial and parallel transmission where they are differentiated in sending a serial bit or multiple bit at the same time [14]. Data stored in flash can be divided into Single-Level Cell (SLC) and Multilevel Cell (MLC). SLC can store a single bit of information, whereas the MLC can store more bits. SLC flash is used in industry-grade devices due to high performance and stable reliability, whereas MLC flash is often used in consumer products, such as MP3 players, USB flash drives, and flash memory cards [15]. Two major types of flash memory in the current market are NAND and NOR flash. NOR is best used for code storage and execution, usually in small capacities, and it offers eXecute In Place

(XIP) capabilities and high read performance. It is cost-effective in low capacities but suffers from extremely low write and erase performance. On the other hand, NAND architecture offers extremely high cell densities and high capacity combined with fast write and erase rates [16]. NAND obviously offers many advantages in term of duplication performance due to its fast write and erase rates. Due to writing operation is at page level which is smaller than erase operation on block level and have a bigger size than page, all the page will be erased as a block needs to be erased before it can be written [10].

2.1 Flash Translation Layer (FTL)

FTL [17] can be described as a block device interface that emulates the hard disk drives operations using flash memory and helps to translate the logical sector address from the operating systems into the physical flash address used by flash memory or address translation. The address translation can be done by some mapping algorithm, which can be basically classified into page level mapping, block, and hybrid of both level mappings. Because both page and block have limitation, hybrid is expected to overcome both methods. Unfortunately, the hybrid has faced the same problem due to erasing before write characteristics as successfully explained in detail in work by [18–20]. Therefore, rather than focus on the FTL, some improvement can be done by focusing more on the data itself and how data is arranged and transferred during data duplication process across flash devices, for example, data transfer of Universal System Bus (USB). USB is an example of MLC flash memory application. Benefits of studying previous work of USB duplicator will help to understand duplication process for flash technology application especially for environment that uses USB.

2.2 Flash Application: Universal Storage Bus (USB)

The market today provides many accessories to access flash storage. Like any other memory card, eMMC can be accessed by using adapter or any memory card reader. Memory card reader commonly used in Secure Digital (SD), CompactFlash (CF), and MultiMedia Card (MMC) to access the data in the memory [21].

Universal System Bus (USB) is a flash application function to store data [22]. USB duplication algorithm can simply be explained by using signal generation circuit and bit stuffing that has been explained in detail in [23]. Synchronization of received bits is managed by bit stuffing. Basically, different versions of USBs would give different performances. The performance is improved with the use of the latest version of USB. Some flash duplicator provides different versions of USB port interfaces that are compatible with the USB flash drive to prevent bottleneck.

Previous works that involve with the use of USB for duplication were studied. Stand-alone and portable USB duplicator was proposed in [22]. Serial UART of the PIC16f877A was connected with PuTTY using the TTL-232 USB cable where sample data was sent to the PuTTY and received from it. The data rate using UART mode is 12 Mbits/s. On the other hand, the authors of [24] proposed flash USB duplicator which consists of a bridge that introduces USB to USB bridge concept that can carry out mass data transfer everywhere, secure and at high speed. This research concisely has explained how duplication and data transfer of a USB is performed. By transferring a 700 MB file in around 12 s and a 2.1 GB movie file in around 40 s, the result of data rate is 480 Mbps approximately.

Their work is so much closer to us; moreover, it involves the use of USB. Even though it is stand-alone which not requires any PC in the environment, our research will focus more on data transfer improvement; therefore, the duplication approached can be adopted as the basic duplication in this research study. Moreover, their method allows the user to select any files or folders to be copied from the source drive to user selected destination drive with a diversity of file format transfer where they transfer different data types including data with bigger size such as movie.

2.3   Duplication Algorithm Enhancement: Sorting Technique

As explained before, some common measures of performance really depend on the input size or on the arrangement of data (e.g., sorting) [13]. Three sorting algorithms have been discussed and experiment in [25] such as bubble sort, selection sort, and gnome sort. The aim of this comparison between the three algorithms is to know the running time of each one according to the number of input elements. The analysis of these algorithms is based on the same data and on the same computer. The result shows that gnome sort algorithm is best for sorted data, but selection sort outperforms gnome and bubble for unsorted data. As the selection algorithm is easy for implementation and useful for sorting small data as it does not occupy a large space of memory, we decide that it is very suitable to be used in this research study as sorting, and the use of heterogeneous data has not been experimented by most of the research.

2.4   Duplication Algorithm Enhancement: Compression Technique

In general, data duplication works by duplicating bit by bit of data and gave some affect to the overall performance of the duplication process. Logically, reducing the data bit or the data size might shorten the time taken to duplicate, and this process can be done by using the compression method. Compression works by reducing the data bits by eliminating the redundancies in each data, thus reducing the data size. Data compression can not only save space, but it can also save time and optimize resources [26]. Data can be compressed in a lossless or lossy manner.

Lossless compression means that the original data can be reproduced exactly after decompression. In contrast, lossy compression, which often results in much higher compression ratios, can only approximate the original data. This is typically acceptable if the data are meant for human consumption such as audio and video. However, program code and input, medical data, email, and other text generally do not tolerate lossy compression [27]. It is important to cater to all types of data to prevent any misfortune or corruption in the future because the goal and purpose of having data backup is not limited to a certain type of data. Therefore, this research will focus more on lossy data compression as our concern is on the performance because logically, the reduction of data size would quicker the naïve duplication algorithm because lossless will not reduce the data size as much as lossy.

3   Research Framework

Overall methodology that has been conducted consists basic duplication that we called as naive duplication algorithm as it performs general duplication from the source to the target destination. Based on the previous work, we were able to simulate basic duplication algorithm that was adopted from [24]; the use of selection sort and compression technique for improvement.

In this research, the methodology encompasses three major stages: the first stage is to do a background study, literature review and conduct a competitive analysis of existing eMMC duplicators market players especially in terms of speed. Hardware that will be used in the research are selected based on the product analysis where the performance in terms of speed between duplicators is set as the benchmark. In stage 2, the first step is to successfully emulate the duplication algorithm by using PC adapter and then enhance the naive duplication process by hybrid with other algorithms such as sorting and data compression. Lastly, the performance evaluation will be done based on the speed, accuracy, and format diversity. Besides that, the performance of data duplication will experiment between the master and the slaves. Figure 1 illustrates the proposed framework during the experiment where naïve duplication will be enhanced with sorting based on data size and reduce the data size. We implement it stage by stage where until this stage, we manage to perform 1 to 1 without parallel technique to speed up the process more.

Datasets are selected to be representative of the target platform for the studies and framework evaluation. This is to cover the feasibility of different digital data that usually being stored in flash or any memory card. The framework evaluation is conducted with representative of selected data (ranges of file sizes and types). Besides, the research is focused mainly on the roughly 3.2–3.4 GB data file size as the dataset. This is because of the previously benchmarking result of duplicator using 3.4 GB data for testing. Due to the limitation of the device used, the eMMC card reader device is only able to store up until 3.2 GB despite using 4 GB size of data. The division of individual dataset has been set up and adopted by [24] where they use approximately 700 MB file and a 2.1 GB movie file. In this research, the dataset

Fig. 1 Research framework

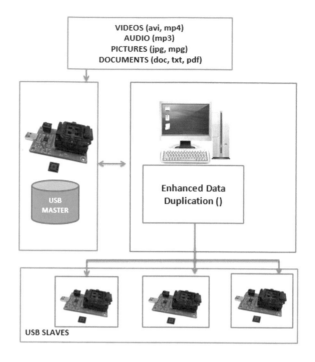

consists of videos, files, image, audio, and normal documents with the same format but different contents were used in the experiments. This research borrows concept brought by USB duplication and video streaming where data transfer which is fast and accurate is needed. This greatly shows that digital data borrows the same concept during transfer. The dataset that will be used in experiments will be divided into the first and second rounds, where for the first round, the same set of the data type being duplicated separately. This is to observe the movement of different data types with the same size across local PC and from the devices during naive duplication process because different data characteristics would give different performances [27] prior to the compression process. All categories of dataset would be experimented one by one according to the data type. Therefore, for the first experiment, the dataset that is used is roughly 800 MB taken from each data type that gives a total of 3.2 GB and 800 MB that mix all the data types altogether which roughly takes 200 MB from each data type for the second round as clearly illustrated in Tables 1 and 2.

By conducting experiments, the process would enable this research to analyze problems and to prove some concept as more gaps to be found and to fill in the gap with an effective solution. For validation and observation, three experiments will be conducted. Each experiment has different purposes. The goal of experiment 1 is to observe different characteristics of different data types during naive duplication, especially when we only duplicate the same group of data type and mix different data types during duplication. In experiment 2, the goal is to improve the naïve duplication process with sorting algorithm from the previous experiment. Other than

Table 1 Dataset division for the first round

Data	Type	File size (MB)
Documents	MS Word, PDF, Excel, Text file	800
Videos	Avi, mp4, mkv	800
Pictures	Jpg, mpeg	800
Audio	mp3, mp4	800

Table 2 Dataset division for the second round

Data	Type	File size (MB)
Mixture of all data type in one folder (doc, vid, image, and audio)	MS Word, pdf, Excel, Text File	200
	Avi, mp4, mkv	200
	Jpg, mpeg	200
	mp3, mp4	200

that, in experiment 3, the implementation of basic compression algorithm will be done to observe the difference between compressed and uncompressed data from all the experiments using the division dataset and later to perform data compression during data duplication process. The benchmark of this research has been done based on human judgment and experiment of the duplication process duration of the existing duplicator. This is to study and investigate data characteristic, when it is sorted and the data size is reduced, therefore to observe its impact during the duplication process. All the experiments, especially in 2 and 3, will be compared with naive duplication that has been performed in experiment 1. The experiments are conducted throughout this research were implemented using the high-level language of C# programming language resides in Visual Studio, which is run on an Intel® Core™ i7 CPU 1.73 GHz with 4.00 GB of RAM under 64-bit Windows 7 Professional operating system and the use of USB eMMC Reader as the PC adapter.

4　Preliminary Result and Analysis

For the first experiment, we will record the time taken for the duplication process to complete with and without eMMC reader device to know the overall process time taken. In Table 3, the results and analysis of the results are discussed.

Other than that, as shown in Table 3, time for duplication process for the first round shows that video is the fastest compared to other data even though the size of the dataset is the same and document shows the slowest duplication process. This was the same as our expectation where the same data type would give many differences compare when we combine all the data even though the size is the same. Therefore, after the second round of experiment, for a dataset that is combined, the overall

Table 3 Result from
experiment 1 (Naïve)

Data	Size (MB)	Duration (locally) (hh:mm:ss:ms)	Duration (with device) (hh:mm:ss:ms)
Documents	800	00:00:22:92	00:03:36:31
Pictures	800	00:00:09:45	00:02:27:94
Videos	800	00:00:09:45	00:02:27:94
Audio	800	00:00:15:59	00:02:28:30
Overall	800	00:00:18:50	00:02:31:50

Table 4 Result from
experiment 2 after sorting

Data	Size (MB)	Duration (locally) (hh:mm:ss:ms)	Duration (with device) (hh:mm:ss:ms)
Documents	800	00:00:19:12	00:02:40:50
Pictures	800	00:00:15:71	00:02:29:97
Video	800	00:00:21:61	00:02:37:32
Audio	800	00:00:00:74	00:02:29:74
Overall	800	00:00:00:96	00:02:31:51

process took 18 s and 50 ms. Next, we experimented our dataset across USB device; we are aware that there might have a slight change in the performance due the use of USB reader compares to perform duplication locally due to flash characteristic, transportation cost, device overhead, etc. Even though there is a big difference, the delay is much worsened during the process, but the performance is much better when we compare with the benchmark of the duplicator performance; Dediprog which record 11 min at worst case to duplicate the data. Plus, the performance would not be the same due to the version of devices that were used. USB interface that was used for the experiment is a 2.0 version. Therefore, we assume that the performance might improve if we used the latest and more advanced version such as USB 3.0 and high-end performance computer.

A long flash mounting time was observed when a USB flash device is plugged in. We believe that this is complementary with research done in [28], which stated there is a process taken by a flash memory to scan for a free space in flash, and this delays the flash mounting time significantly. The result after applying selection sort is shown in Table 4.

It is proven that the sorting algorithm helps to speed up the duplication process. Then we apply the same method using devices where we try to duplicate data from PC to USB devices. Even though there were not so many differences compare to the previous result, sorting method still have an impact on some data and reduce the duration time for the duplication process when the device was used. Overall, by using the sorting algorithm, the result shows that this method able to reduce the time taken to duplicate all the data separately except for videos and pictures. Previously in experiment 1, videos show the fastest among the data during the duplication process

Table 5 Result of data size reduction for an individual dataset

Category	Doc	Picture	Video	Audio
Before compression (MB)	800	800	800	800
Before compression (bytes)	839,932,668	839,289,135	839,170,434	839,045,636
After compression (MB)	731	796	800	774
After compression (bytes)	767,557,790	835,319,222	838,643,175	811,981,701

(locally) but the performance worsens after being sorted. Surprisingly, it takes a long time compared to other data which gives a reverse result when it has been sorted. The same goes for pictures, the delay is increased from 19 to 21 s longer. It is assumed that sorting might not give satisfactory results for some data type such as video and pictures and due to flash characteristics. Overall, it has a big impact on document, and the combination of all datasets shows a big improvement after sorting.

Next, with the use of flash devices, surprisingly, the results are the same as before sorting except for document that gives a big difference. Able to reduce by 1 min is also an achievement and assume if we backup data that consist of a lot of document this can be a big contribution. This concludes that sorting might help to improve for certain type of data, and it is performed locally but not across devices. We assume that flash transportation cost and sorting might increase overhead and therefore increase the delaying process. Because of this result, the urge to try other sorting techniques that are suitable might help to speed up flash performance during duplication increase. In the future, we will explore another sorting method to the dataset and observe the performance. In experiment 3, we manage to perform reduction of data size until this stage. This work will not stop here but will continue to adopt compression with sorting and parallel method in a later stage. Table 5 shows the result of the data size reduction.

Referring to the table above, we have shown that the previous benchmark of our dataset is 800 MB; each data would have the same number of byte size. Unfortunately, it is not the same when we check the file system properties. This is because, in the file system, each data shows different bytes. Therefore, we decided to consider the bytes of our individual data in the table above for our next observation to ease in tracking the differences.

The analysis of the data is shown after we had successfully compressed each individual categories of our dataset. It is shown in the table that despite the same amount of data which are 800 MB, a different byte is shown. After compression has been done, the method successfully reduces the size roughly from 69 MB. The measurement of the data either is measured by MB and bytes. Except for videos category, it surprisingly maintains the size of data in MB but the size in bytes was reduced. Due to this, we are ought to maintain the measurement in bytes because it will help to track the differences. Therefore, overall, compression method thus helps in reducing digital data size up to 2% and more. Even though it seems like a small percentage, successfully reducing the size up to 69 MB is quite an achievement for an improvement. This is because as we previously mentioned, due to the size of data, the bigger data size would slower the performance during the duplication process. Therefore, the performance later expected to be faster due to the reduction of data size and would improve more with the parallel method.

5 Conclusion

The proposed framework uses compression and sorting to improve naive duplication method. This paper only presents the basic proposed framework and some of the improvements as a preliminary result. For overall framework, we will include parallel method at later stage, but up to this stage, we manage to do till the reduction of data, and it shows some positive results. Logically, the reduction of data size will improve the performance more. Due to some performance greatly depending on the data structure, for example, on how the data are stored, we came across the sorting technique. Selection sorting shows some improvement on certain data even though there is an increase in delay when using the device, due to the characteristic of flash that is unavoidable. It is proven that sorting works for our environment during the duplication process and is really affected by the data type and data size. Moreover, compression works differently to different data types. As mentioned above, if time is not our limitation in this study, we will try to explore more sorting algorithm and compression strategy combining with the parallel method in the future.

Acknowledgements This work was supported by Collaborative Research in Engineering, Science and Technology (CREST) and Sophic Automation Sdn Bhd under P13C2-2014 Grant.

References

1. Schoenherr E, Steven (5 May 2004) The digital revolution. Archived from the original on October 2008
2. Chiao ML, Chang DW HybridLog: an efficient hybrid-mapped flash translation layer for modern NAND flash memory
3. Zhang C, Wang Y, Wang T, Chen R, Liu D, Shao Z (2014) Deterministic crash recovery for NAND flash based storage systems. In: Proceedings of design automation conference, Dresden, Germany, 24–28 March 2014, ACM, Dresden, Germany, pp 1–6
4. Xie W, Chen Y, Roth PC (2016) ASA-FTL: an adaptive separation aware flash translation layer for solid state drives. Parallel Comput https://doi.org/10.1016/j.parco.2016.10.006
5. Shim H, Kim JS, Maeng S (2012) BEST: best-effort energy saving techniques for NAND flash-based hybrid storage. IEEE Trans Consum Electron 58(3)
6. Tseng HW, Grupp LM, Swanson S (2013, May) Underpowering NAND flash: profits and perils. In: Proceedings of the 50th annual design automation conference. ACM, New York, p 162
7. Jimenez X, Novo D, Ienne P (2014, February) Wear unleveling: improving NAND flash lifetime by balancing page endurance. In: FAST vol 14, pp 47–59
8. Luo Y, Cai Y, Ghose S, Choi J, Mutlu O (2015, May) WARM: improving NAND flash memory lifetime with write-hotness aware retention management. In: 2015 31st symposium on mass storage systems and technologies (MSST). IEEE, New York, pp 1–14

9. Jeong J, Song Y, Hahn SS, Lee S, Kim J (2017) Dynamic erase voltage and time scaling for extending lifetime of NAND flash-based SSDs. IEEE Trans Comput 66(4):616
10. Gal E, Toledo S (2005) Algorithms and data structures for flash memories. ACM Comput Surv (CSUR) 37(2):138–163
11. Parthey D (2007) Analyzing real-time behavior of flash memories. B. Smith
12. Longley PETAL (2001) Geographic information systems and science. Wiley, Chichester
13. Kammer F Algorithm design under consideration of different resources and a faulty world
14. Retrieved   https://www.quantil.com/content-delivery-insights/content-acceleration/data-transmission/
15. Micheloni R, Marelli A, Commodaro S (2010) NAND overview: from memory to systems. In: Inside NAND flash memories. Springer Netherlands, pp 19–53
16. Iniewski K (Ed) (2010) CMOS processors and memories. Springer Science & Business Media
17. Na GJ, Moon BK, Lee SW (2011) IPL B+ -tree for Flash Memory Database Systems. J Inform Sci Eng 27:111–127
18. Kale MN, Jahagirdar AS (2012) An innovative algorithm for flash memory. Int J Comput Sci Inf Technol (IJCSIT) 3(3):4371–4376
19. Suh YK, Moon B, Efrat A, Kim JS, Lee SW (2012) Extent mapping scheme for flash memory devices. In: 2012 IEEE 20th international symposium on modeling, analysis & simulation of computer and telecommunication systems (MASCOTS). IEEE, New York, pp 331–338
20. Lee S, Shin D, Kim YJ, Kim J (2008) LAST: locality-aware sector translation for NAND flash memory-based storage systems. ACM SIGOPS Oper Syst Rev 42(6):36–42
21. Lin T (2016) U.S. Patent No 6654841B2. U.S. Patent and Trademark Office, Washington, DC
22. Rashid MU, Khan ZH (2012) Design and implementation of a low-cost USB duplicator. Bull Electr Eng Inform 1(4):263–278
23. Intel microprocessors: 8086/8088, 80186/80188, 80286, 80386, 80486 Pentium, Pentium Pro Processor, Pentium II, Pentium 4 and Core2 WITH 64-bit extensions Architecture, Programming and Interfacing, Eight Edition Barry B. Brey
24. Sawant T, Parekh B, Shah N (2013, December) Computer independent USB to USB data transfer bridge. In: 2013 6th international conference on emerging trends in engineering and technology (ICETET). IEEE, New York, pp 40–45
25. Hammad J (2015) A comparative study between various sorting algorithms. Int J Comput Sci Netw Secur (IJCSNS) 15(3)
26. Null L, Lobur J (2014) The essentials of computer organization and architecture. Jones & Bartlett
27. Dzhagaryan A (2016) A framework for optimizing data transfers between edge devices and the cloud using compression utilities
28. Yim KS, Kim J, Koh K (2005, March) A fast start-up technique for flash memory based computing systems. In: Proceedings of the 2005 ACM symposium on applied computing. ACM, New York, pp 843–849

A Design of Automated Surgical Illumination System

Kee Soek Fuan, A. R. Syafeeza, Mohd Fitri Alif Mohd Kasai
and Saleha Mohamad Saleh

Abstract Surgical lights which consist of a single- or multiple-light head assembly attached to a suspension arm are designed to illuminate surgical site. However, it was found out that the need of nurse or surgeon to move the surgical light manually might obstruct the surgical flow and cause contamination of instruments. Thus, this project aims to design an illumination system that will automatically track the movement of the hand of surgeon with specific color of glove. Yet, the project will mainly focus on the automation movement of light with the tracking of hand glove, not related to the specifications of light, operating table, operating room condition and design of suspension arm. An algorithm that could track the surgeon's hand movement is designed using image processing. The project will also include designing a surgical light which consists of a single light head attached with camera to illuminate surgical site. Raspberry Pi is used with the Pi camera to track the movement of the hand of surgeon through the program called OpenCV and the programming language, Python. At the end of this project, the algorithm of color object tracking and the prototype of automation surgical illumination system will be produced. The significant finding of this study is in terms of the capturing process of camera at each of the eight positions of the prototype after each movement of the light.

Keywords Surgical illumination system · Surgical light · Object color tracking · Raspberry Pi

1 Introduction

Surgery is an important medical procedure carried out in operating room (OR) to treat injuries or diseases by incision, especially with equipment and instrument. There are many types of surgeries such as elective surgery, cosmetic surgery, reconstructive

K. S. Fuan · A. R. Syafeeza (✉) · M. F. A. M. Kasai · S. M. Saleh
Fakulti Kejuruteraan Elektronik dan Kejuruteraan Komputer, Center for Robotics and Industrial Automation (CeRIA), Universiti Teknikal Malaysia Melaka, 75450 Durian Tunggal, Melaka, Malaysia
e-mail: syafeeza@utem.edu.my

© Springer Nature Singapore Pte Ltd. 2019
V. Piuri et al. (eds.), *Intelligent and Interactive Computing*, Lecture Notes in Networks and Systems 67, https://doi.org/10.1007/978-981-13-6031-2_15

surgery, transplant surgery, and more. The operating room is a specialized environment, which is strict adherence to standards and guidelines of practice [1]. Every equipment, supply, and instrument use in the operating room must be accurate and differentiate depending on the level of surgery complexity. OR must be spacious enough to accommodate all the facilities such as the illumination system, operating table, anesthesia machine, and more [2]. Thus, the recommended size of OR is 6.5 m × 6.5 m × 3.5 m [3]. As one of the basic requirements of OR, illumination system is used to provide high light intensity and true color illumination of a wound. The light must illuminate the surgical site constantly even though the head or hands of surgeon are in between the surgical site and light source [4]. The system can be adjusted in different aspects such as color temperature, color rendering index, light spot, and illumination. However, in the real situation of surgery, the surgeon faces many inconveniences such as inadequate lighting and shadowing effect. The need to reposition the illumination system manually is also a major distortion to the surgeon. In the study carried out by [5] with 98 OR staff members, it was found out that the need for repositioning the illumination system is high and that repositioning is cumbersome.

The surgeon may sometimes need to operate the light by releasing one hand from the operation area. If the illumination system is controlled by nurse, communication is necessary to locate the light spot. Both methods may distract surgeon and his or her attention might be lost on the operation. Consequently, the period of surgery and anesthesia duration of patient is prolonged which is not encouraged. It is essential to have an automated surgical illumination system that applying the method of color tracking to detect the color of surgeon's glove to alleviate the problem.

2 Background Studies

The illumination system applied in [6] was voice recognition to convert the selection commands into corresponding selection signals and controlled signals. Incorrect process of recognizing voice may occur when there are more than one staff in the operation room. In [7], robot and ultrasonic sensors were used to locate the position of the surgeon's head. However, interference of ultrasonic may occur during surgery which affects the performance of automation illumination system. In [8], the intelligent lighting system used light intensity sensors, Inertial Measurement Unit (IMU), and Light-Dependent Resistors (LDR). This method tracks the position of surgeon's hand accurately as IMU and LDR were used to take the position of the sensor panel. The light intensity can be adjusted to avoid burning of illumination on surgical site. However, this method is not suitable for all types of surgeries. The surgeon has to obey the rule: "bare below the elbow" in which the surgeon is not allowed to wear anything below the elbow to avoid any contamination. Based on these reviews, the most practical method is based on color tracking approach which refers to the design in [8].

3 Methodology

3.1 Software Development

In general, the algorithm for this study is based on these eight steps as shown in Fig. 1. In this project, the method of image processing is applied to track the specific color of surgeon's glove before converting the image from BGR to HSV. Mask is created to filter the image with the required color. The morphological operation, namely, opening, is applied to eliminate the dots appearing randomly in the frame to reduce the noise. The operation of bitwise is used to compare the mask and the original image to show the filter part in the specific color. Lastly, the required part with the specific color will be bounded in a box.

3.2 Hardware Development

The connection of components is shown in Fig. 2. The duty cycle of desire angle can be calculated using Formula (1).

$$(\text{Desired Angle})/18 + 2.5 = \text{Duty Cycle} \tag{1}$$

Step 1: Start
Step 2: Image Identification to camera
Step 3: Image processing in OpenCV
Step 4: Convert RGB into HSV color space
Step 5: Filtering the mask
Step 6: Operation of bitwise_and
Step 7: Tracking the object and control the motor
Step 8: End

Fig. 1 Algorithm of surgical illumination system

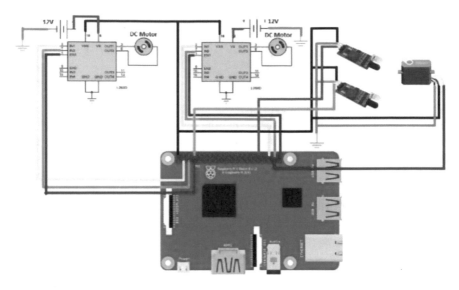

Fig. 2 Connections of components with Raspberry Pi

4   Result and Discussion

4.1   *Operation of the System*

When the system is started, the light will move downward automatically. Surgeon can determine the distance of the patient to the light and stop the system from moving downward by putting the hand just below the light as shown in Fig. 3a. This step is important to protect the patient from burn risk. Next, the system will track the color glove. In Fig. 4a, the yellow dot at pixel 320,240 is the center of the frame, while red dot is the center of color tracking. If there is more than one hand appearing in the frame, the system will bound both of the hands in green box (Fig. 4b).

By comparing the red and yellow dots in x-axis, the motor will decide either moving the light to the left or right direction (Fig. 2b). If the difference of the dots is in between 20 pixels, the motor will stop rotating. Next, the system will compare the dots in y-axis. If the pixel of red dot in y-axis is less than 80 pixels, the servomotor will rotate the light to 72°. When the red dot is more than 400 pixels, the servomotor will rotate to 99° (Fig. 3c). After the rotation of servomotor, the system will compare again the dots in x-axis to ensure the accuracy of light position. The motor will move the light upward to the standby position and servomotor will rotate to 90° (Fig. 3d).

(a)   (b)

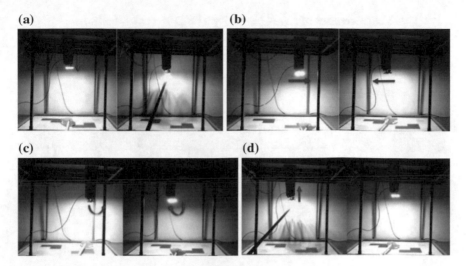

(c)   (d)

Fig. 3 Movement of motor when system is started

(a)   (b)

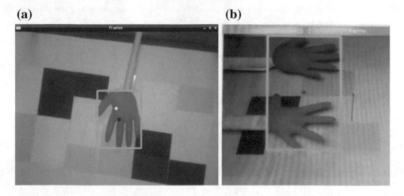

Fig. 4 Result of color tracking

4.2 Result Analysis

In order to analyze the performance of the system, light sensor module is used to calculate the difference of light intensity before and after the movement of light. The measurement of light intensity is taken at eight different locations of the prototype.

Based on the graph in Fig. 5, it was found out that the increment of the light is not constant as some of the points may get up to 500%, while others may increase only 130%. At point 2 and 5, the percentage of increment is the highest as the light is illuminating straight to the points. The servomotor maintains at 90° and the light is only moving in x-axis. Thus, the rotation of the servomotor is the major reason which influences the accuracy of the light's position. In this system, the servomotor is set to rotate in two angles only which are 72° and 99°. However, these two angles are

(a) **Measurement of Light Intensity at 8 Different Locations**

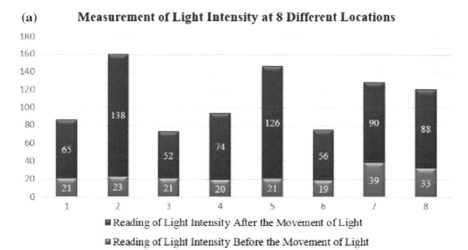

■ Reading of Light Intensity After the Movement of Light

■ Reading of Light Intensity Before the Movement of Light

(b) **Increment of Light Intensity**

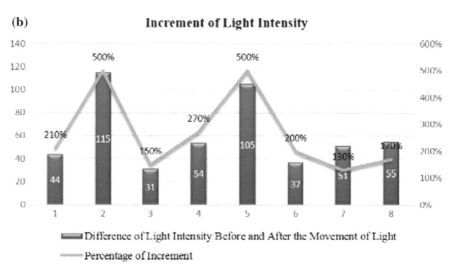

▬▬▬ Difference of Light Intensity Before and After the Movement of Light

▬▬▬ Percentage of Increment

Fig. 5 Graph of light intensity's measurement

not enough to cope with the movement of hands at different points. Besides, when the light is rotating to specific angles, the displacement of light to the hand increases which affects the value of light intensity.

5 Conclusion

This project has been successfully completed by achieving three objectives. Method of converting RGB to HSV is applied to the system to track the specific color. A prototype had been created to demonstrate the movement of light to the color glove.

The performance of the system is analyzed in the factor light intensity. Although the system is demonstrating well with prototype, the appropriateness of applying the method of color tracking with the surgical illumination system is still unsure. More research and improvements should be done as there are many circumstances and qualifications need to be considered.

Acknowledgements Authors would like to thank Universiti Teknikal Malaysia Melaka (UTeM) and Ministry of Education for supporting this research under PJP/2018/FTK(9D)/S01603.

References

1. Eskola S, Roos M, McCormack B, Slater P, Hahtela N, Suominen T (2016) Workplace culture among operating room nurses. J Nurs Manag 24(6):725–734
2. Burlingame B (2014) Operating room requirements for 2014 and beyond. FGI Guidel Updat Ser 1–6
3. Copenhaver MS et al (2017) Improving operating room and surgical instrumentation efficiency, safety, and communication via the implementation of emergency laparoscopic cholecystectomy and appendectomy conversion case carts. Perioperative Care Oper Room Manage 8:33–37
4. Knulst AJ, Stassen LPS, Grimbergen CA, Dankelman J (2009) Standards and performance indicators for surgical luminaires. Leukos 6(1):37–49
5. Mooijweer R (2011) The surgical lighting problem: manipulation problems with the surgical lighting system during surgical procedures
6. Wang Y, Goleta C, Jordam CS, Santa Barbara C, Uecker DR, Wooters CC (2003) General purpose distributed operating room control system. US 6,642,836 B1
7. Choi DG, Yi BJ, Kim WK (2007) Automation of surgical illumination system using robot and ultrasonic sensor
8. Roshan TAU, Sachinthana LMN, Senarathna PMNK, Jayathilaka WADM, Welgama WPD, Amarasinghe YWR (2016) Design and development of an intelligent lighting system for operation theatres, no. October

A Cloud Computing Framework for Higher Education Institutes in Developing Countries (CCF_HEI_DC)

Ban Salman Shukur, Mohd Khanapi Abd Ghani and M. A. Burhasnuddin

Abstract In order to satisfy the rapid development in ICT, high educational institutes including members who are involved and interested in research's field began to investigate and seek new tools, ways, and technologies to support their mission. On the other hand, cloud computing technology, as a new technology have emerged the education and learning scene lately and the cloud computing service providers (CSPs) are getting more interested in this field to offer their productions and services. Nevertheless, the association of cloud computing within high educational institutes is mostly directed to learning, teaching, and other student's activities area of interest. In most developing countries, there are countable obstacles and problems in their high educational institutes, above-mentioned one before, such as shortness in funds and necessary IT infrastructure and resources which are required to establish, develop and maintain an ICT environment for their students, lecturers and researchers in such institutes. In this paper, we propose a Cloud Computing Framework model for Higher Education Institutes in Developing Countries (CCF_HEI_DC) of six layers which could be implemented and developed in any developing country besides a modified hybrid cloud model which is suggested to host and support the proposed cloud computing framework model.

Keywords Developing countries · Cloud computing · Cloud framework · Cloud model · High education institutes

B. S. Shukur (✉) · M. K. A. Ghani · M. A. Burhasnuddin
Faculty of Information and Communication Technology, Universiti Teknikal
Malaysia Melaka, Durian Tunggal, Malaysia
e-mail: Ban_s_s_daoud@yahoo.com

M. K. A. Ghani
e-mail: khanapi@utem.edu.my

M. A. Burhasnuddin
e-mail: burhanuddin@utem.edu.my

© Springer Nature Singapore Pte Ltd. 2019   397
V. Piuri et al. (eds.), *Intelligent and Interactive Computing*, Lecture Notes in Networks
and Systems 67, https://doi.org/10.1007/978-981-13-6031-2_24

1   Introduction

The power in building and developing the society in any country is the higher education. Information and Communication Technologies (ICT) have emerged and affected
the education field as anything else in today's world. In today's world universities,
colleges and other higher educational institutes are seeking new technological fields,
methods, and instruments to use and adopt in order to satisfy the rapid development in ICT and other needs in their institutes [1]. They began to investigate new
technological solutions and use them for the benefits of their institutes.

Gradually, education, in a way or another becomes associated with information
technology (IT) in terms of communicating and delivering materials.

Cloud computing is relatively a new technology phenomena which enables an on-
demand network access by sharing a pool of computers and resources, has become
the new technology that emerged the education and learning landscape lately [2].

As a result of the growth in the cloud computing services and its involvements
in this area, many high educational institutes have already began in adopting cloud
computing ready services like Microsoft Live@edu which ensures that students can
access Microsoft products from anywhere without having to purchase them besides
providing students with direct access email and office packages among other services.
Google Apps Education (GAE) is another option which holds Google facilities like
Mail, Talk and Docs have similar benefits as those of its equivalent within Microsoft.

Such services have proved to be good practice in enhancing students' online
collaboration and improve their learning experiences [3]. Other cloud computing
services like documentation applications and storage space applications are available
to be used by all high institutes' members [4].

Cloud computing delivered services also influence positively the on online education and face-to-face learning in higher education, in case they were adopted successfully [5].

Besides all abovementioned usages of cloud computing within the higher educational institutes' area, there is the hosting of learning management systems (LMSs),
which could be considered as an evolving example because of the high costs involved
in the creation and maintaining these systems without the cloud's use [4, 8].

Muli and Kimutai mentioned that there are successful adoptions for cloud computing services in the US, the UK, Africa, and Asia's universities. They stated that
in Africa only, there are more than 30 institutes for high education which already
using one or more of cloud computing services [4].

This paper is into seven main sections as follows: Sect. 1 gives an introduction
to issues and information discussed by the paper, while Sect. 2 discusses the main
problems in higher educational institutes in developing countries. Section 3 gives an
idea of literature review for depth knowledge and analyzes strengthens and Sect. 4
discuss cloud computing as a solution for high educational institutes in developing countries. Section 5 provides the proposed framework model CCF_HEI_DC,
describing its main functions, while Sect. 6 discusses the service models and utilities

for the framework. The last section, however, describes a modified model which could be hosted in any developing country to host the proposed framework model.

2 Problems of Higher Education Institutes in Developing Countries

In many developing countries, high educational institutes suffer from poor or limited Information Technology (IT) infrastructure and resources to establish or implement their own necessary ICT requirements for teaching, learning, researching and other educational activities [6–9]. Even big high educational institutes could have lack of funds regarding their scientific researches and shortage in available resources [9].

Karim and Rampersad stated that the studies about examining the factors which affect the adoption of cloud computing in developing countries' universities and pointed out that there is a gap in the practical studies concerning this particular field [10].

Another problem within high educational institutes in developing countries is the laboratories' availability. Truong, Pham, Thoai, and Dustdar mentioned that in developing countries, the facilities of laboratories in different kinds of learning objects are not available out of class hours which themselves could be insufficient for all students and/or all high educational institutes [11].

Masud and Huang pointed out another problem which some high educational academic institutes suffer from in developing countries, that is not having the required expertise in IT field to operate and support their own infrastructure [1].

Within this context, another important issue was mentioned by Kurelovic, as he said clearly that the association of cloud computing within high educational institutes is mostly with e-learning and m-learning. This means that most cloud computing services and applications which are delivered by cloud computing providers for high education institutes are dedicated for students, learning and education purposes not for research requirements [12].

3 Literature Review

Almajalid [8] proposed a hybrid cloud model (Fig. 1) for universities in Saudi Arabia under the supervision of Ministry of Higher Education (MOHE) in Saudi Arabia for the purpose of simplifying the learning in Saudi Arabia universities and colleges in the context of knowledge sharing [8].

The model consists of a public cloud owned by MOHE which contains all data generated or used by all colleges and universities and private cloud and private cloud owned by relevant university or college. Each university and/or college can give its own authorities and privileges for users.

Fig. 1 Hybrid cloud model
proposed by Almajalid [8]

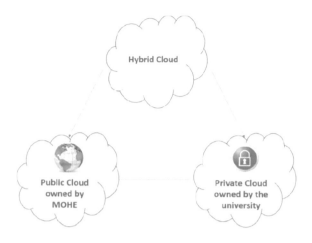

Pardeshi [13] proposed an architecture for higher education institute in India. The architecture composes of the three layers of cloud, IaaS is the first layer followed by PaaS built upon it and lastly the SaaS at the top of this architecture which should provide applications to users by including several functions for presentation and management capabilities as illustrated in Fig. 2. He pointed out that trading off between the three layers which compromises this architecture is made by taking into considering that features are integrated with one layer are integrated also with other layers and this is done by respect of complexity versus security. Development and extensibility are also considered without forgetting the aforementioned featured [13].

Wang et al. [2] said that e-learning methods are no longer sufficient and need to be improved by using cloud computing solutions and they proposed a framework that integrates an existing e-learning contents which has object-based facilities with cloud computing where learners can reuse, share, and access for future knowledge [2].

While Gital et al. [14] proposed a framework for a Collaborative Virtual Environment (CVE) which could be used for educational purposes. The framework comprises of cloud computing infrastructure layer, application layer, virtual environment, and access entries. The researchers claimed that this framework was able to improve the usual CVE effectiveness [14].

On the other hand, Encalada and Castillo [15] presented a model by utilizing "social cloud" to enable the universities in teaching practical skills of IT by implementing ecosystems. The model is based on the idea of providing virtualized labs based on cloud computing by using massive open online courses (MOOCs). Students are able, within this model, to enhance their basic educational pillars which are learning to live together, learning to be, and learning to do with IT training from a MOOCs with the researchers' opinion [15]. The (CVE) model has been experienced and statically evaluated through a course of computer operating systems as researchers' claimed.

Waga et al. [16] suggested a framework of a government cloud computing with a repository of data which could be accessed by any terminal for educational purposes

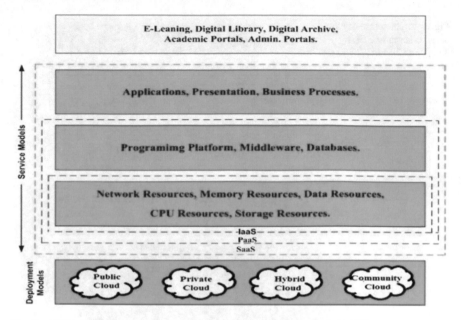

Fig. 2 Architecture for cloud implementation in HE institutes by Pardeshi [13]

so that the researchers and students can collaborate with each other by using this framework and centralized e-learning components with necessary resources. The researchers proposed the idea of using Virtual desktop infrastructure (VDI) to meet the development goals for Kenya's vision 2030 [16].

For e-learning purposes also, Kaur and Chawla [7] proposed a cloud architecture and framework which is specialized in advance java e-learning applications on cloud and they called it the Cloud-based E-Learning (CEL) using Web 2.0 based on cloud to develop, this e-learning application could be used. Figure 3 shows CEL, which consists of three layers which are cloud model layer, service model layer, and learning application model layer [7].

4 A Cloud Computing as a Solution for High Educational Institutes

Any high education institute is interested in improving teaching, learning, and research activities within its own academic institutional' perspectives. Within this context, investigating new learning tools and mechanisms, cooperating with other research institutes and organizations and other researches has become a high interest for every high educational institute. This is to improve the scientific and research levels of their lecturers as well as students.

Fig. 3 Architecture of cloud
based e-learning CEL by
Kaur and Chawla [7]

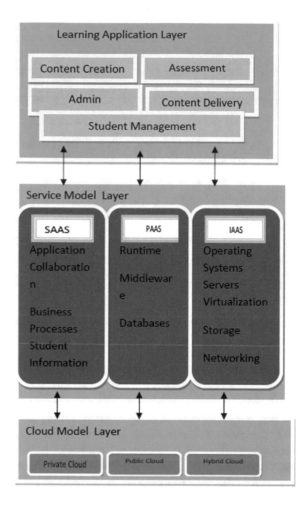

According to that, such institutes require the use of IT infrastructure to provide a groundwork for the abovementioned context, which in turn costs a lot of money and requires specialized IT people to implement, develop and maintain such projects.

A large number of educational services have already migrated to be available online for students and high educational institutes and started to be largely used because of their noticeable benefits in helping them to achieve their goals with suitable cost such as cloud computing which can successfully decrease the investment cost for required IT infrastructure [13].

Kurelovic mentioned that there are countable educational institutes around the world which are already using cloud computing services to store data. At the same time, many of these institutes have already adopted cloud computing solutions for their students' basic requirements such as email [12].

Indeed, cloud computing, as a new concept has created new ways and established new paths to think about software, hardware, services, resources, and infrastructure in the new world.

Although benefits of using cloud computing within high institutes are visible and most of them are immediate like creating a collaborative environment and sharing information between students, still moving to a cloud computing technology in the developing countries for the high educational institutes is a big decision to make. A decision which should be investigated and studied carefully to be supported on factual bases before migration steps happened.

All involved stakeholders in the educational process should be aware of cloud computing services and solutions so that to be a part of the decision-making strategy to adopt and migrate towards the cloud computing solutions [12].

Stein, Ware, Laboy, and Schaffer said that education efficiency and quality could be increased with the use of cloud computing [17].

Vanjulavalli agreed that cloud computing can afford amazing application for high educational institutes, but he stated that it is more beneficial to use its application for e-learning and educational purposes [18].

Today, in developed countries, the adoption and use of cloud computing in high educational institutes is a reality. Many institutes and students are already common with most of its services and using it widely. Mircea and Andreescu investigated higher education IT leaders and concluded that about 70% recommend using cloud computing rather than traditional computer technologies [19].

In developing countries, because of financial constraints, most universities suffer from recourses' limitation, lack in IT infrastructure and few IT experts [6, 8, 9].

Usually, in higher educational institutes, teaching is the centric importance matter, while researches' field comes in second place although it is encouraged.

Lecturers, researchers, and other involved people in researching's field are working in fragmented and small groups and are unable to communicate easily with other researchers outside the academic comity very easily. Small research groups could be founded but with no coordination between them.

Waga, Makori, and Rabah said that, in Kenya, as an example, and because of the high cost, researchers are working in isolation with lower novelty [16].

Even when there are investments in the developing countries regarding high educational institutes, it is mostly oriented towards teaching and learning activities, they are rather far from research considerations.

Fragmented, on-demand, not connected and unorganized groups may be working at the same subject or field without being aware of each other and not able to benefit from each other.

5 The Proposed Cloud Computing Framework for Higher Education Institutes in Developing Countries (CCF_HEI_DC)

The proposed framework model (Fig. 4) consists of six layers, each of which consists of its own components to suit the high educational institutes for research purposes and researchers' cooperation which should be ready to use by relevant users of the cloud within high educational institutes. The layers incorporate the cloud's services (SaaS, PaaS, IaaS, and DaaS). DaaS which was added to the model as a revolutionary new added service to the cloud's three known services is able to provide raw data as well as generated data in an efficient manner for users.

As each layer consists of its own components; management, security, and monitoring are also available within all layers to ensure quality of service of the cloud.

Layers could be described as follows:

1. User interface layer: users can login to high educational institute cloud by using various enabled devices such as (smartphones, iPad, laptop, desktop) supported by some needed applications and protocols such as, Hypertext Transfer Protocol (Http/Https), Lightweight Directory Access Protocol (LDAP) and Remote Desktop Protocol (RDP).

2. Software application layer: this layer provides the SaaS to end users. A licensed third-party software is installed in the high educational institute cloud and such software is provided to the researchers and lecturers as a service as desired from Cloud Service Providers (CSP). Web 2.0 tools should be available here and the session management is used as an application in this layer to arrange sessions between relevant researchers to discuss a matter of interest. Identification application for registry and user account management applications are also should be available within this layer.

3. The platform Layer: this layer provides a platform to the end users of the high educational institutes cloud to develop, design, test and deploy their applications and managements. This is done with the help of applications such as the Software Developing Kit (sdk) and Integrated Development Environments (IDE's). It also includes a platform to design the arrangement for team collaboration and session management which is necessary for researchers, lecturers, and other involved members in developing and working on academic researches in order to have the required cooperation between them towards achieving their targets in research field.

   Data management and integration is also needed to be available in this layer to maintain database in the cloud besides library. Scheduling algorithms also needed to manage available resources for end users.

4. Virtualization Resource Management Layer: this layer should be capable of managing and maintaining the resources of the virtualized high educational institutes' infrastructure including their components which participants in the cloud. All components such as live migration, templates, resource's scheduler, recovery

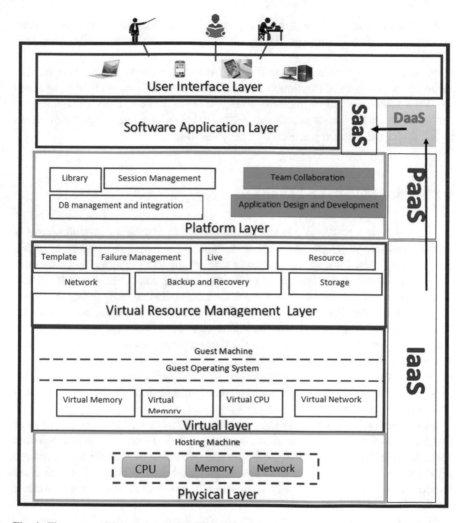

Fig. 4 The proposed framework (CCF_HEI_DC)

and backup manager, and all other resources which should be available for participants in the high educational institutes cloud.

5. Virtual Layer: A number of virtual machines (VM) are created here, each of which has its own CPU, Memory, Storage and Network. Above the VM, the Operating System (OS) is placed and acts as a guest OS within a quest machine.

6. Physical Layer: The actual physical devices, actual data centers which is the host machine exist within this layer.

Layers are integrated with the cloud service model (SaaS, DaaS, Paas, and Iaas).

For each faculty, college, researcher, research scholar, and lecturer their must be an Identification Document (ID) and password to login to the (CCM_HEI_DC) cloud.

Each user should have an authorization level to use the cloud, that is to be able for using one or more of the cloud's functions and applications, such as using ready applied services, develop a platform, arrange a session of common interest, upload new software or any available functions within the cloud, the user must be authorized to do so, and this service is within the user's authority.

6 Service Models and Utilities

DaaS: (Data as a Service) means that the service is capable of providing raw or restructured data which could be understandable by machine and will be available to different applications, system and users who are involved in using online services and various applications unrelatedly of their belonging to organizations. It enables businesses to have remote access to database beside reading and writing data [20]. Figure 5 shows the DaaS correlation within other cloud computing services. DaaS can provide the many benefits for the higher education institutes within the use of cloud computing because it goes beyond other services provided by the cloud [21]. Accessing a wide range of external data sources can bring positively affects for institutes; especially for developing countries which suffer from a lot of difficulties to acquire information and cooperation between institutes to share the same or related data and information.

SaaS: (Software as a Service) here the researchers and lecturers can use some of the cloud's on-demand ready software according to their needs. Educational software could be included as well. Office packages and other free/paid application software provided by the cloud which need not be installed on the physical machine are provided as a service to consumers. It provides the ready-made application and their capabilities to end users.

Besides that, it provides different options which could be customized as required to end users by using important known protocols, such as Representational State Transfer (REST) and Simple Object Access Protocol (SOAP).

PaaS: (Platform as a Service) the high educational institutes' authorized members, researchers and lecturers within the cloud can use the application development platforms like software developing kit (sdk) and software applications like the integrated development environment (IDE). Data is stored to develop, test, deploy, host and maintain applications by developers in java and PHP on demand. It provides a programming environment for its users to develop and build their applications on a cloud platform.

By means of multitenancy, their own applications can also be shared with their colleagues, other researchers, lecturers, and faculties.

IaaS: (Infrastructure as a Service) which provides and manage the infrastructure such as storage, memory, network to users on demand. It should be able to manage the

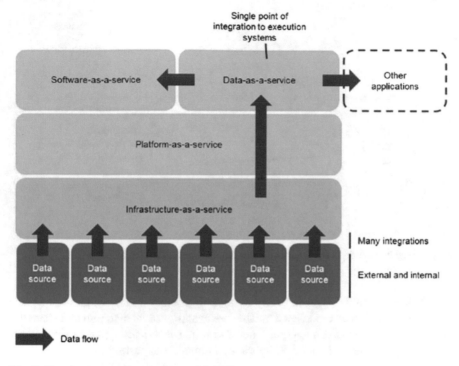

Fig. 5 The cloud computing service models [21]

infrastructure with different forms of users' interaction within the cloud environment. It should contain application programming interfaces (APIs) to allow and control this management to happen [22].

7 A Modified Hybrid Cloud Model Suggestion

The proposed cloud computing framework model (Fig. 4) was proposed in the first place to be specialized and exclusively used by faculties, universities, lecturers, and other academic community members for scientific and researches' objectives. For that, and for one country consideration, a community cloud resident within this country would be suitable.

Still, and as expected users of such cloud will require to access and use other necessary information, applications, data and other requirements outside these specifications which are delivered only via a public cloud by CSPs.

For that Almajalid [8] suggestion to adopt a hybrid cloud (private and public) in Saudi Arabia, as a developing country, under the supervision of Ministry of Higher Education (MOHE) there (Fig. 1) [8].

This model could be expanded and modified to include other developing countries.

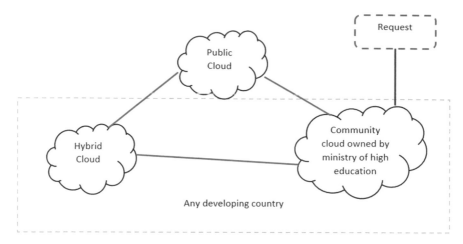

Fig. 6 Modified hybrid cloud for higher education institutes in developing countries

The model could be hosted in any developing country and for other high educational institutes which are located outside the country and wish to join and share the facilities of this cloud can apply a request to join, share, and be part of this hybrid cloud. Figure 6 shows a modified hybrid cloud model suggestion.

8 Conclusion

In this paper, a cloud framework was proposed for higher educational institutes in developing countries in an innovation perspective by including the DaaS cloud service so that raw and structured data could be available for researchers, lecturers and other concerning people in the field of research can get the benefit from this service besides other services provided by the cloud. In this sense, this framework could be established in any developing country as a base to be expanded after that to include other countries. A modified hybrid cloud model was suggested also to support the framework model.

References

1. Masud AH, Huang X (2012) A novel approach for adopting cloud-based e-learning system. In: Proceedings—2012 IEEE/ACIS 11th international conference on computer and information science, ICIS 2012, pp 37–42
2. Wang L, Lau SH, Lew SL, Leow MC (2015) Proposed object-based e-learning framework embracing cloud computing. In: International conference on e-commerce (ICoEC 2015), pp 8–13

3. Lakshminarayanan R, Kumar B, Raju M (2014) Cloud computing benefits for educational institutions. In: Second international conference of the omani society for educational technology, vol 8, No 2, pp 104–112
4. Muli E, Kimuai J (2015) Adoption of cloud computing for education in kenyan universities: challenges and opportunities. Int J Adv Res Comput Eng Technol (IJARCET) 4(6)
5. Alharthi A, Yahya F, Walters RJ, Wills GB (2015). An overview of cloud services adoption challenges in higher education institutions. In: Proceedings of the 2nd international workshop on emerging software as a service and analytics. SCITEPRESS—Science and Technology Publications, pp 102–109
6. Alghali M, Najwa HMA, Roesnita I (2014) Challenges and benefits of implementing cloud based e-learning in developing countries. In: Proceedings of the social science research ICSSR, vol 2014, pp 1–9
7. Kaur G, Chawla S (2014) Cloud e learning application : architecture and framework. In: SSRG international journal of computer science and engineering (SSRG-IJCSE), 1 June, pp 1–5
8. Almajalid RM (2017) A survey on the adoption of cloud computing in education sector. In: CoRR, pp 1–12
9. Odeh M, Perez G, Warwick A (2017) Cloud computing adoption at higher education institutions in developing countries: a qualitative investigation of main enablers and barriers. Int J Inf Educ Technol 7(12):921–927
10. Karim F, Rampersad G (2017) Factors affecting the adoption of cloud computing in Saudi Arabian universities. Comput Inf Sci 10(2):109
11. Truong HL, Pham T-V, Thoai N, Dustdar S (2012) Cloud computing for education and research in developing countries. Cloud computing for teaching and learning: strategies for design and implementation
12. Kurelovic EK (2013) Cloud computing in education and student's needs. MIPRO
13. Pardeshi V (2014) Architecture and adoption model for cloud in higher education: Indian perspective. Bonfring Int J Ind Eng Manage Sci 4(2):43–47
14. Gital AY, Ismail AS, Chen M, Chiroma H (2014) A framework for the design of cloud based collaborative virtual environment architecture. In: Lecture notes in engineering and computer science, vol 2209
15. Encalada WL, Castillo Sequera JL (2017) Model to implement virtual computing labs via cloud computing services. Symmetry 9(7):117
16. Waga D, Makori E, Rabah K (2014) utilization of cloud computing in education and research to the attainment of millennium development goals and vision 2030 in Kenya. Univers J Educ Res 2(2):193–199
17. Stein S, Ware J, Laboy J, Schaffer HE (2013) Improving K-12 pedagogy via a cloud designed for education. Int J Inf Manage 33(1):235–241
18. Vanjulavalli N (2015) An effective tool for cloud based e-learning architecture 6(4):3922–3924
19. Mircea M, Andreescu AI (2011) Using cloud computing in higher education: A strategy to improve agility in the current financial crisis. Commun IBIMA 2011:1–15
20. Rida B, Ahmed E (2015) Multiview SOA: extending SOA using a private cloud computing as SaaS and DaaS. In: Proceedings of 2015 international conference on cloud computing technologies and applications, CloudTech 2015. Institute of Electrical and Electronics Engineers Inc.
21. Pringle T (2014) Data-as-a-service: the next step in the as-a-service journey, p 10
22. Brunette G, Mogull R et al (2009) Security guidance for critical areas of focus in cloud computing V2.1, cloud security alliance, pp 1–70

Wireless Wearable for Sign Language Translator Device with Android-Based App

Tan Ching Phing, Radzi Ambar, Chew Chang Choon
and Mohd Helmy Abd Wahab

Abstract Sign language translator device is an assistive device that helps hearing impaired and normal people communicate with each other. The device can be either a camera vision-based or wearable glove-based system that translates hand gestures into text or voice. However, vision-based devices are susceptible to low light conditions and expensive. In contrast, glove-based devices are popular and low cost, but many designs utilized cable system that limits hand motions. The main focus of this work is to develop a wireless wearable sign language translator device, so that free-hand motions can be achieved. An Android-based smartphone application (app) has been developed to receive and display translated gestures from the wearable device in text form. The usage of smartphone app significantly reduces development cost, simplifies the design and improves the ease-of-use for hearing impaired communities. In this article, the developed hardware, circuit diagrams, app development as well as experimental results are presented.

Keywords Wearable device · Sign language translator · Accelerometer · Flex sensor · Android-based app

1 Introduction

Sign language plays an important role in deaf person's daily life. It has become a main communication method for them to communicate with others. For people that do not understand sign language, they can hire sign language interpreter to help them translate sign language to verbal language when they communicate with the deaf person. Sign language interpreter is a person that is able to translate sign language to spoken word and vice versa. However, deaf persons may find it difficult to find an interpreter that is able to fully translate their messages especially those with complex meanings and personal feelings. Furthermore, there are less than 100 certified sign

T. C. Phing · R. Ambar (✉) · C. C. Choon · M. H. A. Wahab
Department of Computer Engineering, Faculty of Electrical and Electronic Engineering,
Universiti Tun Hussein Onn Malaysia, Batu Pahat, Johor, Malaysia
e-mail: aradzi@uthm.edu.my

© Springer Nature Singapore Pte Ltd. 2019
V. Piuri et al. (eds.), *Intelligent and Interactive Computing*, Lecture Notes in Networks
and Systems 67, https://doi.org/10.1007/978-981-13-6031-2_27

language interpreters in Malaysia [1]. The limited amount of interpreters are not sufficient to provide service for the deaf communities in the country [2].

A sign language translator device is able to substitute sign language interpreter with the use of technology. In today's technology, there are devices that are able to translate sign language into voice, text, and vice versa. Sign language signers are able to communicate with non-signers by using these devices.

There are basically two methods in designing a sign language translator device: vision-based system and wearable devices. Vision-based systems utilize image processing method through feature extraction techniques to identify hand and finger movements [3–5]. Whereas, wearable devices for sign language recognition usually utilize sensors attached on the user or glove-based approach [6–8]. Vision-based systems may not involve wearing sensory devices that can be uncomfortable, but, the system is complex, expensive and usually not portable. On the contrary, wearable devices can easily be a portable system and low cost. But, wearables usually consist of data cables that limit hand movements. Therefore, a wireless wearable can easily permit free-hand movement that can be portable and reduce cost. Furthermore, the devices that have been developed previously consists of dedicated display devices that may increase development costs. Thus, taking into account the ease-of-use and cost reduction, the wearable should be connected wirelessly to other available devices such as a smartphone app.

This work describes the development of an Android-based smartphone app called Sign Language Translator (SLT) app that can display translated sign language gestures. The SLT app is connected to a wireless wearable device via Bluetooth communication. The wearable can translate sign language gestures into texts. This paper also describes the method to develop the app, and how to interface the wearable device with the app.

The remainder of the paper is as follows: Sect. 2 describes the design of the system. Section 3 gives detail description of sensor calibration and evaluation including experimental results and analysis, followed by a brief conclusion and future recommendation in Sect. 4.

2 Design Description

Figure 1 shows an overview of the proposed system for the sign language translator device which is consists of a glove-based wearable that translates sign language gestures into text and sends it to an Android-based smartphone via Bluetooth communication. The text is displayed on the proposed smartphone application (apps). As shown in the figure, the wearable is consists of five (5) flex sensors, an accelerometer sensor, an Arduino Nano, and an HC-05 Bluetooth module.

Figure 2a shows the flow chart of the developed sign language translator device. First, the wearable measures the movements of the wrist using the accelerometer, while flex sensors measure the amount of bending for each finger. When a user creates a sign language gesture, the sensors produce raw data that are conveyed to the

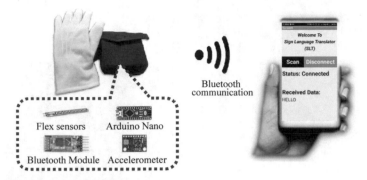

Fig. 1 Overview of the wearable sign language translator device with Android app

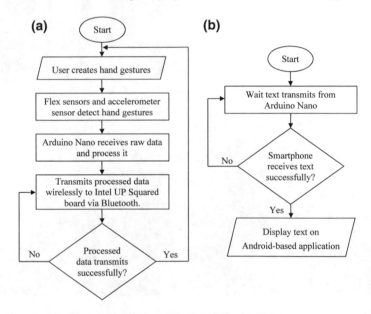

Fig. 2 Flow chart for **a** wearable device and, **b** Android-based SLT apps

Arduino Nano microcontroller. Arduino Nano translates the gesture into appropriate text based on a pre-determined gesture-sensor values mapping. Then, the text is transmitted wirelessly from Arduino Nano to the Sign Language Translator (SLT) apps via Bluetooth communication. As shown in Fig. 2b, if the translated gesture is successfully transmitted to the apps, the system will wait for another hand gesture input. At the same time, the smartphone displays the text on the apps. However, if the text data transmission failed, the system will continue to wait for subsequent data transmission.

Fig. 3 Circuit diagram for
the wearable device

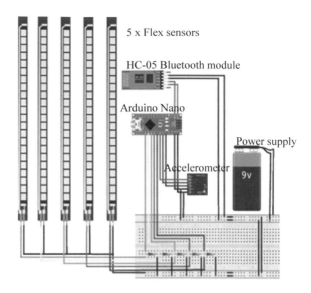

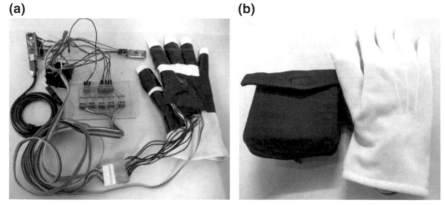

Fig. 4 Glove-based wearable device. **a** Actual circuit. **b** Circuit is packed inside a pouch, while a
clean glove covers the glove attached with sensors (right image)

2.1 Hardware Design for the Wearable Device

Figure 3 shows the circuit diagram for the glove-based wearable device. The device
is consists of an Arduino Nano, five (5) units of 4.5" Spectra Symbol's flex sensors,
a GY-61 3-axis accelerometer module, an HC-09 Bluetooth module, 220 Ω resistors,
jumper wires and a 9 V battery. Figure 4a shows the actual wearable that has been
developed. Figure 4b shows the circuit of the wearable is packed inside a small pouch
bag for portability, while a clean glove, covers the glove attached with sensors.

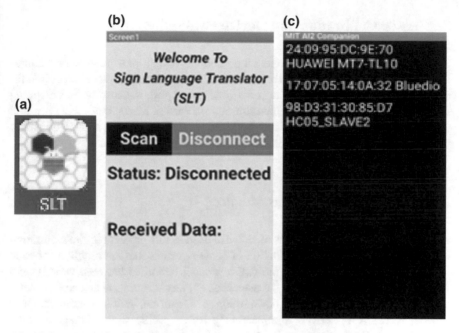

Fig. 5 Sign language translator (SLT) app: **a** SLT app icon, **b** SLT main user interface and, **c** user interface that shows a list of Bluetooth devices

2.2 Sign Language Translator (SLT) App Design

In this work, a Sign Language Translator (SLT) app has been developed using an open source web application called MIT App Inventor 2 for Android operating system (OS). It allows users to build apps easily using interactive drag-and-drop graphical user interface. Android is chosen because it has been dominating the smartphone market worldwide. This app development software requires a Google email account to create a MIT App Inventor 2 account.

Figure 5 shows the developed SLT app user interfaces (UI). Figure 5a shows the icon for SLT app. This icon will appear on the user's smartphone screen, and the user requires to click the icon to launch the app. Figure 5b shows the main UI for the app. In this interface, the user can choose to scan for existing devices to be connected or disconnected with an already-connected device. This interface also displays the text of a translated gesture received from the wearable. Figure 5c shows the UI that list out existing devices that can be connected via Bluetooth. The detail on how to utilize the app with the developed wearable is described in the next section.

3 Sensor Calibration and Device Evaluation

This section describes the experimental procedures to calibrate the sensors. Calibration of sensors is necessary steps to determine the sensor values that are required for each gesture. The calibration produced a table called gesture-sensor values mapping table that will be explained in the next subsection. Furthermore, experiment setup and results to show the usefulness of the proposed wearable for sign language translator with Android-based app is demonstrated at the end of this section.

3.1 Gesture-Sensor Values Mapping

As previously described, the sensors that are used in this project require calibration in order to determine the combination of sensors values that are required for each gesture. It is important so that the device can differentiate between various sign language gestures performed by the user. Five (5) flex sensors and an accelerometer are used in the wearable device shown in Fig. 4. Therefore, in this experiment, these sensors values were derived simultaneously for a specific sign language gesture. Then, the combination of recorded sensor values is used to create a gesture-sensor values mapping table that will be described later in this subsection.

Experimental steps. In this work, nine (9) sign language gestures based on Malaysian Sign Languages have been determined for experimentation. Initially, the wearable device shown in Fig. 4a was worn by a user. As shown in the figure, the cables for the flex sensors and accelerometer are connected to an Arduino Nano board that is powered by a 5 V power supply. The board has been uploaded with a source code that records flex sensors and accelerometer values. Then, the user was asked to create the first pre-determined sign language gesture for three (3) times. The sensor values were recorded for each movement. The user was asked to repeat these steps until all nine (9) pre-determined gestured were done. The pre-determined gestures are "THANK YOU", "HELLO", "BYE", "YES", "SURE", "NO", "YOU", "GOOD", and "PLEASE".

Experimental results. Figures 7a–i show the experimental results. As shown in the figures, there are nine (9) gestures that were tested, which are "THANK YOU", "BYE", "YES", "NO", "YOU", "HELLO", "SURE", "GOOD", and "PLEASE". As seen in each of the figures, the figures on the right are screen captures of the example of the result for a trial for all sensors values (there are three trials as explained in experimental steps). The values for "$f0$", "$f1$", "$f2$", "$f3$", and "$f4$" are values for five (5) flex sensors, and "x", "y", and "z" are data values from an accelerometer. As described in the experimental steps, all gestures were trialed for three (3) times to determine the raw data value range (minimum-maximum) for each sensor. Based on the values in these three trials, the value range for each sensor was determined as shown in Table 1, which is also called gesture-sensors values mapping table.

Table 1 Gesture-sensors values mapping table

Gestures	Flex sensors data					Accelerometer data		
	f0	f1	f2	f3	f4	x	y	z
Thank you	645–672	870–881	866–871	826–834	853–860	344–353	300–315	406–410
Hello	684–692	860–879	866–870	850–853	852–860	279–289	306–315	350–360
Bye	476–483	880–882	867–873	840–855	833–859	300–320	295–300	334–340
Yes	506–519	765–777	781–790	714–718	627–689	358–369	375–380	283–289
Sure	480–486	756–761	853–854	679–726	844–846	392–398	290–298	319–325
No	570–580	847–853	838–843	769–773	772–777	356–362	300–304	275–290
You	497–508	870–876	760–775	692–705	690–720	280–291	348–351	329–335
Good	663–672	770–774	778–786	704–707	727–733	299–306	324–328	300–303
Please	570–589	876–882	867–874	850–854	845–851	284–287	349–352	304–310

As seen in the table, the combination of these sensors values is used to determine the appropriate text for a specific gesture. For example, in order to translate "THANK YOU" gesture, by referring to the table, the raw data values for flex sensor $f0$ (thumb) is between 645 and 672, flex sensor $f1$ (index finger) is between 870 and 881, flex sensor $f2$ (middle finger) is between 866 and 871, flex sensor $f3$ (ring finger) is between 826 and 834 and flex sensor $f4$ (little finger) is between 853 and 860. In addition, the accelerometer raw data on x, y, and z axes are between 344 and 353, 300 and 315, and 406 and 410 respectively. When a user makes a gesture and each sensors data value falls within the ranges shown in the table, then, the appropriate text will be displayed on the app. Similarly, the rest of eight (8) gestures utilize a different combination of eight (8) raw data values from the sensors as shown in the table.

3.2   Wearable Device and App Evaluation

In this section, the experiment to show the usefulness of the developed wireless wearable sign language translator device with Android-based app is described. Initially, Bluetooth communication must be established beforehand. Once communication is established, the wearable will starts to send text data to the app.

Experimental setup to establish Bluetooth communication. First, a user was required to wear the wearable and connects the Arduino Nano with a 5 V power supply. Then, the user launched the SLT app by clicking the SLT app icon on the smartphone screen as shown in Fig. 7. As seen in the figure, the screen promptly showed the main user interface of the SLT app. In this situation, Bluetooth connection is not established yet, thus, the status shows "Disconnected". Furthermore, no text is received from the wearable. Then, the user clicked the "Scan" button to show a list of existing Bluetooth devices. Here, the user selected the Bluetooth device named "HC05_SLAVE2" to establish communication with the HC-05 Bluetooth module in the wearable. Once the Bluetooth communication is established, the main UI reappeared and the status has changed into "Connected". This means Bluetooth communication between the wearable device and SLT app has been established. Now, the user can make any gesture based on the nine (9) sign language gestures as shown in Fig. 6. For example, in the experiment, the user made "HELLO" gesture. The wearable translates the gestures into text, and Arduino Nano transmits it to the smartphone app via Bluetooth. Finally, the text "HELLO" is displayed on the app as shown in Fig. 7. User can disconnect Bluetooth communication by clicking "Disconnect" button on the app UI.

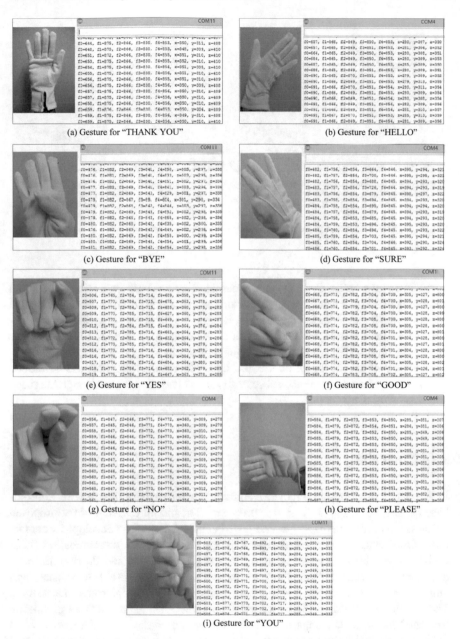

(a) Gesture for "THANK YOU"

(b) Gesture for "HELLO"

(c) Gesture for "BYE"

(d) Gesture for "SURE"

(e) Gesture for "YES"

(f) Gesture for "GOOD"

(g) Gesture for "NO"

(h) Gesture for "PLEASE"

(i) Gesture for "YOU"

Fig. 6 Experimental results

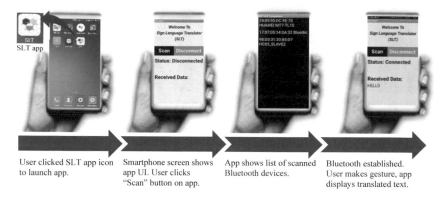

User clicked SLT app icon to launch app.

Smartphone screen shows app UI. User clicks "Scan" button on app.

App shows list of scanned Bluetooth devices.

Bluetooth established. User makes gesture, app displays translated text.

Fig. 7 Android-based sign language translator (SLT) app launching steps

Experimental results. Once the experimental setup has been completed, Bluetooth communication between the wearable device and SLT app will be established. In this experiment, the user made nine (9) sign language gestures as described in subsection 3.1 to demonstrate the effectiveness of the developed system. Figure 8 shows the results on nine (9) types of sign language gestures and the corresponding translated texts that are displayed on the SLT app. The results show that the developed sign language translator device successfully translated nine (9) basic sign language gestures using the wearable.

4 Conclusion

In this paper, a wireless wearable for sign language translator device with Android-based smartphone app (SLT app) is proposed. The circuit diagram of the proposed wireless wearable and display device have been explained. The developed wearable has been successfully connected to the SLT app via Bluetooth communication. Steps on how to establish Bluetooth connection between Arduino and the app has been described. Experiment results show that the developed device is able to translate sign language gestures into text, and displays it on the app's user interface successfully. The design described in this work presents some advantages such as permitting free hand movements compared to a wired glove-based device, and a low-cost and portable wearable design. Currently, works are underway to upgrade the app so that it just not only displays translated text, but also produces speech for the translated gestures.

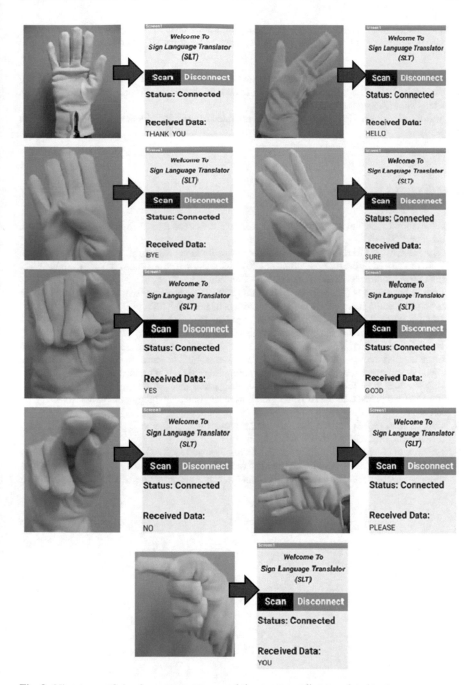

Fig. 8 Nine types of sign language gestures and the corresponding translated texts

Acknowledgements The authors would like to thank the Research Management Center (RMC), UTHM and Ministry of Higher Education for sponsoring the research under Tier 1 Research Grant (H161).

Ethical Approval. All procedures performed in studies involving human participants were in accordance with the ethical standards of the Human Research Ethics Committee of Universiti Tun Hussein Onn Malaysia (UTHM).

Informed Consent. Informed consent was obtained from all individual participants included in the study.

References

1. Acute shortage of sign language interpreters in M'sia. Retrieved from https://www.malaysiakini.com/news/376165
2. Mohid SZ, Zin NAM (2011) Accessible courseware for kids with hearing impaired (MudahKiu): a preliminary analysis. In: Proceedings of 2011 international conference on pattern analysis and intelligent robotics ICPAIR 2011, vol 2, pp 197–202
3. Igari S, Fukumura N (2016) Recognition of Japanese sign language words represented by both arms using multi-stream HMMs. In: Proceedings of IMCIC-ICSIT, pp 157–162
4. Dong C, Yin Z, Leu MC (2015) American sign language alphabet recognition using microsoft kinect. In: IEEE conference on computer vision and pattern recognition workshops, pp 44–52
5. Hai PT et al (2018) Automatic feature extraction for Vietnamese sign language recognition using support vector machine. In: 2018 2nd international conference on recent advances in signal processing telecommunications and computing (SigTelCom), pp 146–151
6. Rishikanth C et al (2014) Low-cost intelligent gesture recognition engine for audio-vocally impaired individuals. In: 2014 IEEE global humanitarian technology conference (GHTC), pp 628–634
7. Gałka J et al (2016) Inertial motion sensing glove for sign language gesture acquisition and recognition. IEEE Sens J 16(16):6310–6316
8. Ambar R et al (2018) Preliminary design of a dual-sensor based sign language translator device. In: Ghazali R, Deris M, Nawi N, Abawajy J (eds) Recent advances on soft computing and data mining. SCDM 2018. Advances in intelligent systems and computing, vol 700. Springer, Cham

Part IV
Multimedia and Immersive Technologies

ARMyPat: Mobile Application in Learning Malay Historical Patriots Using Augmented Reality

Mudiana Mokhsin, Amer Shakir Zainol, Emma Nuraihan Mior Ibrahim,
Mohd Husni Mohd Som, Kamarul Ariffin Abdul Basit
and Amir Afwan Azman

Abstract In the era of evolving technology, the generation who was born in 1995 onwards or known as Gen Z in Malaysia is becoming less interested in knowing and learning about Malay Historical Figures. Currently, the medium or tool applied in learning about Malay Historical Figures is only by using textbook as reference in the classroom. Looking at this scenario, this project attempts to attract their interest in learning historical information, especially about Malay Figure by converting the text or image information into an interactive learning method. The approach for this project is by implementing an augmented reality technology where the interactive fighting game has been embedded. The objectives of this study are to identify the requirements, design and develop Augmented Reality mobile learning application about Malay Historical Figures with the commercial name of *ARMyPat* (Augmented Reality based in Learning about Malay Patriots). This application prototype will be compatible for the Android platform mobile devices and the target users for this application are students aged between 10 and 15 years old who learn the historical subject at school. The project methodology is ADDIE model that covered analyses, design, develop, implement and evaluation phase since this methodology is suitable for the

M. Mokhsin (✉) · E. N. M. Ibrahim · K. A. A. Basit · A. A. Azman
Faculty of Computer & Mathematical Sciences, Universiti Teknologi MARA,
40450 Shah Alam, Selangor, Malaysia
e-mail: mudiana@tmsk.uitm.edu.my

E. N. M. Ibrahim
e-mail: emma@tmsk.uitm.edu.my

A. A. Azman
e-mail: amirafwan94@gmail.com

M. Mokhsin · A. S. Zainol · M. H. M. Som
Institute of Malay Thoughts and Leadership (IMPAK), Universiti Teknologi MARA,
40450 Shah Alam, Selangor, Malaysia
e-mail: amers781@salam.uitm.edu.my

M. H. M. Som
e-mail: husni820@salam.uitm.edu.my

A. S. Zainol
Faculty of Art and Design, Universiti Teknologi MARA, 40450 Shah Alam, Selangor,
Malaysia

© Springer Nature Singapore Pte Ltd. 2019
V. Piuri et al. (eds.), *Intelligent and Interactive Computing*, Lecture Notes in Networks
and Systems 67, https://doi.org/10.1007/978-981-13-6031-2_6

interactive learning development process. The students can obtain the information such as the summary of character, biodata, heritage and contribution interactively. For 3D character model view, they can learn and know the attraction of the martial art which is Silat. Therefore, they can learn a few movements of the Silat martial art. The last part is an interactive fighting game whereby it can bring the students to the fighting action in real environment and encourage them to know about the heroism characters of that Malay Historical Figure. In the end, this application is to encourage young generation patriotism spirits towards Historical Figure in Malaysia by utilizing augmented reality technology.

Keywords Augmented reality · Mobile learning · Mobile game · Malay figures · ADDIE research methodology

1 Introduction

Malay historical figure is famous and very important people. They have significant contribution to the country. In education, the students will learn about Malay Figures such as Tun Mahathir, Tun Hussein Onn and Tun Abdul Razak especially in the historical subject. Some of them struggled to get independence and make efforts to ensure that our country is peaceful. Many people wonder why everyone must know the contribution of Malay Figure. The reason is to create patriotic spirit among the citizen [1]. Hang Tuah is a very popular and legendary warrior in Malacca. He is the most illustrious Malay historical hero and leader of a closely knit band of Melaka's finest warriors—Hang Kasturi, Hang Lekiu, Hang Lekir and Hang Jebat. He was considered as one of the greatest *silat* masters and more powerful than the other warriors at that time. Nowadays, Hang Tuah's legend is still popular in Malaysia. According to Jamil [2], Hang Tuah's story is true and known as a symbol of glory during the reign of Malacca sultanate. He stated that whether it is in the form of manuscript or historical record, the life story of Hang Tuah is the most valuable legacy for today's generation and future. However, the rest four close friends of Hang Tuah are not widely known and receive less popularity as compared to Hang Tuah.

In the school's history learning environment, the current method used by the teachers is by using textbooks which is an unattractive way of learning as it is a one-way learning approach. The content and learning process are supposed to be more interesting for education by including the combination of audio, video, games or other interactive media to attract students to learn history. The idea to combine education with entertainment, or popularly known as edutainment has been used before. Therefore, to visualize the historical content, the use of Augmented Reality application is used as one of the methods to attract students to learn interactively. This effort is in line with the Malaysia government agenda towards education 4.0 where an immersive technology has been embedded in the teaching and learning process. Augmented reality can help the user to interact with real and virtual environment with a tangible interface that consists of combination of words, specific information and 3D models [3]. This technique can deliver the information by using many kinds of multimedia elements. The user will interact with visual object in a real environment. Therefore, it will increase the understanding of the user and at the same time improve the learning process. Billinghurst [3] stated that by learning using Augmented Reality, especially for history subjects, it can offer vivid information and character's emotion compared to traditional ways. By traditional method, the historical situation of figure can only be described by words, images and oral report [4]. Learning Malay Historical Figure using Augmented Reality mobile application is believed to be the way to attract the young generation to learn about it. Apart from that, in Venice, Italy, there is an Augmented Reality application game called *VeGame* which guides the user in learning about the historical places such as churches, monuments and etcetera in giving a new experience to the tourist [5]. It is one of the Augmented Reality applications that expose on the historical information. In London, England, there is also an augmented reality application called *Shakespeare's Globe* 360 by Sam Wanamaker, which had a vision to provide theatre to all children. This mobile app carries out primary purpose of theatre by bringing history to life. Learning about the performing arts, alongside its history will provide new depth to the content of both arts and English classes. AR lets students explore the theatre and feel the atmosphere of it. AR supports education in a way that was previously impossible. For instance, adding a virtual reality headset to this app will take the experience to another level, by bringing the user inside this historic theatre [6].

Malaysia has many historical figures, but most of today's generation is less sensitive about the matter because they were born after independence and do not understand the importance to know about historical figures [7]. Nowadays, young generation in Malaysia is less interested to read or know about the country's historical figures. They only learn about historical figures at school during History subject. Talin [8] stated in her study that History is not a popular subject in schools as students complained that the subject is not interesting as they must memorize a lot of facts. However, to prevent this situation from happening which may cause the younger generations to forget the contribution of historical figures in Malaysia; historical figures need to be presented in a form that can attract students to learn about it. The idea of this study is to convert the historical figure's information into interactive form of knowledge sharing by using Augmented Reality. It is a known fact that the younger generation nowadays is addicted to use digital device such as computer, and mobile phone. Despite the situation, learning through mobile application has not been productively promoted as an alternative learning method because the education system is still confined to the conventional way of learning through media such as newspaper and television [9].

Learning approach using mobile applications is still not very popular in the school environment in Malaysia. Learning approach is still based on the traditional method of learning-centred class which is mostly geared towards examinations. As the technology advances, the students will feel frustrated since there are so many entertainment alternatives out there which are more interesting than learning traditional method [10]. Some previous studies proved that most teachers are still using the traditional way of teaching the subject in the classroom [11, 12]. However, since most students now were born in the digitalized age, naturally they will be more motivated when learning using technologies' application such as Augmented Reality [13, 14]. Therefore, the intention of this project is to use Augmented Reality that can effectively attract students to learn about historical figures in Malaysia.

2 Literature Review

Augmented Reality (AR) is a new technology that integrates the digital information of the real environment to the user [15]. Usually, user only sees the information directly from the book and paper. Nowadays, the technology that view the elements of a physical or real environment are augmented by computer generated sensory input video, sound and image. The objective of Augmented Reality is to capture the meaning and information of the place or real object.

The tremendous importance of AR technology is proven by various investments made all over the world to explore the usage AR. For instance, AR offers engaging entertainment, particularly in the media. It allows designers with the opportunity to develop new kinds of appealing entertainment that offers more emotional and intimate characters. AR technology also assists in facilitating the development of better buildings. In the same manner that AR is transforming the entertainment and media industry, it is also influencing the construction environment. AR shows tremendous potential in benefiting the education and learning environment while applications in other areas such as the medical environment are also feasible [16]. In the gaming environment, AR is showing remarkable potential. It is facilitated in the integration of audio and visual material that provides the user feelings as though they are in the real environment. Compared to virtual reality that demands confined space or separate area to develop immersive environment, AR utilizes the real environment and establishes its own playing field. Virtual reality demands the usage of specific VR headsets while just certain AR systems need them. For the AR games, they expand the playing field on the real world's diversity to ensure that the game remains interesting [17]. Currently in school, traditional ways in education are still used. The students must follow all the knowledge and learning instruction arranged by the teacher face-to-face [13, 18]. Moreover, the learning process still uses paper. This method of learning does not share dynamic information such as motion or movement. Although the conventional method is very useful, there are rooms for improvement to spark the interest in learning. Since technology begins to grow, the combination of technology has affected the learning system. The use of technology in learning and teaching provides the attraction to develop an engaging, authentic, realistic and extremely fun learning environment [19]. It is also found that the student engagement and understanding of the content increase when learning using technology. Therefore, many technology applications have been combined with education such as the use of computer, multimedia, Internet, mobile devices, virtual reality and augmented reality. The increase in usage of mobile devices has led to the introduction of Augmented Reality application for mobile learning.

Mobile AR offers educators and designers with new ways of developing the situation and improving the context of the mobile learner. One of the major concerns about AR is that the augmenting encounters in real-world settings regardless of the location of the learner. Technologies used in AR have the capacity to leverage on location, experience, situation or environment to entirely new avenues of understanding and meaning. AR games are significantly altering people's ways of learning while using mobile devices. When considering a mobile learning game, it is crucial to emphasize several major aspects unique to mobile augmented reality. Utilization of information that supports the learning process using the AR is done by merging mobile together with AR facilities. These comprise of the user's mobility, the geographical position of the user, physical environment for engagement in learning and linkages between informal and formal learning. Furthermore, in the event of mobile learning games using AR, emphasizing on the "interaction modes" is essential since the modes interact with each other to offer learners passive overlaps of information based on their gestures, physical location or movements. AR games can also involve the learners in investigative mode for the purpose of discovering or solving a problem [20].

Nowadays, education field had used Augmented Reality technology for the end user. Furthermore, many people use smart phones that include this technology. The reason why this technology had been used for educational purpose is because the features available in Augmented Reality can make the students easy to understand during the learning process and become more fun. Most of Augmented Reality application currently focuses on personalization area that needs the agent approach to create the best learning experience. According to the Bower et al. [21], learning through Augmented Reality can increase motivation to the learner because it is realistic, fun and interesting. AR has the capability to display the information in various forms of media and above virtual real-world environments. It also helps to improve the student thinking capabilities, problem solving and communication skills. This is due to the reason that the Augmented Reality can do impressive hybrid-learning [22]. Students who have problems to visualize in any learning concepts can be helped using Augmented Reality technique. The students will be able to see the 3D design like a real situation. Besides that, Augmented Reality can increase the level of interaction between the subject's content with the students in terms of physical exploration [23]. By using this technology, the attraction towards learning among the students can be increased. Mobile devices have great features, can be accessed anytime and everywhere and at the same time can access Internet usage with affordable price [24]. Smartphones, tablets and wearables are examples of mobile devices that have the capability in computer vision, cloud computing based on mobile and cooperative in networking which allow the science fiction to be transformed into a reality vision using Mobile Augmented Reality (MAR). Even though the mobile devices have some constraint compared to traditional computer, their systems have many sensors that are very useful to develop MAR applications. The intensive parts execution process can also be assisted by the remote server. The technology in mobile devices had been improved such as the built-in camera, mobile cloud computing, resources for computer and

System	Hardware platform	Software platform	Display	Tracking	Network	Collaborative	Indoor/ outdoor	Application field
MARS	notebook computer	JABAR/ Ruby	optical see-through HMD	RTK GPS, orientation tracker	campus WLAN	multiusers	indoor, outdoor	Tourism, Navigation
ARQuake	notebook computer	Tinmith-evo5	optical see-through HMD	GPS, digital compass	cable network	single	indoor, outdoor	Entertainment
BARS	notebook computer	unknown	optical see-through HMD	inertial sensors, GPS	WWAN	multiusers	outdoor	Training
Medien welten	PDA	Studiers-tube ES	video see-through screen	visual marker tracking	WLAN	multiusers	indoor	Education, Entertainment
MapLens	mobile phone	unknown	video see-through phone screen	nature feature tracking	WWAN	multiusers	indoor, outdoor	Tourism, Entertainment
Virtual LEGO	mobile phone	UMAR	video see-through phone screen	visual marker tracking	unknown	single	indoor	Authoring, Assembly
InfoSPOT	Apple iPad	KHARMA	video see-through pad screen	sensors, geo-reference visual markers	WWAN	single	indoor	Information Management
Pokemon Go	mobile phone	Android & iOS	video see-through screen	GPS, digital compass	Internet Access	multiusers	indoor, outdoor	Entertainment
Ingress	mobile phone	Android & iOS	video see-through screen	GPS, digital compass	Internet Access	multiusers	indoor, outdoor	Entertainment

Fig. 1 The existing application that using augmented reality in mobile [24]

sensors. This is the reason why AR has very high potential usage and development in mobile devices. Since the rapid growth between human–computer interaction, computer vision and mobile cloud computing, the user can feel the new experience of interaction and get the information in various ways using the mobile devices. Currently, Mobile Augmented Reality (MAR) had been used in many fields such as entertainment and advertisement, tourism and navigation, assembly and maintenance and lastly for education and training. Figure 1 shows the existing application using Augmented Reality in mobile devices.

Augmented reality has been used on desktop computing. Some of Augmented Reality applications use the components of website infrastructure, but not every part can be used to develop Augmented Reality applications. AR is a system that enhances the user's view in real-time by using generated image or other inputs by the events with the specification of AR. By using the most popular product for AR such as Oculus Rift and Google Cardboard, it usually makes the content of AR application more interactive and improves the user experience. Previous studies also discussed the usage of Augmented Reality on web technology. For example as an open source, Argon Augmented Reality web browser tries to combine the existing technology for web development. The Argon apps have commercialized the Augmented Reality mobile application to the compatible mobile devices. Figure 2 shows the browser architecture that includes a framework called "Augmented Reality Framework" (AREF), whereby this is the framework that has been referred in developing the application. Modules in the framework are illustrated below.

Even though there are many significant benefits in using Augmented Reality during learning activity, there are some issues that need to pay attention on. Research had shown that some students do not know and have lack of information about this technology. As a way to reduce the problem, the students were given some instructions as exposure before they use Augmented Reality technology. Another

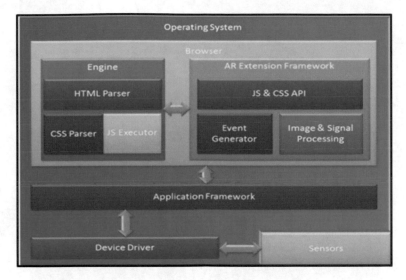

Fig. 2 Extended browser architecture with AR, CSS and JS [25]

reason is that it was difficult for the user to find the 3D for virtual contents of Augmented Reality on their mobile device because of the screen limitation [26].

The Integrated Curriculum for Secondary School (ICSS) announced that History is the core subject for the lower secondary and upper secondary in Malaysian schools. This subject has been included in class timetable. For Sijil Pelajaran Malaysia (SPM) level, the Ministry of Education stated that History subject is one of the compulsory subjects for the students to pass. History is considered as an important subject because according to Johdi et al. [27], learning History subjects can improve the students' understanding of human being in order to make critical judgement to prepare an individual to handle any problems.

3   Research Methodology

The research methodology is described as a logical way to solve a task. The discipline is to research what type of methodology to be chosen. It consists of the process to explain, define and forecast about their task and it is called research methodology. By doing the research methodology, researchers can gain knowledge and describe the task strategy of the project [28]. The methodology used for this research is ADDIE model. The acronym of "ADDIE" was described by Schlegel [29], in his book "A Handbook of Instruction and Training Program Design" as the generic Design Model of Analyse, Design, Development, Implementation and Evaluation (ADDIE) (Fig. 3).

For this application, ADDIE methodology is found suitable because it is an interactive learning application. The step-by-step process methodology during the devel-

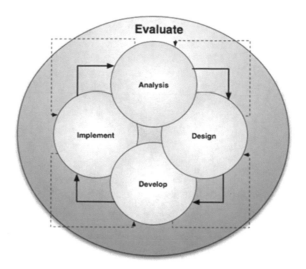

Fig. 3 ADDIE model [29]

opment is very important especially during certain stage when problems occur. By using ADDIE method, this research could access the previous phase and repeat the process again when problems happened during the research process. According to Linda [30], the ADDIE methodology process is an effective and efficient design for teaching which helps with instructional design. Analysing learners' problem and needs, designing the instructional design which includes activities and material are the components of the instructional design.

(i)   Analysis Phase

Firstly, the details related to the project such as problems need to be identified. At this phase, the analysis resources must be not less than five years back to make sure it is up to date, and can be accepted to be applied in the current year. The problem statement in this project is that the young generation is less interested to know about national figures in Malaysia. After the problem had been defined, next step is to make sure the best method to provide solution for the problem. To find the best method, the target user had to be identified. As the project was meant for students who learn the History subject in their schools, the method used for this study is AR. Along the process of researching the resources, objective, and scope are used to ensure that the resources relate to the topic. The resources taken from numerous sites, journals and books were then compared to gain better results (Fig. 4; Table 1).

(ii)   Design Phase

After the completion of the analysis phase, the next step is the design phase which is needed in order to develop this application. In design phase, the use case diagram was first designed using Microsoft Visio. Use case diagram or usually referred as behaviour diagram was used to list all set of actions performed by the system's user. After that, the design for prototype using paper sketching was made for system interface to show the activities and process. The purpose for prototyping process is to make sure that the development phases run easily. The design was derived from the analysis that had been done to develop this application. Besides that, the activities for application design flowchart and storyboard were also covered in this phase. This project application has three components which are the main activities for this project. Firstly, the information on Malay figure was explained at information part. It contained a navigation button to attract the student's interest to learn. The second one was interactive game for the students. Because the nature of students like to play games, the game was designed based on fighting scenes to increase the understanding and help the user to think fast which will activate their brain to act fast based on the fighting necessity in the game. Last but not least is the usage of Augmented Reality technology which was included in the information and fighting game. The users need to focus their phone camera to the picture of Malay figure on the paper and then the animation of character plus the information will appear on the phone screen. This system used fighting game as way to attract students to know about Malay figure using interactive games (Table 2).

This part describes the flow of the application prototype; Fig. 5 shows the application design flow. First, after user click the icon of the application, the home page will appear. The home page contains two buttons which are for navigating to the game section and learning section. At learning section, users can select any Malay figure that they want to know. To play the game, the user can click the game button at the game section page (Figs. 4 and 5).

Table 1 Summary for analysis phase

Research objective	Phase	Activity	Technique	Deliverable
To identify requirements for augmented reality, mobile learning application about Malay historical figures	Analysis	• Identify the problem statement • Identify the objective • Identify user requirement • Identify the projects significant • Define functional and nonfunctional requirement	Conducting observation on previous similar system that focus on historical figures Read from article, journals and previous research Produce Gantt chart	List of problem statement List of objectives List of user requirement List of project significant
		• Estimated project planning	Produce project schedule tools using microsoft project	Gantt chart

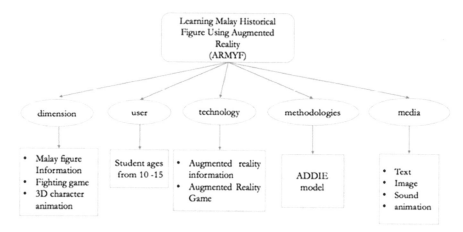

Fig. 4 Brief description of project analysis

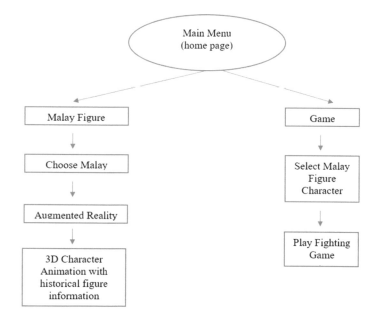

Fig. 5 ARMyF application design flow

(iii) Development

After the analysis and design phase completed, the results were used as guidelines to proceed to development phase. In this phase, software and hardware were both needed to build the project using Augmented Reality. The technique used for this development was marker techniques to display the virtual information directly to the user using mobile devices. The types of coding used to build this application were C# and Java. The most important part is to develop the main functionalities of the application (Table 3).

Before getting started, there are a lot of matters that need to be considered for Augmented Reality (AR) application such as to extend the functionality of a mobile software offering. The truth is, there are many ways to approach the design and development of an AR app and the best approach depends on the requirements of the project. It is always best to start with the end result in mind, establish the key measures of success and ensure a mechanism by which to measure the performance of the application that has been established. Measuring success can be tricky and therefore having the right data and analysis is essential. To make it successful, there are five steps or processes in the development of an augmented reality game as shown in Fig. 6.

(iv) Implementation

For implementation phase, after the development of the application is completed, it needs to be installed and transferred to smartphone using Android Studio as a platform to run it. Therefore, the user would be able to test the application and then the responses were taken to make sure the application achieves the project objective.

(v) Evaluate

Table 2 Summary for design phase

Research objective	Phase	Activity	Technique	Deliverable
To design augmented reality, mobile learning application about Malay historical figures	Design	• Design use case diagram • Design storyboard • Design 3D character model with an animation • Design interface and button • Producing the flow of application design	• Microsoft Visio which provides tools to design use case diagram • Using JustInMind software • Using Blender software • Using Unity 3D software • Using Microsoft PowerPoint	Use case diagram Storyboard prototyping 3D character model with an animation Menu, game, information, interface and button Application design flow

Table 3 Summary of development phase

Research objective	Phase	Activity	Technique	Deliverable
To develop augmented reality, mobile learning application	Development	• Develop the project using augmented reality and game • Development of augmented reality application	• Use augmented reality marker techniques to display the virtual information directly to the user using mobile devices • Use unity 3D to develop the augmented reality game • Before and Unity 3D software to create mobile applications that using augmented reality	Marker-based augmented reality application 3D game augmented reality fighting game Augmented reality, mobile learning application about Malay historical figures (ARMyF)

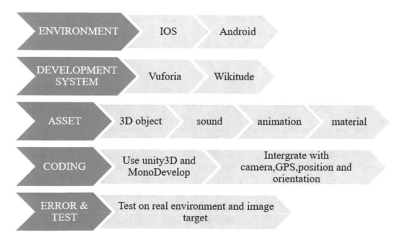

Fig. 6 Augmented reality game development process

Table 4 Summary for implementation and evaluation phase

Research objective	Phase	Activity	Technique	Deliverable
To develop augmented reality, mobile learning application about Malay historical figures	Implement Evaluate	• Transfer and install to smartphone as a platform • Feedback and response from the user is taken to improve the application	• Using android studio • Select 5–10 students to do the testing of the system	An application prototype completed install User feedback Result of testing

Finally, the feedback and response from the user were taken to improve the application. To complete this evaluation process, a few students were selected to do testing of the system. The results of the test and user feedback are very important in order to improve this application (Table 4).

4 Analysis and Findings

This section discusses discoveries of the entire stage which covered three objectives that have been achieved which are; user requirements, design and development of Learning Malay Historical Figure using augmented reality (ARMyF). In this section, the point of convergence of the paper is shown. User requirements are needed for any project, especially when developing mobile application projects. Use case, functionality and nonfunctionality interact with the use cases. Figure 7 shows the use case diagram for the development of ARMyF.

Functional Requirements are a description of the facility or feature required for development of a system or application. It deals with what the system should do or provide for users. A use case diagram is used to describe the functional requirement. Use case diagram is a depiction of a user's interaction and relationship within a system or application. It contains use cases and role where use case is a list of functionality that the user can perform actions whereas role is an actor (Tables 5 and 6).

Several steps of the design process were done to develop this mobile application. It started with producing a use case diagram, designing storyboard, 2D model, 3D model and user interface design. All these processes used different types of tools in order to complete, but the whole design is related to each other. Figure 8 shows the design phase processes in the development of the ARMyF.

The first step was to sketch the process flow of the application by using the storyboard. The storyboard explained the short description about this system. It was developed in the early planning for the design phase and used as a reference to develop the complete application prototype. Next step was creating the application graphic. The application graphic covered application background, game character,

Table 5 Functional requirements for ARMYF mobile

Function	Description
Select Malay Figure	Students can choose Malay figure that they want to know about
View 3D model and animation	Student can view 3D character model and animation with an interactive sound
Play an interactive game	Student can play an interactive fighting game and select which character they want to use before starting the game
Implement of augmented reality	Student can watch information and play the game by using augmented reality technology

Table 6 Non-functional requirements for ARMYF

Characteristic	Description
Availability	Uptime less than 5 s
Performance	Good performance with minimum constraint
Flexibility	Capability to increase or extend the functionality
Portability	The ease to view on all necessary platform
Reliability	High capability to maintain its performance over time
Usability	Easy to be understood, learned and used by intended user

buttons and icons. This was followed by the process of designing the 2D model of the character where Adobe Photoshop was used as a tool for this activity. Next was to create the 3D model. The design was based on the same 2D character design, and

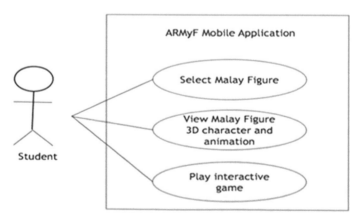

Fig. 7 Use-case diagram for ARMyF

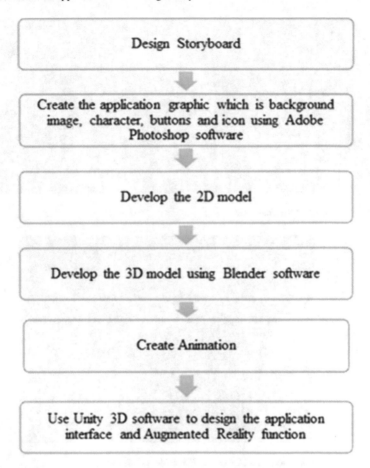

Fig. 8 Design phase process

then it was converted into 3D character. In order to model the 3D character, this study used a tool or software called Blender. After the 3D models had been created, the process to animate the character was executed. As for the augmented reality function and to design the application interface, a tool called Unity 3D was used. Figures 9 and 10 show the process of creating and animating the 3D model.

As for the user interface, there are three parts in this mobile application prototype. First is a Malay Figure section which displays information about the Malay Figure such as the summary of the character, the biodata and their contribution. Second is the 3D character part which shows 3D character model with animation. The final one is the game part where user can play interactive fighting game and use the 3D Malay Historical Figure as the fighting character. Figure 14 shows the game animation flow for all the processes involved at each character. The changes of states from one process to another process are linked to each other to create the reaction during the

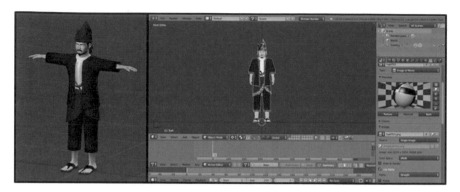

Fig. 9 The 3D model of the character

Fig. 10 Putting the animation elements of the 3D model of the character

fighting game. Every fighting button such as kick, punch, walk forward and backward were set at the animation flow as illustrated below (Refer Figs. 11, 12, 13 and 14).

In the development process, Unity 3D was used as the tools to develop the application. All the completed process in the previous stage namely graphics, audio and 3D models were imported into this process. The next process was to search and edit the audio. To create the best experience for the user, a couple of audio systems were used. Internet browser was used as search engine and the audacity as tool to edit the audio. The final process was to test and evaluate the application prototype. To make sure the application runs smoothly, several tests and evaluations were completed before the application was given to end user (Fig. 15).

Fig. 11 Interface for home
page of the application

Fig. 12 Malay Figure section

Fig. 13 Interface for the
game section

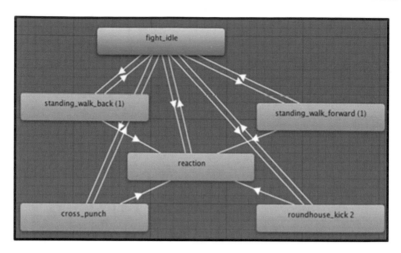

Fig. 14 Game animation flow

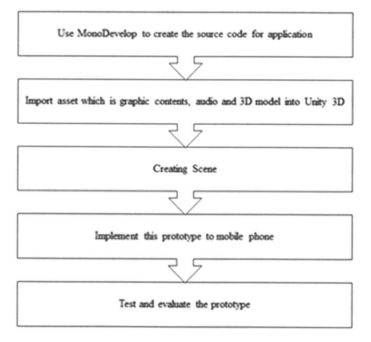

Fig. 15 The development process

5   Conclusion and Recommendation

There are many limitations that occurred during the development process. The main issue was when gathering the information about historical figures. It was very difficult to find storybooks about the historical figure so the researchers opted to use the internet. After that, need to find the suitable part to convert the information into 3D design and make it interactive to the user. Secondly, the process to design the 2D character was quite complicated using Adobe Illustrator and then need to convert it into 3D character modelling. This process needs a good skill to imagine the 2D drawing and sketch it using the software. Lastly, the limitation was during the process to create the augmented reality game. The game developed must be relevant to the historical figure and gain interest from the user to know more about it. Even though this project gives a lot of benefits to the user, there is room for improvements that need to be done in the future to improve the application efficiency and effectiveness. The further development of this application prototype needs to be continued after the positive survey and result from the participants. To understand user requirement, the prototype needs to involve an expert person such as the teacher and academician. The cooperation between these two groups can make this application more useful. Second, better design, interactive animation and smooth model development can be achieved with good computer specs and development skills. Clear audio sound can deliver better understanding of user. With better design and development of the application, the user will become more attracted to use the application. Lastly, more user testing is required to identify the weaknesses for this application prototype. The numbers of participants especially students need to be increased, for example, select one secondary school and focus on students from 13 to 15 years old. The main focus is to attract the user's attention about Malay Historical Figures. There are many historical figures in Malaysia who can be used as the next development of the application. Therefore, the students can learn History in school by downloading the application on their mobile phones.

This application prototype exposed the user about Malay Historical Figure, which may be forgotten in the next generation since there were born after the Independence Day. The idea to develop an interactive learning version of Malay Historical Figure to bring competitive advantages for the user to learn, remember and know about historical person. The interactive element in this application is a 3D character model with animation and another section is fighting game. Besides that, Augmented Reality technology was used in this application. This technology enables the user to interact with the application and view the 3D object in the real world using their smartphone camera. The user can get the opportunity to know about how Augmented Reality works in mobile application. The fighting game helps to gain user's interest as younger generation nowadays like to play games from their smartphone. The result of this study showed that the application is efficient for the user, which proved that it is successfully delivered and meet the objective.

Acknowledgements We would like to express our high gratitude to University Teknologi MARA, Shah Alam, Selangor, Malaysia for funding this paper submission via an internal grant called

BESTARI (Ref. No: 600-IRMI/DANA 5/3/BESTARI (P) (109/2018)). The highest appreciation also goes to the Faculty of Computer and Mathematical Sciences (FSKM) and Institute of Malay Thoughts and Leadership (IMPAK), Universiti Teknologi MARA, Shah Alam, Selangor, Malaysia.

References

1. Razliqah R (2010) Cara2 meningkatkan semangat patriotisme dalam Kalanganremaja. Impian Rika
2. Jamil M (2015) Hang Tuah Bukan Mitos. Sinar Online 8 Aug 2015
3. Billinghurst M (2012) Augmented reality in education. New horizon for learning 12
4. Duveen J, Solomon J (1994) The great evolution trial: use of role-play in the classroom. J Res Sci Teach 31(5):575–582
5. Schrier KL (2005) Revolutionizing history education: using augmented reality games to teach histories. Master of Science thesis in comparative media studies at the Massachusetts Institute of Technology
6. Kim G (2017) Living history: education through augmented reality. Research Center of the Master of Arts Management Program Heinz College, Carnegie Mellon University Pittsburgh, Pennsylvania
7. Suliati A (2015) myMetro. Hayati Perjuangan Tokoh Negara. https://www.hmetro.com.my/node/47468. Published on Thursday, 30th April 2015
8. Talin R (2014) The teaching of history in secondary schools. Int J Soc Sci Humanit Res 2(3):72–78
9. Majid NAA, Husain NK (2014) Mobile learning application based on augmented reality for science subject: Isains. ARPN J Eng Appl Sci 9(9):1455–1460
10. Parhizkar B, Gebril ZM, Obeidy Wk, Ngan MNA, Chowdhury SA, Lashkari AH (2012) Android mobile augmented reality application based on different learning theories for primary school children. In: Proceedings of the international conference on multimedia
11. Ahmad A, Abdullah AG, Ahmad MZ, Abd Rahman Abd Aziz (2005) Kesan efikasi kendiri guru Sejarah terhadap amalan pengajaran berbantukan teknologi maklumat dan komunikasi (ICT). J Penyelidikan Pendidikan 7:15–24
12. Ahmad A, Rahman SHA, Abdullah NAT (2009) Tahap keupayaan pengajaran guru Sejarah dan hubungannya dengan pencapaian murid di sekolah berprestasi rendah. Jurnal Pendidikan Malaysia 34(1):53–66
13. Liu TY, Chu YL (2010) Using ubiquitous games in an English listening and speaking course: Impact on learning outcomes and motivation. Comput Educ 55(2):630–643
14. Di Serio A, Ibáñez MB, Delgado C (2013) Impact of an augmented reality system on students' motivation for a visual art course. Comput Educ 68:586–596. https://doi.org/10.1016/j.compedu.2012.03.002
15. Siltanen S (2012) Theory and applications of marker-based augmented reality. VTT Science 3. ISBN 978-951- 38-7449-0 (soft back ed.)
16. Aukstakalnis S (2016) Practical augmented reality: a guide to the technologies, applications, and human factors for AR and VR. Addison Wesley Professional, Boston
17. Lanham M (2017) Augmented reality game development. Packt Publishing Ltd., Birmingham
18. De Freitas S, Rebolledo-Mendez G, Liarokapis F, Magoulas G, Poulovassilis A (2009) Learning as immersive experiences: using the four-dimensional framework for designing and evaluating immersive learning experiences in a virtual world. Br J Educ Technol 41(1):69–85
19. Kirkley BSE, Kirkley JR (2004) Creating next generation blended learning environments using mixed reality. Video Games Simul Techtrends 49(3):42–53
20. Haag J (2013) Using augmented reality for contextual mobile learning. Retrieved 23 Jan 2018, from https://www.learningsolutionsmag.com/articles/1310/using-augmented-reality-for-contextual-mobile-learning

21. Bower M, Howe C, McCredie N, Robinson A, Grover D (2013) Augmented reality in Education; cases, places, and potentials. In: 2013 IEEE 63rd annual conference international council for education media (ICEM), pp 1–11. https://doi.org/10.1109/CICEM.2013.6820176

22. Dunleavy P, Bassili-Gallo JF, Bocci M, Brieba D, Kim T, Tinkler J (2009) Delivering decentralised public services: how e- government adds value. Paper for OECD, Paris: OECD

23. Matcha W, Rambli DRA (2011) Preliminary Investigation on the Use of Augmented Reality in Collaborative Learning. Int Conf Inform Eng Inf Sci 254:189–198

24. Chatzopoulos D, Bermejo C, Huang Z, Hui P (2017) Mobile augmented reality survey: from where we are to where we go. IEEE Access PP(99) 1. https://doi.org/10.1109/ACCESS.2017.2698164

25. Ramanujam RSU, Veluchamy P, Joy B (2016) Augmented reality adaptive web content. In: IEEE consumer communications & networking conference, Las Vegas USA. https://doi.org/10.1109/ccnc.2016.7444740

26. Lin TJ, Duh HBL, Li N, Wang HY, Tsai CC (2013) An investigation of learners' collaborative knowledge construction performances and behavior patterns in an augmented reality simulation system. Comput Educ 68:314–321

27. Johdi MS, Baharom M, Said AA (2013) The significant of History curriculum in the development of active citizen: a critical analysis of the ICSS History, Malaysia. Int J Sci Res Publ 3(9):1–8

28. Rajasekar S, Philominathan P, Chinnathambi (2013) Research methodology. Retrieved June 12, 2013, from http://arxiv.org/abs/physics/0601009

29. Schlegel MJ (1995) A handbook of instructional and training program design. ERIC Document Reproduction Service ED383281

30. Linda K (2015) 5 steps to developing an effective training plan. Adobe systems incorporated. USA

Augmented Reality Model to Aid Al-Quran Memorization for Hearing Impaired Students

Hayatunnufus Ahmad, Norziha Megat Mohd. Zainuddin, Rasimah Che Mohd Yusoff, Nurulhuda Firdaus Mohd Azmi and Wan Azlan Wan Hassan

Abstract Industrial Revolution 4.0 becomes the driver of innovation like Augmented Reality. Nowadays, Augmented Reality is applied in the education area to prompt and enhance students' learning ability. Islamic education is not spared from this; innovations in Islamic educational materials should be enhanced so that students are able to learn better, especially among the special needs students. Hence, this research work is aimed at developing an integrated software model, which can assist the Quran memorization among the hearing-impaired students. This proposed prototyped model is based on Augmented Reality Based Content (ARBC). It is called mAR-Quran and it meets the Augmented Reality environmental development requirement. It enables the students to arrange the sequence verses of surahs in the correct order so that eventually, they memorize verses of Quran better and easier.

Keywords Augmented Reality · Hearing impaired · Al-Quran memorization

H. Ahmad · N. M. Mohd. Zainuddin (✉) · R. C. M. Yusoff · N. F. M. Azmi
Razak School of Technology and Advance Informatics, Universiti Teknologi Malaysia, Jalan Sultan Yahya Petra, 54100 Kuala Lumpur, Malaysia
e-mail: norziha.kl@utm.my

H. Ahmad
e-mail: hayatunnufus2@graduate.utm.my

R. C. M. Yusoff
e-mail: rasimah.kl@utm.my

N. F. M. Azmi
e-mail: huda@utm.my

W. A. W. Hassan
Faculty of Communication, Visual Art & Computing, UNISEL, 45600 Bestari Jaya, Selangor, Malaysia
e-mail: wanazlan@unisel.edu.my

© Springer Nature Singapore Pte Ltd. 2019
V. Piuri et al. (eds.), *Intelligent and Interactive Computing*, Lecture Notes in Networks and Systems 67, https://doi.org/10.1007/978-981-13-6031-2_12

1 Introduction

Industrial Revolution 4.0 has changed the direction of an organization in achieving its goal [1]. It has also introduced new innovation such as Augmented Reality (AR) in the education sector. Thus, this has attracted many researchers and industrial players to deeply discover the potential of AR technology toward students learning environment [2]. The same goes with Islamic education, there is a need for Islamic teaching materials and approaches to be integrated with technology especially among the Hearing Impaired (HI) students [3]. The Quran is a book that is compulsory for each Muslim to learn including the disabled. Nevertheless, Al-Quran education is less common among the HI [4–6] as many people believe that HI community is being given the exception to learn Al-Quran. This belief has made learning the Quran among the HI children to be less important and to a certain extent, taken for granted.

A memorization method called Tahfz Akhyar has been introduced to enable HI children to learn Al-Quran [7]. However, these students faced difficulty in arranging the verses of surahs in a correct sequence and made mistakes while arranging the verses of surahs [8]. Thus, the use of technology can help HI students to learn Al-Quran better and eventually improve the students' learnability [9] as well as the memorization quality [10]. Therefore, the objective of this article is to develop the software development model to aid the HI students to memorize Al-Quran based on AR technology.

1.1 Industry 4.0

Since the early nineteenth century, the world has experienced several modernizations in the way things were done and now, the world is experiencing the Industrial Revolution 4.0 [11]. The term Fourth Industrial Revolution (FIR) and Industry 4.0 refer to the same concept, in which many industries choose FIR as it is more familiar than Industry 4.0 [11, 12]. Industry 4.0 is considered as the next phase in the digitalization of manufacturing sector and it is driven by four changes which are: (i) the astonishing rise in data volumes, (ii) computational power, (iii) connectivity, especially new low power wide-area network and new forms of human–machine interaction such as touch interface and AR system and (iv) improvement in transferring digital instructions to the physical world, such as the use of advanced robotics and 3D painting [12]. Figure 1 shows the digital compass for Industry 4.0 that consists of new technologies with their value drivers.

Based on Fig. 1, we can see that Industry 4.0 paradigms are becoming the driver in the development of a new generation of digital technical instructions, such as AR and Virtual Reality(VR) that exploit more graphical and visual elements [13]. Indeed, the implantation of AR technology has been adopted in SmartFactory Lab hosted by German Research Center for Artificial Intelligence (DFKI), in which they used mobile devices and AR to demonstrate the visual computing visualization for

Industry 4.0 levers

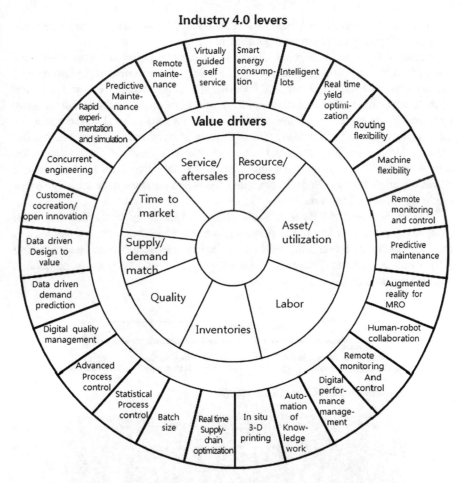

Fig. 1 Digital compass of Industry 4.0 Levers with its Value Drivers (*Source* [12])

accessing and analyzing the information generated in an integrated intelligent factory [14]. Not only that, AR technology also has been used in other factories to support workers in the rapid changing production environment [11].

1.2 Augmented Reality (AR) Technology

AR refers to the augmentation or meditate digital information like graphics, video, GPS into the real environment by utilizing the device camera [15]. Milgram et al. [16] states that AR refers to the mixed reality that combines the real environment and virtual environment that is in a single display. Figure 2 shows the reality–virtuality continuum, AR is known as Mixed Reality (MR) that can be an interchange between the real environment and virtual environment.

Industry 4.0 is more than just a change in a technology-driven environment. The innovations that are due to Industry 4.0 positively impact our core industries and sectors such as education, health, and business. It has changed the focus and direction of the educational sector by redefining conventional learning methods and teaching by bringing new technologies into classrooms [1]. According to [17], the quality of teaching and learning in the future year of 2025 in Malaysia can be achieved through the use of virtual learning and creation of the virtual university. Figure 3 shows the four future scenarios of teaching and learning in 2025 based on the anticipatory workshop attended by over 50 lectures and deans and conducted by the Ministry of Higher Education, Malaysia.

The establishment of VR and AR universities will enable lectures doing their work from anywhere and at any time. Moreover, it has been proven that AR can help to increase motivation among students [18, 19], enhance students' learning satisfaction [20], increase students' retention of knowledge [18, 19], and provide ubiquitous learning experience [2].

Fig. 2 Reality–Virtuality Continuum (*Source* [16])

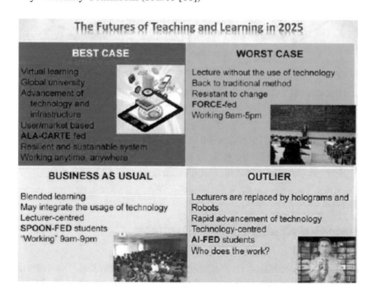

Fig. 3 The Futures of Teaching and Leaning in 2025, Malaysia (*Source* [17])

1.3 Background of Problem

The use of AR in the education system from preschool to tertiary can be seen from the implementation in classrooms such as in preschool science subject [21], science subject in secondary school [22, 23], astronomy [24], arithmetic [25], geometry [26], and computer science [27]. However, according to [3], there is a need for Islamic learning materials to be incorporated with modern methods and technologies.

Learning the book of Quran is compulsory for all Muslims. It is a gift from Allah to all Muslims as it has explanations, warnings, and guidelines for Muslims and others on how to live life here and in the world hereafter. All Muslims regardless of age, gender, race, ability, or wealth are required to learn the Quran. However, the method in learning Quran among HI students needs to be improved [4–6]. This is due to the notion that many people believe that having hearing impairment is an acceptable excuse to not read the Quran [9]. Moreover, people with a hearing problem have difficulty to hear and utter sounds as Quran is usually taught by reading repeatedly and hearing how the word is pronounced with good makhraj and Tajweed [9]. Not only that, they also use sign language, writing, and gesture acts as the communication medium [4, 28] that requires all words to be translated into the sign language. However, many of the HI are not able to understand or grasp something abstract and some of the Islamic terms [29]. Therefore, Tahfiz Akhyar, the Al-Quran memorization method was introduced by Hj Norakyairee Hj Raus and Zaharatul Sophia Mohamed Amir Abas in 2014 that is based on Prophet Muhammad's teaching method to memorize *doa, zikr, selawat,* and *solah* [7]. The method is based on arranging the pieces verses of surah in Al-Quran in the correct sequence. However, according to [8], HI students faced difficulty in arranging the verses of the surah in the correct order and made mistakes when arranging the verses of the surah. Thus, the use of technology can help the HI students to learn Al-Quran better and eventually can improve the students' learnability [9] as well as the memorization quality [10].

2 Methodology

There are many software development methodologies that have been developed. In this research work, the prototyped model which is based on AR Based Content (ARBC) Model is discussed.

2.1 Prototype Model

A prototype model is a software that is developed and partially produced and it is examined by both user and developer [30]. Figure 4 depicts the prototype model that consists of sequence phases, in which the next phase can be executed after the current phase has been completed [31]. The four main phases are the following:

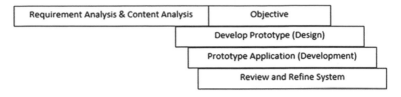

Fig. 4 The Sequence Prototype Model (*Source* [31])

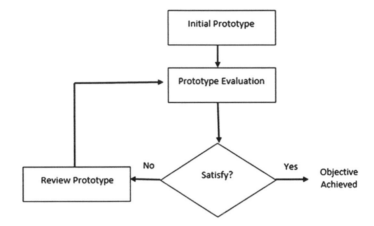

Fig. 5 Prototype Model Process (*Source* [32])

  i. Determining the basic requirement.
 ii. Developing the initial prototype.
iii. Reviewing the prototype by users.
 iv. Revising and enhancing the prototype by developers.

Figure 5 shows the process of the prototype model. It is iterative in which the version of the proposed prototype will be updated based on the users' feedback until all the requirements have been fulfilled [30]. Once the proposed prototype has met the users' satisfaction, the iterative cycle will be stopped as it has already achieved the development objective.

This prototype model has several advantages. It takes less time to be developed. Hence, it is economical and takes a shorter time to be delivered. In addition, it is able to satisfy users' learning and it also helps developers to define and refine the system requirement. Moreover, it can be integrated with other model and the training would then start earlier in the life cycle [30].

2.2 AR Based Content

ARBC model was introduced by [33] specifically for AR development process that consists of five phases which are analyze, determine, produce, use, and evaluate as shown in Fig. 6.

A brief explanation of each step is given in the following paragraphs:

i. Analyze

In this step, a problem needs to be identified and the system requirements need to be gathered to understand the purpose of employing AR technology.

ii. Determine

Developers need to clearly define the objectives of the content in order to understand the purpose of AR development, determine teachers' or students' role in using the application. In addition, the application's interface such as the object and its colors, dimension and the movement in the system need to be described. Besides, software and hardware requirement needs to be determined in order to produce the objects that are needed for the application, the type of software authoring (Adobe Illustrator, 3DMax, AR builder, and Aurasma or Unity) and the types of hardware needed (personal computer, tablet, projector, high-definition camera, or Kinect camera) should be well clarified. Moreover, the learning environment that includes the lighting, electricity supply, safety features, and ventilation should also be described.

iii. Produce

In this step, the AR objects will be produced to develop the AR application. The objects can be in either 3 Dimension (3D) or 2 Dimension (2D). They need to be designed by using suitable software that has been identified. Also, in this phase, there is a need to design the marker which is related to the objects and this is done by employing suitable software to connect both marker and objects.

iv. Use

The developed application is applied in the classroom to measure or observe students' improvement in their learning. Teachers then need to train students to use the application, observe their interactions, monitor their learning strategies, do a lot of reinforcements, and provide ample motivations to students.

v. Evaluate

The evaluation or feedback from users is taken into account in evaluating the effectiveness of the software application and its implementation. In this step,

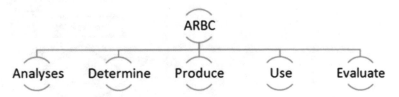

Fig. 6 AR Based Content (ARBC) Model (*Source* [33])

the AR evaluation needs to be carried out to determine whether it runs smoothly without an error or bug, the objects appear in the correct manner and students understand the whole content or otherwise.

3 Findings

Based on the process of that the prototype model went through, the findings are presented in this section. Figure 7 illustrates the development process of this prototyped model, and the findings are discussed in the following paragraphs.

In this mAR-Quran model, the ARBC model was adapted according to the name of phases and four phases were employed which were analysis, determine, produce, and use. While for the evaluation phase, the prototype model was assessed to determine users' satisfaction in using the model. During the analysis phase, five hearing impaired students have been involved in the preliminary study to identify the students' behavior toward Al-Quran education. The results from the preliminary study have been used for the next phases to produce the prototype. It follows the prototyping model where each phase is in the sequential order, in which the coming phase can only be executed after the present phase has been completed.

Furthermore, the iterative cycle is embedded in this model to allow the process to repeat. This enables the developers to make changes easily and in accordance with the user requirement. The feedback for users was prepared by using smiley and sad icons. When users were arranging the verses of surahs, they would get the feedback. If the arrangement was in the right order, they would get a smiley icon and if the arrangement was in the wrong order, they would receive a sad icon. The use of icons would definitely help users to identify whether they had the correct order of verses and surahs. This would make their learning easier, fresher and more effective. Figure 8a, b show the mentioned feedback. Therefore, we can say that this mAR-

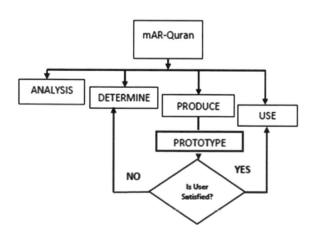

Fig. 7 Proposed Model, mAR-Quran Model

(a)   **(b)**

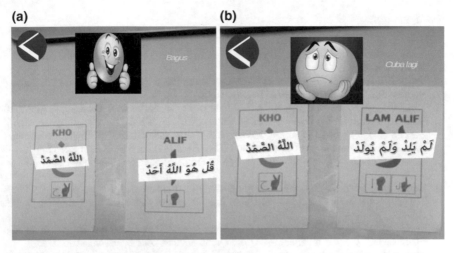

Fig. 8 **a** Feedback answer for the correct arrangement. **b** Feedback answer for the wrong arrangement

Quran model can be used to fully develop the AR application to aid the Al-Quran memorization for HI students.

4 Conclusion and Future Works

Revolution Industry 4.0 has brought about great innovations in technology and the education sector is also experiencing the benefit of these innovations. In line with the birth of these innovations, a new learning approach is needed in order to help students to learn better. Therefore, this research work has developed a prototype model using the AR technology to assist the special need students to learn Quran. It has been discovered that this prototype model can be further developed in producing software to aid Al-Quran memorization among the hearing impaired students.

For future work, the proposed model will be evaluated using heuristics evaluation and end-user testing. The scaffolding learning approach will also be embedded in the application to enable the students to have a rough idea of AR technology before they can start with the content. Moreover, reward and punishment learning approach can also be employed in improving their level of motivation and the quality of their memorization.

Acknowledgements This research was supported by Universiti Teknologi Malaysia under Research University Grant, (Q.K 130000.2538.15H95). We thank Madam Norazmah Suhaila for comments that greatly improved the manuscript.
We would also like to show our gratitude to the teachers and students in Yayasan Faqeh for their involvement in the preliminary study.

References

1. Xu M, David JM, Kim SH (2018) The fourth industrial revolution: opportunities and challenges. Int J Financ Res 9:90–95
2. Santos MEC et al (2016) Augmented reality as multimedia: the case for situated vocabulary learning. Res Pract Technol Enhanc Learn 11.1–23
3. Yasin AM, Ali M, Isa M, Endut NA (2016) Interactive Prophet's storybook using Augmented Reality. In: International conference on e-learning 2015, pp 391–399
4. Hussain A, Jomhari N, Kamal FM, Mohamad N (2014) mFakih: modelling mobile learning game to recite Quran for deaf children. Int J Islam Appl Comput Sci Technol 2:8–15
5. Jaafar N et al (2014) Quran education for special children: teacher as murabbi. Creative Educ 5:435–444
6. Sabdan MS, Alias N, Yusof M, Yakub M @ Zulkifli, Jomhari N, Daud M, Aziah N (2016) Tinjauan Pelaksanaan Pendidikan al-Quran Bagi Golongan Pekak Di Malaysia. J al-Turath 1:43–51
7. Mohd Rashid SM (2017) Kemahiran Asas Serta Kaedah Pengajaran dan Pembelajaran Al-Quran Kepada Golongan Istimewa. J-QSS J Quran Sunnah Educ Spec Needs, no. June
8. Ahmad H, Zainuddin NMM, Ali R, Maarop N, Yusoff RCM, Hassan WAAW (2018) Augmented reality to memorize Al-Quran for hearing impaired students: a preliminary analysis. J Fundam Appl Sci 10:91–102
9. Ghadim N, Jomhari N, Alias N (2013) Mother's Perspective toward Al-Quran education for hearing impaired children in Malaysia. Malays Online J Educ Technol 1:26–30
10. Hashim A, Tamuri AH, Jemali M (2013) Latar Belakang Guru Tahfiz Dan Amalan Kaedah. Online J Islam Educ 1:28–39
11. Lu Y (2017) Industry 4.0: a survey on technologies, applications and open research issues. J Ind Inf Integr 6:1–10
12. Sung TK (2017) Industry 4.0: a Korea perspective. Technological forecasting and social change
13. Wally G, Gattullo M, Fiorentino M, Ferrise F, Bordegoni M, Emmanuele A (2018) Converting maintenance actions into standard symbols for Augmented Reality applications in Industry 4.0. Comput Ind 98:68–79
14. Posada J et al (2015) Visual computing as a key enabling technology for industrie 4.0 and industrial internet. Visual computing challenges, pp 26–40
15. Agarwal A et al (2014) Mobile application development with Augmented Reality. Int J Comput Sci 5:20–25
16. Milgram P, Takemura H, Utsumi A, Kishino F (1994) Mixed Reality (MR) Reality-Virtuality (RV) continuum. Syst Res 2351:282–292
17. Inayatullah S, Milojević I (2014) Augmented reality, the Murabbi and the democratization of higher education: alternative futures of higher education in Malaysia. Horiz 22:110–126
18. Radu I (2014) Augmented reality in education: a meta-review and cross-media analysis. Pers Ubiquit Comput 18:1533–1543
19. Diegmann P, Schmidt-kraepelin M, Van Den Eynden S, Basten D (2015) Benefits of Augmented Reality in educational environments—a systematic literature review. In: 12th International conference on Wirtschaftsinformatik, pp 1542–1556
20. Akcayir M, Akcayir G (2017) Advantages and challenges associated with augmented reality for education: a systematic review of the literature. Educ Res Rev 20:1–11
21. Trnova E, Trna J (2015) Formation of science concepts in pre-school science education. In: 7th world conference on educational sciences (WCES-2015), pp. 2339–2346
22. Gopalan V, Zulkifli AN, Abu Bakar JA (2016) A study of students' motivation using the augmented reality science textbook. In: Proceedings of the international conference on applied science and technology, pp. 20040-1–20040-6
23. Cheng K-H, Tsai C-C (2013) Affordances of augmented reality in science learning: suggestions for future research. J Sci Educ Technol 22:449–462

24. Yen J-C, Tsai C-H, Wu M (2013) Augmented Reality in the higher education: students' science concept learning and academic achievement in astronomy. In: 13th international educational technology conference augmented, pp 165–173

25. Young JC, Kristanda MB, Hansun S (2016) ARmatika: 3D game for arithmetic learning with Augmented Reality technology. In: International conference on informatics and computing, ICIC 2016, pp 355–360

26. Olalde K, García B, Seco A (2013) The importance of geometry combined with new techniques for augmented reality. In: International conference on virtual and Augmented Reality in education 25:136–143

27. Kose U, Koc D, Yucesoy SA (2013) An augmented reality based mobile software to support learning experiences in computer science courses. In: International conference on virtual and Augmented Reality in education, vol 25, pp 370–374

28. Hidayat L, Hidayatulloh F (2017) Multimedia based learning materials. Eur J Spec Educ Res 2:77–87

29. Awang MH, Zakaria HB, Mohd Shafie BH, Talib NHF, Kassim N (2010) Persepsi Pelajar Bermasalah Pendengaran terhadap Pembelajaran Fardu Ain: Cabaran terhadap Guru. In: Proceedings of the 4th international conference on teacher education: join conference UPI & UPSI, pp 240–250

30. Ben-zahia MA, Jaluta I (2014) Criteria for selecting software development models, pp 1–6

31. Amin REM (2007) Perisian Kursus Multimedia Pembacaan Awal bagi Kanak-kanak Sindrom Down (MEL SindDown). Universiti Kebangsaan Malaysia

32. Ubaidullah NH (2007) Perisian Kursus Multimedia dalam Literasi Matematik (D-Matematika) untuk Pelajar Disleksia. Universiti Kebangsaan Malaysia

33. Aqel MS (2017) A new model for an Augmented Reality Based Content (ARBC) (A case study on the Palestinian curriculum). IOSR J Res Method Educ 7:95–100

Augmented Reality: A Systematic Literature Review and Prospects for Future Research in Marketing and Advertising Communication

Fatrisha Mohamed Yussof, Sabariah Mohamed Salleh
and Abdul Latiff Ahmad

Abstract The combination of digital technology and the reality has further strengthened the aspect of modern society life to cater for new developments in the industrial revolution 4.0. Augmented reality is a technology that superimposes computer-generated images on a user's view of the real world. However, the knowledge of augmented reality in the context of marketing and advertising communication is still scarce that needs to be discovered to identify the necessities in the study. Thus, by conducting a systematic literature review, this article seeks to answer four main research questions: (1) Which areas of augmented reality research have been conducted in? (2) What methodologies are being utilized to research augmented reality in advertising and consumer behaviours? (3) What theories or construct are being utilized to research augmented reality in advertising and consumer behaviours? (4) What are the research gaps in augmented reality in advertising and consumer behaviours? The study was a review of the 10 years period of publication from 2009 until 2018.

Keywords Augmented reality · Technology · Marketing · Advertising · Consumer behaviour

1 Introduction

In the context of marketing, according to Kotler and Armstrong [1], there are four P's components; the product, place, promotion (advertising) and price that are geared to successful marketing of brand companies. The proliferation of digital technologies and advanced media offer many opportunities to advertisers, marketers and promoters for generating advertising to promote products that engage consumers to notice, spend time, and get their messages known and absorbed. However, advertisers have difficulties in grabbing consumers' attention and alertness due to an increasing competitive message environment that urge the advertising to be significant [2, 3]. Based

F. M. Yussof (✉) · S. M. Salleh · A. L. Ahmad
Media and Communication Department, Social Sciences & Humanities Faculty,
National University of Malaysia, Bandar Baru Bangi, Malaysia
e-mail: fatrishayussof@yahoo.com

© Springer Nature Singapore Pte Ltd. 2019
V. Piuri et al. (eds.), *Intelligent and Interactive Computing*, Lecture Notes in Networks
and Systems 67, https://doi.org/10.1007/978-981-13-6031-2_36

on reports from 'Advertising Expenditure' (ADEX) in early 2017, there has been a steady decline over the past few years towards the advertising expenditure of traditional media such as advertisement in magazines, newspapers and televisions [4, 5]. Thus, the technology change has also given marketers access to think of new alternatives for targeting and interacting with audiences such as augmented reality or virtual reality [6].

In addition, augmented reality has the strong advantage for contributing of integrated marketing programmes and offering various potential to implement experience marketing to enhance brand communication [7, 8]. In addition, it helps the marketers to quickly gain advantages on their competitors [9].

Thus, augmented reality could be used in marketing to motivate people in order to grab their attention on advertisement appeal and to improve perception and recall of the advertising message that led to purchasing intention behaviour. Unfortunately, although we are in the wave of advancing technology, there is often lack of empirical study available regarding consumer behaviours towards advertising that adopt augmented reality as compare to other fields such as education, game and science health. Thus, the aim of this paper is to discover the situation in augmented reality in terms of consumer behaviour of technology usage in marketing and advertising communication, or any possible future research within the context.

The author has conducted a Systematic Literature Review (SLR) research to get the overview. SLR refers to a deep reviewing of previous literature that addresses the problem by identifying, critically determining and synthesizing, the findings of all relevant, high-quality individual studies addressing one or more research questions [10]. By mapping what is known, this review will be providing insights or the comprehensive view of prospects for future research in augmented reality marketing and advertising communication which is increasingly important field of studies. Therefore, the study sought to answer the research questions as stated (1) Which areas of augmented reality research have been conducted in? (2) What methodologies are being utilized to research augmented reality in advertising and consumer behaviours? (3) What theories and construct are being utilized to research augmented reality in advertising and consumer behaviours? (4) What are the research gaps in augmented reality in advertising and consumer behaviours?

2 Literature Review

2.1 Augmented Reality Overview

Augmented Reality (AR) is a merger of reality and virtual objects in a real environment that runs in actual time and interactive to immerse in realistic experience [11, 12]. Basically, AR was created and developed by Ivan Sutherland in 1960 to be adopted for health applications, engine maintenance and information systems [13]. However, since the rise of the consumption of smartphones and digital devices, AR technology has been extended to several new functions such as advertising, gaming, tourism and education. There are various ways to experience AR technology through devices such as head-mounted display, laptops, smartphones, tablets and most recently is the use of smart glasses.

AR is also known to have two categories which are marker-based and marker-less AR [14]. For based-markers AR, it requires an image labelled specifically to locate 3D object positions on real-world images [15] while marker-less AR using a combination of compass and location (GPS) and electronic devices by pointing to a position in the physical world for transmitting information in 3D [9]. The technology can provide users with information that is relevantly significant and pleasurable. Interacting with different environments through AR could provide a novel and vast experiences [16]. The interest of the developers and companies in using AR has significantly increased due to the ubiquitous adoption of smartphones and other mobile devices. Consequently, AR is shifting from the laboratory into consumer markets [17].

2.2 Marketing and Advertising Using 'Augmented Reality' Technology

The advertising industry is now in a critical condition. It has experienced a period of unprecedented change due to the emergence of new media that challenges business models and industrial structures [18]. AR seems to provide advantages to the industry that should be adopted to win consumers' decision and positively impact consumers [6, 19]. According to Scholz and Smith [8] AR has great opportunities in contribution to integrated marketing communications (IMC) programmes by stimulating the communication among consumers in fulfilling advertising objectives to showcase different and interesting ideas so that consumers are more likely to share in social media, and indirectly promote word of mouth. However, underdeveloped approach initiatives will lead to failure to connect consumers and adversely affect branding images, resource waste and spoil programmes that uses AR in the future [8]. Despite being in the world of high technology, there is lack of awareness among the consumers regarding the AR technology [20]. Additionally, the majority of consumers have expressed their desire to use AR as it is innovative and innovative to attract attention.

3 Methodology

The Systematic Literature Review (SLR) that has been conducted is a category of descriptive. The descriptive review has variations in how data are determined, analysed and integrated which is consisted of narrative review (Green et al. 2001, as cited in Xiao and Watson [21], textual narratives synthesis (Popay et al. 2006; Lucas et al. 2007, as cited in Xiao and Watson [21], meta-summary (Sandelowski et al. 2007, as cited in Xiao and Watson [21], metanarrative (Greenhalgh et al. 2005, as cited in Xiao and Watson [21], and scoping review (Arksey and O'Malley 2005, as cited in Xiao and Watson [21]. However, the author aims at focusing scoping review, which is extracting as many relevant data from each piece of literature as possible including methodology, finding, variables or any important inclusion to provide a complete overview of what has been done [21]. However, the scoping reviews need to be comprehensive even though the research quality is not emphasized [22]. Therefore, the method was deemed the most suitable for the purposes of this paper because there are scarce in the knowledge of augmented reality in marketing and advertising communication.

The author also tends to include the iterative assortment process to ensure there is no data will be an exception. According to Irshad and Rambli [23], the iterative assortment is used to get the most related results by analysing and refined iteratively to maintain the quality of research. However, the author has set the limitation to exclude 'grey literature' such as pattern, thesis, news report and blogs to ensure the SLR is reliable and it used appropriate sources. The keyword filters were applied to discard any vague or ambiguous search results by carefully reviewed the articles during the search. The articles containing augmented reality marketing or advertising studies were identified. Finally, the researcher categorized AR publications based on consumer behaviour in marketing and advertising communication.

3.1   The Process

The main aim is to identify the comprehensive view of prospects for future research in augmented reality marketing and advertising communication hence provides evidence to fill the gaps in research. The six steps are utilized in the SLR will be described and design in the flow chart at the following subsections.

3.1.1   Step 1—Planning

At the initial phase, planning is a crucial step to conduct SLR. Google Scholar and Mendeley Desktop database have been utilized for the SLR due to its rigorous search. The duration of the publication search was set and identified within ten years from 2009 to 2018.

3.1.2   Step 2—Research Scope

Second, by conforming the research scope will lead to answerable research questions. An initial brainstorming, focusing context and literature search has been defined in the early stage as an iterative process. Then, the research questions have been defined as:

Q1 Which contexts of augmented reality research have been conducted in?
Q2 What methodologies are being utilized to research augmented reality in advertising and its consumer behaviours?
Q3 What theories and construct are being utilized to research augmented reality in advertising and its consumer behaviours
Q4 What are the research gaps in augmented reality in advertising and its consumer behaviours?

The research questions were set in order to help the researcher to give an overview for future prospects in Augmented Reality Advertising.

3.1.3 Step 3—Searching

The search step consists of browsing the databases by utilizing the keyword of 'Augmented Reality' AND 'Advertising', 'Augmented Reality' AND 'Advertising' AND 'Marketing'. The idea was selected based on the research questions and context stated in the previous section. The word 'AND' is used to provide an accurate screening. The search started on 13 February 2017. The time allocation was based on the availability of researcher until it has reach to an exhaustive search. From the search, it was found that 12 600 documents on Google Scholars and 512 733 from Mendeley Desktop.

3.1.4 Step 4—Assessing

The assessing phase aims to scrutinize and narrow down the articles by finalizing the quantity of documents that are relevant to context study. The inclusion and exclusion criteria have been set up as below:

First, the screening was done by reviewing the title, journals and abstract by the following inclusion criteria which were:

(1) The primary study that represents augmented reality in advertising communication
(2) The study that represents the augmented reality in marketing
(3) The context study is in consumer behaviours and perception
(4) Peer-reviewed journals or proceeding

Next, the deep screening was done to exclude the number of articles as the following criteria:

(1) The article that is not written in English
(2) The document is older than 2009
(3) A redundancy articles

(4) Inappropriate sources such as website pattern, thesis, news report, and blogs.

Thus, there were 89 articles found as a relevant document at the first screening. However, after following the exclusion criteria, the number of articles has reduced to 22 articles.

4 Data Analysis and Results

4.1 Related Research in Augmented Reality Advertising and Marketing

After assessing the documents, the next step was synthesizing and analysing the data in order to answer the RQ1. The data was tabulated into tables to calculate and view the trend of study (Table 1). There are the four areas of topic in the context of marketing and advertising that is related to consumers' behaviour and perception that has been studied previously from 2009 until 2018. In 3 recent years, the 'consumer decision-making' area of the topic was the most studied by researchers which is 42% followed by area of the topic 'consumer engagement' and 'consumer perception' which is 25%, respectively. There is only 8% of managerial perception study was done along the three recent years. However, it has shown that 2015 was the peak year for past researchers in understanding marketing and advertising that adopting AR technology. There is no study was conducted in 2009 among these four topics in the SLR. It is due to the limitation of the study that only considering the peer-reviewed articles.

4.2 Methodologies Usage in Augmented Reality Advertising and Marketing

The previous studies were solely an empirical data that have been through quantitative or qualitative research method. The majority of the research was conducted in a quantitative manner 63.6% while qualitative 36.3%. The usual type of quantitative research is an experimental design which is to test the theory that applied in topic areas of 'consumer engagement', 'consumer perception' and 'consumer decision-making' in marketing and advertising communication research. Qualitative research was conducted mostly around the year 2011, until 2015 that is at the early stage of understanding consumers' behaviour in the adoption of AR technology or other possibilities such as lacking of previous information on AR advertising and marketing.

Table 1 2009–2018 augmented reality advertising and marketing research

No.	Author and article title	Year	Topic area
1	Brengman, M., Willems, K & Kerrebroeck, H. M.,- Can't touch this: the impact of augmented reality versus touch and non! touch interfaces on perceived ownership	2018	Consumer decision-making
2	Phua, J., Kim, J.- Starring in your own snapchat advertisement: influence of self-brand congruity, self- referencing and perceived humor on brand attitude and purchase intention of advertised brands	2018	Consumer decision-making
3	Pantano, E., Rese, A., Baler, D., Online decision- making process by using augmented reality: a two country comparison of youth markets	2017	Consumer decision-making
4	Mark Yi-Cheon Yim & Shu-Chuan Chu & Paul L. Sauer, Is augmented reality technology an effective tool for E-commerce? An interactivity and vividness perspective	2017	Consumer decision-making
5	Scholz, J & Smith, A.N, Augmented reality: designing immersive experiences that maximize consumer engagement	2016	Consumer engagement
6	Hopp, T. & Harsha Gangadharbatla, Novelty effects in augmented reality advertising environments: the influence of exposure time and self-efficacy	2016	Consumer engagement
7	Syarifah Nurleyana Wafa, Ellyna Hashim, adoption Of MAR advertisements by brands In Malaysia	2016	Managerial perception
8	Rese, A., Baier, D., Schulz, A.G, Schreiber, S. -How augmented reality apps are accepted by consumers: a comparative analysis using scales and opinions	2016	Consumer decision-making
9	Shafaq Irshad and Dayang Rohaya Bt Awang, User perceptions oil MAR as a marketing tool	2016	Consumer perception
10	Ross, H.F, Harrison, T., Augmented reality apparel: an appraisal of consumer knowledge, attitude and behavioral intentions	2016	Consumer perception

(continued)

Table 1 (continued)

No.	Author and article title	Year	Topic area
11	Baratali, E., Mohd Helmi Abd. Rahim, Behrang Parhizkar, Zahra Mohana Gebril- A case study on brand interactive advertising effective of augmented Reality (Ar) in marketing communication	2016	Consumer perception
12	Mauroner, O., Le, L., Best, S., Augmented reality in advertising and brand communication: an experimental study	2016	Consumer engagement
13	Tony Liao-Augmented or admented reality? The influence of marketing on augmented reality technologies	2015	Managerial perception
14	Jasmina Stoyanova, Pedro Quelhas Brito, Petia Georgieva, Mariofanna Milanova-Comparison of consumer purchase intention between interactive and augmented reality shopping platforms through statistical analyses	2015	Consumer decision-making
15	O' Mahony, S., A proposed model for the approach to augmented reality deployment in marketing communications	2015	Managerial and consumer perception
16	Stockinge, H., Consumers' perception of augmented reality as an emerging end user technology: social media monitoring applied	2015	Consumer perception
17	Chao-Hung Wanga, Yi-Chen Chianga. Mao-Jiun Wanga-Evaluation of an augmented reality embedded on-line system shopping	2015	Consumer engagement
18	Djamasbi et al., Augmented reality and print communication	2014	Consumer engagement
19	Jungyeon Sung and Kwangsu Cho. User experiences with augmented reality advertising applications: focusing on perceived values and telepresence based cu the experiential learning theory	2012	Consumer perception
20	Ezgi EyuboGLu, Augmented reality as an exciting online experience: is it really beneficial for brands?	2011	Consumer perception

(continued)

Table 1 (continued)

No.	Author and article title	Year	Topic area
21	Marias Bulearca and Daniel Tamarjan, reality: a sustainable marketing tool?	2010	Consumer engagement
22	Connolly, P. Chambers, C. Eagleson, E. Matthews, D. and Rogers, T., augmented reality effectiveness in advertising	2010	Consumer engagement

4.3 Variables in Augmented Reality Advertising and Marketing Research

In recent years, the 'consumer decision-making' area of topic became crucial in the field because of technology advancement especially the industrial revolution 4.0 being introduced and discussed worldwide by scholars and practitioners. In the Brengman et al. [24] research, he wanted to discover the impact of augmented reality on users' perceived ownership of touchscreen mobile phone as compared to a laptop that affected product attitudes and purchase intentions. He measured the perceived ownership, interactivity, product type and product attitude on purchasing intention. While Phua and Kim [25] used self-referencing, self-brand congruity, perceived humour and attitude variables to measure purchasing intention in examined the self-endorsed brand advertisements on consumers' brand-related preferences. In Yim et al. [20] studies, the variables measured were interactivity, vividness, immersion, media usefulness and enjoyment, attitude on purchase intention. It was to identify the effectiveness of AR-based presentation compared to traditional media to understand the consumer's evaluation or attitude that leads to purchase intention. Other studies have concurrently conducted by Pantano et al. [26] which measures the aesthetic quality, interactivity, response time, quality of information, ease of use, usefulness, enjoyment and attitude on behavioural intention. While, Rese et al. [27] study measured the perceived informative, perceived enjoyment, ease of use usefulness, enjoyment and attitude on behavioural intention. The study was to investigate the effect of AR technologies on consumer behaviour within the online retail environments.

However, there are few studies on the 'consumer engagement' conducted in recent years. In Scholz and Smith [8] research suggested eight actionable recommendations which is ENTANGLE, the augmented reality initiatives. They suggested E-experiences, N-nourishing engagement, T-target audiences, A-aligning AR with the marketing programme, N-neutralizing threats, G-goals, L-leveraging brand meanings and E-enticing consumers. On contrary to Hopp and Gangadharbatla [28] studies which conducted an experiment to measure the product relevancy, perceived technological, self-concept, familiarity, gender and self-efficacy on attitudes towards the brand. In addition, Mauroner et al. [7] study measured the brand recall, brand recognition, ad credibility and perceived interactivity on attitude towards ads and attitude towards the brand.

In 'managerial perception' area of the topic, Wafa and Hashim [14] study was focusing on brand utilizing AR in newspaper ads. A study on 'consumers' perception' was also conducted that are measured the variables of control in your hands, enjoyments, experience, digitalizing product, pleasure, recommendation, easy shopping on AR [29]; perceived informative, perceived enjoyment, ease of use, usefulness, enjoyment on AR [30]; design, usability, performance, user friendly and popularity on AR [9].

4.4 Consumer Decision-Making, Engagement and Perception

In prior studies on 'consumer decision-making', Brengman et al. [24] found that perceived ownership positively affects product attitudes and purchase intentions. In addition, there is no interaction effect according to product type and insignificant difference in perceiving ownership between the mobile phone and the laptop interfaces. Phua and Kim [25] stated that greater perceived self-brand congruity through seeing oneself portrayed next to a brand in the ad can lead to a significantly more positive attitude and purchase intention of the advertised brand. He also found that greater perceived humour in the ads resulted in significantly more positive brand attitude and intent to purchase the advertised brand as compared to lower perceived humour. He also emphasized that advertisers should implement creative messaging strategies that effectively represent consumers alongside advertised brands to maximize positive brand outcomes. Furthermore, in Yim et al. [20] stated that AR is a powerful tool for supporting the decision-making process. Pantano et al. [26] found that interactivity and vividness generate diverse positive consumer that makes AR is a success medium as an information source to persuade consumers.

In 'consumer engagement' area of topic, Scholz and Smith [8] stated that AR has strong potential in an integrated marketing programme. In other related research, it was found that there were positive relationship exposure time and brand among high self-efficacy respondents [28]. They also stated that interactive examination with an advertisement leads to a better and more intense cognitive processing of the brand message. Furthermore, AR combines printed ads with virtual reality and has the potential to provide the consumer with product relevant experiences [7] and it has a positive impact on engagement [31]. It is increasingly important to understand how the consumer perceives AR. Ross and Harrison [30] suggests that the mainstream requires to inform the potential market of the possibilities that AR technology presents. It is because little known on augmented reality in relation to innovating business process and improving the value of ads [9]. Furthermore, it helps organizations to quickly gain advantages on their competitor.

5   Conclusion and Future Works

SLR research aims to answer the key research questions. The findings in context fields of marketing and advertising communication have been described based on the SLR. The recent technology adopted has been outlined among the 22 articles of this review has been provided. In general, AR technology in marketing and advertising communication is still immature and lacking of research that requires further discovery to stay consistent with the industrial revolution 4.0. The data extraction process has been explained and applied systematically. Therefore, the author believes this SLR provides a contribution to AR in marketing and advertising communication. The study could be used for anyone approaching AR at an industrial level as well as academic research. The future study that has been suggested by previous researchers may fill in the gap in the future study of marketing and advertising communication (Table 2).

Table 2 Recommendation of future study by previous researcher

ID	Author	Topic area	Recommendation
1	Brengman et al. [24]	Consumer decision-making	(1) Impact of AR for different types of products, such as apparel. (2) Determine the exact return on investment of the technology
2	Pantano et al. [26]	Consumer decision-making	(1) Older consumer. (2) Online consumer. (3) Measuring diffusion technology
3	Yim et al. [20]	Consumer decision-making	(1) Timing of purchase. (2) Consumer cognition and emotion. (3) Consumer learners oriented. (4) Relationship confidence and product infonnation. (5) AR increased sales and improved brand image

(continued)

Table 2 (continued)

ID	Author	Topic area	Recommendation
4	Hopp and Gangadharbatla [28]	Consumer engagement	(1) AR Familiarity. (2) Behavioural intention (3) Individual differences as means of identifying (he optimal conditions for the application and use of persuasive AR messaging
5	Wafa and Hashim [14]	Managerial perception	(1) Newspaper ads in AR. (2) Novelty effect of AR
6	Rese et al. [27]	Consumer decision-making	(1) The adoption of AR apps may vary depending on the cultural context. (2) Older consumer
7	Ross and Harrison [30]	Consumer perception	(1) Potential engagement of gender
8	Mauroner et al. [7]	Consumer engagement	(1) Low and high involvement conditions. (2) Age and gender differences. (3) Usability, acceptability and consumer habits
9	Liao [32]	Managerial perception	(1) AR stakeholders, implications of their decisions
10	Stockinge [33]	Consumer perception	Analysis of arousal, sentiments and emotionality
11	Djamasbi et al. [31]	Consumer engagement	Comparison of conventional media with technology to provide proof of concept for managers
12	Sung and Cho [34]	Consumer perception	User' s responses such as perceived hedonic, utilitarian and motivational value

(continued)

Table 2 (continued)

ID	Author	Topic area	Recommendation
13	Bulearca and Tamarjan [35]	Consumer engagement	(1) Findings on other cultures. (2) Emotional and functional value, and positive customer satisfaction
14	Connolly et al. [36]	Consumer engagement	(1) Impact of virtual reality technology. (2) Include user-controlled manipulation capability of the augmented reality image, including text-based information

References

1. Kotler P, Amrstrong G (2008) Principles of marketing. Pearson Prentice Hall, United States of America
2. Bakar MHA, Desa MAM, Mustafa M (2015) Attributes for image content that attract consumers' attention to advertisements. In: World conference on technology, innovation and entrepreneurship, Procedia—social and behavioral sciences, pp 309–314. https://doi.org/10.1016/j.sbspro.2015.06.349
3. Kim S, Haley E, Koo G-Y (2009) Comparison of the paths from consumer involvement types to ad responses between corporate advertising and product advertising. J Advertising: 67–80. doi: https://doi.org/10.2753/joa0091-3367380305
4. Malaysian Advertiser Associations (2017) 2017 full year ADEX report, http://www.malaysiaadvertisers.com.my/malaysia-adex-2017-full-year-adex-report/
5. Statista (2017) Change of advertising spending in 2016 and 2017 by medium. https://www.statista.com/statistics/240679/global-advertising-spending-growth-by-medium/
6. Yaoyuneyong G, Foster J, Johnson E, Johnson D (2016) Augmented reality marketing: consumer preferences and attitudes toward hypermedia print ads. J Interact Advertising:16–30. https://doi.org/10.1080/15252019.2015.1125316
7. Mauroner O, Le L, Best S (2016) Augmented reality in advertising and brand communication: an experimental study. Int J Inf Commun Eng: 422–425
8. Scholz J, Smith AN (2016) Augmented reality: designing immersive experiences that maximize consumer engagement. Bus Horiz: 149–161. https://doi.org/10.1016/j.bushor.2015.10.003
9. Baratali E, Helmi M, Rahim A, Parhizkar B, Gebril ZM (2016) Effective of augmented reality (AR) in marketing communication: a case study on brand interactive advertising. Int J Manag Appl Sci:133–137. doi: 4379
10. Siddaway A (2014) What is a systematic literature review and how do i do one? https://www.semanticscholar.org/paper/What-Is-a-Systematic-Literature-Review-and-How-Do-I-Siddaway/22142c9cb17b4baab118767e497c93806d741461
11. Azuma RT (1997) A survey of augmented reality. Presence: Teleoperators Virtual Environ 6(4):355–385
12. Craig AB (2013) Chapter 1–what is augmented reality? Understanding augmented reality. Concepts Appl 1–37. doi:0.1016/C2011-0-07249-6

13. Sutherland IE (1968) A head-mounted three dimensional display. In: Proceedings of the December 9–11, 1968, fall joint computer conference, part I (757–764). ACM. https://doi.org/10.1145/1476589.1476686
14. Wafa SN, Hashim E (2016) Adoption of mobile augmented reality advertisement by brands in Malaysia. In: 3rd global conference on business and social sciences 2015, Procedia—social and behavioral sciences, pp 762–768. https://doi.org/10.3402/rlt.v20i0.19189
15. Imbert I, Vignat F, Kaewrat I, Boonbrahm P (2013) Adding physical properties to 3D models in augmented reality for realistic interactions experiments. Procedia Comput Sci 25:364–369. https://doi.org/10.1016/j.procs.2013.11.044
16. Olsson T, Lagerstam E, Karkkainen T, Väänänen-Vainio-Mattila K (2013) Expected user experience of mobile augmented reality services: a user study in the context of shopping centres. Personal Ubiquitous Comput. 17(2):287–304. http://dx.doi.org/10.1007/s00779-011-0494-x
17. Daponte P, De Vito L, Picariello F, Riccio M (2014) State of the art and future developments of the augmented reality for measurement applications. Measurement 57:53–70. https://doi.org/10.1016/j.measurement.2014.07.009
18. Pfeiffer M, Zinnbauer M (2010) Can old media enhance new media. J Advertising Res: 43–49. https://doi.org/10.2501/s0021849910091166
19. Vidal E (2014) Are marketers seeing the whole picture? The rise of augmented reality. http://www.marketingprofs.com/opinions/2014/24758/are-marketers-seeing-the-whole-picture-the-rise-of-augmented-reality
20. Yim MY-C, Chu S-C, Sauer PL (2017) Is augmented reality technology an effective tool for E-commerce? An interactivity and vividness perspective. J Interact Market 39:89–103. https://doi.org/10.1111/j.1460-2466.1995.tb00748.x
21. Xiao Y, Watson M (2017) Guidance on conducting a systematic literature review. J Plann Educ Res. 0739456X1772397. https://doi.org/10.1177/0739456x17723971
22. Peters MDJ, Godfrey CM, Khalil H, McInerney P, Parker D, Soares CB (2015) Guidance for conducting systematic, scoping reviews. Int J Evid-Based Healthc 13(3):141–146
23. Irshad S, Rambli DRA (2014) User experience of mobile augmented reality. A Rev Stud:125–130. https://doi.org/10.1109/iuser.2014.7002689
24. Brengman M, Willems K, Kerrebroeck HV (2018) Can't touch this: the impact of augmented reality versus touch and non-touch interfaces on perceived ownership. Virtual Reality:1–12. https://doi.org/10.1007/s10055-018-0335-6
25. Phua J, Kim J (2018) Starring in your own snapchat advertisement: influence of self-brand congruity, self-referencing and perceived humor on brand attitude and purchase intention of advertised brands. Telematics Inform. https://doi.org/10.1016/j.tele.2018.03.020
26. Pantano E, Rese A, Baier D (2017) Enhancing the online decision-making process by using augmented reality: a two country comparison of youth markets. J Retail Consum Serv 38:81–95. https://doi.org/10.1016/j.jretconser.2017.05.011
27. Rese A, Baier D, Geyer-Schulz A, Schreiber S (2017) How augmented reality apps are accepted by consumers: a comparative analysis using scales and opinions. Technol Forecast Soc Change 124:306–319. https://doi.org/10.1016/j.techfore.2016.https://doi.org/10.010
28. Hopp T, Gangadharbatla H (2016) Novelty effects in augmented reality advertising environments: the influence of exposure time and self-efficacy. J Curr Issues Res Advertising 37(2):113–130. https://doi.org/10.1080/10641734.2016.117117
29. Irshad S, Rambli DRA (2016) User perception on mobile augmented reality as a marketing tool 109–113. https://doi.org/10.1109/iccoins.2016.7783198
30. Ross HF, Harrison T (2016) Augmented reality apparel: an appraisal of consumer knowledge, attitude and behavioral intentions. In: 2016 49th Hawaii international conference on system sciences (HICSS), pp 3919–3927. https://doi.org/10.1109/hicss.2016.487
31. Djamasbi S, Wyatt J, Luan X, Wang H (2014) Augmented reality and print communication. In: Proceedings of the 20th, Americas conference on information systems (AMCIS), Savannah, Georgia, pp 1–9
32. Liao T (2015) Augmented or admented reality? The influence of marketing on augmented reality technologies. Inf Commun Soc 18(3):310–326. https://doi.org/10.1080/1369118x.2014.989252

33. Stockinge H (2015) Consumers' perception of augmented reality as an emerging end user technology. Soc Media 29:419–439. https://doi.org/10.1007/s13218-015-0389-5
34. Sung J, Cho K (2012) User experiences with augmented reality advertising applications: focusing on perceived values and telepresence based on the experiential learning theory. Lect Notes Electr Eng 182:9–15. https://doi.org/10.1007/978-94-007-5086-9_2
35. Bulearca M, Tamarjan D (2010) Augmented reality: a sustainable marketing tool? Global Bus Manag Res 2&3:237
36. Connolly P, Chambers C, Eagleson E, Matthews D, Rogers T (2010) Augmented reality effectiveness in advertising. In: Presented at the 65th midyear conference on engineering design graphics division of the American Society of Engineering Education, Houghton, Michigan, October
37. O' Mahony S (2015) A proposed model for the approach to augmented reality deployment in marketing communications. Procedia—Soc Behav Sci 175. https://doi.org/10.1016/j.sbspro.2015.01.1195

Mobile Games for Children with Autism Spectrum Disorder to Support Positive Behavioural Skills

Shin Wey Tan, Muhammad Haziq Lim Abdullah
and Nor Farah Naquiah Mohd Daud

Abstract Autism Spectrum Disorder (ASD) is one of the fastest growing disorders around the world. The increasing rate of occurrence in Malaysia is estimated to occur in 1 of every 625 children and has since become a great concern to the community. Due to insufficient resources to cater for the autism education services, ineffective teaching strategy, and inadequate good positive behavioural support, this work is initiated as a means of solution to aid this issue. Findings of how a mobile games application, "Safe and Sound" featuring personalisation avatar can foster autistic children's positive behaviour in social practices aspect are reported in this paper. Mobile game applications are used to trigger the children's imagination for cognitive development as it is expected to become reliable to support content for the children to learn and apply in real situations. This analysis is entirely focused on interactions in schools by examining data from classroom activities. Three findings were revealed: (i) Personalised avatar can trigger children with ASD interest, (ii) Personalised avatar can promote emotions and feelings among children with ASD, and (iii) Avatar can facilitate social interaction among children with ASD. Thus, mobile games that incorporate personalised avatar can support children with ASD, especially to facilitate their positive behavioural skills.

Keywords Autism spectrum disorder · ASD · Children · Mobile games · Socialisation · Behaviour · Personalisation · Emotion · Avatar

S. W. Tan (✉) · M. H. L. Abdullah · N. F. N. Mohd Daud
Fakulti Teknologi Maklumat Dan Komunikasi (FTMK), Universiti Teknikal
Malaysia Melaka (UTeM), Hang Tuah Jaya, Melaka, Malaysia
e-mail: b031410178@student.utem.edu.my

M. H. L. Abdullah
e-mail: haziq@utem.edu.my

N. F. N. Mohd Daud
e-mail: b031410279@student.utem.edu.my

© Springer Nature Singapore Pte Ltd. 2019
V. Piuri et al. (eds.), *Intelligent and Interactive Computing*, Lecture Notes in Networks
and Systems 67, https://doi.org/10.1007/978-981-13-6031-2_43

1   Introduction

As the world keeps evolving towards modernisation, the prevalence of autism is becoming increasingly common today. Autism Spectrum Disorder (ASD) is a neurological disorder which impacts a child's social interaction skills and imagination [1]. ASD and autism are common terms used to describe a group of complex disorders of brain development. It is identified that autism occurs four time more frequent in boys than girls [2]. Despite that, the symptoms tend to be more severe among girls [3].

In Malaysia, the rate of occurrence is also worrying. Azizan [4] stated that the ratio of individuals diagnosed with autism in Malaysia is estimated to be 1 in 625 children. The number shows that it is important to support these children. Hence, resources in the country need to be doubled or tripled to tackle the situation [2]. Awareness programmes and better services for autism can be of great help to the society [3].

With this issue arising, it leads to the initiation of this project which aims to cater to the needs of children with autism in social practices aspect. The use of mobile games is to enhance the effectiveness of educating these children on road safety learning. Mobile technologies such as smartphones, tablets, and PDAs are able to support patients in mental decay to solve some of their behaviours and physiological problems [5]. Boyd et al. [6] also revealed a positive result of using collaborative assistive technologies including video games on iPad to facilitate social relationships for children with ASD.

Thus, the use of mobile games is promising to support behavioural skills for children with ASD. In fact, the Safe and Sound application can potentially benefit both special education teachers and children with ASD in terms of the educating process. By incorporating mobile games in teaching, children with ASD will be able to learn the provided content faster and easier. Hence, this paper specifically explores how a mobile game with personalised avatar was used in the classroom activity of 8 autistic children to support positive behavioural skills.

2   Related Work

The literature review section investigates mobile games that best engage children with ASD and discuss ways to support positive behavioural skills. Prior work on mobile games and personalised avatar are discussed. Opportunities and implications for design in the field are also presented in this section. The details of each aspect will be discussed below.

2.1 Use of Games Technology in Autism Intervention

The term games, in the special education field, is often defined as important and highly dynamic educational motivators which enable users to be involved in immersive experiences while provoking reflection and improving cognitive capacity [7]. As games had been used to integrate learning since the early century, it had become a tool for teaching and learning since then. This soon leads to the acceptance and prevalence of game-based learning applications such as Khan Academy where games began to grow in parallel with technology [8].

Hence, the potential of games to be used as an intervention for children with special needs is evident because the current learning and interpersonal communication tools are getting more digital and focused on logics of production, exchange, and sharing of contents [9]. This is also supported by the fact that individuals with ASD usually find computers and electronics less intimidating to work with for many reasons [10]. One of the reasons is that it is easier for them to focus visually on the material illustrated on the screen. As a result, many game-based learning applications can be found currently in the market. Therapeutic games including Second Life developments make up almost 65% of the entire games for health market in 2008 [11].

According to Boyd et al. [6], games are found to have great potential as an intervention for children with ASD due to the ability to interweave thoughtfully applied design elements with the naturally occurring contingencies (i.e., having fun together). Children with special needs can benefit from the use of computer-based technologies such as games to facilitate their educational activities in the learning aspect [12]. For example, word processing, communication, and multimedia projects can incorporate these technologies to help students with specific learning and emotional disorders to keep up with their non-disabled peers.

Boyd et al. [6] revealed that intervention through computer games have been shown to be effective for children with learning disabilities in terms of learning cognitive and social skills improvement. In fact, children are trained using games because they enjoyed playing games rather than undergoing the conventional learning process. The main benefits from the use of games technology as an intervention for children with ASD are fostering social skills, increasing attention span, and easily accessible by users.

2.1.1 Games Technology Can Support Teaching Social Skills for Children with Autism

Many researchers had investigated games technology as part of the intervention for people with autism. The feedback is deemed encouraging to foster social skills development. Results from a study on collaborative game indicated that games can be designed to support the teaching of social skills at all three levels of intimacy; membership, partnership, and friendship without the help of a human mediator [6]. Through membership, children are able to be in close physical proximity to one another, leading to new relationships formation and maintenance of the existing ones. It brings encounters and chances for interaction while allowing them to become jointly engaged in a team or pair. Although some of the players in the study expressed discomfort due to the small space shared, this form of factor may help them to overcome their challenges by maintaining close proximity with others.

On the other hand, a partnership involves collaboration through turn-taking and simultaneous gestures to achieve mutual goals [6]. This level of intimacy resulted in players to demonstrate non-verbal interactions such as eye contact and touching partner on the arm. Lastly, friendships built on mutual interest usually involved the sharing of joy and demonstrating empathy. Boyd et al. [6] also found that games can support friendship by occasioning celebration and commiseration through the wins and losses in the game. Players interacted by expressing their reactions and feelings by glancing, commenting, or coordinating gestures during these situations.

Another research employed an interactive social story mobile game called Friend-Star to teach the social skill of greeting and making new friends in school for children with ASD between 9 and 13 years old [13]. Based on the findings, it was found that all of the participants showed a positive opinion towards FriendStar. The participants showed a high acceptance and willingness to try the game. By using the game to learn social skills, the children were able to learn how to initiate a conversation and find a better way to act after making a mistake. Furthermore, comments from the professional described the game as attractive and engaging despite several lacking features such as voice over and limited topics. Since games can be used to train social skills and collaboration to children in a fun way in which they learn by doing, this makes intervention using game approach suitable for children with ASD to be treated or being assisted [14].

2.1.2 Games Technology Provides an Accessible Platform for Children with Autism

The growing number of games deployed in mobile platform affirmed that games are preferred to be used in mobile devices such as smartphones and tablets rather than computer-based. According to Hani and Wandi [5], a mobile-based application is considered as an exploitation of a new application which utilised the advancement of smartphone technology, targeting the increased number of children with ASD in society along with their parents and teachers. Researchers also revealed that the technology used in these devices is able to aid in the educational process among children with ASD. Technology devices such as tablets and smartphones which contain inherently collaborative features including touchscreens and other interactive surfaces are used in the study because they are able to teach a variety of social skills [6]. In their study, tablet is their preferred research tool due to its benefits of being a smaller scale collaborative system which can be used outside of the research setting.

A research interview by Eder et al. [15], revealed that pupils with ASD nowadays can play well on mobile devices and able to learn while playing games. They are also attracted to bright and lively colours which are yet to be found in the existing applications in regards to science. Following this, Eder and her colleagues designed a mobile game application where students with ASD can learn and enjoy while being guided by their guardians. Furthermore, Hiniker et al. [16] also found that mobile devices support gestures that allow small fine motor precision which is essential for users to engage in player-controlled scenarios game. Therefore, it can be proposed that mobile games technology has high potential to be used in educating social practice about road safety for children with ASD.

2.2 The Aid of Avatar in Fostering Positive Behaviours Among Children with ASD

Recent studies revealed that the use of avatar integrated into assistive technologies for children with ASD shows quite promising results in supporting positive behavioural skills. These results include promoting social interaction, expression of emotion and feelings, and triggering children's engagement while facilitating their educational process. The details of each aspect will be discussed below.

2.2.1 Avatar Can Promote Social Interaction Among Children with Autism

A study by Ying et al. [17] embedded personalisation avatar in social stories and digital storytelling and showed positive result in fostering positive behavioural skills for children with autism. Their study found that social interaction between the characters is promoted when the personalisation avatar is involved in the content. Children are able to interact and follow the storytelling activity by showing imitation pattern and gestures repeatedly. They also kept pointing to the avatar with their own picture, probably to seek the teacher's attention on the respective character.

Additionally, Cheng and Ye [18] found that collaborative virtual environments consisting of a 3D expressive avatar, an animated social situation, and verbal and text communication allowed children with ASD to improve their social interactions and understanding of social interactions. Hence, from these findings, it can be deduced that the avatar feature is helpful in promoting social interactions among children with ASD.

Carter et al. [19] examined the outcome of using interactions with various avatars or a video and their study showed a contrary result. Appearance complexity of the avatar was revealed to have not significantly affected any social interaction behaviours. It was identified that out of the three appearance conditions, one received increased responses from a subset of children but there was not a uniform pattern of performance across participants.

The participants showed a higher verbal score towards computer generated (CG) avatar and higher non-verbal score towards video avatar. Video appearance avatar was described to be the most complex appearance as compared to CG and cartoon avatar. Therefore, the proposed hypothesis that predicted lower visual complexity appearance would receive increased responses was not supported. Despite that, it is suggested that computer-based intervention (CBI) provides options for various appearances of avatars to allow parent or therapist choose the best-suited one for a specific child. This limitation shows a room of improvement that can be integrated into this project.

2.2.2 Avatar Allows Expression of Emotion and Feelings in Children with ASD

Children's emotion could be motivated while allowing them to express their own feelings via an avatar. Ying et al. [17] found that children tend to show their interest and emotions when following the storytelling activity. This finding revealed that the use of personalisation avatar using the children's own picture could induce their emotion when they participate and engage in the activity. Abdullah and Brereton [20] found that empowering the child interest can aid in the engagement of preliminary discussions across settings (home and school) and across child, teachers, classmates, and parents. In addition, they found that any photo activity that is within the child's interests is able to support the development of interactions between parent, teacher, and child. According to Konstantinidis et al. [21], affective avatar is used in the studies due to its high distinction and clear recognition of emotions expressed which enable people with autism to understand and use them better. Most of the time, people with autism could recognise the avatar's mental and emotional state based on its facial expression. This indicated that interactive avatars can help children with autism to express or communicate their feelings through these characters.

Since most of the children with ASD usually possessed social anxiety and lack social skills, a 3D system called "Invirtua" has been developed to cater to this aspect. Invirtua features live and animated 3D characters which demonstrate positive behaviours in a fun and entertaining way [22]. With this feature, the children will feel less anxious and embarrassed when interacting with these cartoon characters as compared to interacting with a real person. Contributing to this finding, FaceSay, an intervention which uses interactive, realistic avatar assistants to promote eye gaze, facial expressions, and emotion showed a positive result [23]. From the outcome, it was found that students with high-functioning autism (HFA) had successfully exhibited emotion recognition and social interaction skills.

2.2.3 Avatar Can Trigger Children with Autism's Engagement While Facilitating Their Educational Process

The integration of avatar in technologies is also found to trigger engagement on the learning content and facilitate the educational process among children with ASD. Kuan, Mat Sah, and Abdullah [17] revealed that children positive engagement are triggered when personalised avatar is adapted in the activity. This is because the avatar is able to keep children's interest to engage in the digital storytelling and prolong their attention span which in turn provide a positive opportunity for their learning. Based on the Fabri et al. [24], the positive result of their research supports the optimism that emotionally expressive avatars could be used effectively to educate individuals with autism by becoming a "teacher" or tutor in the context. The use of such avatar is important in the educational process as it enhances the quality of tutor-learner and learner-learner interaction.

Findings from research which uses the engagement of an affective speaking avatar for teacher-child learning scheme indicate that implementing avatar in the context can advance the educational process [21]. Educators suggest that integrating native voice in an avatar is more effective than an avatar without a voice. It is believed that the educational processes are facilitated in a more proper way when the avatar is integrated with a voice function in the autistic native language. Based on the results, the content should be presented in a simple and attractive way for people with autism to engage and not lose interest.

As suggested in other research, avatar feature could play an important role in cultivating positive behaviours, especially in terms of mental and emotion display, social interactions, and engagement in the learning process. While avatar character can foster the child interest and increase their attention span, it is crucial for a more customised avatar character to be integrated into the children's life. According to Ardamerinos et al. [10], children with ASD preferred games that contain images of themselves or individuals who they are familiar with.

Results from another research using mobile game find that the participants liked the ability to customise their own avatars along with the look of the game, the story, and star rewards [13]. It was specifically mentioned that the characters are likeable due to its "not too many details" and "cartoon". Hence, our project is extending the features introduced in these research but with a different approach by integrating the avatar feature which allows more options for customising the avatar according to the children's preferences. In this paper, we report how Safe and Sound application with personalised avatar was used in the classroom activity with children on the autism spectrum to support positive behavioural skills.

3 Safe and Sound Application

The first author designed and developed the Safe and Sound application in accordance with a guideline designed by a special education teacher who teaches children with autism. As mentioned above, Safe and Sound is a serious game which is intended to train the players in fostering positive behavioural skills. Thus, before the development process of Safe and Sound was made, a preliminary investigation is conducted to collect data for avatar design using paper prototypes as shown in Fig. 1.

Hence, the digital version is designed specifically to cater to children with autism needs in social practice. Therefore, Safe and Sound application was developed and

Fig. 1 Prototype of male and female avatar

Fig. 2 Game scene which players have to travel from one destination to another destination

run on smartphones with Android OS 6.0 (JellyBean). In addition, it is recommended to deploy the game on Android devices which have a screen size of 1280×720 pixels. Besides that, the storyline of the game uses road safety content of which player would be travelling from one destination to another as shown in Fig. 2. During the process, a player needs to overcome the obstacles to cross the road safely in order to reach the final destination.

All game characters are developed based on cartoon themes along with the feature of personalised avatar as shown in Fig. 3. This feature allowed them to use the picture of themselves which will then be mapped on to the avatar to be used throughout the entire game (see Fig. 4). Participants can choose either male or female avatar to start playing the game.

Three episodes are available in the Safe and Sound application and each of it has a different level of difficulty (see Fig. 5.). Participants can play any episode without having to complete the episodes in sequence. The goal of all these episodes is the same which is to guide the avatar to reach the opposite end safely.

Fig. 3 Sound application's interface design for main menu screen

Fig. 4 Personalisation avatar screen on safe and sound application

Fig. 5 Episode selection scene

4 Method

The Human Research Ethics Committee of Universiti Teknikal Malaysia Melaka approved the study protocol, and written consent was obtained from the coordinator of The National Autism Society of Malaysia (NASOM Melaka) to involve each participant prior to the trial. In fact, the testing was conducted in a classroom and helped by the special education teacher. Two groups of users participated in the trial. The first group of users has a total of 8 participants consisting of 4 females and 4 males. All participants in this group are children with autism of different

Fig. 6 Special education teacher assisting the participant as she got anxious during the trial

age range and level of ASD. The participants are selected by the special education teacher from the centre. This group of participants will be tested mainly on their level of understanding towards the game and their capability of playing the game. Meanwhile, their behaviours and reactions will be observed throughout the trial.

The second group of participants is the special education teachers who are specialised in teaching children with autism. As they are more familiar with the participants' behaviours, they will check whether the content used is suitable and in accordance with the target user's level of understanding. The teachers will also provide their opinions and suggestions on the application's relevancy towards children with autism.

A total of 3 scenarios will be conducted in sequence, consisting of cartoon scenario, own scenario, and peer scenario. Firstly, participants are required to choose one cartoon card for them to take a picture for their avatar. Next, they have to take a picture of their own for their avatar set up. Lastly, peer scenario will take place where participants have to take their peer's pictures for their avatar customisation. For every scenario, participants are encouraged to explore the game by themselves. However, special education teachers may step into assist the participants if required or when there are no further interactions between the tester and the game (see Fig. 6).

During the trial, observations are performed on children with ASD while they are playing the game. Participants' emotion and interaction with the game are being observed and recorded. The data were collected and analysed using manual coding. Then, interview was conducted by interviewing the special education teacher to confirm or to refine the identified themes from findings. The thematic analysis of Safe and Sound application used in classroom has identified three additional themes in relation to teachers and children interactions. The emerging themes from the analysis are discussed in the next section.

5 Findings and Discussion

Safe and Sound application is demonstrated to foster positive behavioural skills among children with autism spectrum disorder. Out of all scenarios, playing using avatar with own picture recorded the highest in the total duration spent but participants were found to play more frequent using cartoon picture. This shows that participants prefer to play using the cartoon and own picture rather than a peer's picture. Meanwhile, the lowest average of duration spent is recorded at 1.5 min while the highest is recorded at 7.2 min. The result indicated that the game is able to capture the participants' interest and attention in engaging the game. Throughout the game, participants were also found to be able to express their basic emotions such as happiness, anger, and fear. All these reporting are tagged as skills in socialisation, expression of emotion and behaviour which leads to the identification of three themes. These themes are discussed below.

Theme 1: Personalised avatar can trigger children with ASD's interest
It was found that the avatar can trigger children's with ASD interest when interacting with the application. From the observation, many participants showed their interest in playing the game, especially when taking pictures for their avatar. One of the participants kept returning to the taking picture scene instead of completing the episodes when he is exploring the game on his own. The teacher commented in the interview on how taking pictures can trigger the children interest: *"Most of the children enjoy photography session but it doesn't apply to all. It's true that self-created pictures can increase their interest and encourage them to play for those who enjoy taking pictures"*.

When comparing the three different avatars, the avatar with cartoon picture and their pictures are found to be most fond to them compared to the avatar with peer's picture. This is because children with ASD are more focus primarily on themselves in their own space rather than their surroundings. Furthermore, these children love cartoons, especially local cartoons such as Didi & Friends and Upin Ipin. Since participants spent more time on own scenario than cartoon scenario, it can be concluded that they like to play using the avatar with their own picture. The teacher responded towards the findings: *"Children with autism are living in their own world. So their focus is more on themselves than other things. They are not usually aware of their surroundings or people around them. As for cartoons, they really love these cartoons because they always watch Upin & Ipin at home while The NASOM Melaka usually played Didi & Friends music or videos for their school activities"*.

In addition, the avatar is also found to build attention skills among children with ASD. Averagely, participants spent a minimum duration of 1.5 min and maximum duration of 7.2 min in a scenario. According to the teacher, she commented: *"I think that the avatar used in the game can prolong the participants' attention span because they keep playing the game for quite some time during the session. For example, one of the participants played the game for quite long but he usually can't focus on one task for long. Furthermore, most of the children with autism love using gadgets which are causing them to stay on task much longer"*. The finding was also supported by

Fig. 7 Participant laughed
when looking at her own
picture on the game avatar

Fig. 8 Participant danced
along with the music when
he completed the level

Ying, Sah and Abdullah [17]. They found that the avatar can keep children's interest
to engage in the activity and prolong their attention span, thus, providing a positive
opportunity for their learning. In summary, avatar can help to trigger children's
interest in interacting with the application, especially for their learning process.

Theme 2: Personalised avatar can promote emotions and feelings in children with ASD.

From the class observations, this study found that the avatar used in the application
can promote emotions and feelings in children with ASD. The results showed that the
majority expressed basic emotions such as happiness, anger, and fear when playing
the game. Since many participants showed the happy emotion including smiling and
laughing, this suggests that they enjoy playing the game. Some even laughed when
looking at their own picture (see Fig. 7). The finding is supported by the teacher's
statement: "*Since they couldn't verbally express themselves, this might be one of
their ways to express their emotion when playing using avatar. You can see most of
them smiled and some even danced along with the music because they feel happy
when they completed the episode and listen to the music played (see Fig. 8). Some
laugh at own picture probably because they like to look at own picture in the game*".

Besides the happy emotion, one participant showed fear and said "Tidak, nanti!"
which means "No, wait!" when her avatar almost got hit by the vehicles during the
own scenario as shown in Fig. 6. This is because she actually recognised herself in

Fig. 9 Peer looks interested
to play the game

the game. Therefore, she felt as though she was getting hit by a car so she brings her emotions into the game. However, there is one participant who expressed anger when playing the game. It was being clarified by the teacher that: *"Her behaviour is the type tends to get frustrated when something happens. Also, her avatar got hit continuously at that time, making her unable to accept the situation, thus she got frustrated"*. This result is also supported by Hughes [23]. Based on Hughes [23], the use of interactive and realistic avatar assistants is able to promote eye gaze, facial expressions, and emotion among students with autism. In conclusion, avatar can help to promote children's expression of emotion throughout the process of interacting with the application. In a similar vein, Abdullah and Brereton [25] concluded that photos and videos of children with autism personal activity are able to foster the children's expressive language that can be used to help children to understand their own emotions.

Theme 3: Avatar can facilitate social interaction among children with ASD
This study also found that the avatar can facilitate social interaction among children with ASD when they played with the application. Based on the observation, the peers of participants showed interest to join in the activity when they saw their friends playing the game. The peers tend to come over and look at the game when the participant is playing or taking pictures of his or her peer (see Fig. 9). Some even tried to take hold of the application as shown in Fig. 10. This shows that these children with ASD are seeking to share their interest and enjoyment with other people by playing the game. Through this approach, they will also be able to sustain their friendships even better. During the interview, the teacher commented: *"I agree that the avatar can promote social interaction because these children don't know the right way to communicate. So when they look at their peers, it caused them to become interested and wanted to interact along"*.

Furthermore, the finding also showed that participants are able to use non-verbal communication such as eye gaze when interacting with the application. Although these children have difficulties in making eye contact with others, they can look directly at the avatar and the surrounding when playing this game. Other factors also contribute to the result which are the use of interactive music and graphics

Fig. 10 Peer trying to join
in the game play

that maintain their attention. Hence, it can be deduced that avatar with the help of
other interactive elements is able to encourage eye contact in children with ASD.
The teacher also mentioned in the interview that: *"It's also fortunate to see that the
participants can have eye contact with the game because the avatar together with
the music and graphics trigger them to interact with the game"*.

The participants are also seen to be imitating the voice instruction that is played
throughout the game. They repeat after what the instruction said such as "Wait"
and "Oh no". This result showed that the participants are capable of interacting
with the content. Through this imitation behaviour, these children with ASD will be
able to improve their socialising skills by learning the languages. Our findings are
supported by Cheng and Ye [18]. In their research, they found that the avatar allowed
children with ASD to improve their social interactions and understanding of social
interactions. In short, the use of avatar can facilitate social interaction among children
with ASD which is important for them to cope in various social situations. Abdullah
and Brereton [26] also found that the visual of each individual child's life is able to
prompt communication and support social interaction of children with autism in the
school setting.

6   Conclusion

This project had demonstrated how the Safe and Sound mobile game is designed and
supported positive behavioural skills among children wi≫th ASD. Safe and Sound
was also determined as a potentially useful game for children to learn and engage
in social practice for them to apply in real-life situations. It is simple to play and
accessible anytime for teachers to use as part of their classroom activity or for parents
to educate their children about social practice at home. Through this approach, children who are equipped with these skills are believed to become more self-motivated
and able to adapt to the physical environment more easily. Nonetheless, suggestions
for improvement of the application which have been gathered through the analysed
result and feedback obtained will be discussed next.

A few future enhancement suggested for this application are to add other language options, provide a video tutorial on button controls, using facial recognition technology, and change the current sliding menu to fixed buttons. Language option such as native language, Malay is suggested since not every user is familiar with English. This is important for users to understand the instruction easier. Meanwhile, a video tutorial can be added at the beginning of the episode to help the user understand the control better. The tutorial can be guided along with music for them to stay focus and follow the tutorial. In addition, facial recognition technology can be added to the camera capture during the avatar personalisation so that user will be capturing faces instead of empty canvas or plain backgrounds. It was also being suggested to change the sliding menu into buttons where all the episodes are fit into one screen instead. This interface design is easier to control in terms of navigation for children who are not familiar with using touch inputs.

Nevertheless, this project had made a contribution towards the autism community by developing a mobile game to support children with ASD"s positive behavioural skills. Through this approach, the children's behaviour and learning process are expected to improve, especially in terms of their social practices. This project also played a role in investigating the importance of personalisation avatar in fostering the children's positive behaviour. The results of this project can be used as a reference for future works in the related topic.

Acknowledgements I would like to acknowledge my faculty, Fakulti Teknologi Maklumat dan Komunikasi (FTMK) and express my deepest gratitude to the academic staff from The National Autism Society of Malaysia (NASOM Melaka) for their time and assistance in the trial process of this project. Not to forget, the participants who are involved and their willingness to take part in this project trial. The authors would also like to acknowledge Universiti Teknikal Malaysia Melaka (UTeM), Center for Advance Computing Technology (C-ACT), Specialists in Special Needs Awareness and Research (SPEAR) Group and Applied Oriented Research Grant (OARG)-PJP/2017/FTMK-CACT/S01567 for supporting and encouraging this research.

References

1. PERMATA Kurnia. http://www.programpermata.my/en/kurnia/autism
2. National Autism Society of Malaysia (NASOM). http://www.nasom.org.my/
3. Gomez R (2016) Autism Awareness in Malaysia. Early Autism Project. http://www.autismmalaysia.com
4. Azizan H (2008) The burden of autism. The star online. http://www.thestar.com.my/story/?file = %2F2008%2F4%2F27%2Ffocus%2F21080181&sec = focus
5. Hani H, Abu-Wandi R (2015) DISSERO mobile application for AUTISTIC children's. In: Proceedings of the international conference on intelligent information processing, security and advanced communication pp 90
6. Boyd LE, Ringland KE, Haimson OL, Fernandez H, Bistarkey M, Hayes GR (2015) Evaluating a collaborative iPad game's impact on social relationships for children with autism spectrum disorder. ACM Trans. Accessible Comput. (TACCESS) 7(1):3
7. Saridaki M, Mourlas C (2016) Playing in the special education school: from gamers to game designers

8. Teach Thought (2012) A brief history of video games in education. http://www.teachthought.com/uncategorized/a-brief-history-of-video-games-in-education/
9. Bertolo M, Mariani I (2013) Game and play as means for learning experiences. INTED 2013 Proceedings. pp 698–707
10. Ardamerinos G, Shevchuk H, Smith J (2015) Investigating the use of collaborative technology in autism therapy
11. Cannon-Bowers J (Ed.) (2010) Serious game design and development: technologies for training and learning: technologies for training and learning. IGI global
12. Hasselbring TS, Glaser CHW (2000) Use of computer technology to help students with special needs. Future Child. pp 102–122
13. Zhu J, Connell J, Kerns C, Lyon N, Vecere N, Lim D, Myers C (2014) Toward interactive social stories for children with autism. In Proceedings of the first ACM SIGCHI annual symposium on Computer-human interaction in play, pp 453–454
14. Andrade A, de Carvalho CV (2013) Gamifying a serious games community. In: International conference on computer, networks and communication engineering, pp 249–252
15. Eder MS, Diaz JML, Madela JRS, Mag-usara MU, Sabellano DDM (2016) Fill me app: an interactive mobile game application for children with Autism. Int J Interact. Mobile Technol. (iJIM) 10(3):59–63
16. Hiniker A, Daniels JW, Williamson H (2013) Go go games: therapeutic video games for children with autism spectrum disorders. In: Proceedings of the 12th international conference on interaction design and children. pp 463–466
17. Ying KT, Sah SBM, Abdullah MHL (2016) Personalised avatar on social stories and digital storytelling: Fostering positive behavioural skills for children with autism spectrum disorder. In: The 4th international conference on user science and engineering (i-USEr) 2016, pp 253–258
18. Cheng Y, Ye J (2010) Exploring the social competence of students with autism spectrum conditions in a collaborative virtual learning environment–the pilot study. Comput Educ 54(4):1068–1077
19. Carter EJ, Hyde J, Williams DL, Hodgins JK (2016) Investigating the influence of avatar facial characteristics on the social behaviors of children with autism. In: Proceedings of the 2016 CHI conference on human factors in computing systems, pp 140–151
20. Abdullah MHL, Brereton M (2012) A child led participatory approach for technology-based intervention. In: 2012 participatory innovation conference digital proceedings, pp 1–5
21. Konstantinidis EI, Hitoglou-Antoniadou M, Luneski A, Bamidis PD, Nikolaidou MM (2009) Using affective avatars and rich multimedia content for education of children with autism. In: Proceedings of the 2nd international conference on pervasive technologies related to assistive environments, pp 58
22. Autism Speaks (2015) Animator's digital avatars to help kids with autism ease anxiety. https://www.autismspeaks.org/news/newsitem/animator039s-digital-avatars-help-kids-autism-ease-anxiety
23. Hughes D (2014) The design and evaluation of a video game to help train perspective-taking and empathy in children with autism spectrum disorder
24. Fabri M, Elzouki SYA, Moore D (2007) Emotionally expressive avatars for chatting, learning and therapeutic intervention. In: International conference on human-computer interaction, pp 275–285
25. Abdullah MHL, Brereton M (2017) Mycalendar: supporting children on the autism spectrum to learn language and appropriate behaviour. In: Proceedings of the 29th Australian conference on computer-human interaction, pp 201–209
26. Abdullah MHL, Brereton M (2015) MyCalendar: fostering communication for children with autism spectrum disorder through photos and videos. In: Proceedings of the annual meeting of the Australian special interest group for computer human interaction, pp 1–9

A Study on Gamification for Higher Education Students' Engagement Towards Education 4.0

Rafidah Ab Rahman, Sabrina Ahmad and Ummi Rabaah Hashim

Abstract An evolution of teaching and learning experience in higher education witnesses many techniques being introduced towards student-centered learning. One of the most popular techniques is game-based learning or known as gamification to enrich students experience in classroom. However, the awareness among educators is lacking and the anxiety whether gamification benefits the learning process is prominent. This paper is written to provide an insight into gamification and its ability to instill students' engagement. The questionnaire set and model to evaluate the students' engagement following gamification during teaching and learning process are also studied and presented. Literature review method is used to investigate the relevant subjects related to the gamification for higher education and ways to evaluate its effectiveness. This study showed that there exist several studies on students' engagement and gamification which are deemed useful to promote students' engagement. Besides, the study discovered that improved Technology Acceptance Model (TAM) is a good measure to investigate its effectiveness.

Keywords Gamification · Students' engagement · Higher education

1 Introduction

One of the main challenges confronted by educators today is to engage students during teaching and learning process in the classroom. Students' preceding learning

R. Ab Rahman
Politeknik Muadzam Shah, Lebuhraya Tun Abdul Razak, 26700 Muadzam Shah, Pahang, Malaysia
e-mail: rafidah@pms.edu.my

S. Ahmad (✉) · U. R. Hashim
Faculty of Information and Communication Technology, Universiti Teknikal Malaysia Melaka, Hang Tuah Jaya, 76100 Durian Tunggal, Melaka, Malaysia
e-mail: sabrinaahmad@utem.edu.my

U. R. Hashim
e-mail: ummi@utem.edu.my

© Springer Nature Singapore Pte Ltd. 2019
V. Piuri et al. (eds.), *Intelligent and Interactive Computing*, Lecture Notes in Networks and Systems 67, https://doi.org/10.1007/978-981-13-6031-2_5

(readiness), enthusiasm for learning, and the way the input is introduced to them are some of the factors that influence their ability to learn [1]. Diverse learning styles among the students also contribute to the way they engage in the activities conducted by the educators. Sustaining the students' interest and participation is a struggle that leaves the educator in a quandary. This is because students' involvement plays an important factor in their achievement and performance [2] measured during either the formative or summative assessment. Mohd et al. [3] discovered that active students have more tendency to perform well as they retained more knowledge during learning activities.

Several studies have been done referring to gamification as a technique to improve the students' engagement [4–6]. Technological developments allow the use of game elements in a non-game environment by extending the methods that can be employed by educators in developing lesson plans. Not all educators are creative enough to include gamification in their lessons, thus online platforms such as Kahoot!, Quizizz, Socrative and Quizalize provide excellent options for educators to choose from in diversifies lesson plans and activities that can captivate and inspire students' motivation and increase students' engagement during lessons in the classroom.

In recent years, gamification has been gaining interest as a point of study in education development in the effort of enriching students experience in their learning journey, especially in a classroom. Educators keep on searching for ways to enhance their students' engagement during lessons with the hope that the students will be able to gain more knowledge and consequently help them to perform better in their assessment. Some of the problems that have been identified are:

1. Lack of awareness of the online platform available to implement gamification during lessons among the educators.
2. The arising question whether the students are susceptible to participate in gamification during lessons in an effort to increase their engagement.

The purpose of this paper is to provide an insight into the feasibility of the gamification technique to engage students during teaching and learning process in the classroom. Following Introduction in Sect. 1, the knowledge on students' engagement including the factors, influences and its instruments are compiled and summarized in Sect. 2. This is then followed by Sect. 3 which presented the options of gamification techniques and online gamification platforms available. Subsequently, Sect. 4 explains the evaluation instruments which covers Student Course Engagement Questionnaire (SCEQ) and Technology Acceptance Model (TAM). Section 5 concludes the paper with recommendation for future works.

2 Students' Engagement

There are many definitions arose from numerous studies regarding the term of students' engagement. Most literature defined students' engagement as activities performed either physically or mentally by the students in their pursuit to gain knowledge

[3, 6, 7]. In a different perspective, a study by Hu et al. [8] defines them as engagement that occurred when the students used the online learning platform in their lessons, as in this context; the learning materials that can only be accessed by the students themselves.

There are studies that identify the factors contributed to the students' engagement. Mohd et al. [3] and Hu et al. [8] stated that the students' engagement comprises three dimensions; cognitive, behavioral, and emotional engagement. However, other studies by Handelsman et al. [2], Dixson [6] and Marx et al. [7] categorized the students' engagement into four factors which are skill engagement (represented by skill displayed by the students), emotional engagement (represented by the students feelings), participation (represented by activities performed by the students during lessons) and performance engagement (represented by the result of assessments conducted on the students). In addition, Marx et al. [7] also listed another engagement which is the total engagement to measure the students' perception of their overall engagement in one of the courses taken in the college.

Based on the literature, several significant influences were identified. The relationship between the students and educator plays an important factor in students' engagement [3, 7]. Furthermore, Marx et al. [7] also stated that educators' expectation on the students' engagement during lessons can be met if the educators themselves reciprocate this expectation towards the students.

Table 1 shows a summary of the four studies in the context of students' engagement conducted in this section.

3 Gamification and the Online Platforms

In recent years, the concept of gamification in education is gaining attention as an area of study among researchers. Researchers describe gamification as infusing game component into a non-game context [4, 9] which can be used as a mean to promote the students' engagement during lessons [4, 5, 10].

Leaderboard, badges, points and levels are some of the game elements employed in previous studies [4, 5, 9–11]. Based on Table 2, badges are the top choices among researchers to be used in the gamification, followed by the leaderboard. This is due to the users' perception that badges will highlight their social status to their peers [10]. Although most studies found that gamification learning has a constructive influence on the students' engagement [5, 9–11], the discovery from Hanus and Fox [4] contradicts this finding. Hanus and Fox [4] found that as time elapsed, the students who experienced gamification learning show a decline in their motivation, and consequently affect their final exam scores. This led to the conclusion that any gamification employed must be considered with great care, as to it may be detrimental to the students as opposed to helping them in their lessons.

Table 1 Summary of previous studies on student engagement

References	[3]	[6]	[7]	[8]
Definition	Refers to the amount of physical and emotional energy that the student applies to the academic practice	Student engagement is the extent to which students enthusiastically participate by thinking, talking, and interacting with the lesson, their peers in the class, and the educator	Engagement is student's connection to their learning and the learning environment which incorporates behavioral, emotional, and cognitive aspects	Student engagement in online learning is engagement when using online learning platform to learn
Factors of engagement	Cognitive engagement	Participation engagement	Total engagement	Cognitive engagement
	Behavioral engagement	Skill engagement	Skill engagement	Behavioral engagement
	Emotional engagement	Performance engagement	Participation engagement	Emotional engagement
		Emotional engagement	Emotional engagement	
			Performance engagement	
Influences on engagement	Lecturer support		Teacher immediacy behaviors	Multidirectional communication
	Blended learning		Class size	Involvement
				Support
				Control
Instrument to measure engagement	Adapted from motivated strategies for learning questionnaire (MLSQ) and several others	Online student engagement Scale (OSE) adapted from SCEQ	Student course engagement questionnaire (SCEQ)	
			Nonverbal immediacy behaviors (NIB)	
			Verbal immediacy behaviors (VIB)	

Table 2 Comparison of game elements used in previous studies

	References				
Game mechanics	[4]	[5]	[9]	[10]	[11]
Badges	●	●	●	●	●
Leaderboard	●	●	●		●
Challenges					●
Levels			●		●
Points		●	●		
Online activity			●		
Incentive	●		●		
XP					●

3.1 Online Gamification Platform

Studies conducted by Wang [12], Sawang [13], Wang and Lieberoth [14] and Chaiyo and Nokham [15] are some of the few literature found which discuss the online gamification platform in the form of students' response system, particularly regarding Kahoot! and Quizizz. In Wang [12], the diminishing effect of using Kahoot! in a different situation (Event and Semester) played by two different groups of students was investigated. It was found that the group of students who used Kahoot! in a longer period agree that the game is still engaging. Both groups also agree that using helped their engagement during lessons as it provides a meaningful, interactive, and fun way of learning, besides the determination to get ahead in the game. Thus, it was proven that longer period of gamification during lessons has no diminished effect on the students' engagement.

In another study, Wang and Lieberoth [14] expanded the research by investigating on how the elements of the game in Kahoot!, especially the points and audio elements, affects the students' engagement. The study indicates that audio has the greatest impact on students' interaction when combined with the use of points. On the other hand, Chaiyo and Nokham [15] studied the effects of three different gamification tools to the students' engagement, enjoyment, concentration, perceived learning, satisfaction and motivation in lessons. The study revealed that, although all three tools, namely Kahoot!, Quizizz and Google Form, did not show any difference on how the students perceived their learning, the students show preference towards Kahoot! and Quizizz, compared to Google Form in other factors such as concentration, engagement, enjoyment, motivation and satisfaction [15]. Table 3 shows the comparison of the features between Kahoot! and Quizizz, compared by Pilakowski [16].

Table 3 Comparison of online gamification platforms [16]

	Kahoot!	Quizizz
Student log-In	Game pin	Game code
Question length	Less than 95 characters for each question and 60 characters for answer	No characters limit
Timer for questions	Changeable from 5 to 120 s	Changeable from 5 s to 15 min
Disarm timer	Cannot be disarmed	Can be disarmed
Pacing	Based on teacher's speed	Based on student's speed
Multimedia in questions	YouTube and image can be inserted into the question. Music played during game cannot be muted	Image or GIFs appears behind the question text and limited to 16:9 format. Image can also be added to answer options. Music can be turned on or off by student
Scoreboard	Shown after each question. Highest climber also shown	Yes, but may be disabled
Type of feedback to students	Immediately after each question whether the answer is correct or not	Funny meme reveals either student's answer was correct or not
Feedback of correct answers	After each question	After each question and after the quiz ended but can be turned off
Can be assigned as homework	Not assignable as the questions and answer are not displayed on students' screen	Yes but also can be played live
Public library of quizzes	Able to access other quizzes shared by others and have option to share or not own quizzes	Yes, can access other quizzes shared by others and also share own quiz via social media platform, email and shareable link
Randomized questions	Can randomize the questions and the answers	Can randomize questions and answers using Jumble
Main purpose	Questions that require rapid answer	To encourage more thinking depending on student's pace

4 Evaluation Instruments

4.1 Student Course Engagement Questionnaire (SCEQ)

To measure the students' engagement, a few survey instruments have been proposed, for example, Online Student Engagement Scale (OSE), National Survey of Student Engagement (NSSE), and the most notable one is Student Course Engagement Questionnaire (SCEQ). Handelsman et al. [2] developed an instrument to measure the students' course engagement as there is no existing survey capable of emphasizing the students' engagement on individual courses offered rather than as the whole program of the institution which is undertaken by the National Survey of Student Engagement (NSSE).

Handelsman et al. [2] ended up with a 23-items instrument from the initial 27-items after performing an exploratory factor analysis which produces a 4-factors survey instrument denoted as skill engagement (9 items), emotional engagement (5 items), participation/interaction engagement (6 items), and performance engagement with 3 items. The SCEQ was then validated with another study conducted on another group of participants. It was found that there was a relationship between every two factors of the SCEQ. These findings were supported by several studies which adapted the SCEQ items in their research [6, 7].

4.2 Technology Acceptance Model (TAM)

TAM by Davis et al. [17] is an information systems model that theorize the process of acceptance and the usage of technology where the users' decision to accept and use a new technology relied on several factors. Introduced by Davis et al. [17], TAM is adapted from the Theory of Reasoned Action (TRA) [18] as shown in Fig. 1. Legris et al. [19] stated that TRA was proposed to clarify and foresee the practices of individuals in a particular circumstance.

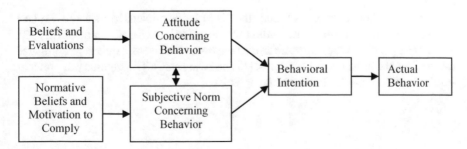

Fig. 1 Theory of reasoned action (TRA) [18, 19]

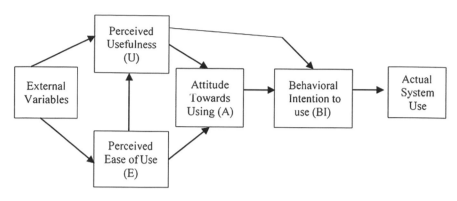

Fig. 2 TAM [17]

Davis et al. [17] have proposed two new constructs, namely the Perceived Use-fulness (U) and the Perceived Ease of Use (E) representing the users' belief on the beneficial of the technology used to his/her work performance and the users' believe on the effortless usage of technology. The subjective norm constructs as the contributing factor to behavioral intention was eliminated which in turn can also be influenced by external variables as shown in Fig. 2 [17].

Based on this model, Davis et al. [17] concluded that the Perceived Usefulness (U) and the Perceived Ease of Use (E) are the major factors in the acceptance of technology among the users. The instrument consists of four items in each factor, which were finalized from 14 items (for each factor) during the development process and further finalized to 10 items per construct after pretesting and wording refinement [17]. After another item analysis has been conducted, the items were streamlined to 6-items per construct as used in [20]. As the validity and reliability of these construct were found to be very high, another item analysis was done to produce the 4 items per construct which were then used in [20]. In his study, Davis [20] found that the usefulness construct is more important towards the user acceptance of technology compared to the ease of use of a system. The difficulty in using the technology may be overlooked by the user but not in the case of a useless system however easy it is to use it.

However, other researchers found that TAM is insufficient to measure the user acceptance in certain situations, leading to attempts of TAM enhancement [21, 22]. Table 4 (Appendix 1) depicts the different constructs used by researchers as compared to the original constructs by Davis et al. [17] and Davis [20]. The summary of previous studies is also shown in Table 5 (Appendix 2).

Table 4 Constructs found in previous studies using TAM

References	Tool	Sample size	Model used	Method
[13]	KeyPad (Interactive student response system)	First-year students of Queensland University of Technology's Business School	TAM (without perceived usefulness, perceived ease of use) + social pressure + lecture engagement + facilitating condition for KeyPad + extraversion	Survey targeted at a large-sized classroom and done at the end of weekly lecture
[17]	WriteOne	107 MBA students	TAM + TRA	Survey done over after introduction and after 14 weeks usage
[20]	Chart Master, Pendraw	40 MBA students, Boston University	TAM (without Attitude), use self-predicted usage	Survey done after one hour of hands-on experience
[21]	Canvas LMS	560 faculty members and graduate teaching assistants from two universities in the USA	TAM + system quality + perceived self-efficacy + facilitating conditions	Survey targeted at individuals with teaching responsibilities and had used Blackboard as their LMS before adopting Canvas
[22]	Interactive Mobile Maps	65 respondents at University Utara Malaysia	TAM (without attitude and behavioral intention) + perceived enjoyment	Survey targeted participants who have used mobile map on their mobile device

5 Conclusion

A thorough study has been conducted to provide an understanding of gamification technique definition and its related components to promote students' engagement. The study also shows the feasibility of the gamification in a higher education teaching and learning process as many options and platforms are available to be utilized. In addition, the survey technique and model for evaluating students' engagement are also discussed to be adopted by educators. These evaluations are important as they can

Table 5 Summary of previous studies using TAM

References	[17]	[20]	[21]
Perceive ease of use	Learning to operate W would be easy for me	Learning to operate C would be easy for me	My interaction with CANVAS is clear and understandable
	I would find it easy to get W to do what I want it to do	I would find it easy to get C to do what I want it to do	Interacting with CANVAS does not require a lot of my mental effort
	It would be easy for me to become skilful at using W	My interaction with C would be clear and understandable	I find CANVAS to be easy to use
	I would find W easy to use	I would find C to be flexible to interact with	I find it easy to get CANVAS to do what I want it to do
		It would be easy for me to become skilful at using C	
		I would find C easy to use	
Perceive usefulness	Using W would improve my performance in the MBA program	Using C in my job would enable me to accomplish tasks more quickly	Using CANVAS improves my performance as a faculty member
	Using W in the MBA program would increase my productivity	Using C would improve my job performance	Using CANVAS in my job increases my productivity
	Using W would enhance my effectiveness in the MBA program	Using C in my job would increase my productivity	Using CANVAS enhances my effectiveness in my job
	I would find W useful in the MBA program	Using C would enhance my effectiveness on the job	I find CANVAS to be useful in my job
		Using C would make it easier to do my job	

give an insight to educators on what are the suitable game elements which are able to be applied in gamification and whether the gamification will give positive impact in the students' engagement in the classroom. An exciting prospect for future research would be to study the effect of using existing gamification platforms available online as opposed to using a specially designed gamification for a specific course.

Acknowledgements The publication of this paper is sponsored by Universiti Teknikal Malaysia Melaka through a research grant numbered PJP/2017/FTMK-CACT/S01573.

Appendix 1

See Table 4.

Appendix 2

See Table 5.

References

1. Eltegani N, Butgereit L (2016) Attributes of students engagement in fundamental programming learning. In: Institute of Electrical and Electronics Engineers (ed) 2015 international conference on computing, control, networking, electronics and embedded systems engineering (ICCNEEE 2015). Red Hook, NY, Curran Associates, Inc., Khartoum, Sudan, 7–9 September 2015, pp 101–106
2. Handelsman MM, Briggs WL, Sullivan N, Towler A (2005) A measure of college student course engagement. J Educ Res 98:184–192. https://doi.org/10.3200/JOER.98.3.184-192
3. Mohd IH, Aluwi AH, Hussein N, Omar MK (2016) Enhancing students engagement through blended learning satisfaction and lecturer support. In: Engineers Institute of Electrical and Electronics (IEEE) (ed) 2016 IEEE 8th international conference on engineering education (ICEED2016): "Enhancing engineering education through academia-industry collaboration." Red Hook, NY, Curran Associates, Inc., Kuala Lumpur, Malaysia, 7–8 December 2016, pp 175–180
4. Hanus MD, Fox J (2015) Assessing the effects of gamification in the classroom: a longitudinal study on intrinsic motivation, social comparison, satisfaction, effort, and academic performance. Comput Educ 80:152–161. https://doi.org/10.1016/j.compedu.2014.08.019
5. Sanmugam M, Zaid NM, Abdullah Z, et al (2016) The impacts of infusing game elements and gamification in learning. In: 2016 IEEE 8th international conference on engineering education (ICEED), pp 131–136
6. Dixson MD (2015) Measuring Student Engagement in the online course : the online student engagement scale (OSE). Online Learn J 19
7. Marx AA, Simonsen JC, Kitchel T (2016) Undergraduate student course engagement and the influence of student, contextual, and teacher variables. J Agric Educ 57:212–228. https://doi.org/10.5032/jae.2016.01212

8. Hu M, Li H, Deng W, Guan H (2016) Student engagement: one of the necessary conditions for online learning. In: Institute of Electrical and Electronics Engineers (IEEE) (ed) 2016 international conference on educational innovation through technology (EITT). Red Hook, NY, Curran Associates, Inc., Tainan, Taiwan, 22–24 September 2016, pp 122–126

9. Kuo MS, Chuang TY (2016) How gamification motivates visits and engagement for online academic dissemination—an empirical study. Comput Human Behav 55:16–27. https://doi.org/10.1016/j.chb.2015.08.025

10. Hamari J (2015) Do badges increase user activity? A field experiment on the effects of gamification. Comput Human Behav 71:469–478. https://doi.org/10.1016/j.chb.2015.03.036

11. Barata G, Gama S, Gonçalves D, Jorge J (2013) Improving participation and learning with gamification. In: Proceedings of the first international conference on gameful design, research, and applications—gamification 2013. New York, ACM, Stratford, Ontario, Canada, 2–4 October 2013, pp 10–17

12. Wang AI (2015) The wear out effect of a game-based student response system. Comput Educ 82:217–227. https://doi.org/10.1016/j.compedu.2014.11.004

13. Sawang S, O'Connor P, Ali M (2017) IEngage: using technology to enhance students' engagement in a large classroom. J Learn Des 10:11–19. https://doi.org/10.5204/jld.v9i3.292

14. Wang AI, Lieberoth A (2016) The effect of points and audio on concentration, engagement, enjoyment, learning, motivation, and classroom dynamics using Kahoot! In: Connolly T, Boyle L (eds) Proceedings From the 10th European conference of game based learning. Academic Conferences and Publishing International Limited, Paisley, UK, 6–7 October 2016, pp 737–748

15. Chaiyo Y, Nokham R (2017) The effect of Kahoot, Quizizz and Google Forms on the student's perception in the classrooms response system pp 178–182

16. Pilakowski M (2015) Comparing student-response systems. http://technologypursuit.edublogs.org/2015/11/25/comparing-student-response-systems/. Accessed 28 May 2017

17. Davis F, Bagozzi R, Warshaw PR (1989) User acceptance of computer technology: a comparison of two theoretical models. Manage Sci 35:982–1003. https://doi.org/10.1287/mnsc.35.8.982

18. Fishbein M, Ajzen I (1975) Belief, attitude, intention and behaviour: an introduction to theory and research. Addison-Wesley Pub. Co, Reading, MA

19. Legris P, Ingham J, Collerette P (2003) Why do people use information technology? A critical review of the technology acceptance model. Inf Manag 40:191–204. https://doi.org/10.1016/S0378-7206(01)00143-4

20. Davis FD (1989) Perceived usefulness, perceived ease of use, and user acceptance of information technology. MIS Q 13:319–340

21. Fathema N, Shannon D, Ross M (2015) Expanding the technology acceptance model (TAM) to examine faculty use of learning management systems (LMSs) in higher education institutions. MERLOT J Online Learn Teach 11:210–232

22. Hussain A, Mkpojiogu EOC, Yusof MM (2016) Perceived usefulness, perceived ease of use, and perceived enjoyment as drivers for the user acceptance of interactive mobile maps. In: Abdul Nifa FA, Mohd Nawi MN, Hussain A (eds) Proceedings of the international conference on applied science and technology 2016 (ICAST'16). AIP Publishing, Kedah, Malaysia, 11–13 April 2016

Learning Style on Mobile-Game-Based Learning Design: How to Measure?

Hanif Al Fatta, Zulisman Maksom and Mohd. Hafiz Zakaria

Abstract Some researchers have recognized the importance of pedagogical aspects in designing digital game-based learning. It is believed that learning style, learning content will affect the form of the game genre, game activities, and mechanics for delivering suitable game-based learning. This study aimed to analyze the relation of an individual's learning style to their game preference. Learning style of samples is measured by Learning Style Questionnaire (LSQ) by Honey and Mumford. In addition, the measurement to determine if selected samples choose the appropriate game according to their learning style is proposed. As a contribution to knowledge, this study introduced an instrument to asses the relationship of learning style with the way learner using the game. The result indicated the samples did not follow the trend that preference for a game they played is affected by their learning style. Still, it confirmed the previous studies where learning style is a significant factor that should be addressed during the game design process.

Keywords Game-based learning style · LSQ · Pedagogical aspect

1 Introduction

Nowadays, we have reached a new level of the learning system. According to [1], the students from all levels of education institutions are grouped into the so-called "Digital Natives". Their daily life is surrounded by computer games, digital devices, and smartphones. This situation changes the abilities and learning style of the Digital

H. A. Fatta (✉)
Faculty of Computer Science, Universitas AMIKOM, Yogyakarta, Indonesia
e-mail: hanif.a@amikom.ac.id

Z. Maksom · Mohd. H. Zakaria
Faculty of Information and Communication Technology, Universiti Teknikal
Malaysia Melaka, Durian Tunggal, Malaysia
e-mail: zulisman@utem.edu.my

Mohd. H. Zakaria
e-mail: hafiz@utem.edu.my

© Springer Nature Singapore Pte Ltd. 2019
V. Piuri et al. (eds.), *Intelligent and Interactive Computing*, Lecture Notes in Networks and Systems 67, https://doi.org/10.1007/978-981-13-6031-2_8

Natives. This is the significant reason why digital game-based learning arises and evolves. Furthermore, a digital game-based learning or educational games [2] played an important role in defining new methods of education because of radical change in the learner and appropriate ways to motivate this kind of learners [1].

As the role of digital game-based learning is highly recognized, the study to improve the quality of this game grew. The nature of learning in the physical world will apply as well in this digital world. The importance of recognizing the variation among students' learning style has been studied by [3]. According to [3], the teachers' understanding of the variation of student's reaction to a particular classroom setting and pedagogical practices increased the possibility of accommodating the students' diversity of learning demand. Moreover, [4] showed the value of learning style in delivering learning content to the student in some level. If addressing learning style improved learning outcome in real-world learning, it is also possible for game-based learning to adopt it and gain similar advantages [5].

This study aimed to propose empirical evidence on the effect of the user's learning style on their game preference. This study also focuses on whether the user with the particular type of learning style tends to play and interact with the game differently with the user with different learning styles.

2 Learning Style

During the deliverance of learning the material through educational games, some educational aspects such as learning style and learning methods should be considered as it is the nature of learning in the real world. It is significant to understand an individual learning style to anticipate how the learner responds and feel in the various conditions. In addition, different people will learn and digest information with their different approach [6]. Numerous learning styles models have been used to measure. However, there are few models which have become prominent models to study in the education literature. They are the following:

a. **The Myers–Briggs-Type Indicator** [7]

Based on Jung's Theory of Psychological Types, Myers–Briggs-Type Indicator (MBTI) classifies people into four categories: Extraverts, Sensors, Thinkers, and Judgers.

b. **Kolb's Experiential Learning Model** [8]

This model classified students based on their preference on (1) concrete experience or abstract conceptualization and (2) active experimentation or reflective observation. And four types of learners are classified as follows: (a) concrete-reflective, (b) abstract-reflective, (c) abstract-active, and (d) concrete-active.

c. **The Felder–Silverman Model** [9]

Based on this model, learners are classified as: sensory or intuitive, visual-verbal, active-reflective, and sequential-global.

d. **The VARK Model** [10]

Based on this model learners are classified as Visual, where learner experienced the best learning by seeing it. Aural where the best learning is by hearing. Read/Write-learner that prefer to look at the written word. Kinesthetic-experience and practice are the best way to learn. And the Multimodal category for a learner with multiple types of learning style.

However, this study employed Kolb's model for the measurement basis. According to [4], Kolb's model has a broad equivalent orientation with learning style instrument develop by Honey and Mumford [11]. So in this study, Learning Style Questionnaire (LSQ) by Honey and Mumford will be used to determine the learning style of chosen samples. Briefly described, this model divided learning style into four categories: activists, reflectors, theorists, and pragmatists.

1. **Activists** learn by doing, and like a new experience and sudden assignment, they evolve through challenge and are bored by the routine implementation.
2. **Reflectors** learn through reflection. Before taking a decision, they analyze carefully and can delay decision-making.
3. **Theorists** learn from systems, concepts, and models. They are adapting observation into theories. They are tidy and rational learners.
4. **Pragmatists** learn by the practical application of theory; they are ideal learners who put balance to theory and practice, decision-making, and problem-solving.

3 Previous Research

Research about the relationship between pedagogical aspect and design of educational games is insufficient. Luckily, some studies have revealed the significance of learning content, learning style, and technique for the deliverance of learning through the utilization of game-based learning.

A study conducted by Rapeepisarn et al. [6] proposed a relationship between learning style and game style chosen by their user. This study underlined the study undertaken by Prensky [12], concerning the association between three concepts: learning activity, learning content, and learning type. And at the end of the study, Rapeepisarn et al. concluded the relation of learning style, behavior when interacting with a game, behavior when using a computer and desirable, suitable game genre. It means that a learner with a specific learning style will perform different behaviors while playing a game. By knowing the way a user is playing a game, we can develop a suitable game genre for delivering learning content. The proposal of the relationship between learning style, behavioral on playing a game and using a computer, and game genre are presented in Table 1.

Table 1 Relationship of pedagogical aspects and game genre according to Rapeepisarn et al. [6]

Learning style	Behavior when playing the game	Behavior when using the computer	Learning activities	Possible game genre
Activists	Would better work in a team, choose to become group's leader, capable of deliberating problem-solving	Get used to shortcut key but able to employ button available on toolbars	Practice. Imitation work with other Deal with problem "head on"	Multiplayer interaction, action game, role-playing game
Reflectors	Inspects the crucial data in the game, stick to the instructions, allocating a long time before making a decision, does not take a position as a leader during game	Usually choose drop-down menu but in the end, find what is best, for example, SEARCH FOR HELP in the HELP menu	Attentive. Feedback. Measured task. Work solo at their motion	Concentration game, adventure game, a simulation game
Theorists	Inspects data and follow the instruction before starting the game, be able to give careful thoughts when choosing the game elements, develop a positive strategy to overcome the opponents	Usually use drop-down menus to explore application's ability, like to browse through the INDEX or SEARCH FOR HELP in the HELP menu	Logic. Understanding the principle. Analyze and formulate the plan Examine the relationship between things	Strategies game, a simulation game. Puzzle
Pragmatists	Follow the direction and scenario carefully during the briefing, believe they can play better with sufficient guidance Demonstrate a strong interest in the puzzle game and dislike a role-playing game	Apparently, utilize the toolbars buttons to complete things, often find HELP menu to get things done	Trial and error. Asking question, try things out. Constructing a plan with obvious purpose	Puzzle game, building game, constructing game, reality testing game, detective game

The relation presented in Table 1, however, triggered several questions. First, does a learner with a specific dominant learning style, always perform the same behavior when playing a game or using a computer? If this question is not solidly answered, the conclusion that connects learning style with possible game genre should be reconsidered. Second, learning style can be determined by several models, in this case, if we replace the LSQ with another model, VARK model, for example, will it show a similar relationship?

To answer the first question, this study proposed measurement tools to indicate the relation of learning style and behavior of learner while playing a game. The measurement tool tested on selected samples and then the result is analyzed. Although Rapeepisarn et al. only indicated a small number of behaviors performed by the learner when playing a game, it is possible to construct a measurement tool using those indicators.

4 Research Methods

4.1 Proposed Measurement Instrument

This study develops a measurement instrument to identify the learning style of an individual based on their behavior when playing a game. The measurement instrument is designed based on the relationship of learning style and behavior while playing games proposed by Rapeepisarn as depicted in Table 1. From the information in Table 1, it is possible to determine the learning style of a learner by identifying their behavior when playing the game, including educational games. If the participant is a mature individual, they can use this instrument as a self-assessment. But if the learner is a very young aged learner, an observer can conduct observation while children were playing some educational games and filled the tool based on the observation condition, and sometimes can be followed by interviews to confirm several things found during observation.

The detailed assessment is available in Appendix. In this form, the participant will answer 14 questions available related to behavior type based on the kind of behavior proposed by Rapeepisarn et al. The participants respond from the most suitable description by circling "1" score and to the least suitable type of behavior by choosing "4". Consecutively, Type A will represent activists learning Type, B for reflectors, C for theorists, and D for pragmatists type. Detailed procedure to use and produce the result of the measurement is presented as follows: (1) Examine all the answers to the given question. (2) Put attention on the question with a "1" score as an answer. (3) Calculate the number of "1" answers in each type. (this question has 4 types A, B, C, and D). (4) For example, a respondent has "1" score for question number 1, 2, 3, (5) calculate the percentage for each type, for example, Type A consists of question no 1,2, and 3 and all the questions are scored "1". So, it has

100% Type A result. According to this result, the participant showed Type A. So his/her dominant behavior is activists learning the style.

From their answer, a leaner's learning style can be drawn. The result from this measurement will be compared to the prior result taken using LSQ questionnaire. If the trend follows the concept proposed by the Rapeepisarn et al., the result from LSQ and proposed measurement should inform the same learning type of an individual, and a learner's behavior while playing a game can be predicted based on their learning type.

4.2 Study Sample

Participants were taken randomly from undergraduate students in the informatics department, at Universitas AMIKOM Yogyakarta, Indonesia. Because this study is intended to analyze learning style and behavior playing a game, a confirmation to ensure respondents are familiar with the digital game is carried out. To ensure they also play educational games, two educational games are installed and played by the participants before data retrieval. 40 participants are given a questionnaire, and 29 questionnaires are returned and filled correctly. Because the samples are undergraduate students and they are supposed to know precisely their behavior in accessing the game, the survey is given in the form of self-assessment.

4.3 Evaluation Methods

Before data retrieval is started, the process that should be done is to ensure that participant is familiar with digital games or educational games, two types of mobile educational games is given to the participant, and they play it about around 10 min duration for each game. Next, the LSQ questionnaire is distributed among selected samples. Providing about 10–15 min, the participant had their form filled and evaluated. This is an entirely self-assessment process so the honesty of the participants will be the most challenging part to support the valid result. As soon as they finished the LSQ questionnaire, the second questionnaire as presented in Appendix is distributed. The participant and then selected the suitable behavior type according to the available type on the form.

Statistical analysis is employed before the comparison stage is conducted. Reliability test is applied for both questionnaires using SPSS. The calculation showed that the value of Cronbach's alpha for the LSQ questionnaire is 0.636. And the Cronbach's alpha value for the proposed survey is 0.824. From these results, the internal consistency for both surveys is considered acceptable.

And finally, the comparison stage is initiated. From the result of two types of questionnaires distributed to the participant, the data reflecting the dominant learning style and the dominant behavior while using a game is acquired. From this result, we can only match the result of the LSQ with the consequence of behavior prediction from the second questionnaire. And the accuracy of the projection can be made with a just a simple comparison. If the precision is high, then we can say that learning style based on LSQ questionnaire does have a significant role in predicting the game player's behavior.

5 Result

From the result of LSQ questionnaire, participants were classified into four categories: activists, reflectors, theorists, and pragmatists according to their score. For example, a participant showed these following scores: activist/10, reflector/17, theorists/15, and pragmatist/8; then this participant will be categorized as a theorist. The result indicated that 3% of the students are classified into activists, 66% into reflectors, 17% into theorists, and 14% into pragmatists.

A different result is showed in the second questionnaire. The proposed measurement tools are designed to predict learning style characteristics of a participant based on gaming behavior as presented in Table 1. The result indicated that 45% of the students are categorized into activists, 38% into reflectors, 3% into theorists, and 14% into pragmatists. Figure 1 displayed the graphic showing the result of both questionnaires.

An interesting fact also appeared if the result of a participant is compared. According to Table 1, the ideal result should show that a participant will have the same learning style measured in both questionnaires. This will be the justification that learning style can be used to predict user's gaming behavior. But the result shows the accuracy of this matching is less than 50%. Figure 2 indicates that only 45% of the participant have the same result in both questionnaires. 55% of the participant showed they have different result indicating from their questionnaire answer. This

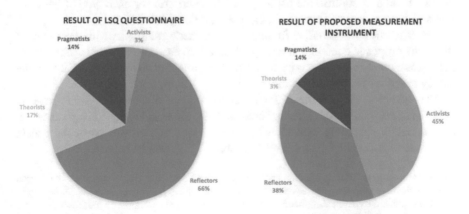

Fig. 1 The chart showed the different results collected from two questionnaires in categorizing learning style among the same participants

Fig. 2 Result comparison of
LSQ questionnaire and
gaming behavior
questionnaire with one
dominant learning style

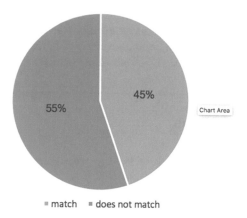

study only used simple data presentation to show that the finding did not follow the
trend proposed by Rapeepisarn et al.

6 Conclusion and Discussion

The finding confirmed several limitations in the current study. First, the use of LSQ
questionnaire to perform categorization on learning style should be reconsidered.
Using the different model of learning style to determine participants' learning style
is worth trying. The different model may reveal the perfect learning style of the
learner. The limitation of the proposed measurement instrument is also needed to be
reconsidered. The instrument is written based on criteria developed by Rapeepisarn
et al. No additional detail for the gaming behavior is added to the requirements
proposed by Rapeepisarn et al. The simple questionnaire suggested by this study may
lack in detail and confuse the participant. So the criteria for each type of behavior
when playing games as shown in Table 1 should be updated.

The limitation also applied to data retrieval methods. Perhaps the best way of
filling the proposed questionnaire is by observation. Participant is set to play certain
games, and close inspection is made during the gameplay and observer filled out the
survey based on the behavior performed by the participants.

Nevertheless, 45% of the matching result indicated there is a weak relation
between learning style and gaming behavior. Still, the learning style is a good aspect
to consider in designing educational games, so research on this particular aspect
should be made to enhance the model proposed by Rapeepisarn et al.

Acknowledgements Before the data acquisition is conducted, the ethical approval is required
to ensure the data gathering process does not violate the ethical policy in Indonesia, especially in
AMIKOM University. Ethical approval is granted from the ethical committee in the university level,
that is in charge of examining the research permission and giving the ethical approval according
to the national standard. The participant filled up consent form assuring their willingness as a

volunteer. The department where the participants enrolled also filled up a consent form to make sure the activity is under the supervision of the faculty of computer science.

We would like to mention special thanks to AMIKOM University for their contribution on funding the research published in this paper, and also for the students of Information Technology Department as they are the voluntary participants for primary data retrieval. And we also would like to mention special thanks to Universiti Teknikal Malaysia Melaka for their contribution in providing experts to review, discuss, and revise the content of this paper.

Appendix: Proposed Measurement Instrument

Instructions:						
	From the following question, circle the most suitable behavior according to your experience while playing a game					
	1 indicates the most suitable, 4 indicates the least suitable					
My Behavior while playing a game:						
Type	Behavior		Circle your score here			
	1	Prefer working as a team	1	2	3	4
Type A	2	Being a group leader	1	2	3	4
	3	Be able to brainstorm to solve the problem	1	2	3	4
	4	Go through important data in the game	1	2	3	4
	5	Follow the instructions	1	2	3	4
Type B	6	Spend a long time before making a decision	1	2	3	4
	7	Not to lead the game	1	2	3	4
	8	Go through the data and follow the instruction before starting the game	1	2	3	4
Type C	9	Be able to give careful thoughts when choosing the game elements	1	2	3	4
	10	Formulate a good strategy to defeat the enemy	1	2	3	4
	11	Follow closely the instructions and strategies that were mentioned in the briefing	1	2	3	4
Type D	12	Believe they can play better if they were given proper instruction	1	2	3	4
	13	Show a great interest in the puzzle game	1	2	3	4
	14	Dislike role-playing game	1	2	3	4

References

1. Prensky M (2001) Digital game-based learning, 1st edn. McGraw Hill, New York
2. De Freitas S (2006) Learning in immersive worlds a review of game-based learning. Prepared for the JISC e-Learning Programme. JISC e-Learning Innov 3(3):73
3. Felder RM, Brent R (2005) Understanding student differences. J Eng Educ 94(1):57–72
4. Price L (2004) Individual differences in learning: cognitive control, cognitive style, and learning style. Educ Psychol 24(5):681–698
5. Soflano M, Connolly TM, Hainey T (2015) An application of adaptive games-based learning based on learning style to teach SQL. Comput Educ 86:192–211
6. Rapeepisarn K, Wong KW, Fung CC, Khine MS (2008) The relationship between game genres, learning techniques and learning styles in educational computer games. In: Pan Z, Zhang X, El Rhalibi A, Woo W, Li Y (eds) Technologies for e-learning and digital entertainment, vol 5093. Springer, Berlin, pp 497–508
7. Pittenger DJ (1993) The utility of the Myers-Briggs type indicator. Rev Educ Res 63(4):467–488
8. Kolb DA (2015) Experiential learning: experience as the source of learning and development, 2nd edn. Pearson Education, Hoboken
9. Felder R, Silverman L (1988) Learning and teaching styles in engineering education. Eng Educ 78(June):674–681
10. Fleming N, Baume D (2006) Learning styles again: VARKing up the right tree! Educ Dev 7(4):4–7
11. Honey P, Mumford A (1982) The manual of learning style. Peter Honey, Berkshire
12. Prensky M (2005) Computer games and learning: digital-based games. Handbook of Computer Game Studies, pp 97–124

Evaluation of Production Model for Digital Storytelling via Educational Comics

Farah Nadia Azman, Syamsul Bahrin Zaibon, Norshuhada Shiratuddin
and Mohamad Lutfi Dolhalit

Abstract The paradigm of Education 4.0 empowers learners to produce innovations, the follow-on substantiations of knowledge production in a form of learner-generated content. Originated from this trend, there are a growing number of Digital Storytelling online platforms that grant learners interactive tools for producing educational comics. Thus, a conceptual production model of learner-generated comic is proposed to serve as a guideline for learners to design and develop digital educational comics. To evaluate the quality of the proposed model, experimental testing is carried out where participants quantitatively rate their experience of using the model. Nonparametric findings from statistical test disclose that the proposed production model learner-generated comic is significantly generalizable, flexible, complete, usable, and understandable. Therefore, results conclude that the production model for learner-generated comic has significantly served as guideline for learners to design and develop digital educational comics.

Keywords Educational comic · Digital Storytelling · Learner-generated content · Media production

F. N. Azman (✉) · S. B. Zaibon · M. L. Dolhalit
Faculty of Information and Communication Technology, Universiti Teknikal Malaysia Melaka,
Hang Tuah Jaya, 76100 Durian Tunggal, Melaka, Malaysia
e-mail: farah@utem.edu.my

S. B. Zaibon
e-mail: syamsulbahrin@uum.edu.my

M. L. Dolhalit
e-mail: lutfi@utem.edu.my

N. Shiratuddin
School of Multimedia Technology and Communication, Universiti Utara Malaysia,
06010 Kedah, Malaysia
e-mail: norshuhada@uum.edu.my

© Springer Nature Singapore Pte Ltd. 2019
V. Piuri et al. (eds.), *Intelligent and Interactive Computing*, Lecture Notes in Networks and Systems 67, https://doi.org/10.1007/978-981-13-6031-2_45

1 Education 4.0 and Learner-Generated Content

Resulting from the Fourth Industrial Revolution, the paradigm of Education 4.0 empowers learners to produce innovations, the follow-on substantiations of knowledge production [1]. These innovations also identified as "student-generated content" and "students as producers", which are some of the other related terms currently in use for learner-generated content [2]. In prior, teachers were more or less the solitary source who provided information; today the learners rather passive users of information, the role is switched to creators of knowledge in networked structures [3].

The production of learner-generated content may include producing e-portfolios [4], e-books [5], blogs [6], computer games [7], videos [8], podcasts [9], animation [10], multimedia poster [11], models [12], as well as comics [13–15]. Hence, the focus of this paper is on comics as the learner-generated content.

2 Production Model for Learner-Generated Comic

Digital Storytelling elevates understanding across difference, communication, and critical reflection [16]. A Digital Storytelling project serves a tool for communication, observation, analysis, and metacognition [17]. In the effort to converse the narrative to the audience through technology, the digital story's creator also learns how to incorporate their perspectives, beliefs, and needs, in order to better present the story [18]. As a result, there is a growing number of Digital Storytelling online platforms that grant users interactive tools for creating comic narrative such as Comicshead, TechnoToon, Madefire, Storyboard That, and more. Digital stories in a form of learner-generated educational comics may consist of illustrative characters [13] or sorted collection of annotated photos [14, 15] in comic panels.

One of the various types of digital narratives is stories that meant to inform and instruct [19]. Instructional messages are integrated with learning activities to create more engaging and exciting learning environments [20] by allowing learners to self-reflect on their thinking in their digital story [21]. Despite the importance of Digital Storytelling via educational comics, most scholars do not emphasize on quality, theoretically supported, and strategic learner-generated comic production methodology that accommodate to interrelated key elements and production methods of digital educational comics. Thus, a conceptual production model of learner-generated comic is proposed. The proposed model was intrinsically constructed through content analysis, comparative study of learner-generated comic classroom strategies, Digital Storytelling models and frameworks, expert consultation, user participation [22] and validated through expert review [23]. Primarily, the core components of learner-generated comic production model are categorized into phases and supported by task, activity, and flow subcomponents as illustrated in Fig. 1.

Fig. 1 Production model for learner-generated comic [23]

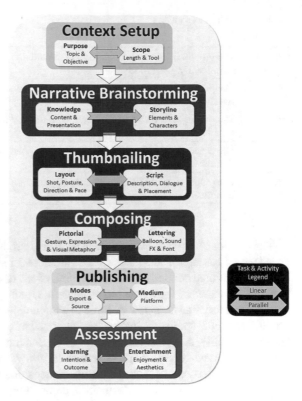

It is hypothesized that the production model for learner-generated comic would significantly serve as a guideline for learners to design and develop digital educational comics. To verify this claim, the proposed model is evaluated through experimental testing as explained in the next section.

3 Methodology

The experimental testing adopted quasi-experimental nonequivalent control group design methods in order to evaluate the proposed production model for learner-generated comic. Unequivocally, the users of the production model are learners. Consequently, convenience sampling was adopted involving 57 students of an institution as the participants in experimental testing activity. This number is adequate for learner-generated comic production model quality evaluation since at least 30 datasets should be employed for obtaining reliable result in statistical tests [24]. Plus, the apparent homogeneity of university students enhances research validity where they tend to be similar on dimensions such as age and education [25]. Accordingly, the designated participants are students from the same discipline and course.

Table 1 Research hypotheses

Hypothesis		Sources
H_0	The proposed learner-generated comic production model is not significantly generalizable	[26, 27]
H_0	The proposed learner-generated comic production model is not significantly flexible	[26, 27]
H_0	The proposed learner-generated comic production model is not significantly complete	[26, 27]
H_0	The proposed learner-generated comic production model is not significantly usable	[26, 27]
H_0	The proposed learner-generated comic production model is not significantly understandable	[26, 27]

The quality of the proposed model was measured in order to evaluate if the proposed model has significantly served as a guideline for learners to design and develop digital educational comics. Based on the dimensions of model characteristics in [26], five hypotheses were formulated to evaluate the quality of learner-generated comic production model as depicted in Table 1. Fundamentally, learner-generated comic production model quality evaluation is categorized into five dimensions, namely, generality, flexibility, completeness, usability, and understandability. Generality dimension assesses at what level the proposed model allowed its user to utilize it in diverse educational purposes while the flexibility dimension evaluates on how flexible the proposed model to its users. Meanwhile, completeness dimension measures if the components proposed in learner-generated comic production model are entirely required to successfully to create digital educational comics. Usability dimension on the hand refers to how usable the proposed model for the users in constructing digital educational comics. Lastly, understandability dimension evaluates at what level the proposed model is comprehensible to its users.

To carry out the Digital Storytelling session, BitStrips for School online application was selected as the comic authoring tool for the experimental testing. The participants were instructed to develop digital educational comic that presents their understanding on selected topics in web programming subject [28]. Based on what they have learned so far in official lecture and lab sessions, the participants referred to the proposed model help them reflect and organize their overall idea about the topic into their digital educational comic. They may also refer to additional books and online resources on the respective topic. Divided into three members per group, the participants began the project by discussing and generating ideas based on the activity stated in the learner-generated comic production model. Using instructor account in BitStrips, the participants' progress in finishing the comics digitally was inspected. Incomplete comic pages, character usage, access time, and more were monitored as displayed in Fig. 2.

The Digital Storytelling session ended after the participants submitted their completed comics to the instructor. As satisfaction survey comparison is one of the

Fig. 2 BitStrips for school instructor interface

evaluation items to measure the functionality of the artifact with the solution objectives [29], the 57 participants were provided with questionnaire instrument [30] to rate their experience in using learner-generated comic production model. Informed consent was obtained from all individual participants included in the study.

4 Results and Findings

Initially, data normality was assessed both visually and through Shapiro–Wilk test for lower than 100 sample size [31]. Table 2 displays the results of data normality test for each dimension.

Since the p value was less than 0.05, it was assumed that the data were not normally distributed. Thus, in analyzing the descriptive results for each item in a specific dimension, the median ($\tilde{x}$) score is featured. Consequently, to analyze and

Table 2 Test of normality results

Dimension	W	df	Sig.
Generality	0.873	57	0.000
Flexibility	0.964	57	0.045
Completeness	0.864	57	0.000
Usability	0.948	57	0.015
Understandability	0.921	57	0.001

Lilliefors significance correction

descriptively interpret the data from the 9-point semantic scale instrument items, a six-scale measurement with the range of interval 1.33 was formulated as the scale (see Fig. 3). This number was achieved by dividing the range of scale with a number of scales as suggested in [32].

As shown in Table 3, the results exhibited that there was a comprehensible positive tendency of agreement toward the statements stated in the questionnaire. This is because item values greater than 6.33 indicate that the 57 respondents agreed with the statement for each measurement item.

Next, all data from the items were pulled together to describe a more accurate reflection on participants' perception toward the dimension [33] where they are visualized in Fig. 4 and summarized in Table 4.

So far, the findings had hinted participants' positive acceptance toward the proposed model. These numbers implied that majority of the participants have come to an agreement that the production model for learner-generated comic is generalizable, flexible, complete, usable, and understandable.

To strengthen the validity of the results, evaluation of learner-generated comic production model quality in a mode of hypothesis testing was conducted (refer to Table 1). Based on the established method in measuring users perceived quality of the proposed conceptual model [34, 35], the $(\tilde{x}) = 6.33$ measurement was used as the primary indicator to determine whether the proposed production model for learner-generated comic is significantly generalizable, flexible, complete, usable, and understandable. Essentially, a score of 6.333 and above is the cutting point (hypothesized median) for statistical significance for each dimension. In compliance to the positive interval value, the decision to accept or to reject the null hypothesis (H_0) depends on the value. Particularly, this study rejects H_0 when $(\tilde{x}) \geq 6.33$ and fails to reject H_0 when $(\tilde{x}) < 6.333$. Accordingly, one sample Wilcoxon signed rank nonparametric test was run and the results are displayed in Table 5.

The results of one sample Wilcoxon signed rank test is given in Table 5 which indicates that the null hypothesis is rejected. This finding discloses that the proposed production model learner-generated comic is significantly generalizable, flexible,

Interval   = (Highest score – lowest score) ÷ number of scale
= (9-1) ÷ 6
= 1.333

List of Scale (Level of Agreement)
Highly Disagree   =   1.000 – 2.322
Disagree   =   2.333 – 3.655
Slightly Disagree   =   3.666 – 4.988
Slightly Agree   =   4.999 – 6.322
Agree   =   6.333 – 7.655
Highly Agree   =   7.666 – 9.000

Fig. 3 Conversion of numerical scale for data analysis

Table 3 Proposed model dimension results

Statement	Median ($\tilde{x}$)	Mean	Standard deviation
Generality			
The model enables me to summarize what I had learnt in the form of digital educational comic	7.000	7.460	1.151
The model enables me to elaborate and organize my knowledge the form of digital educational comic	8.000	7.560	1.035
During the digital educational comic development, the model enables me to relate my learning towards essential theories/ideas/information/knowledge	8.000	7.670	0.893
During the digital educational comic development, the model enables me to apply knowledge to other situation/scenario/context	8.000	7.720	0.996
During the digital educational comic development, the model enables me to reflect prior knowledge and connect it to new knowledge	8.000	7.600	1.178
Flexibility			
Using the model fits well with the way I like to work	8.000	7.390	1.082
The model enables me to produce digital educational comic according to my own taste and preferences	7.000	7.490	1.037
I have the options to follow or deviate from the phases and activities suggested in the model	7.000	7.320	1.152
The model enables me to make alterations towards certain phases and activities in digital educational comic development process	7.000	7.330	1.041
Completeness			
All the concepts and components included in the model are strictly necessary for digital educational comic development	7.000	7.440	1.053
All components in the model are pertinent for the depiction of the digital educational comic development process	7.000	7.610	0.881
The model provides a thorough exemplification of the digital educational comic development process	8.000	7.470	0.966
The model enables me to accomplish tasks in digital educational comic development more thoroughly	8.000	7.610	0.940
The model allows me to intelligently check the relevance and completeness of my digital educational comic	7.000	7.470	0.966
Usability			
Using the model produces the digital educational comic, for which it is intended for	8.000	7.560	0.945
The model is useful in providing information I need on digital educational comic development	8.000	7.680	1.003

(continued)

Table 3 (continued)

Statement	Median ($\bar{x}$)	Mean	Standard deviation
Using the model enhances the quality of my digital educational comic	8.000	7.860	0.915
The model would be an enhancement to a textual portrayal of the digital educational comic development process	8.000	7.700	0.944
Understandability			
The model is clear and understandable	8.000	7.960	0.999
Understanding the model does not entail a lot of mental exertion	8.000	7.610	0.978
The model as a whole is workable	8.000	7.880	0.927
The phases and activities in the model can be followed	8.000	8.020	0.834

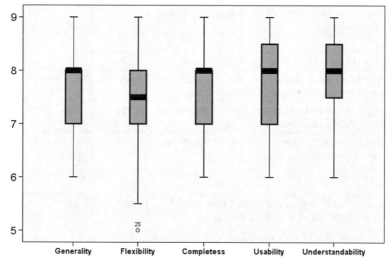

Scale: Highly Disagree = (1.000 – 2.333), Disagree = (2.333 – 3.655), Slightly Disagree = (3.666 – 4.988), Slightly Agree = (4.999 – 6.322), Agree = (6.33 – 7.65), Highly Agree = (7.666 – 9.000)

Fig. 4 Boxplot of generality, flexibility, completeness, usability, and understandability dimensions

Table 4 Descriptive statistics of composite factors

Dimension	Summary statistics				
	Mean	Median ($\bar{x}$)	Std. deviation	Skewness	Kurtosis
Generality	7.597	8.000	0.863	0.036	−0.634
Flexibility	7.351	7.500	0.925	−0.309	0.213
Completeness	7.597	8.000	0.821	0.081	−0.516
Usability	7.737	8.000	0.791	−0.160	−0.573
Understandability	7.921	8.000	0.784	0.508	0.429

Table 5 One sample Wilcoxon signed rank analysis on generality, flexibility, completeness, usability, and understandability dimensions

Attribute	Sig.	Decision
The median of generality equals 6.333	0.000*	Reject the null hypothesis
The median of flexibility equals 6.333	0.000*	Reject the null hypothesis
The median of completeness equals 6.333	0.000*	Reject the null hypothesis
The median of usability equals 6.333	0.000*	Reject the null hypothesis
The median of understandability equals 6.333	0.000*	Reject the null hypothesis

*Asymptotic significance is displayed. The significance level is 0.05

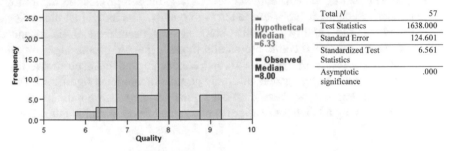

Fig. 5 Summary of one sample Wilcoxon signed rank analysis on generality, flexibility, completeness, usability, and understandability collectively (quality)

complete, usable, and understandable. Finally, one sample Wilcoxon signed rank nonparametric was run another round (including all five dimensions) to measure the overall quality of the proposed model. Results are displayed in Fig. 5.

The results revealed that the quality median [$(\tilde{x}) = 8.000$] is higher from the hypothesized median [$(\tilde{x}) = 6.333$]. Therefore, it is confirmed that users perceived the proposed learner-generated comic production model as having quality. This concludes that the production model for learner-generated comic has significantly served as guideline for learners to design and develop digital educational comics.

5 Limitations

For the purpose of homogeneousness, convenience sampling was adopted. Hence, the conclusions of learner-generated comic production model quality may not be generalizable to a broader cross section of the population because this study was restricted to a particular group of learners, which were 57 Malaysian undergraduate students. Consequently, replication studies of measuring the quality of learner-generated comic production model are encouraged as part of the adoption effort so as to add to the body of knowledge. Plus, the demographic background of participants did not consider their past experience in reading or producing comics. This indicates that in what way existing and previous experience in digital educational comics influence users' perception and attitude toward production model for learner-generated comic should be explored in extended research.

6 Conclusion

This paper presents the analysis and evaluation results of the production model for Digital Storytelling via educational comic. The proposed production model for learner-generated comic is measured in terms of quality. The individual dimension results convey that generality, flexibility, completeness, usability, and understandability of learner-generated comic production model evaluation score high overall median and mean. Testing of hypotheses is also conducted, and the results demystify that the quality of the learner-generated comic production model is significant. This implied that the proposed production model for learner-generated comic is significant in serving as a guideline for learners to design and develop digital educational comics.

Acknowledgements All procedures performed in studies involving human participants were in accordance with the ethical standards of the institutional and/or national research committee. The authors would like to express our high gratitude to the Center for Advanced Computing Technology (C-ACT) and Centre for Research and Innovation Management (CRIM) Universiti Teknikal Malaysia Melaka (UTeM) for funding this paper via an internal grant (Ref. No: PJP/2018/FTMK(5D)/S01633). The highest appreciation also goes to Universiti Utara Malaysia for supporting the data collection. The authors declare that they have no conflict of interest.

References

1. Harkins AM (2018) Leapfrog principles and practices: core components of education 3.0 and 4.0. Futures Res Q 24(1):19–31
2. Sener J (2007) In search of student-generated content in online education. E-Mentor 4:1–8
3. Schuster K, Groß K, Vossen R, Richert A, Jeschke S (2016) Preparing for industry 4.0–collaborative virtual learning environments in engineering education. In: Automation, communication and cybernetics in science and engineering 2015/2016. Springer International Publishing, pp 417–427
4. De Jong T, Van Joolingen WR, Giemza A, Girault I, Hoppe U, Kindermann J, Van Der Zanden M (2010) Learning by creating and exchanging objects: the SCY experience. Br J Educ Technol 41(16):909–921

5. Tsai CW, Shen PD, Lu YJ (2015) The effects of problem-based learning with flipped classroom on elementary students' computing skills: a case study of the production of ebooks. Int J Inf Commun Technol Educ (IJICTE) 11:32–40
6. Chang Chi-Cheng, Liang Chaoyun, Tseng Kuo-Hung, Tseng Ju-Shih (2014) Using e-portfolios to elevate knowledge amassment among university students. Comput Educ 72:187–195. https://doi.org/10.1016/j.compedu.2013.10.015
7. Baytak Ahmet, Land Susan M, Smith Brian K (2011) Children as educational computer game designers: an exploratory study. Turk Online J Educ Technol-TOJET 10(4):84–92
8. Omar Hanan, Khan Saad A, Toh Chooi G (2013) Structured student-generated videos for first-year students at a dental school in Malaysia. J Dent Educ 77(5):640–647
9. Johnson CG (2008) Student-generated podcasts for learning and assessment. In: Proceedings of the 8th international conference on computing education research. ACM, pp 84–87. https://doi.org/10.1145/1595356.1595371
10. Chang H-Y, Quintana C (2006) Student-generated animations: supporting middle school students' visualization, interpretation and reasoning of chemical phenomena. In: Proceedings of the 7th international conference on learning sciences. International Society of the Learning Sciences, pp 71–77
11. Howell Emily, Reinking David, Kaminski Rebecca (2015) Writing as creative design: constructing multimodal arguments in a multiliteracies framework. J Literacy Technol 16(1):2–36
12. Schwarz C, Reiser B, Fortus D, Shwartz Y, Acher A, Davis B, Hug B (2009) Models: defining a learning progression for scientific modeling. In: Learning progression in science (LeaPS) conference, Iowa City, IA, USA
13. Vassilikopoulou M, Retalis S, Nezi M, Boloudakis M (2011) Pilot use of digital educational comics in language teaching. Educ Media Int 48(2):115–126. https://doi.org/10.1080/09523987.2011.576522
14. Schäfer L, Valle C, Prinz W (2004) Group storytelling for team awareness and entertainment. In: Proceedings of the third Nordic conference on Human-computer interaction. ACM, pp 441–444
15. Suwardy Themin, Pan Gary, Seow Poh-Sun (2013) Using digital storytelling to engage student learning. Acc Educ 22(2):109–124
16. Stewart Kristian D (2017) Classrooms as 'safe houses'? The ethical and emotional implications of digital storytelling in a university writing classroom. Crit Stud Teach Learn 5(1):85–102
17. Schuck Sandy, Kearney Matthew (2006) Capturing learning through student-generated digital video. Aust Educ Comput 21(1):15–20
18. Bratitsis T (2017) Contextualized educators' training: the case of digital storytelling. In: Research on e-learning and ICT in education. Springer, Cham, pp 31–43
19. Robin B (2006) The educational uses of digital storytelling. In: Society for information technology & teacher education international conference. Association for the Advancement of Computing in Education (AACE), pp 709–716
20. Smeda Najat, Dakich Eva, Sharda Nalin (2014) The effectiveness of digital storytelling in the classrooms: a comprehensive study. Smart Learn Environ 1(1):6
21. Gakhar S, Thompson A (2007) Digital storytelling: engaging, communicating, and collaborating. In: Society for information technology & teacher education international conference. Association for the Advancement of Computing in Education (AACE), pp 607–612
22. Azman FN, Zaibon SB, Shiratuddin N (2015) Modelling learner-generated comic production: an initial design. J Eng Appl Sci 13, 2237–2241. Springer, Cham
23. Azman FN, Zaibon SB, Shiratuddin N (2018) A revised production model of learner-generated comic: validation through expert review. In: MATEC web of conferences, vol 150. EDP Sciences, p 05044
24. Sekaran U, Bougie R (2016) Research methods for business: a skill building approach. Wiley & Sons
25. Peterson Robert A, Merunka Dwight R (2014) Convenience samples of college students and research reproducibility. J Bus Res 67(5):1035–1041

26. Matook Sabine, Indulska Marta (2009) Improving the quality of process reference models: a quality function deployment-based approach. Decis Support Syst 47(1):60–71
27. Syamsul Bahrin Z (2011) Mobile game-based learning (mGBL) engineering model. Ph.D. diss., Universiti Utara Malaysia
28. Zaibon SB, Azman FN, Shiratuddin N (2018) Enhancing performance of student in web programming using digital educational comics. J Telecommun Electron Comput Eng (JTEC) 10(2–4):161–165
29. Peffers K, Tuunanen T, Gengler CE, Rossi M, Hui W, Virtanen V, Bragge J (2006) The design science research process: a model for producing and presenting information systems research. In: Proceedings of the first international conference on design science research in information systems and technology (DESRIST 2006), pp 83–106. sn
30. Azman FN, Zaibon SB, Shiratuddin N (2016) Toward the development of an instrument to evaluate learner-generated comics. Int J Interact Digit Media 4(2):28–32
31. Ghasemi Asghar, Zahediasl Saleh (2012) Normality tests for statistical analysis: a guide for non-statisticians. Int J Endocrinol Metab 10(2):486
32. Zakaria Z, Hishamuddin Md (2001) Analisis data menggunakan SPSS Windows. Penerbit Universiti Teknologi Malaysia (UTM)
33. Melnick Steven A (1993) The effects of item grouping on the reliability and scale scores of an affective measure. Educ Psychol Measur 53(1):211–216
34. Maes Ann, Poels Geert (2007) Evaluating quality of conceptual modelling scripts based on user perceptions. Data Knowl Eng 63(3):701–724
35. Shiratuddin N, Kuan TH (2014) Quality evaluation of a digital storytelling (DST) conceptual model. In: 2014 International conference on multimedia computing and systems (ICMCS). IEEE, pp 690–695

Comparing DBpedia, Wikidata, and YAGO for Web Information Retrieval

Sini Govinda Pillai, Lay-Ki Soon and Su-Cheng Haw

Abstract Knowledge graphs serve as the primary sources of structured data in many Semantic Web applications. In this paper, the three most popular cross-domain knowledge graphs (KGs), namely, DBpedia, YAGO, and Wikidata were empirically explored and compared. These knowledge graphs were compared from the perspectives of completeness of the relations, timeliness of the data and accessibility of the KG. Three fundamental categories of named entities were queried within the KGs for detailed analysis of the data returned. From the experimental results and findings, Wikidata scores the highest in term of the timeliness of the data provided owing to the effort of global community update, with DBpedia LIVE being the next. Regarding accessibility, it was observed that DBpedia and Wikidata gave continuous access using public SPARQL endpoint, while YAGO endpoints were intermittently inaccessible. With respect to completeness of predicates, none of the KGs have a remarkable lead for any of the selected categories. From the analysis, it is observed that none of the KG can be considered complete on its own with regard to the relations of an entity.

Keywords Semantic web · Knowledge graphs · DBpedia · YAGO · Wikidata

1 Introduction

Semantic Web, an extension of the current web, was introduced by Tim Berner's Lee in 2001 [1]. According to him, "The Web was designed as an information space, with the goal that it would be useful not only for human–human communication, but also

S. G. Pillai · L.-K. Soon (✉) · S.-C. Haw
Faculty of Computing and Informatics, Persiaran Multimedia, Multimedia University,
63100 Cyberjaya, Selangor, Malaysia
e-mail: lksoon@mmu.edu.my

S. G. Pillai
e-mail: sinisanjay@gmail.com

S.-C. Haw
e-mail: sucheng@mmu.edu.my

© Springer Nature Singapore Pte Ltd. 2019
V. Piuri et al. (eds.), *Intelligent and Interactive Computing*, Lecture Notes in Networks and Systems 67, https://doi.org/10.1007/978-981-13-6031-2_40

that machines would be able to participate and help." Semantic Web is designed to include information which are expressed in machine-readable form in files without modifying the existing web structure. Thus, Semantic Web is the Web of Data and provides a common framework that shares and reuses this data from any part of the word.

Linked data is one of the core parts of Semantic Web. This is a way of publishing data on the Web. Berner's Lee outlined the principles of linked data in 2006. Linked data publishes structured data on the Web using resource description framework (RDF) and hypertext transfer protocol (HTTP) and connects data between different sources, effectively allowing data in one data source to be linked to data in another source [2]. Linked open data is linked data which is openly available on the web.

Semantic Web uses RDF standards to publish data. Knowledge graphs which are based on RDF standards are increasingly popular. Zaveri et al. [3] define knowledge graph as RDF graph. The structured data in the knowledge graphs can be interlinked so that artificial intelligence (AI) applications can make use of this through semantic queries. In recent years, semantic data sources have become increasingly popular with AI applications. Various knowledge graphs (KGs) are available on the web like DBpedia, YAGO, Wikidata, OpenCyc, NELL, and ConceptNet.

As more and more AI applications start to use the linked open data, it is highly desired to get as much facts of an entity as possible. When an entity is linked to a KG for extracting information, missing information in the queried KG can have great impact on the applications, depending on the nature of the application. In this paper, the comprehensiveness of three selected KGs along with certain dimensions is empirically analyzed. This paper discusses the KGs based on specific dimensions like accessibility of the dataset, timeliness of the data, and completeness of the relations of an entity. To the best of our knowledge, previous researches do not focus on the specific dimensions addressed in this paper.

Accessibility is defined as "the extent to which data are available or easily and quickly retrievable". Accessibility of a knowledge graph refers to how easy and flexible the data in the knowledge graph can be accessed. Zaveri et al. [3] describe this as one of the intrinsic categories of data quality. Timeliness is defined as how up-to-date the data is with respect to the task it is used for [4]. Completeness can have many definitions. In our experiment, completeness is meant as how complete properties/relations of a KG compared to the total predicates available on this entity across all the three KGs.

The remainder of this paper is organized as follows: Sect. 2 highlights the background study, which mainly focuses on the knowledge graphs as well as previous work done. Section 3 presents the experimental setup. Section 4 presents the experimental results and findings, while Sect. 5 presents findings and concludes the paper.

2　Literature Review

There are several semantic data sets on the web, which are publicly available to human users as well as machines. In this paper, we focus on these public knowledge graphs and do a comparative study. In the field of knowledge graphs, there are some previous research works done to analyze and compare the prominent KGs like assessment of overall global properties (e.g., number of instances and facts), queriability and informativity of the data set, and the quality of the data set on the LOD cloud.

For a given setting for AI applications, Zaveri et al. [3] proposed a framework for identifying the most appropriate KG for a particular domain. Authors have identified 11 data quality dimensions and 34 data quality criteria in them for analyzing linked data quality. KGs which are openly available under linked open data (LOD) cloud were evaluated along these criteria. Ten key statistics were also used to examine each KGs.

The work of Li et al. included two more dimensions for usability of a data set such as queriability and informativity and analyzes DBpedia and YAGO along this [5]. It studies and highlights the importance of schema for the ease of use of data sets and the requirement of balances between number of classes and number of properties among others.

Paulheim discussed that high coverage and the correctness of a KG is very difficult to achieve at the same time [6]. It studied various approaches for type and class prediction of entities and predicting relation for various KGs. It also discusses methods for finding erroneous types, relations, literals, interlinks, etc. and the evaluation methodologies.

Kern-Isberner et al. have concluded that the prominent KGs like DBpedia, Wikidata and YAGO are comparable, each has its own advantages and disadvantages, which cannot be replaced by one or another for diverse AI applications [7]. 25 popular classes were compared and the conclusions are detailed in their paper. They highlight the need for a careful analysis of the KGs based on the task and the domain before finalizing a KG for applications. Based on the complementarity observed on KGs, they hint on the need for combining more than one KG for considered applications.

In this paper, in continuation to these recent researches, we explore different KGs with the intention of identifying unresolved issues for the directions of future research. We have analyzed the same entity from three popular classes from different KGs with respect to different aspects.

2.1　Open Knowledge Graphs

This paper presents our preliminary experiments that compare some knowledge graphs from few dimensions. Since the focus is on open knowledge graphs, Google's

Knowledge Graph and Facebook's Open Graph are excluded from this comparison. Google's Freebase is also not considered since it was shut down in 2017, and the data has been transferred to Wikidata. Hence, instead of Freebase, Wikidata is included. Wikipedia is one of the largest public knowledge bases. It contains textual and structured data. It does not support direct access to data using any query language or as downloadable dump. Since data is drowned in 30 million Wikipedia articles in 287 languages [8], data extraction is comparatively difficult. Therefore, this is not included. At this stage, we have eliminated KGs which do not store data as RDF models. RDF is stored as triples subject–predicate–object (SPO) formats. DBpedia, YAGO, and Wikidata are chosen for the analysis. Most recent versions of these selected KGs at the time of analysis are chosen for analysis. The basic information of the three selected KGs is presented below.

DBpedia[1]

DBpedia is one of the most popular knowledge graphs in research community. DBpedia extracts information from Wikipedia. DBpedia acts as the hub of linked data. DBpedia contains links to most of the other knowledge bases including Wikipedia, YAGO, Wikidata, Freebase, Cyc, etc. DBpedia is a static KG that is updated periodically (roughly 6 to 18 months). DBpedia is updated from the static dump of Wikipedia and the information extraction process is heavy. The latest version of DBpedia is 2016-10, which is based on Wikipedia dumps from October 2016. DBpedia 2016-10 has 13 billion RDF triples. This is extracted from English and other language editions of Wikipedia and Wikipedia commons and Wikidata, as 1.7 billion, 6.6 billion, and 4.8 billion, respectively. DBpedia datasets can be downloaded as N3/TURTLE serialization format.

DBpedia LIVE: Since Wikipedia content is getting updated every second, a dynamic version of DBpedia exists, which is DBpedia Live.[2] This is built with the objective to provide DBpedia in synchronization with Wikipedia. DBpedia LIVE presents up-to-date information.

YAGO[3]

YAGO is developed by Max Planck Institute for Computer Science in Saarbrücken. YAGO is automatically constructed from Wikipedia, GeoNames, and WordNet [9]. It also attaches temporal and special dimension to the facts and entities, if it is available. It extracts data from infoboxes, categories, redirects, etc. from Wikipedia, synsets, hyponymy from WordNet and GeoNames. One notable advantage of YAGO is its high accuracy of 95% which has been manually evaluated. This manual evaluation is one of the key factors that affect YAGO releases. YAGO is updated roughly less than once a year. YAGO3 was released in 2015, while YAGO2 in 2011, with one more update of YAGO2 in 2013. Latest version of YAGO, which is YAGO3, is multilingual [9, 10]. This has been extracted from ten different language versions of Wikipedia.

---

[1] http://wiki.dbpedia.org/.

[2] http://wiki.dbpedia.org/online-access/DBpediaLive.

[3] https://www.mpi-inf.mpg.de/departments/databases-and-information-systems.

Wikidata[4]

Wikidata is a collaborative knowledge graph hosted by Wikimedia Foundation. This KG is built through crowdsourcing. It also imports facts from different primary sources and includes their references. Freebase has been integrated with Wikidata after it has been shut down in June 30, 2015. Attaching references is helpful for verifying the validity of facts. Purpose of Wikidata is to provide a central storage which can be used by Wikipedia, other Wikimedia projects and other external applications. Wikidata is multilingual—labels, aliases, and descriptions of the entities are provided in more than 350 languages. For the purpose of experiments, Wikidata was last accessed on 2018-06-14. A page has been created in Wikidata for every Wikipedia article. These pages are called "items". Items are uniquely identified by a name starting with "Q". For example, Q22686 is used for Donald Trump. Properties have a "P" followed by a number. For example, Property P21 is "sex or gender". A property and value are together known as statements.

2.2 Accessibility of Knowledge Graphs

Public SPARQL Endpoints All of the selected KGs have official public SPARQL endpoints to access the data sets on the web. This method of accessing does not require any prior setup to access the dataset. This will also enable the applications to use the latest data set provided by the KGs. Public SPARQL endpoint for DBpedia [11] is provided by Virtuoso server. This can be accessed via the URL http://dbpedia.org/sparql. Other URLs to access the DBpedia data set is http://dbpedia.org/snorql. YAGO is also provided with a SPARQL endpoint. Data Science Center of Paris-Saclay University has issued a SPARQL interface, which is hosted using Virtuoso Universal server to access the YAGO3 dataset. This SPARQL endpoint can be accessed via URL https://linkeddata1.calcul.u-psud.fr/sparql. For Wikidata, the graph database platform Blazegraph provides public SPARQL endpoint. This Wikidata Query Service is accessed via URL https://query.wikidata.org.

Graphical Methods Max Planck Institute has provided a handful of visualization tools to access YAGO data set. Users can use these tools provided in their website like graph browser[5] and SPOTLX browser.[6] In graph browser, the triple is displayed in the shape of a star, with entity being queried at the center with in and out going edges forming the shape of the star. Origin of facts is shown using flags. SPOTLX (subject, predicate, object, time, location, conteXt) browser allows querying YAGO data set with spatial and temporal visualizations. As of writing this paper, graph browser and SPARQL endpoint provide access to YAGO3 data set, while SPOTLX browser lets you query YAGO2 data set.

---

[4]https://www.wikidata.org/wiki/Wikidata:Main_Page.
[5]https://gate.d5.mpi-inf.mpg.de/webyago3spotlxComp/SvgBrowser.
[6]https://gate.d5.mpi-inf.mpg.de/webyagospotlxComp/WebInterface.

Private SPARQL Endpoints If there is RDF dump provided by the KG, then this can be downloaded and can be loaded into a triple store. Clients have to host their own private SPARQL endpoint to read data from the triple stores [12]. All of the selected KGs, DBpedia, YAGO, and Wikidata provide RDF dump. There are many triple stores available; some of the popular ones are Apache Jena, BigData RDF Database, OWLIM Lite, and OpenLink Virtuoso. For a more detailed comparison on these triple stores, you can refer to this paper [13]. Entire YAGO3 data set can be downloaded as TSV or TURTLE serialization format. The entire data set is divided into nine portions, which are called theme. Each theme is allowed to download separately. DBpedia data set is also downloadable as RDF dumps. The new DBpedia data set is available in N3/TURTLE serialization format. Wikidata provides several kinds of dumps like RDF dumps, JSON dumps, and XML dumps.

3 Experimental Design

The overall flow of the processes of the experiments conducted is detailed in the following subsections.

3.1 Analyzed Entities

In order to analyze the completeness of entities from the selected KGs, we have selected two diverse entities each from the three most popular classes. We chose general classes, person, organization, and location. Here, we use the following criteria to choose entities from each class. First criterion is entities which have some activities (modification in their data) in their life in the last one and a half years. Second criterion is entities which can cause disambiguation.

We have chosen two entities from person class, the incumbent President of United States Donald Trump and an English theoretical physicist and cosmologist Stephen Hawking. Donald Trump is in office since January 20, 2017. Next entity from Person class, Stephen Hawking died on March 14, 2018. The first condition thus satisfies these two entities. Next, we chose other entities which can cause disambiguation upon selection. We chose Apple Inc. and Blackberry Limited for organization and the island Java in Indonesia and the country China for location. All of these selected entities can cause disambiguation. Other than the mentioned entities, apple and blackberry are names of fruits, java is coffee as well as a programming language. China is used interchangeably with porcelain and there are people, music[7] which can be referred as China. This issue of entity disambiguation is not addressed in this paper but can be explored for further research work.

---

[7]https://en.wikipedia.org/wiki/China_(disambiguation).

In any knowledge graph representation, entities are connected by relations, where entities are the nodes of the graphs and relations are the edges of the graph. For this experiment, if there is at least one triple exists for a given entity relation (e.g., *has children*), it is considered as the existence of that relation for that entity. For example, *Stephen Hawking has child (Lucy Hawking)* is given by predicate *children* in DBpedia, which *hasChild* in YAGO and *child(P40)* in Wikidata. Few predicates like *s*, *v*, *n* from DBpedia are excluded from this measurement due to lack of clarity in meaning. A percentage score of a KG is measured as the number of distinct relations identified for that entity in that KG to the total number of relations gathered from all the three KGs for that entity. Overlap between two KGs is measured as the number of relations which are common to those particular KGs under consideration for the chosen entity.

3.2 Evaluation Process

Data is extracted from each knowledge graph for the selected entities for the classes of person, organization, and location. Data extraction is done via public SPARQL endpoint. We used a Python wrapper to access the SPARQL endpoint. Data is fetched in JSON format. We performed a detailed comparison of the data extracted from different knowledge graphs. Purpose of the comparison is to further explore research prospects of this, with machine learning methods. At this stage, due to the manageable size of data in experiment, the results are manually verified.

4 Experimental Result and Findings

4.1 Accessibility of the Knowledge Graphs

Public SPARQL Endpoints All of the selected KGs have official public SPARQL endpoints to access the data sets on the web. This method of accessing does not require any prior setup to access the dataset. Thus, do not require additional hardware and software or skilled man power to do the installation. This will also enable the applications to use the latest data set provided by the KGs. Because of these advantages listed here, in this experiment, we have used public SPARQL to access the data sets. Public SPARQL can be accessed via a browser or by many of the popular programming languages like Python, Java, etc., and thus can be accessed by both humans and machines. While using the SPARQL interface, it has been observed that DBpedia and Wikidata public SPARQL interfaces have high availability. In the last 2 months of observance, there is hardly any downtime noted with DBpedia. The same has been observed with Wikidata. There is no noted downtime for Wikidata in the observed time of 2 months. But for YAGO, this is not the case. In the last 2 months, it has been observed that the public SPARQL endpoint was down two times with each lasting up to many days. Hence, unlike DBpedia and Wikidata, it is not recommended to use public SPARQL to access YAGO for critical applications.

RDF Dumps DBpedia, YAGO, and Wikidata provide RDF dumps, which are in N-triples and N3/Turtle serialization formats. Downloaded dump can be loaded into a triple store for accessing locally. This data set can be queried using SPARQL queries by setting up a private SPARQL endpoint. Some of the popular triple stores available are Apache Jena, BigData RDF Database, OWLIM Lite, and OpenLink Virtuoso. Entire YAGO3 data set can be downloaded in turtle format. The data set is divided into nine portions, which are called theme. Each theme is allowed to be downloaded separately.

Data Dumps (Other formats) All of the selected KGs provide dumps in formats which can be loaded into SQL databases. This data set can be queried using SQL. YAGO can be downloaded into TSV format, whereas DBpedia provides dump in CSV/JSON formats. Wikidata provides in JSON format, which is the most stable format of Wikidata.

4.2   Timeliness of Data

While checking with data of different entities from these knowledge graphs, we found that Wikidata has the most up-to-date information, while DBpedia and YAGO do not cover the latest activities/actions of that entity. Knowledge from DBpedia and YAGO is not updated, and hence when any AI applications, you can get erroneous data or data which is not correct at the time of retrieval.

For person class, for the selected entity, Stephen Hawking, the death details were not found in DBpedia and YAGO. In Wikidata, it is shown as "Most significant event," while this is completely missing in DBpedia and YAGO. But DBpedia LIVE is updated with this fact. Therefore, DBpedia LIVE is used for this comparison. Another entity from person class is Donald Trump, who became the President of United States on January 20, 2017. If we query DBpedia for this entity, the closest information you can get is occupation is given as a Politician. This information is not correctly available in YAGO. In Wikidata, the value of predicate "position held" is given as President of United States of America. Here, we can see that retrieving entity from any knowledge graphs other than Wikidata conveys partial or wrong information. This can result in wrong information supplied to the AI applications. In order to solve this issue by machines, there is a need to systematically explore and discover a method to obtain the measure of timeliness as well as how to solve this for time-critical applications. Since timeliness is one of the critical data quality features for industrial applications which ensures the authenticity of the data, solving the issue of timeliness greatly improves the quality of the data set.

Fig. 1 Relation completion
of selected entities

Completeness of Relation of selected
entities

Class	Entity	KG D	W	Y
Location	China	47.71	48.62	64.22
	Java	80.65	70.97	74.19
Organization	Apple Inc.	72.97	89.19	75.68
	BlackBerry Limited	72.41	65.52	79.31
Person	Donald Trump	67.92	79.25	47.17
	Stephen Hawking	54.72	90.57	45.28

Percentage of complet..

45.28   90.57

4.3 Completeness and Overlapping of the Data

The experiment conducted on six entities from different classes and the result is
shown in a heat map in Fig. 1. Relation completion of selected entities. For Person
class, Wikidata consistently scored highest for both entities. For Stephen Hawking,
total number of distinct relations identified from three KGs was 53, with DBpedia
having 29, YAGO having 25 and Wikidata having 45. For the location class, for
People's Republic of China, YAGO achieves highest score, while for the island of
Java, DBpedia has greater scores. For organization, Wikidata scores highest for Apple
Inc. and YAGO scores for BlackBerry Limited. From these experiments, it can be
seen that relations in none of the KG is complete on its own and we cannot zero
down to any particular KG for a particular domain.

Overlapping of properties of entities among different KGs is analyzed. For example, for Stephen Hawking, relation *citizen* exists in DBpedia and Wikidata, but missing in YAGO. Number of relations which overlap among different KGs are presented
in heat map Fig. 2. Overlap of relations among different KGs. This indicates that if
data from different KGs can be combined, we shall be able to get a more comprehensive data about each entity.

5 Findings and Conclusion

In this paper, the three most popular cross-domain KGs were empirically explored.
These KGs were assessed from end user's perspectives, which included the accessibility, timeliness, and completeness of the relations of the KGs. Three fundamental
categories of named entities like people, organization, and location were queried
within the KGs for detailed analysis of the data. From the designated categories,
we choose two celebrities Donald Trump and Stephen Hawking for people class,
Apple Inc. and BlackBerry Limited for the class of organization, and island Java in

Fig. 2 Overlap of relations
among different KGs

Overlap of Relations of selected
entities

		KG		
Class	Entity	D2W	D2Y	W2Y
Location	China	33.94	45.87	41.28
	Java	51.61	70.97	45.16
Organization	Apple Inc.	62.16	56.76	67.57
	BlackBerry Limited	41.38	62.07	48.28
Person	Donald Trump	47.17	43.40	41.51
	Stephen Hawking	50.94	41.51	43.40

Percentage of overlap

33.94   70.97

Indonesia and People's Republic of China for the location. The completeness of the relations and timeliness of the data of these selected entities in KG were assessed.

In terms of timeliness, Wikidata scores the highest because of global community update. DBpedia LIVE is a derived version of DBpedia that is continuously updated from Wikipedia. It also benefits from the crowdsourcing nature of Wikipedia and will tend to have up-to-date information. However, the extent of relations existing in each KG varies from entity to entity. From the comparative study given above, it is evident that none of the KG is sufficient on its own. Overlapping of relations between KGs suggests that machine learning methods for integration of KGs would be a promising direction for future work. It is desirable to provide the accurate information which is not stale to the AI applications or humans who extract entity information from these knowledge graphs. In order to achieve that, for future work, an automated solution to integrate the KGs for complete and updated information will be proposed.

Acknowledgements This work is partially funded by Fundamental Research Grant Scheme (FRGS) by Malaysia Ministry of Higher Education (Ref: FRGS/1/2017/ICT02/MMU/02/6).

References

1. Tim B, Lee B, Hendler J, Lassila O (2001) The semantic web will enable machines to comprehend semantic documents, no. May, pp 1–5
2. Bizer C et al (2008) Linked data on the web. WWW2008 Work. Linked Data Web, pp 1265–1266
3. Zaveri A, Kontokostas D, Leipzig U, Hellmann S (2017) Linked data quality of DBpedia, freebase Semant. Web, 0(0):1–53
4. Pipino LL, Lee YW, Wang RY, Lowell Yang Lee MW, Yang RY (2002) Data Quality Assessment. *Commun.* ACM 45(4):211
5. Li TRBY, Wang H, Zhao L (2016) The semantic web. Latest advances and new domains 9678:52–68
6. Paulheim H (2015) Knowledge graph refinement: a survey of approaches and evaluation methods. Semant. Web, 0:1–0
7. Ringler D, HPB (2017) KI 2017: advances in artificial intelligence 10505:366–372

8. Vrandečić D, Krötzsch M (2014) Wikidata: a free collaborative knowledgebase. Commun ACM 57(10):78–85
9. Hoffart J, Suchanek FM, Berberich K, Weikum G (2013) YAGO2: a spatially and temporally enhanced knowledge base from Wikipedia. IJCAI Int Jt Conf Artif Intell 3161–3165
10. Demner-Fushman D et al (2013) YAGO3 : a knowledge base from multilingual wikipedia. J Biomed Inform 46(SUPPL):129–132
11. Auer S, Bizer C, Kobilarov G, Lehman J, Cyganiak R, Ives Z (2007) DBedpia: a nucleus for a web od open data, Emantic Web Lect Notes Comput Sci 4825:722
12. Verborgh R et al (2014) Querying datasets on the web with high availability. Lect. Notes Comput. Sci. (including Subser. Lect. Notes Artif. Intell. Lect. Notes Bioinformatics) 8796(Iswc 2014):180–196
13. Voigt M, Mitschick A, Schulz J (2012) Yet another triple store benchmark? Practical experiences with real-world data. CEUR Workshop Proc, 912(Sda):85–94

Snake-Based Boundary Search for Segmentation of 3D Polygonal Model

Kok-Why Ng, Junaidi Abdullah, Sew-Lai Ng and Hau-Lee Tong

Abstract Many existing 3D model segmentation methods require user input to accurately segment the virtual models. This paper applies snake-based (or contour) approach to automatically search for the boundary between the two connected functional features. The search is guided by the skeleton model. The proposed approach courageously modifies the geometric meshes with respect to each sectional skeleton to ensure the snake meets end to end to form a complete ring. The result of the ring is then interpolated to fit the local geometric feature model. This step eradicates the non-related feature and invalid rings. In the end, the created ring is reverted to its original coordinates so that the meshes are non-distorted. The entire process is done automatically without any user input. The proposed method is compared to two well-known methods: Shape Diameter method and Core Extraction method. The results generated by the proposed method turn out to be more accurate than the two existing methods.

Keywords 3D model segmentation · Snake-based approach · Skeletonization · Clustering · Mesh decomposition

K.-W. Ng (✉) · J. Abdullah · S.-L. Ng · H.-L. Tong
Faculty of Computing and Informatics, Multimedia University, 63100 Cyberjaya, Selangor, Malaysia
e-mail: kwng@mmu.edu.my

J. Abdullah
e-mail: junaidi.abdullah@mmu.edu.my

S.-L. Ng
e-mail: slng@mmu.edu.my

H.-L. Tong
e-mail: hltong@mmu.edu.my

© Springer Nature Singapore Pte Ltd. 2019   537
V. Piuri et al. (eds.), *Intelligent and Interactive Computing*, Lecture Notes in Networks and Systems 67, https://doi.org/10.1007/978-981-13-6031-2_35

1   Introduction

Segmentation or decomposition of Three-Dimensional (3D) models is greatly debated in recent years. Many new methods are proposed to achieve a reasonable result. Technically, a model can be segmented either based on geometric direction (called batch-type segmentation) or meaningful shape (called part-type segmentation) [8]. Each method offers unique advantages to the extended research works and applications. This paper researches on part-type segmentation as the segmented features are expressive to human eyes. In this paper, the segmented features will be termed as functional features as the results always turn out to be a functional part.

Prior to segmentation, many existing methods needed to determine how many functional features were in a model and approximated the boundary to split two connected features. They located the associated edge meshes, model volume or through human observation input to determine the significant functional features. Some researchers extended the well-defined 2D segmentation methods to 3D segmentation to retrieve the functional features. Each method displayed different strengths and results. In this paper, it does not need any prior knowledge of the functional features.

The rest of the paper is organized as follows. Section 2 will discuss and analyse the related works on 3D model segmentation. Section 3 will present the proposed method. Section 4 will show the experimental results, analysis and comparison. Section 5 will summarize this paper.

2   Related Work

Many 3D model segmentation methods have been proposed. Though, they can be tabled into three broad categories: clustering method, skeleton-based segmentation and snake-based segmentation.

Clustering method is a greedy approach. It commences with several groups of different properties and assigns the similar components (e.g., vertices or faces) to each group. Refinement is required to find the boundary features. The different properties make the proposed clustering methods output different results. Among the popular clustering methods are iterative clustering [3, 4, 6, 20, 21], region growing [5, 7], feature point-based clustering [9], fuzzy clustering with graph cut [8], watershed [15] and spectral clustering [18, 19]. These techniques are well accepted; however, they are very sensitive to noises and mesh features.

Skeleton-based segmentation is a rational approach [13, 14, 16]. It provides a good representation to a model shape in one-dimensional array. It reveals the significant features of a model and ignores the minor irregular meshes. However, there is no robust definition on how exactly a skeleton is to be constructed. One may apply any of the four methods below to generate skeleton for segmentation purpose: Thinning and boundary propagation [1, 16], distance field based [13, 14], geometric [2] and general-field function [17]. These methods, in general, are complex and slow.

Snake-based or contour segmentation [10–12] in common defines the cutting path of ridges or textures. It divides the model into parts along the concave discontinuity of the tangent plane. The snake is evolved around the model and forms a closed loop for cutting the individual feature. However, if a model contains noisy textures, the contour will be deviated.

The proposed approach takes one step ahead to make use of both snake and skeleton methods to form a robust and unbroken loop.

3 Proposed Method

Below is the proposed method for segmenting a model.

3.1 Input of 3D Model and Generate Skeleton Features

The input 3D model can be a point cloud model or polygonal meshes, which each point/vertex provides only the Cartesian coordinate. The algorithm starts by first generating the skeleton. As mentioned earlier that the skeleton is just a stepping stone for the segmentation process, this paper (borrows the idea) generates the skeleton through *Laplacian* contraction method [1]. However, it does not follow the narrow threshold proposed by Cao et al. [1]. The generated skeleton though is less accurate, it is simple and the quality is acceptable.

Due to the contraction method is not the original work in this paper, readers may approach [1] for more detail. In this work, the generated skeletons will be computed for *Significant cost C_{fs}* in (Eq. 1):

$$\text{Significant cost, } C_{fs} = f(i) - \frac{\psi}{n} \sum_{i=0}^{n-1} f(i) \tag{1}$$

where $f(i)$ is a distance function, $i = 0, 1, 2, …, n - 1$ of all the edges, n is the number of skeleton edges and ψ is a positive coefficient for tuning the average distance. The coefficient can be set to value one if tuning is unnecessary. In this paper, the coefficient is set to 0.5. This is an experimental value to preserve smaller skeletons.

Next, the skeleton features are sought by locating the skeleton nodes that makeup of three or more than three branches (see Fig. 1). Each branch of the node will form a skeleton feature. This feature will be sent for segmentation in next section.

Fig. 1 Skeleton node (the red dot) with four branches

Fig. 2 **a** The nearest surface points v_1 and v_2 to the first and second skeleton node s_1 and s_2, respectively. **b** A repeated jagged lines. **c** A reverse ring makes unnecessary rounds and extends to the other non-related feature

3.2 Initiate Surface Point and Relocation

One may initiate the nearest surface point (say v_1) to the first skeleton node (s_1), to be the first opening point in a ring. However, from the finding, it is only necessary at second skeleton node (s_2) because the first created ring at s_1 is generally too short for a functional feature (see Fig. 2a).

There are two types of boundary edges: (i) along a skeleton feature and (ii) at the end of a skeleton feature. We need the first type to identify each continuous sectional feature; whereas the second type is needed to adaptively learn the connected boundary feature edges. A special attention will be applied to second type as it will easily create a weird ring. In Fig. 2b, the red jagged line does not meet end to end and creates repeating jagged lines; Fig. 2c shows the ring reverses and makes a number of rounds.

The main problem to these outcomes is down to the jagged geometric meshes which are likely to reverse the ring. The problem will persist if the applied metric is exclusively based on the meshes. Many existing works introduced extra metrics (e.g., volume measure) to prevent the problem. This paper takes a brave act to modify the original mesh by relocating the vertex coordinate orthogonal to the skeleton node. The relocation neither base on the geometric direction of the meshes nor cutting across the meshes as introduced by Lee et al. [12]. It bases on the direction of the skeleton edge. For example, in Fig. 3a, the vertex v is not orthogonal to s_2. When the angle $\theta = 90 - \alpha$ is less than zero (Fig. 3c), it moves the vertex v in the direction of $-\mathrm{d}t * \hat{v}_s$ where the distant error $\mathrm{d}t = d \sin \theta$; d is the distant from v to s_2; $\hat{v}_s$ is the unit skeleton vector at s_2.

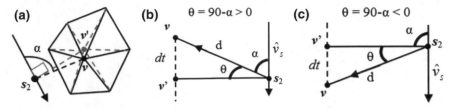

Fig. 3 **a** Center vertex v is not orthogonal to skeleton node s_2. It is relocated to v'. **b**, **c** are the measures for the relocation

If the angle $\theta = 90 - \alpha$ is more than zero (Fig. 3b), the vertex v will be moved in opposite direction of the skeleton edge. The relocation is shown in (Eq. 2).

$$\text{New coordinate, } v' = \begin{cases} v + dt * \hat{v}_s & \text{if } \theta = 90 - \alpha > 0 \\ v - dt * \hat{v}_s & \text{if } \theta = 90 - \alpha > 0 \\ v & \text{if } \theta = 0 \end{cases} \qquad (2)$$

3.3 Computation of Minimum Cost and Relocation

With the initiated orthogonal vertex v, next, it is to find the adjacent vertex that is close or orthogonal to vertex v by computing the cost in (Eq. 3).

$$\text{Costv}_{\text{ortho}} = \min_{i \in v_{\text{adj}}}(\text{abs}[90 - (\cos^{-1}(\hat{u}_i \cdot \hat{v}_s))]) \qquad (3)$$

where v_{adj} is the adjacent vertices of v; $\hat{u}_i$ is the unit vector to the surface point of s_2; $\hat{v}_s$ is the unit skeleton vector. The minimum cost is computed so that the geometric alteration is minimal.

Once the next vertex is determined, Eq. (2) is applied to relocate the vertex so that it is orthogonal to s_2.

3.4 Computation of Prevention Cost

The same measure of Eq. (3) is applied to the rest of the adjacent vertices to form the contour. However, from the third vertex and onwards, a constraint of (Eq. 4) is set to prevent the reverse of the contour. If the selected v_3 with *Prevention cost* (T_{cstr}) is less than 60°, it will be set as the third vertex. Otherwise, it will be flagged and a new adjacent vertex will be redetermined. The value of 60° is made to be the threshold by considering that most of the regular generated model meshes have six triangles incident on each vertex. The third edge can vary within 60° from its adjacent edges.

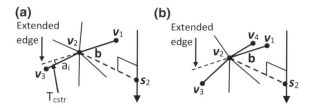

Fig. 4 Computation for third vertex (v_3) and onwards. **a** Vertex v_3 is the closest vertex to the extended edge or which the T_{cstr} is minimal. **b** If T_{cstr} is not applied, vertex v_4 would be selected and this would produce a reverse ring

While computing the constraint, if more than one edge fulfils the constraint, the minimum of T_{cstr} will be considered.

$$\text{Prevention cost, } T_{cstr} = \min_{i \in v_{2adj} \cap (\neg v_1)}(\text{abs}[180 - (\cos^{-1}(a_i \cdot b))]) \qquad (4)$$

where $v_{2adj} \cap (\neg v_1)$ is the adjacent vertices of v_2 but not including v_1; a_i is the adjacent edge to vertex v_2; **b** is the vector of $v_1 v_2$ as shown in Fig. 4.

3.5 Computation for Contour at Feature End

As mentioned in Sect. 3.2, the second type of boundary edges occurs at the end of skeleton feature. In this stage, the same metrics in Eqs. 2, 3 and 4 are applied to find the contour of the feature model.

In Fig. 5a, if a contour is continuously constructed after the node s_{11}, the contour will include the other non-related features. This is proven in Fig. 5b. Visually speaking, the finger is defined from s_0 to region close to s_{11}, and is not until s_{13}. This paper computes and compares the circumference of the last contour to the previous contour. If the error is more than 0.5, the node s_{11} will be interpolated by 0.1 percentage from the last node. This will stop when the error is equal or less than 0.5 (refer to Fig. 5c). The error is an experimental threshold.

3.6 Restoration to Original Coordinate

In earlier sections, the geometric meshes are modified to ensure the computed contour meets end to end to form a complete ring. Keeping the modified meshes for visualization will spoil the original look of the model. Furthermore, the modified surface meshes are typically distorted and unappealing. Therefore, after a complete contour is computed, the original coordinates will be restored to preserve the original look of the model.

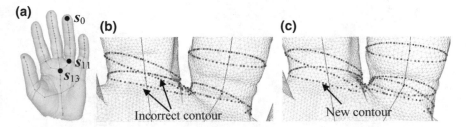

Fig. 5 **a** Node s_0 to node s_{11} makes the actual feature model regardless of the skeleton ends at node s_{13}. **b** This shows the incorrect contours created after s_{11}. **c** A new contour is adaptively computed

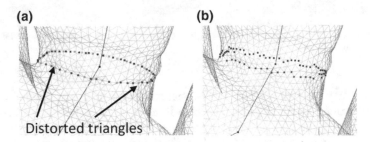

Fig. 6 **a** A complete contour is computed at the end of a feature model. **b** Restoration of the contour to the original coordinates

In Fig. 6a, a complete contour is computed. One can notice that the triangles are distorted. Figure 6b shows the original geometric coordinates of the contour. The jagged contour is naturally seen. The above steps will be repeated for each individual branch of a node illustrated in Fig. 2 to finish segmenting the entire model.

4 Results, Analysis and Comparison

The proposed method in this paper is computed by Intel® Core™i3 CPU, of 2 GB RAM, 64-bit Operating System. The hand model discussed in the previous section is fully segmented and shown in Fig. 7. Figure 8 shows the results of various models segmented by the proposed method and two existing segmentation methods for comparison purpose.

In Fig. 7a, rings are computed along a skeleton finger. The exact feature model is determined. The last ring stops at the end of the finger. Figure 7b shows the entire vertices of the feature to be assigned to red colour. This is done immediately after the last ring of each functional feature is determined. Figure 7c shows the complete result of the automatic segmenting the entire hand model.

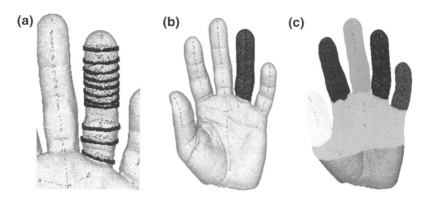

Fig. 7 a A number of rings are formed along a skeleton finger. **b** A segmented finger. **c** A fully segmented hand model

Reader may be doubted that how to segment the palm if the feature skeleton is started from the center palm. The clue is to keep those feature skeletons (or branches) which the node is joined by three or more than three branches to be the last for segmentation process. Meanwhile, for the remaining surface points that are not belonging to any of the features, they can be assigned to the nearest feature which is made by three or more than three branches. This way is simple, fast and reasonable.

For comparison purpose, Shape Diameter (SD) [14] and Core Extraction (CE) [9] have been chosen to be compared with the Proposed Method (PM) in this paper. The reason to choose them for comparison is because all the three methods predict the number of feature segmentation automatically. No user input is required to indicate the number of features in each model. Therefore, they are fair to be compared.

In Fig. 8, the segmented longhand by SD shows more than one colour in two fingers; meanwhile, the middle finger segmented by CE does not accurately cover the entire finger. The result from the PM is more accurate than CE to segment the entire finger precisely.

In crooked fingers model, both SD and CE introduce extra segmentation on the palm, whereas PM segments perfectly on this. The segmented pliers by SD are again with extra segmentation at one of the sharp ends. Both CE and PM have segmented the pliers accurately.

The segmented birds by the three methods look fine. However, only SD and PM are able to detect and segment the mouth of the bird accurately. For octopus model, SD and PM methods produce accurate segmented results at all the legs. The front-left leg of the segmented octopus by CE method is not accurate.

Overall, the proposed method (PM) offers a most convincing result.

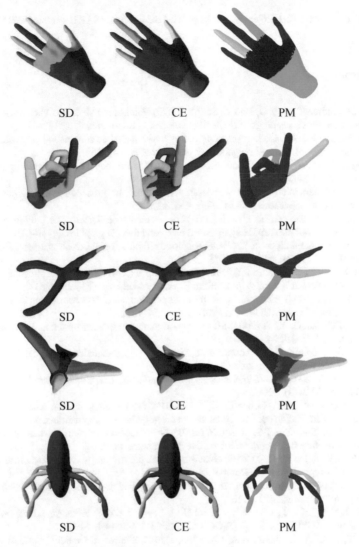

Fig. 8 (Top row) Segmented long hand; (Second row) segmented crooked fingers; (Third row) segmented pliers; (Fourth row) segmented bird and (Last row) segmented octopus

5 Conclusion

The proposed method makes use of the skeleton model to guide the snake-based (contour) approach to search for the boundary between two functional features. The results show that the proposed method is more accurate and more convincing. It is neither over-segmenting nor under-segmenting the models.

Acknowledgements This project is funded by MMU Mini fund (SAP ID: mmui/180152).

References

1. Cao JJ, Andrea T, Matt O, Hao Z, Su ZX (2010) Point cloud skeletons via Laplacian-based contraction. In: Proceedings of IEEE shape modeling international, pp 187–197
2. Dey TK, Sun J (2006) Defining and computing curve skeletons with medial geodesic function. In: Proceedings of symposium geometry processing, pp 143–152
3. Dong X, Lin HW, Xian CH, Gao SM (2011) CAD mesh model segmentation by clustering. Comput Graph 35(3):685–691
4. Frank L, William WC (2010) Power iteration clustering. In: Proceedings of international conference on machine learning, ICML Citeseer, vol 10
5. Garland M, Willmott A, Heckbert P (2001) Hierarchical face clustering on polygonal surfaces. In: Proceedings of 2001 ACM symposium on interactive 3D graphics, pp 49–58
6. Golovinskiy A, Funkhouser T (2008) Randomized cuts for 3D mesh analysis. ACM Trans Graph (Proc SIGGRAPH ASIA) 27(5)
7. Inoue K, Takayuki I, Atsushi Y, Tomotake F, Kenji S (2001) Face clustering of a large-scale CAD model for surface mesh generation. Comput Aided Des 33(3):251–261
8. Katz S, Tal A (2003) Hierarchical mesh decomposition using fuzzy clustering and cuts. ACM Comput Graph (Proc SIGGRAPH 2003) 22(3):954–961
9. Katz S, Leifman G, Tal A (2005) Mesh segmentation using feature point and core extraction. Vis Comput 21(8–10):649–658
10. Kiefer W (2004) Intelligent scissoring for interactive segmentation of 3D meshes. Doctoral dissertation, Princeton University
11. Lee Y, Lee S (2002) Geometric snakes for triangular meshes. Comput Graph Forum (Eurographics 2002) 21(3):229–238
12. Lee Y, Lee S, Shamir A, Cohen-Or D, Seidel HP (2004) Intelligent mesh scissoring using 3D snakes. In: 12th Pacific conference on computer graphics and applications, pp 362–371
13. Li X, Woon T-W, Tan T-S, Huang Z (2001) Decomposing polygon meshes for interactive applications. In: Symposium on interactive 3D graphics, pp 35–42
14. Lior S, Ariel S, Daniel C-O (2008) Consistent mesh partitioning and skeletonization using the shape diameter function. Vis Comput 24(4):249–259
15. Nassir S (2006) Image segmentation based on watershed and edge detection techniques. Int Arab J Inf Technol 3(2):104–110
16. Oscar K-CA, Tai C-L, Chu H-K, Daniel C-O, Lee T-Y (2008) Skeleton extraction by mesh contraction. ACM Trans Graph (SIGGRAPH 2008) 27(3):44:1–44:10
17. Qian Z, Sharf A, Tagliasacchi A, Chen BQ, Hao Z, Sheffer A, Daniel C-O (2010) Consensus skeleton for nonrigid space-time registration. Computer graphics forum (Special Issue of Eurographics), pp 635–644
18. Rong L, Hao Z (2007) Mesh segmentation via spectral embedding and contour analysis. Comput Graph Forum 26:385–394
19. Rong L, Hao Z (2004) Segmentation of 3D meshes through spectral clustering, pp 298–305
20. Toony Z, Laurendeau D, Giguere P, Gagne C (2013) 3D-PIC: Power Iteration Clustering for segmenting three-dimensional models. In: 3DTV-conference: the true vision-capture, transmission and display of 3D video (3DTV-CON), pp 1–4
21. Xiaobai C, Aleksey G, Thomas F (2009) A benchmark for 3D mesh segmentation. ACM Trans Graph (Proc SIGGRAPH) 28(3)

An Investigation of Learner Characteristics that Use Massive Open Online Courses (MOOC) in Learning Second Language

Hasmaini Hashim, Sazilah Salam and Siti Nurul Mahfuzah Mohamad

Abstract Learner characteristics are an essential element, in order to design and create an instructional platform/education/assessment according/customized to the target group. In this paper, types of learner characteristics that were using Massive Open Online Courses (MOOC) in the second language were discussed. Literature have emphasized on learner characteristic roles in enhancing learning. A mixed-method approach which involved both quantitative and qualitative research methods was adopted in this study. The results from the survey analysis revealed that the highest dimension among the eight learning styles was the visual learner at 76% while the result from the interview session was also visual (24.24%). The results from the survey analysis revealed that the highest dimension among the eight cognitive styles was thinking learner at 70%, while the result from the interview session was also thinking (21.95%). This study aimed to investigate learner characteristics using a Mandarin Massive Open Online Course (MOOC). This study determines that there are twofold: (i) learning styles of learners that used Mandarin MOOC, (ii) cognitive styles of learners that used Mandarin MOOC. Finding of study aimed to propose a content development model in MOOC based on learner characteristics. Future study will propose an effectiveness assessment model in MOOC based on learner characteristics.

Keywords Learner characteristics · Learning styles · Cognitive styles · MOOC · Second language

1 Introduction

Learner characteristics are important in all forms of online learning [1]. According to [2], learner's characteristics are defined as an individual mental factor, suggested to impact on the learning activities of students. Most researchers focused predominantly

H. Hashim (✉) · S. Salam · S. N. M. Mohamad
Pervasive Computing and Educational Technology (PET) Research Group, Centre for Advanced Computing Technology (C-ACT), Faculty of Information and Communication Technology, Universiti Teknikal Malaysia Melaka, Hang Tuah Jaya, Durian Tunggal, Melaka, Malaysia
e-mail: hasmainie76@gmail.com

© Springer Nature Singapore Pte Ltd. 2019
V. Piuri et al. (eds.), *Intelligent and Interactive Computing*, Lecture Notes in Networks and Systems 67, https://doi.org/10.1007/978-981-13-6031-2_7

on user perception of MOOC features, rather than individual learner characteristics such as exploring the factors, which affect MOOC completion or learner retention [3]. The authors mentioned that many previous studies of MOOCs considered the completion rate to be an important indicator of learner achievement [4]. However, the authors listed three types of learner characteristics that are effective factors for student learning: (i) learning style, (ii) cognitive style and (iii) multiple intelligence [5]. Figure 1 shows the types of learner characteristics.

A good MOOC is based on learner characteristics; however, individual learners come from many different backgrounds [1, 6]. Researchers' have mentioned that learners have different characteristics, preferences, and needs [7]. Previously, many studies were conducted on learner characteristic roles in enhancing learning. However, there is still a lack of studies on learner characteristics for a specific job [8]. In addition, there are certain professional jobs that need to identify the effective or appropriate learner characteristic [9]. Learner's characteristics are important for the instructor to identify the suitable learning contents for the purpose of content reusability [10]. Previous researchers stated learner characteristics are important components for instructional designers who in turn utilize those factors to; (i) design and (ii) create tailored instructions for a target group [3, 4]. According to L. Paepe, learner characteristics are required in an online course, as (i) they underpin the pedagogy of the instructional design, (ii) task development, (iii) learner support and (iv) decisions regarding technology [11]. This study aimed to investigate learner characteristics using a Mandarin Massive Open Online Course (MOOC) and the research questions were constructed as follows:

(i)   What are the learning styles of learners that used Mandarin MOOC?
(ii) What are the cognitive styles of learners that used Mandarin MOOC?

Based on the findings of this study, the researchers proposed a content development model in MOOC based on the learner characteristics.

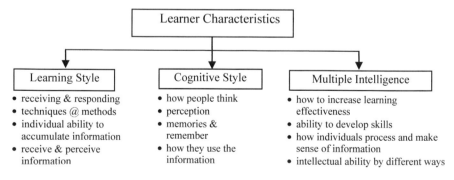

Fig. 1 Types of learner characteristics

2 Related Work

2.1 Learning Styles

Learning styles are authentic as they are the appropriate techniques or methods in which learners learn, comprehend and get information [12]. The authors highlighted that learning styles are an approach to help enhance the rate and nature of learning. Learning styles are the ways of receiving and responding to a learning stimulus with (i) unique psychological, (ii) affective and (iii) cognitive composition [13]. Learning styles refer to the variations in an individual's ability to accumulate as well as assimilate information, sensory preferences that have impact on learning and related to personality [13–15]. Previous researcher mentioned that learning styles refer to the manner in which learners receive and perceive information [16]. Learning styles are one of the most important and can motivate students, enhancing achievement and/or satisfaction [7].

According to S. N. M. Mohamad, each student has his or her own learning style to be considered during the learning process [17]. Researchers [13] listed three important elements for learning styles: (i) academic achievements, (ii) attitudes towards learning and (iii) multimedia technology. The previous study [10] listed a few learning style models to build up their pedagogical hypothesis: (i) Kolb Experiential Learning Theory, (ii) VARK Model, (iii) Felder and Silverman Learning/Teaching Style Model and (iv) Dunn and Dunn Learning Style Mode. Researchers [14] highlighted that the Felder and Silverman Learning Style Model has been selected as the most appropriate model for open learning. Therefore, in this study, we focus this model and investigate the types of learner dimension for student using MOOC in second-language learning.

Previous studies have found four different dimensions of learning styles: (i) processing (active/reflective), (ii) perception (sensory/intuitive), (iii) input (visual/verbal) and (iv) understand (sequential/global). Researchers stated the major differences in learning styles are (i) the way people perceive (sensation versus intuition); (ii) the way they made decisions (logical thinking versus imaginative feelings) and (iii) how active or reflective they were while interacting (extroversion versus introversion) [18].

2.2 Cognitive Styles

Previous research has shown that cognitive styles are (i) the way people think; (ii) the accuracy of their perception; (iii) how they process and remember information; (iv) how they use the information in problem solving or (v) how they organize and process information [7, 19]. The previous study [20] listed three parts of cognitive style: (i) individual's stability, (ii) characteristic of mental approach or (iii) cognitive preferences. The author has proposed three elements of the cognitive style dimension: (i) processing information, (ii) solving problems and (iii) making decisions.

Cognitive styles are preferences or strategies used by individuals that influence functions such as: (i) perceiving, (ii) remembering, (iii) thinking, and (iv) problem solving [21]. Previous researcher has used a cognitive style questionnaire proposed by Ancona et al. [22], and the cognitive style dimensions are: (i) energy (extroversion/introversion), (ii) information (sensing/intuitive), (iii) decisions (thinking/feeling) and (iv) lifestyle (judging/perceiving) [23]. Researchers conclude that the student's preference of cognitive style impacts their preference for e-activities and their performance. Therefore, in this study, we focus this questionnaire to investigate the type of learner dimension for the student using MOOC in second-language learning.

2.3 Learning Styles Versus Cognitive Styles

The previous study found the differences between cognitive styles and learning styles, in which the cognitive style involves cognition based on (i) process or tendency to perceive stimuli, (ii) receiving and responding to a learning stimulus and (iii) use information, whereas learning style is (i) rooted in exterior behaviour, (ii) response to learning situation and (iii) assimilate information [14, 15, 19]. Figure 2 concludes the differences in dimension between learning styles and cognitive styles.

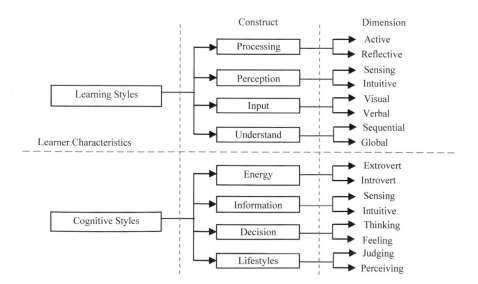

Fig. 2 Dimensions of learner characteristics

Fig. 3 Percentages of participants by faculty

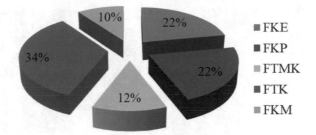

In this paper, the researchers studied two types of learner characteristics using Massive Open Online Courses (MOOC) for a second language: (i) learning style and (ii) cognitive style.

3 Materials and Methods

3.1 Method

A quantitative and qualitative method was adopted in this study. The quantitative data were collected using a survey involving 50 students from a public university in Malaysia. Qualitative data were collected from an interview session, while quantitative data was collected through the students' questionnaires. Moreover, semi-structured interviews were conducted with students in an attempt to obtain their experiences and opinions on MOOC assessments.

3.2 Participants

The participants in the research were 50 technical students in a public university in Malaysia. The study found that the numbers of male students are more than female students. Figure 3 shows the percentages of survey participants by faculty. The total participants for each faculty are FKE (11), FKP (11), FTMK (6), FTK (17) and FKM (5).

In addition to these, the participants for the semi-structured interviews were also selected randomly. 11 students studying at all faculties of the university participated in the interview. Figure 4 shows the percentages of interview participants by faculty. The total participants for each faculty are FKE (2), FKP (2), FTMK (3), FTK (3) and FKM (1).

Fig. 4 Percentages of interview participants by faculty

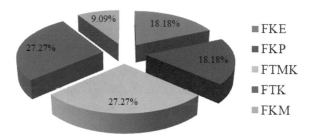

3.3 Research Instruments

Interview and questionnaire methods were used as the research instrument this study. The objective of the interview was to investigate the existing assessment method in a Massive Open Online Course (MOOC). The questionnaire consists of 2 parts, 16 constructs and 24 survey items. The constructs are the eight dimensions of learning styles and eight dimensions of cognitive styles. The survey items were adapted from several research projects with similar research scope [24, 22].

The interview was conducted using 14 items. Based on the students' answers, the researchers conducted an affinity analysis to identify learner characteristics based on the eight constructs and dimensions of learning styles and eight dimensions of cognitive styles. The purpose of this interview was to identify learner characteristics in MOOC for assessment using the OpenLearning platform. The interviews were done to obtain the students' experiences and their opinions on MOOC assessments. The objective of this study was to investigate the existing learner characteristics in learning Mandarin via MOOC.

3.4 Data Analysis

The data were analysed by using frequency and percentages analysis. In the process of data analysis, the collection data are triangulated in the mixed-method approach to study the learning styles in using Mandarin MOOC.

4 Result

In this part, the results of data analysis are presented from both the methods. The findings are presented, analysed and discussed based on the research content.

The findings of the questionnaires indicated visual as the highest percentage of learning style for students based on the Felder and Silverman Learning Style Model in Mandarin MOOC. Figure 5 shows the dimension of learning style by percentages in the second language course. Visual learning style has the highest frequency and percentage, values of 38 and 76.00%, respectively. According to the results of the

Fig. 5 Dimensions of learning styles by percentages

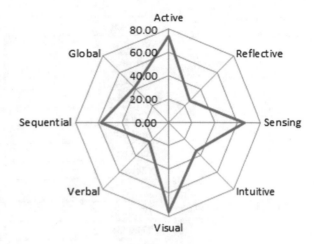

Fig. 6 Dimensions of cognitive styles by percentages

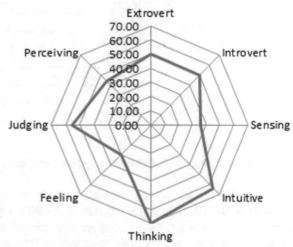

learner characteristics questionnaire, Fig. 6 shows the dimension of cognitive style by percentages in the second-language course. Thinking style has the highest frequency and percentage with the values of 35 and 70.00%.

The results of the learner characteristics interview are presented. Table 1 shows the dimension of learning style by percentages. Visual style has the highest percentage, with the value of 24.24% for the students' experience and opinions on MOOC assessment. Based on the interview, two elements in the visual style were identified: (i) learn best from what they see; and (ii) learners priorities pictorial materials.

According to the results of the learner characteristics interview, Table 2 shows the dimension of cognitive style by percentages. Thinking style has the highest percentages with the value of 21.95% in students' experience and opinions on MOOC assessment. Based on the interview, three elements in the thinking style were iden-

Table 1 Dimensions of learning style by percentages

Dimension	Percentages (%)
Visual	24.24
Active	18.18
Sensing	18.18
Reflective	15.15
Verbal	9.09
Sequential	9.09
Intuitive	6.06
Global	0.00

Table 2 Dimensions of cognitive style by percentages

Dimension	Percentages (%)
Thinking	21.95
Intuitive	19.51
Judging	17.07
Extrovert	12.20
Introvert	12.20
Sensing	7.32
Feeling	4.88
Perceiving	4.88

tified: (i) analysing fact, (ii) structure and function and (iii) logical and rational decisions.

To summarize, the findings of both mixed-method show that visual and active dimension are the highest percentages of learning styles in the second-language course. The results from the survey analysis revealed that the highest dimension among the eight learning styles is visual learner 76%, while the result from interview session is also confirmed visual (24.24%). According to Nicholas [25], for supporting the learners' learning style, they had to provide diagrams and pictures for the visual learners.

The results from the survey analysis revealed that the highest dimension among the eight cognitive styles was thinking learner at 70% while the result from the interview session was also thinking (21.95%). According to MEalor [21], for thinking learner, they think more when see the pictures and more understand pictorial material [26]. This study indicates that visual learner is the type of learning styles of learners and thinking learner is the type of cognitive styles that used Mandarin MOOC.

The findings showed the content development model for learner characteristics using MOOC. This model will be developed based on the interaction between learning styles and cognitive styles dimensions for characteristics of technical students' who learned using MOOC. The researchers identified the learner characteristics based on the two highest percentages dimensions of learning styles and two highest

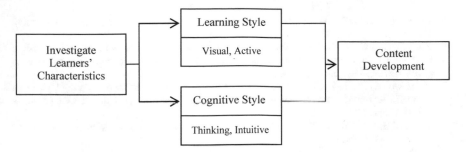

Fig. 7 Content development model for learner characteristics

percentages dimensions of cognitive styles. Figure 7 shows the content development model for learner characteristics.

5 Conclusion

This initial study identified the learning and cognitive styles of learners in using a Mandarin MOOC. The findings of the questionnaire and interview indicated visual and active as the learner characteristics for learning styles, and thinking and intuitive as learner characteristics for cognitive styles. The results from the percentage analysis revealed that out of the eight dimensions of learning styles, the preferred dimensions are visual and active learner. Meanwhile, for the cognitive styles are thinking and intuitive learners. This study also aimed to propose a content development model in MOOC based on learner characteristics. In the future, researchers will further analyse the assessment model in MOOC based on learner characteristics.

Acknowledgements This research is conducted by the Pervasive Computing & Educational Technology Research Group, C-ACT, Universiti Teknikal Malaysia Melaka (UTeM), and supported by the Ministry of Education (MOE). FRGS grant: FRGS/1/2016/ICT01/FTMK-C-ACT/F00327.

References

1. Park Y, Jung I, Reeves TC (2015) Learning from MOOCs: a qualitative case study from the learners' perspectives. EMI Educ Media Int 52(2):72–87
2. Amoozegar A, Mohd Daud S, Mahmud R, Ab Jalil H (2018) Exploring learner to institutional factors and learner characteristics as a success factor in distance learning. Int J Innov Res Educ Sci 4(6):2349–5219
3. Hone KS, El Said GR (2016) Exploring the factors affecting MOOC retention: a survey study. Comput Educ 98:157–168
4. Pursel BK, Zhang L, Jablokow KW, Choi GW, Velegol D (2016) Understanding MOOC students: motivations and behaviours indicative of MOOC completion. J Comput Assist Learn 32(3):202–217

5. Lever-Duffy J, McDonald JB (2009) Teaching and learning with technology, 3rd edn. Pearson, Boston
6. Zhao E, Guo C, Liu L (2017) The construction of MOOC' s curriculum evaluation system model in China. In: Annual international conference on modern education and social science (MESS 2017), 2017, no. Mess, pp 268–275
7. Al-azawei A, Parslow P, Lundqvist K (2017) Investigating the effect of learning styles in a blended e-learning system: an extension of the technology acceptance model (TAM). Australas J Educ Technol 33(2):1–23
8. Baukal CE, Ausburn LJ (2016) Multimedia category preferences of working engineers. Eur J Eng Educ 41(5):482–503
9. Baukal CE, Ausburn LJ (2016) Relationship of prior knowledge and working engineers' learning preferences: implications for designing effective instruction. Eur J Eng Educ 3797(April):1–21
10. Labib AE, Canós JH, Penadés MC (2017) On the way to learning style models integration: a learner's characteristics ontology. Comput Human Behav 73:433–445
11. De Paepe L, Zhu C, Depryck K (2018) Learner characteristics, learner achievement and time investment in online courses for Dutch L2 in adult education. Turkish Online J Educ Technol 17(1):101–112
12. Sadhasivam J, Babu R (2017) MOOC: a framework for learners using learning style. Int Educ Res J 3(2):21–24
13. Ali R (2016) Learning style construct in student' s learning. J Indones untuk Kaji Pendidik 1(2):213–222
14. Abante MER, Almendral BC, Manansala JE, Mañibo J (2014) Learning styles and factors affecting the learning of general engineering students. Int J Acad Res Progress Educ Dev 3(1):16–27
15. Hmedna B, El Mezouary A, Baz O, Mammass D (2016) A machine learning approach to identify and track learning styles in MOOCs. In: International conference on multimedia computing and systems (ICMCS'16)
16. Fasihuddin H, Skinner G, Athauda R (2017) Towards adaptive open learning environments: evaluating the precision of identifying learning styles by tracking learners' behaviours. Educ Inf Technol 22(3):807–825
17. Mohamad SNM (2014) Model for online teaching tools based on interpersonal, visual and verbal intelligence. Universiti Teknikal Malaysia Melaka
18. Rohaniyah J (2017) Integrating learning style and multiple intelligences in teaching and learning process. J Pemikir Penelit Pendidik dan Sains 5(1):19–27
19. Simuth J, Sarmany-Schuller I (2014) Cognitive style variable in E-learning. Procedia—Soc Behav Sci 116:1464–1467
20. Jablokow K, Defranco JF, Jablokow K, Defranco JF, Richmond SS, Piovoso MJ, Bil SG (July 2015) Cognitive style and concept mapping performance
21. Mealor AD, Simner J, Rothen N, Carmichael DA, Ward J (2016) Different dimensions of cognitive style in typical and atypical cognition: new evidence and a new measurement tool. PLoS ONE 11(5):1–21
22. Ancona K, Scully VM, Westney, "Managing For the Future," 1997. [Online]. Available: http://www.analytictech.com/mb021/cogquest.htm. Accessed 30 Apr 2017
23. Mukherjee S (2016) Learning style of humanities, commerce and science students: a study on higher secondary students from West Bengal. Int J Indian Psychol 3(3):20–29
24. Kintakaningrum T (2012) A study of consequences on individual and group learning performance using a web-based and mobile supported learning management system. Universiti Teknikal Malaysia Melaka, Malaysia
25. Nicholas JS, Francis FS (2016) MOOCs and the need for POOCs. Int J Mod Comput Sci 4(4):211–219
26. Koc Januchta M, Hoffler T, Thoma G-B, Prechtl H, Leutner D (2017) Visualizers versus verbalizers: effects of cognitive style on learning with texts and pictures—an eye-tracking study. Comput Human Behav 68:170–179

Cultural Metadata Enhancement by Using Community-Driven Approach to Create Digital Holistic Culture Conservation System

Djoni Setiawan Kartawihardja, Ahmad Naim Che Pee
and Mohd. Hafiz Zakaria

Abstract The biggest problem when conserving cultural information in digital form is figuring out when the stored information has been well structured and meet the needs of the cultural information seeker. Several researches have shown community involvement in creating metadata as the cultural data structure. Now, the problem is how the metadata should be constructed to meet the cultural information seeker's needs that can be changed over time. The structure should cover the holistic items of the culture, not one cultural item only. This preliminary research is to determine how far a community-driven effect can enhance cultural metadata structure processes as the conservation standard for the cultural information structure through study processes of literature. By knowing current cultural metadata structures and community involvement experiences, holistic metadata structure and digital holistic cultural conservation system can be built.

Keywords Culture · Metadata · Conserve · Structure · Community driven

1 Introduction

Cultural conserving processes are carried out by various countries in the world. Their aim is to conserve and preserve their own traditional and national cultural heritage for future generations. These countries work hard to conserve and preserve their culture because culture already became the identity of the nation and is integrated into the

D. S. Kartawihardja (✉) · A. N. Che Pee · Mohd. H. Zakaria
Faculty of Information & Communication Technology, Universiti Teknikal Malaysia Melaka,
Melaka, Malaysia
e-mail: p031710027@utem.edu.my; djoni.setiawan@itmaranatha.org

A. N. Che Pee
e-mail: naim@utem.edu.my

Mohd. H. Zakaria
e-mail: hafiz@utem.edu.my

D. S. Kartawihardja
Faculty of Information Technology, Maranatha Christian University, Bandung, Indonesia

© Springer Nature Singapore Pte Ltd. 2019   557
V. Piuri et al. (eds.), *Intelligent and Interactive Computing*, Lecture Notes in Networks
and Systems 67, https://doi.org/10.1007/978-981-13-6031-2_10

daily activities of the citizen. These conditions are in line with Koentjaraningrat [1] who states that *"culture is the whole system of ideas, actions, and human's work result in community life framework which belong to self-made man with learning."*

Countries have made efforts to conserve and preserve their traditional and national culture from documenting in printed materials until building a special museum to store the cultural items. Despite all these efforts, there are still several problems such as differences in the culture's detail on information and structure, differences in the information that are provided by information provider and seeker, and difficulties in finding or accessing the information. The differences in detail within culture's information can arise especially in a country made up of many tribes. Each part of cultural information may have some special characteristics caused by the tribe's environment; on the other hand the information could have many similarities to other cultures around it. When talking about information provider and seeker, differences on needed and perceived information always become problematic. There is no standard or model that can be used as a basis for connecting between the information provider and seekers. Even with using printed material or special museum, not every information seeker can find the culture's information easily.

To connecting between information provider and seeker, a specific standard is needed as a bridge between them. One approach that can be used as a bridge is metadata. By using metadata, detailed information of culture can be described. Information provider should not create the metadata alone, but it should involve the whole stockholders in cultural activity and management. One of the stockholders is a cultural community. The problem now is how much contribution the cultural community can give on enhancing cultural metadata and how far the involvement of cultural community contributes to enhancing cultural metadata. There are several research activities that have already been conducted on enhancing cultural metadata around the world.

2 Culture Conserving

Since 1989, UNESCO has defined 'folklore' or 'traditional and popular culture' as *"the totality of tradition-based creations of a cultural community, expressed by a group or individuals and recognized as reflecting the expectations of a community in so far as they reflect its cultural and social identity; its standards and values are transmitted orally, by imitation or by other means. its forms are, among others, language, literature, music, dance, games, mythology, rituals, customs, handicrafts, architecture, and other arts"* [2] A similar definition of 'traditional and popular culture' comes from Koentjaraningrat. Koentjaraningrat defining it as *"the whole system of ideas, actions, and human's work result in community life framework which belong to the self-made man with learning"* [1]. Both definitions show that traditional culture has already become the identity of an individual or community group in everyday life. Therefore, the existence of the culture is very important for the future generation and needs to be conserved and preserved.

As a universal aspect in human life, Kluckhohn in [1] define seven culture elements as: language, knowledge system, society or social organization system, living equipment and technology system, livelihood system, religion system, and art. Hoenigman [1] divides the forms of culture as idea, activities, and artifacts. The cultural idea is represented by opinions, values, norms, and rules from others or the community. The cultural activities are referred to patterned actions of society in daily life. The cultural artifacts are the result from ideas, activities, and works of human in a society that can be touched, seen, and documented. From culture elements and forms of culture, the culture is a collection of parts. Therefore, culture should be seen in a single entity as a holistic part and not seen in independently in separate parts.

Conservation is a process of documenting all traditions within culture [2]. The aim of conservation in folklore or traditional culture is to archive, store, and make access available. Another aim of the conservation is to give an archive to show changes in the ways of life and to keep folklore materials that can be passed down to the future generation. But when this conservation process is linked to the complexity information in culture as holistic information, conservation processes becomes less easy.

The importance of cultural conservation also mentions by International Council on Monuments and Sites (ICOMOS). According to ICOMOS [3], cultural preservation is important and should be supported by the public to ensure the culture's survival from generation to generation. The main reason culture is protected, conserved and managed is to preserve the physical and or intellectual heritage and collections for the host community and culture information seeker. Without public awareness and public support in preservation, a culture's item can be dismissed as insignificant.

3 Related Works

Based on literature review results, cultural conservation process can be divided into two main areas: data structure and data management system. Data structure on cultural conservation system can be realized in metadata form. The data management system can be realized in the centralized or distributed system.

3.1 Data Structure

Chen [4] defines metadata as a digital library or museum environment to conserve historical documents, old maps, photographs, calligraphies, archaeological artifacts, and Buddhist scriptures/paintings based on Dublin Core metadata semantics. The created metadata is used as a standard in digitizing museum artifact's items and presenting them on the web in a uniform method. To achieve maximum results, Chen collaborates with the cultural community, such as National Palace Museum, National Museum of history, National Museum of Natural Science, National Central

Library, National Taiwan University, and Academia Sinica. The research's metadata consists of 15 elements to store and conserve information.

As a part of the EMANI project, Niu [5] tried to develop a digital mathematics resource preservation system and shared the resource to other project participants. They develop a metadata framework to describe resources and encoding standards to store metadata and resource structures in the information system. The metadata framework consists of five modules (descriptive metadata module, rights metadata module, technical metadata module, source metadata module, and digitization process metadata module). As descriptive metadata, they used 12 most common core metadata. These common core metadata give basic information that is necessary to certify the existence of resources.

Another project using metadata as a data structure is used in Patel [6]. In this project, metadata is not only used to store information about cultural items only, but it also used to store information to create, manipulate, manage, and present a virtual exhibition view of each cultural artifact in an online virtual reality environment. The metadata of virtual exhibition is developed in Augmented Representation of Cultural Objects (ARCO) systems. The main categories of metadata that used in this research project are curatorial, technical, resource discovery, thematic grouping, presentation, and administrative. The structure of metadata not stay fixed, but continues to grow according to assessment, evaluation, and feedback from user groups who involved in the project.

Saefurrohman and Dewi [7] used Preservation Metadata: Implementation Strategies (PREMIS) to preserve Indonesian batik patterns. The PREMIS metadata is used as a standard on digitalization heritage documents for research, education, restoration, and preserve batik patterns as cultural heritage. On the PREMIS data model, the metadata can be categorized into four major entities, which are object, event, agent, and right statement.

In research about Malay folk dances' preservation system [8], all team members have tried to find a model for digitizing the dances that are comprised of movements, music, and costumes, especially for virtual reality and augmented reality area. According to the research's result, the information of Malay folk dances can be categorized into several principal components as shown in Fig. 1. Figure 1 shows the complexity of information inside information of Malay folk dances.

In another research, Volk and team [9] tried to find the main element of Dutch songs that can be used as a structure for documenting musical culture. This research project tried to make the digital document of Dutch song culture in the past and present. The digital document built from 173 thousand references of songbooks, manuscripts, and field recording. To standardize data structure, they use metadata as a basis. The earliest metadata elements that they used consisted of song texts, melodies and stanzas, music, and song patterns. This metadata structure never stays the same, because the data structure is always consulted to music and literary scholars and conservationist culture communities. The involvement of music scholars, literary scholars, and conservationist communities is very positive and give good impacts to the evolution of the metadata structure as well as the system functionality.

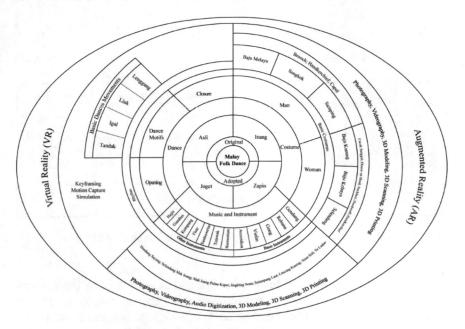

Fig. 1 Categorization of the principal component of Malay folk dance [8]

Digital Library North (DLN) system is a research collaboration among researchers at the University of Alberta, Inuvialuit Cultural Resource, and communities within Inuvialuit Settlement Region in northern Northwest Territories in Canada [10]. This research project developed a digital library infrastructure to support access to Inuvialuit cultural resources. They used the community member to construct, collect, and organized the cultural information which is needed and used by the community. This process ensures the growth of cultural data and development of data structure. The metadata structure that they used was developed under Dublin Core basis. They modified the Dublin Core basis based on community input and information audit. The first metadata structure consists of 14 elements, such as language, dialect, tile, date, type, creator, contributor, subject, description, identifier, is part of, spatial coverage, rights, and audience. Because the project involves members of the community, the metadata structure used in the DLN system is still evolving along with input from members of the community involved.

3.2 Data Management System

Data management system on cultural in Patel [6] is not provided from one stakeholder only but involves more than one stakeholder. The data management process is not only connected to keep the data but also the metadata management. All data management process in Patel [6] can be shown in Fig. 2.

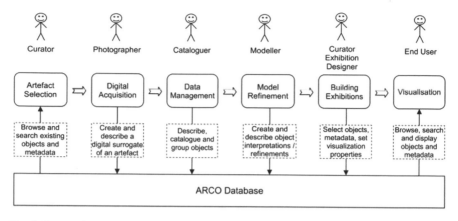

Fig. 2 Data and metadata management process on ARCO metadata element set [6]

From Fig. 2, stakeholders that involved in data and metadata management process on ARCO database and metadata are from curator, photographer, cataloguer, modeler, exhibition curator, until end user. Some of the stakeholders that involved in the project understand the metadata structure, but the others do not. Therefore, the metadata structure that was used in the project was created as simple, easy, and consistent as possible.

Kurniawan and team proposed an approach in managing cultural data through Electronic Cultural Heritage and Natural History (eCHNH) system [11]. The system was created based on Zachman Architecture Framework's approach on web portal form. In their proposed system, all data is centralized within the system. By using a centralized model, anyone can access the information without worrying about incompatible data, technology, and policy (see Fig. 3). Through the eCHNH system, all stakeholders can access, search, and store any cultural information that is available.

Metadata as data structure management is not only used in specific areas, but it can be also used within data itself [12]. The main reason in placing metadata within data itself is to make it useful to discover data. Some implementations of this embedded metadata can be found in digital photography, digital audios, or digital movies. Inside of these digital items, metadata can be found to describe identity, authentication, collection, and copyright. This embedded metadata can be specified by any stakeholder involved or by specific tools that used to create the digital item, for the conservation process. Therefore, this embedded metadata can help a librarian to manage all information related to the query thus increasing the value of the digital data. That information can be harvest, stored or created from within the library catalog system. One barrier in using embedded metadata on daily activity is there are not any standards related to the content of the digital item on local and global levels.

The process of managing data and metadata is not an easy task [13]. One of the researches related to data and metadata management is conducted in Spatial Data Infrastructures (SDIs) management system. The aim of this research is to find

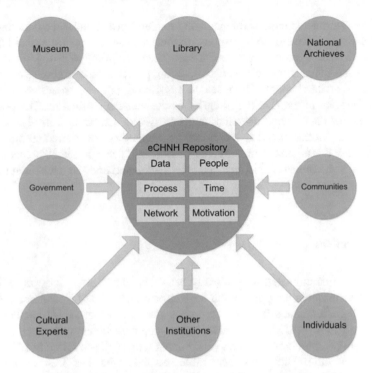

Fig. 3 Electronic cultural heritage and natural history system framework [11]

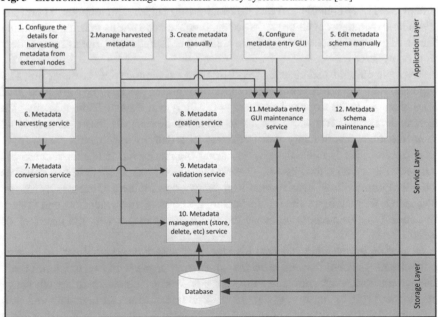

Fig. 4 Architecture of automatic spatial metadata system [13]

effective and efficient methods for managing spatial data demand. Metadata standards have already been developed in response to the need for data custodians and the end users, however, the process remains difficult. The main problem is disjointed data is due to the inefficiency in the metadata updating process. To overcome the growing speed of new data elements, the research has tried to propose an automatic spatial metadata system to manage the integrity of the metadata structure. The proposed architecture of the automatic spatial metadata system can be seen in Fig. 4, which shows that the spatial metadata can be edited manually or converted automatically. Even by using this architecture, the research team still faces problems in recognizing metadata which is contained in received information and finds its equivalent in listed metadata.

4 Discussion

Based on all literature that have already been described above, it appears that all cultural data conservation process still focuses on a small part of the cultural element, such as integration data artefacts that already stored in museums [4], mathematical equations [5], artifacts' data visualization on augmented reality environment [6], batik pattern [7], Malay folk dances [8], Dutch's songs [9], and Inuvialuit dialectical [10]. This condition shows a considerable gap with what has been suggested by Kluckhohn [1]. Kluckhohn mentions that the seven culture elements are a unity within human life. Every element had a close relationship with other elements. Therefore, the cultural metadata development process cannot be done one by one and combining it later. The cultural metadata should be created as a holistic item, not as independent parts of an item.

To create cultural metadata as a holistic item, community involvement plays an important role. Input from conservationist communities will become the direction of cultural metadata development process [4–8, 10]. Any metadata element involved in cultural data comes from conservationist communities' input. They also give input about the main functionality that is needed in a data management system. Input from communities is not only needed during the research and design process but can be continued after the research process has ended [4, 6, 9–11]. These conditions show the communities' involvement in metadata development processes cannot be eliminated. Without input and feedback from the communities, metadata and system functionality development can cause the result to not match with the needs of its users.

In relation to metadata and data management issues, the system should be able to involve all stakeholders and keep the intellectual property of every cultural item. The system framework shown in Fig. 3 already involves all stakeholders who have an important role in conserving data of cultural items. Since all metadata and data is centrally stored, information related to intellectual property right can cause a significant problem, especially for special or valuable items.

One approach that can be used to overcome this problem is making the metadata still centrally stored, but the cultural item data would be distributed in other places. Each data storage owner can manage all its own information and can decide which data can be shared with others. This distributed system, however has a problem in terms of data structure synchronization. The synchronizing process on data structures can be solved through a centralized standardize metadata.

5 Conclusion and Further Works

Cultural metadata development process needs further exploration to get a holistic metadata structure that can fulfill cultural information seekers' needs. The holistic metadata structure can be enhanced by using valuable cultural conservationist communities' inputs and feedbacks. Therefore, a special collaboration model among all conservationist community member should be built to maximize collaborated inputs and feedbacks.

Based on the obtained holistic metadata structure, a centralized metadata and distributed data management systems can be built. Through the systems, the entire process of metadata management based on inputs from conservationist communities and feedbacks from can be developed. The systems can also use to distribute and synchronize metadata structure to any attached data source.

Because cultural data is stored in a distributed manner, an intellectual property right can still be owned by the cultural information owner. Yet, even when intellectual property rights issues are resolved, a distributed data management system can cause other problems such as data accessibility. To ensure all data source can be accessed and recognized by any data management system, a standardized data access model is required to synchronize all connected data sources. By using this access model, all data management systems can exchange information with one another when new data is added.

References

1. Koentjaraningrat: Pengantar Ilmu Antropologi. PT. Rineka Cipta, Jakarta (2015)
2. UNESCO: Recommendation on the Safeguarding of Traditional Culture and Folklore. United Nations Educational, Scientific and Cultural Organization. 15 November 1989. http://portal.unesco.org/en/ev.php-URL_ID=13141&URL_DO=DO_TOPIC&URL_SECTION=201.html. Accessed 21 May 2018
3. ICOMOS (2002) Principles and guidelines for managing tourism at places of cultural, Charenton-le-Pont. International Council on Monuments and Sites, Victoria
4. Chen C-C, Chen H-H, Chen K-H, Hsian J (2002) The design of metadata for the Digital Museum Initiative in Taiwan. Online Inf Rev 26:295–306
5. Jinfang N (2002) A metadata framework developed at the Tsinghua University Library to aid in the preservation of digital resources. D-Lib Mag 8

6. Patel M, White M, Mourkoussis N, Walczak K, Wojciechowski R, Chmielewski J (2005) Metadata requirements for digital museum environments. Int J Digit Libr 5:179–192
7. Saefurrohman S, Ningsih DHU (2015) Metode preservation metadata implementation strategies (Premis) bagi Standarisasi Dokumentasi Digital Batik Tulis Warisan Nusantara. Dinamik 20:140–147
8. Iddris MZ, Mustaffa N, Othman AN, Abdullah MFW (2017) Exploring principle components for digital heritage preservation on Malay folk dances. Int J Acad Res Bus Soc Sci 7:738–747
9. Kranenburg PV, Bruin MD, Volk A (2017) Documenting a song culture: the Dutch Song Database as a resource for musicological research. Int J Digit Libr 18:1–11
10. Farnel S, Shiri A, Campbell S, Cockney C, Rahi D, Stobbs R (2017) A community-driven metadata framework for describing cultural resources: the Digital Library North project. Cataloging Classif Q 55:289–306
11. Kurniawan H, Salin A, Suhartanto H, Hasibuan ZA (2011) E-cultural heritage and natural history framework: an integrated approach to digital preservation. In: International conference on telecommunication technology and applications, Singapore
12. Corrado EM, Jaffe R (2017) Access's unsung hero: the [impending] rise of embedded metadata. Int Inf Libr Rev 49:124–130
13. Kalantari M, Rajabifard A, Olfat H, Petit C, Keshtiarast A (2017) Automatic spatial metadata systems—the case of Australian urban research infrastructure network. Cartography Geogr Inf Sci 44:327–337

Author Index

Printed in the United States
By Bookmasters